Probability and Mathematical Statistics (Conti

RUBINSTEIN • Simulation and The Mo

SCHEFFE • The Analysis of Variance

SEBER • Linear Regression Analysis

SEN • Sequential Nonparametrics: Invariance Principles and Statistical Inference

SERFLING • Approximation Theorems of Mathematical Statistics

TJUR • Probability Based on Radon Measures

WILLIAMS • Diffusions, Markov Processes, and Martingales, Volume I: Foundations

ZACKS • Theory of Statistical Inference

Applied Probability and Statistics

ANDERSON, AUQUIER, HAUCK, OAKES, VANDAELE, and WEISBERG • Statistical Methods for Comparative Studies

ARTHANARI and DODGE • Mathematical Programming in Statistics

BAILEY • The Elements of Stochastic Processes with Applications to the Natural Sciences

BAILEY • Mathematics, Statistics and Systems for Health

BARNETT • Interpreting Multivariate Data

BARNETT and LEWIS • Outliers in Statistical Data

BARTHOLOMEW • Stochastic Models for Social Processes, *Third Edition*

BARTHOLOMEW and FORBES • Statistical Techniques for Manpower Planning

BECK and ARNOLD • Parameter Estimation in Engineering and Science

BELSLEY, KUH, and WELSCH • Regression Diagnostics: Identifying Influential Data and Sources of Collinearity

BENNETT and FRANKLIN • Statistical Analysis in Chemistry and the Chemical Industry

BHAT • Elements of Applied Stochastic Processes

BLOOMFIELD • Fourier Analysis of Time Series: An Introduction

BOX • R. A. Fisher, The Life of a Scientist

BOX and DRAPER • Evolutionary Operation: A Statistical Method for Process Improvement

BOX, HUNTER, and HUNTER • Statistics for Experimenters: An Introduction to Design, Data Analysis, and Model Building

BROWN and HOLLANDER • Statistics: A Biomedical Introduction

BROWNLEE • Statistical Theory and Methodology in Science and Engineering, *Second Edition*

BURY • Statistical Models in Applied Science

CHAMBERS • Computational Methods for Data Analysis

CHATTERJEE and PRICE • Regression Analysis by Example

CHERNOFF and MOSES • Elementary Decision Theory

CHOW • Analysis and Control of Dynamic Economic Systems

CHOW • Econometric Analysis by Control Methods

CLELLAND, BROWN, and deCANI • Basic Statistics with Business Applications, *Second Edition*

COCHRAN • Sampling Techniques, *Third Edition*

COCHRAN and COX • Experimental Designs, *Second Edition*

CONOVER • Practical Nonparametric Statistics, *Second Edition*

CORNELL • Experiments with Mixtures: Designs, Models and The Analysis of Mixture Data

COX • Planning of Experiments

continued on back

Statistical Models and Methods for Lifetime Data

Statistical Models and Methods for Lifetime Data

J. F. LAWLESS

University of Waterloo

JOHN WILEY & SONS

New York Chichester Brisbane Toronto Singapore

Library of Congress Cataloging in Publication Data:

Lawless, J. F. (Jerald F.), 1944–
Statistical models and methods for lifetime data.

(Wiley series in probability and mathematical statistics. Applied probability and statistics section, ISSN 0271-6356)
Bibliography: p.
Includes index.
1. Failure time data analysis. 2. Reliability (Engineering) I. Title. II. Series.

QA276.L328 519.5 81-11446
ISBN 0-471-08544-8 AACR2

Printed in the United States of America

10 9 8 7

To My Family

Preface

The statistical analysis of lifetime or response time data has become a topic of considerable interest to statisticians and workers in areas such as engineering, medicine, and the biological sciences. The field has expanded rapidly in recent years, and publications on the subject can be found in the literatures of several disciplines besides statistics. This book draws together material on the analysis of lifetime data and gives a comprehensive presentation of important models and methods.

My aim is to give a broad coverage of the area without unduly concentrating on any single field of application. Most of the examples in the book, however, come from engineering or the biomedical sciences, where these methods are widely used. The book contains what I feel are the most important topics in lifetime data methodology. These include various parametric models and their associated statistical methods, nonparametric and distribution-free methods, and graphical procedures. To keep the book at a reasonable length I have had to either sketch or entirely omit topics that could have usefully been treated in detail. Some of these topics are referenced or touched upon in the Problems and Supplements sections at the ends of chapters.

This book is intended as a reference for individuals interested in the analysis of lifetime data and can also be used as a text for graduate courses in this area. A basic knowledge of probability and statistical inference is assumed, but I have attempted to carefully lay out the models and assumptions upon which procedures are based and to show how the procedures are developed. In addition, several appendices review statistical theory that may be unfamiliar to some readers. Numerical illustrations are given for most procedures, and the book contains numerous examples involving real data. Each chapter concludes with a Problems and Supplements section, which provides exercises on the chapter material, and supplements and extends the topics discussed. For the reader interested in research on lifetime data methodology I have given fairly extensive references to recent work and outstanding problems.

Chapter 1 contains introductory material on lifetime distributions and surveys the most important parametric models. Censoring is introduced, and its ramifications for statistical inference are considered. In Chapter 2 some methods of examining univariate lifetime data and obtaining nonparametric estimates of distribution characteristics are discussed; life tables and graphical procedures play key roles. Chapters 3, 4, and 5 deal with inference for important parametric models, including the exponential, Weibull, gamma, log-normal, and generalized gamma distributions. This is extended in Chapter 6 to problems with concomitant variables, through regression models based on these distributions. Chapters 7 and 8 present nonparametric and distribution-free procedures: Chapter 7 deals with methods based on the proportional hazards regression model, and Chapter 8 gives distribution-free procedures for single- and many-sample problems. Goodness of fit tests for lifetime distribution models are considered in Chapter 9. Chapter 10 contains brief discussions of two important topics for which it was not feasible to give extended treatments: multivariate and stochastic process models. Several sections in this book are marked with asterisks; these contain discussions of a technical nature and can be omitted on a first reading.

A final remark concerning the methods presented is that the computer is, as always in modern statistics, a useful if not indispensible tool. For some problems, methods that do not require a computer are available, but more often access to a computer is a necessity. I have commented, wherever possible, on the computational aspects of procedures and have included additional material on computation in the Appendices.

Part of the work for the book was done during a sabbatical leave spent at Imperial College, London, and the University of Reading from 1978 to 1979; their hospitality is gratefully acknowledged. I would also like to express my appreciation to two extremely fine typists, Annemarie Nittel and Lynda Hohner, who labored long and diligently in the preparation of the manuscript.

J. F. LAWLESS

Waterloo, Ontario
June 1981

Contents

CHAPTER 1

Basic Concepts and Models

1.1 INTRODUCTION

The statistical analysis of what is variously referred to as lifetime, survival time, or failure time data has developed into an important topic for workers in many areas, especially in the engineering and biomedical sciences. Applications of lifetime distribution methodology range from investigations into the endurance of manufactured items to research involving human diseases. Some methods of dealing with lifetime data are quite old, but many important developments are relatively recent and have to be searched out not only in statistics journals, but also in the literature of various other disciplines. This book is an attempt to draw together some of the basic statistical methods of analyzing lifetime data. The specific techniques and models that underlie them are emphasized over any particular area of application. Life distribution methodology does find its most frequent application in the engineering and biomedical sciences, however, and most of the examples in this book come from one of these two areas.

Throughout the book various types of data will, for convenience, be referred to as "lifetime" data, though the observations may not always refer to lifetimes in the strictest literal sense. Basically, we consider situations in which the time to the occurrence of some event is of interest for some population of individuals; the times to the occurrences of events are termed "lifetimes." Sometimes the events of interest are deaths of individuals in a real sense and "lifetime" is the actual length of life of an individual, or perhaps a survival time, measured from some particular starting point. In other instances "lifetime" is used in a figurative sense. Mathematically, one can think of "lifetime" as merely meaning "non-negative-valued variable." The name given to these variables does not really matter, and although the term "lifetime" is used for general reference, other terms such as "survival time" and "failure time" will also be frequently used.

A few examples will illustrate some typical ways in which lifetime data arise.

Example 1.1.1 Manufactured items such as mechanical or electronic components are often subjected to life tests in order to obtain information on their endurance. This involves putting items in operation, often in a laboratory setting, and observing them until they fail. It is common here to refer to the lifetimes as "failure times," since when an item ceases operating satisfactorily, it is said to have "failed."

Example 1.1.2 Some types of manufactured items can be repaired, should they fail. In this case one might be interested in the length of time between successive failures of an item and refer to these times as "lifetimes."

Example 1.1.3 In medical studies dealing with fatal diseases one is interested in the survival time of individuals with the disease, measured from the date of diagnosis or some other starting point. For example, it is common to compare treatments for a disease at least partly in terms of the survival time distributions for patients receiving the different treatments.

Example 1.1.4 A standard experiment in the investigation of carcinogenic substances is one in which laboratory animals are subjected to doses of the substance and are then observed to see if they develop tumors. A main variable of interest in such experiments is the time to appearance of a tumor, or perhaps the time to death of the animal, measured from when the dose is administered.

In each of these situations there is interest in some aspect of the lifetimes of individuals. The basic problems addressed in this book are those of specifying models to represent distributions of lifetimes and of making statistical inferences on the basis of these models. The part the model plays in the analysis of data varies according to the problem under discussion and the approach taken with the problem. In some situations specific parametric models can be employed to represent lifetime distributions, and inferences based on these. In others, use of a parametric family of models may not be feasible, and nonparametric methods are important. The complexity of the models and related statistical methods called for will also vary greatly from application to application. Sometimes attention is primarily focused on a single lifetime variable in a homogeneous population. Usually, however, the situation is more complicated: different groups of individuals may have different lifetime distributions, there may be several variables associated with each individual, an individual may be liable to suffer any one of several types of "deaths," and so on. In addition, as we shall see, the process by which the data arise is often a complicating factor in problems involving lifetimes.

This chapter covers some preliminaries crucial to the treatment of lifetime data. Section 1.2 introduces concepts associated with statistical models for lifetime distributions. Section 1.3 surveys the most widely used univariate models and briefly discusses more complicated models involving regressor variables or several lifetime variables. These sections provide an introduction to the types of models upon which most statistical analyses of lifetime data are based. Section 1.4 turns to a difficulty frequently associated with the processes by which lifetime data arise. This is the presence of what is usually termed "censoring." Essentially, data are said to be "censored" when there are individuals in the sample for which only a lower (or upper) bound on lifetime is available. Censoring is common in life distribution work because of time limits and other restrictions on data collection. In a life test experiment, for example, it may not be feasible to continue experimentation until all items under study have failed. If the experiment is terminated before all have failed, then for items that are still unfailed at the time of termination only a lower bound on lifetime is available. This is not to say that there is no information available on their lifetimes, but only that the information on them is partial. As we shall see throughout this book, the presence of censoring creates special and interesting problems.

Before continuing, let us consider a few examples of lifetime data to further illustrate and clarify some of the points just mentioned.

Example 1.1.5 Nelson (1972a) describes the results of a life test experiment in which specimens of a type of electrical insulating fluid were subjected to a constant voltage stress. The length of time until each specimen failed, or "broke down," was observed. Table 1.1.1 gives results for three groups of specimens, tested at voltages of 28, 30 and 32 kilovolts, respectively.

Table 1.1.1 Times to Breakdown (in Minutes) at Each of Three Voltage Levels

28 kV	30 kV		32 kV	
68.85	17.05	194.90	0.40	3.91
426.07	22.66	47.30	82.85	0.27
110.29	21.02	7.74	9.88	0.69
108.29	175.88		89.29	100.58
1067.6	139.07		215.10	27.80
	144.12		2.75	13.95
	20.46		0.79	53.24
	43.40		15.93	

The main purpose of the experiment was to investigate the distribution of time to breakdown for the insulating fluid and to relate this to the voltage level. Quite clearly, breakdown times tend to decrease as the voltage increases, and any model for this situation would have to reflect this. In addition to the formulation of a model relating breakdown times and voltage, the estimation of percentiles of the breakdown time distribution, for any given voltage, was also important. The tenth percentile, which is used by engineers as a kind of nominal lifetime measure in this situation, was of special interest.

The experiment in Example 1.1.5 was run long enough to observe the failure of all the insulation specimens tested. Sometimes it may take a very long time for all items in a life test to fail, and it is deemed necessary to terminate the experiment before this can happen. In this case, some of the observations are censored, and one does not know the exact lifetimes of certain items. For example, if a decision had been made in the experiment to terminate testing after 180 minutes had elapsed, then two of the observations in the 28-kilovolt sample and one each in the 30- and 32-kilovolt samples would have been censored. In each case, we would not know the exact failure time of the item, but only that it exceeded 180 minutes.

Censoring arises in lifetime data in a variety of ways and is discussed in more detail in Section 1.4. The remaining examples in this section all involve censoring of some kind.

Example 1.1.6 Bartholomew (1957) considers a situation in which pieces of equipment are installed at different times. At a later date some of the pieces will have failed and the rest will still be in use. The aim is to study the lifetime distribution of this type of equipment and to estimate quantities such as the proportion of the equipment that will fail within a specified time. Bartholomew gives the data in Table 1.1.2, showing results for 10

Table 1.1.2 Operating Times for 10 Pieces of Equipment

Item Number	1	2	3	4	5
Date of installation	11 June	21 June	22 June	2 July	21 July
Date of failure	13 June	—	12 August	—	23 August
Lifetime (days)	2	⩾72	51	⩾60	33
Item Number	**6**	**7**	**8**	**9**	**10**
Date of installation	31 July	31 July	1 August	2 August	10 August
Date of failure	27 August	14 August	25 August	6 August	—
Lifetime (days)	27	14	24	4	⩾21

Table 1.1.3 Lengths of Remission (in Weeks) for Two Groups of Patients[a]

6-MP	6, 6, 6, 6*, 7, 9*, 10, 10*, 11*, 13, 16, 17*, 19*, 20*, 22, 23, 25*, 32*, 32*, 34*, 35*
Placebo	1, 1, 2, 2, 3, 4, 4, 5, 5, 8, 8, 8, 8, 11, 11, 12, 12, 15, 17, 22, 23

[a]Starred quantities denote censored observations.

pieces of equipment. The life test in question was terminated on August 31. At that time three items (numbers 2, 4, and 10) had still not failed, and their failure times are therefore censored; we know for these items only that their failure times exceed 72, 60, and 21 days, respectively.

Example 1.1.7 Gehan (1965) and others have discussed the results of a clinical trial reported by Freireich et al. (1963) in which the drug 6-mercaptopurine (6-MP) was compared to a placebo with respect to the ability to maintain remission in acute leukemia patients. Table 1.1.3 gives remission times for two groups of 21 patients each, one group given the placebo and the other the drug 6-MP.

The starred observations are censoring times. This means that for these patients the disease was still in a state of remission at the time the data were collected. Censoring is very common in clinical trials, since it is often desired to analyze the data before all individuals have "died." In addition, in many trials individuals enter the study at various times and hence may be under observation for different lengths of time. In this experiment, for example, observation continued for 1 year, and individuals entered the study more or less continuously over this period. The observation 10* thus refers to an individual who entered the trial 10 weeks before the end and whose remission was still in effect at the end of the trial.

Example 1.1.8 Table 1.1.4 presents survival data on 40 advanced lung cancer patients, taken from a larger body of data discussed by Prentice (1973). The main purpose of the study from which the data arose was to compare the effects of two chemotherapy treatments in prolonging survival time. All patients in the group represented here received prior therapy and were then randomly assigned to one of the two treatments, termed "standard" and "test." Survival times t from the start of treatment for each patient are recorded in Table 1.1.4. Censored observations are marked with an asterisk and correspond to patients who were still alive at the time the data were collected. A number of concomitant variables were thought to be important in relation to survival time, and these are also shown for each patient. First, different patients can have different types of tumors. Here tumors have been classified into four broad types (squamous, small, adeno, and large). Also given for each patient is a performance status, given at the

Table 1.1.4 Lung Cancer Survival Data[a, b]

t	x_1	x_2	x_3
Standard, Squamous			
411	70	64	5
126	60	63	9
118	70	65	11
82	40	69	10
8	40	63	58
25*	70	48	9
11	70	48	11
Standard, Small			
54	80	63	4
153	60	63	14
16	30	53	4
56	80	43	12
21	40	55	2
287	60	66	25
10	40	67	23
Standard, Adeno			
8	20	61	19
12	50	63	4
Standard, Large			
177	50	66	16
12	40	68	12
200	80	41	12
250	70	53	8
100	60	37	13

t	x_1	x_2	x_3
Test, Squamous			
999	90	54	12
231*	50	52	8
991	70	50	7
1	20	65	21
201	80	52	28
44	60	70	13
15	50	40	13
Test, Small			
103*	70	36	22
2	40	44	36
20	30	54	9
51	30	59	87
Test, Adeno			
18	40	69	5
90	60	50	22
84	80	62	4
Test, Large			
164	70	68	15
19	30	39	4
43	60	49	11
340	80	64	10
231	70	67	18

[a]Starred quantities denote censored observations.
[b]Days of survival t, performance status x_1, age in years x_2, and number of months from diagnosis to entry into the study x_3.

time of diagnosis. This is a measure of the general medical status on a scale of 0 to 100: 10, 20, and 30 mean that the patient is completely hospitalized; 40, 50, and 60, partially confined to hospital; and 70, 80, and 90, able to care for himself. Finally, the age of the patient and the number of months from diagnosis of lung cancer to entry into the study are also recorded.

The main objective is to compare the effects of the two treatments upon survival time, but in doing so one must take into account other factors that may also influence survival. What actually needs to be studied here is the joint effect of the various factors (i.e., treatments and concomitant variables) in the experiment on survival time.

Example 1.1.9 The data in Table 1.1.5 are from an experiment in which new models of a small electrical appliance were being tested (Nelson, 1970b). The appliances were operated repeatedly by an automatic testing machine; the lifetimes given here are the number of cycles of use completed until the appliances failed. There are two complicating factors: one is that there were many different ways in which an appliance could fail, 18 to be exact. Therefore in Table 1.1.5 each observation has a failure code number beside it. Numbers 1 through 18 refer to the 18 different possible causes of failure for the appliances. In addition, some of the observations were censored, since it was not always possible to operate the testing machine long enough for an appliance to fail. Appliances that have censored failure times are indicated in Table 1.1.5 as having a failure code of 0.

The joint distribution of failure times and failure modes is of interest here. This is important knowledge for the testing and development of the appliance, and also for the marketing of the appliance, since a warranty plan has to be determined. One of the reasons why it is difficult to handle data of this type is that the failure time distribution of the appliance will change as the appliance is developed, since product improvements will effectively remove certain causes of failure. This has to be recognized in examining sets of data collected at different times over the development period of the product.

Table 1.1.5 Failure Data for Electrical Appliance Test

Number of Cycles to Failure	Failure Code	Number of Cycles to Failure	Failure Code	Number of Cycles to Failure	Failure Code
11	1	958	10	35	15
2,223	9	7,846	9	2,400	9
4,329	9	170	6	1,167	9
3,112	9	3,059	6	2,831	2
13,403	0	3,504	9	2,702	10
6,367	0	2,568	9	708	6
2,451	5	2,471	9	1,925	9
381	6	3,214	9	1,990	9
1,062	5	3,034	9	2,551	9
1,594	2	2,694	9	2,761	6
329	6	49	15	2,565	0
2,327	6	6,976	9	3,478	9

The examples show some of the ways in which lifetime data arise and some of the questions that such data hope to answer. We now leave the discussion of data for the time being and turn to an examination of statistical models for lifetime distributions.

1.2 BASIC CONCEPTS OF LIFETIME DISTRIBUTIONS

1.2.1 Continuous Models

We begin by considering the case of a single lifetime variable T. Specifically, let T be a nonnegative random variable representing the lifetimes of individuals in some population. Usually T is assumed to be continuous, and we will treat this case first.

All functions, unless stated otherwise, are defined over the interval $[0, \infty)$. Let $f(t)$ denote the probability density function (p.d.f.) of T and let the distribution function be

$$F(t)=\Pr(T\leqslant t)=\int_0^t f(x)\,dx.$$

The probability of an individual surviving till time t is given by the survivor function

$$S(t)=\Pr(T\geqslant t)=\int_t^{\infty} f(x)\,dx. \tag{1.2.1}$$

In some contexts, especially ones involving lifetimes of manufactured items, $S(t)$ is referred to as the reliability function. Note that $S(t)$ is a monotone decreasing continuous function with $S(0)=1$ and $S(\infty)=\lim_{t\to\infty}S(t)=0$.

The pth quantile of the distribution of T is the value t_p such that

$$\Pr(T\leqslant t_p)=p.$$

That is, $t_p=F^{-1}(p)$. The pth quantile is also referred to as the $100p$th percentile of the distribution.

Another useful concept having to do with life distributions is the hazard function $h(t)$, defined as

$$\begin{aligned} h(t)&=\lim_{\Delta t\to 0}\frac{\Pr(t\leqslant T<t+\Delta t\,|\,T\geqslant t)}{\Delta t}\\ &=\frac{f(t)}{S(t)}. \end{aligned} \tag{1.2.2}$$

The hazard function specifies the instantaneous rate of death or failure at time t, given that the individual survives up till t. In particular, $h(t)\Delta t$ is the approximate probability of death in $[t, t+\Delta t)$, given survival up till t. The hazard function also goes under various other names, among them being the hazard rate, the (age-specific) failure rate, and the force of mortality.

The functions $f(t)$, $F(t)$, $S(t)$, and $h(t)$ give mathematically equivalent specifications of the distribution of T. It is easy to derive expressions for $S(t)$ and $f(t)$ in terms of $h(t)$: since $f(t) = -S'(t)$, (1.2.2) implies that

$$h(x) = -\frac{d}{dx}\log S(x).$$

Thus

$$\log S(x)|_0^t = -\int_0^t h(x)\,dx$$

and since $S(0)=1$, we find that

$$S(t) = \exp\left(-\int_0^t h(x)\,dx\right). \tag{1.2.3}$$

For some purposes it is also useful to define the cumulative hazard function

$$H(t) = \int_0^t h(x)\,dx$$

which, by (1.2.3), is related to the survivor function by $S(t) = \exp[-H(t)]$. It can be observed that since $S(\infty)=0$, then $H(\infty) = \lim_{t\to\infty} H(t) = \infty$. Thus the hazard function $h(t)$ for a continuous lifetime distribution possesses the properties

$$h(t) \geqslant 0, \qquad \int_0^\infty h(t)\,dt = \infty.$$

Finally, in addition to (1.2.3), it follows immediately from (1.2.2) that

$$f(t) = h(t)\exp\left(-\int_0^t h(x)\,dx\right). \tag{1.2.4}$$

Example 1.2.1 Suppose T has p.d.f.

$$f(t) = \beta t^{\beta-1}\exp(-t^\beta) \qquad t>0$$

where $\beta>0$ is a parameter; this is a Weibull distribution, discussed in Section 1.3.2. We consider it here to illustrate the formulas just presented. First, it follows easily from (1.2.1) that the survivor function for T is $S(t)=\exp(-t^{\beta})$, and then from (1.2.2) the hazard function is $h(t)=\beta t^{\beta-1}$. Conversely, if the model is specified initially in terms of $h(t)$, then $S(t)$ and $f(t)$ are readily obtained from (1.2.3) and (1.2.4).

1.2.2 Discrete Models

Sometimes, for example, when lifetimes are grouped or when "lifetime" refers to an integral number of cycles of some sort, it may be desired to treat T as a discrete random variable. Suppose T can take on values $t_1, t_2, \ldots,$ with $0 \leqslant t_1 < t_2 < \cdots$, and let the probability function (p.f.) be

$$p(t_j)=\Pr(T=t_j) \qquad j=1,2,\ldots$$

The survivor function is then

$$S(t)=\Pr(T \geqslant t)=\sum_{j:t_j \geqslant t} p(t_j). \tag{1.2.5}$$

Note that for the discrete case, as for the continuous case, $S(t)$ is a monotone decreasing left-continuous function, with $S(0)=1$ and $S(\infty)=0$. The hazard function is now defined as

$$\begin{aligned} h(t_j) &= \Pr(T=t_j \mid T \geqslant t_j) \\ &= \frac{p(t_j)}{S(t_j)} \qquad j=1,2,\ldots \end{aligned} \tag{1.2.6}$$

As in the continuous case, the probability, survivor, and hazard functions give equivalent specifications of the distribution of T. Observe that since $p(t_j)=S(t_j)-S(t_{j+1})$, (1.2.6) implies that

$$h(t_j)=1-\frac{S(t_{j+1})}{S(t_j)} \qquad j=1,2,\ldots \tag{1.2.7}$$

and thus

$$S(t)=\prod_{j:t_j<t}\left[1-h(t_j)\right]. \tag{1.2.8}$$

An analog of $H(t)$ can be defined in the discrete case, but "cumulative hazard function" is not a good name for it. The appropriate analog of the continuous cumulative hazard function is

$$H(t) = -\log S(t)$$

where $S(t)$ is given by (1.2.8). It is easily seen that $H(t)$ does not equal $\Sigma_{j:t_j<t} h(t_j)$ in general, hence the reluctance to use the term "cumulative hazard function" here.

Occasionally situations arise in which one would like T to have both discrete and continuous components. Special notation or definitions will not be introduced for such situations, which will be handled as they occur. No real difficulties are encountered with mixed distributions, especially if one works primarily with the survivor function, which, as usual, is a monotone decreasing left-continuous function on $[0, \infty)$.

1.2.3 Some Remarks on the Hazard Function

The p.d.f. (or p.f.), and the distribution and survivor functions are common representations of a probability distribution, but the reader may not have previously encountered the hazard function. This function is particularly useful with lifetime distributions, since it describes the way in which the instantaneous probability of death for an individual changes with time. Often, in applications, there may be qualitative information about the hazard function, which can help in selecting a life distribution model. For example, there may be reasons to restrict consideration to models with nondecreasing hazard functions or with hazard functions having some other well-defined characteristic.

Figure 1.2.1 shows hazard functions and p.d.f.'s for three continuous distributions. The shapes of the hazard functions are qualitatively quite different; distribution (a) has a monotone increasing hazard function, distribution (b) has a monotone decreasing hazard function, and (c) has a so-called "bathtub-shaped", or U-shaped, hazard function. Models with these and other shapes of hazard function are all useful in practice. If, for example, individuals in a population are followed right from actual birth to death, a bathtub-shaped hazard function is often appropriate. We are, for example, familiar with this pattern in human populations: after an initial period in which deaths result primarily from birth defects or infant diseases, the death rate drops and is relatively constant until the age of 30 or so, after which it increases with age. This pattern also manifests itself in many other populations, including ones consisting of manufactured items.

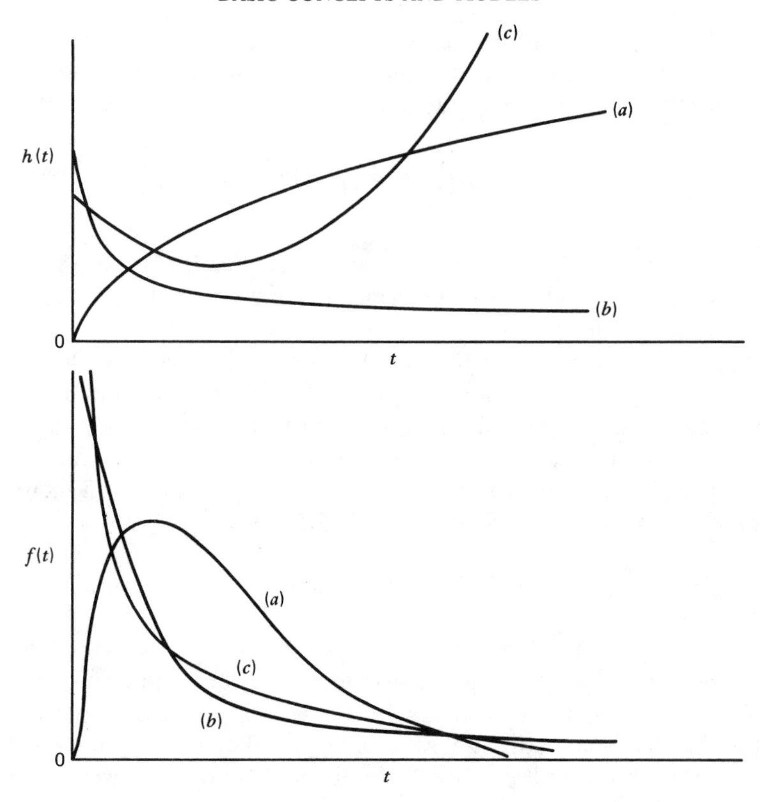

Figure 1.2.1 Some hazard and probability density functions

Models with increasing hazard functions are used the most. One reason for this is that interest often centers on a period in the life of an individual over which some kind of gradual aging takes place, yielding an increasing hazard function. Also, populations that display a bathtub-shaped hazard function are sometimes purged of weak individuals, leaving a reduced population with an increasing hazard function. For example, manufacturers often use a "burn-in" process in which items are subjected to a brief period of operation before being sent to customers. In this way defective items that would fail very early are removed from the population; this frequently leaves a residual population in which individuals exhibit gradual aging, with an increasing hazard function.

Models with a constant hazard function are important and have a particularly simple structure, as we shall see in Section 1.3.1. Models with decreasing hazard functions are less common, but are sometimes used. For example, certain types of electronic devices appear to have decreasing failure rates, at least over some fairly long initial period of use. Nonmonotone hazard functions other than the bathtub-shaped ones are even less common, but possible. All in all, the main point to be remembered is that the hazard function represents an aspect of a distribution that has direct physical meaning and that information about the nature of the hazard function is helpful in selecting a model.

1.3 SOME IMPORTANT MODELS

Numerous parametric models are used in the analysis of lifetime data and in problems related to the modeling of aging or failure processes. Among univariate models, a few particular distributions occupy a central role because of their demonstrated usefulness in a wide range of situations. Foremost in this category are the exponential, Weibull, gamma, and log-normal distributions. This section introduces these models, as well as the generalized gamma distribution, which can be considered as a generalization of each of the others. A few additional univariate models are also noted, and multivariate and regression models are introduced.

Extensive motivation is not provided for the various models. To do this would require a thorough discussion of aging and failure processes and would take us outside this book's intended subject area. Indeed, the motivation for using a particular model in a given situation is often mainly empirical, it having been found that the model satisfactorily describes the distribution of lifetimes in the population under study. This does not, of course, imply any absolute "correctness" of the model. Sometimes there is information about the aging or failure process in a population that suggests a particular distribution, though this information is rarely specific enough to narrow considerations to just one family of models. This situation will no doubt improve as our understanding of aging and failure processes deepens.

Some theoretical motivation for particular models can be found in the references cited below. The interested reader may also consult one of several bibliographic works on univariate distributions. A useful reference is the series by Johnson and Kotz (1970), which extensively catalogs mathematical and statistical properties of most of the distributions presented here and provides additional references concerning their areas of application. An author who examines aging and failure processes is Shooman (1968).

We make two additional remarks before beginning the survey of important models. The first is that models are presented here without the inclusion of a so-called threshold parameter, or "guarantee time." Briefly, this is a time $\mu \geqslant 0$ before which it is assumed that an individual cannot die. Sometimes a situation calls for the inclusion of such a parameter. The distributions considered can all be extended to include a threshold parameter by merely replacing the lifetime t by $t' = t - \mu$, with t' satisfying the restriction $t' \geqslant 0$. For example, we consider in Section 1.3.1 the exponential distribution, in which T has p.d.f. $f(t) = \lambda \exp(-\lambda t)$, with $t \geqslant 0$. If a threshold parameter were introduced, the p.d.f. would be

$$f(t) = \lambda e^{-\lambda(t-\mu)} \qquad t \geqslant \mu.$$

Properties of the latter distribution follow immediately from those of the former, since $T' = T - \mu$ in the latter case has p.d.f. $\lambda \exp(-\lambda t')$, with $t' \geqslant 0$. The second point to be made is that parametric models lead to important, but by no means exclusive, techniques of analyzing lifetime data. Nonparametric procedures are also very important and are discussed throughout this book.

1.3.1 The Exponential Distribution

The exponential distribution has been widely used as a model in areas ranging from studies on the lifetimes of manufactured items (e.g., Davis, 1952; Epstein, 1958) to research involving survival or remission times in chronic diseases (e.g., Feigl and Zelen, 1965). The distribution is characterized by a constant hazard function

$$h(t) = \lambda \qquad t \geqslant 0 \tag{1.3.1}$$

where $\lambda > 0$. The p.d.f. and survivor function are found from (1.2.4) and (1.2.3) to be

$$f(t) = \lambda e^{-\lambda t} \quad \text{and} \quad S(t) = e^{-\lambda t} \tag{1.3.2}$$

respectively. The distribution is also often written using the parameterization $\theta = \lambda^{-1}$, in which case the p.d.f. becomes

$$f(t) = \theta^{-1} e^{-t/\theta} \qquad t \geqslant 0. \tag{1.3.3}$$

The mean and variance of the distribution are θ and θ^2, respectively, and the pth quantile is $t_p = -\theta \log(1-p)$. The distribution where $\theta = 1$ is called the standard exponential distribution; its p.d.f. is shown in Figure 1.3.1. Clearly, if T has p.d.f. (1.3.2), then λT has a standard exponential distribution. Several other properties of the exponential distribution are implicit in

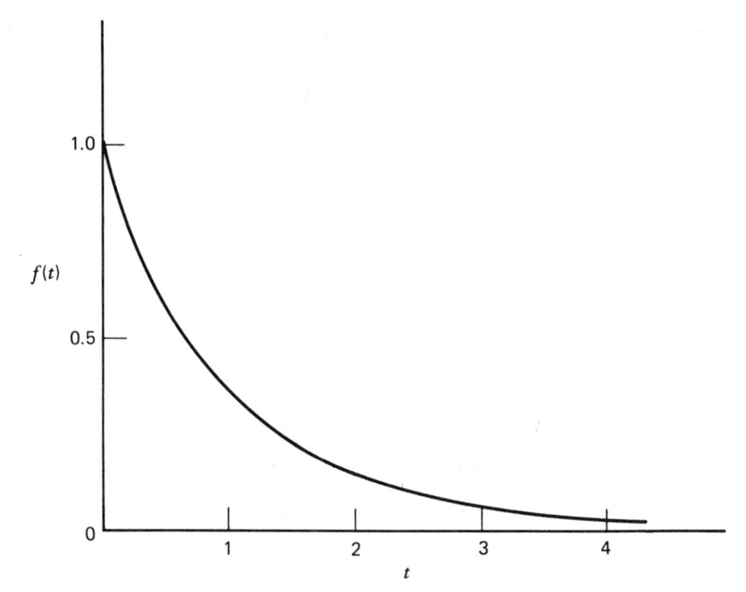

Figure 1.3.1 The standard exponential p.d.f.

the next two sections, since the exponential is a special case of both the Weibull and gamma distributions, which are considered in these sections.

Historically, the exponential distribution was the first widely used lifetime distribution model. This was partly because of the availability of simple statistical methods for it (e.g., Epstein and Sobel, 1953) and partly because the exponential distribution appeared suitable for representing the lifetimes of many things, such as various types of manufactured items (Davis, 1952). The assumption of a constant hazard function is a very restrictive one, however, and the later realization that many inferences are sensitive to departures from the exponential model has led to greater caution in the use of the distribution. Nevertheless, the exponential distribution still proves useful in a wide variety of situations. Statistical inference under an exponential model is considered in Chapter 3.

1.3.2 The Weibull Distribution

The Weibull distribution is perhaps the most widely used lifetime distribution model. Its application in connection with lifetimes of many types of manufactured items has been widely advocated (e.g., Weibull, 1951; Berretoni, 1964), and it has been used as a model with diverse types of items such as vacuum tubes (Kao, 1959), ball bearings (Lieblein and Zelen, 1956), and electrical insulation (Nelson, 1972a). It is also widely used in biomedical applications, for example, in studies on the time to the occurrence of tumors in human populations (Whittemore and Altschuler, 1976) or in laboratory animals (Pike, 1966; Peto et al., 1972) and in many other situations.

The Weibull distribution has a hazard function of the form

$$h(t)=\lambda\beta(\lambda t)^{\beta-1} \tag{1.3.4}$$

where $\lambda>0$ and $\beta>0$ are parameters. It includes the exponential distribution as the special case where $\beta=1$. By (1.2.4) and (1.2.3), the p.d.f. and survivor function of the distribution are

$$f(t)=\lambda\beta(\lambda t)^{\beta-1}\exp[-(\lambda t)^{\beta}] \qquad t>0 \tag{1.3.5}$$

and

$$S(t)=\exp[-(\lambda t)^{\beta}] \qquad t>0. \tag{1.3.6}$$

The rth raw moment $E(X^r)$ of the distribution is easily found to be $\lambda^{-r}\Gamma(1+r/\beta)$, where

$$\Gamma(k)=\int_0^{\infty} u^{k-1}e^{-u}\,du \qquad k>0$$

is the gamma function (see Appendix B). The mean and variance are thus $\lambda^{-1}\Gamma(1+1/\beta)$ and $\lambda^{-2}[\Gamma(1+2/\beta)-\Gamma(1+1/\beta)^2]$.

The Weibull distribution hazard function is monotone increasing if $\beta>1$, decreasing if $\beta<1$, and constant for $\beta=1$. The model is fairly flexible (see Figure 1.3.2) and has been found to provide a good description of many types of lifetime data. This and the fact that the model has simple expressions for the p.d.f. and survivor and hazard functions partly account for its popularity. The Weibull distribution arises as an asymptotic extreme value distribution (see Problem 1.11), and in some instances this can be used to provide motivation for it as a model (e.g., Weibull, 1951; Peto et al., 1972).

The shape of the Weibull p.d.f. depends upon the value of β, in fact, β is sometimes called the shape parameter for the distribution. "Typical" β values vary from application to application, but in many situations distributions with β in the range 1 to 3 seem appropriate. Figure 1.3.2*a* shows some Weibull p.d.f.'s for $\lambda=1$ and several different values of β, and Figure 1.3.2*b* shows the corresponding hazard functions. Note that since λ in (1.3.5) is a

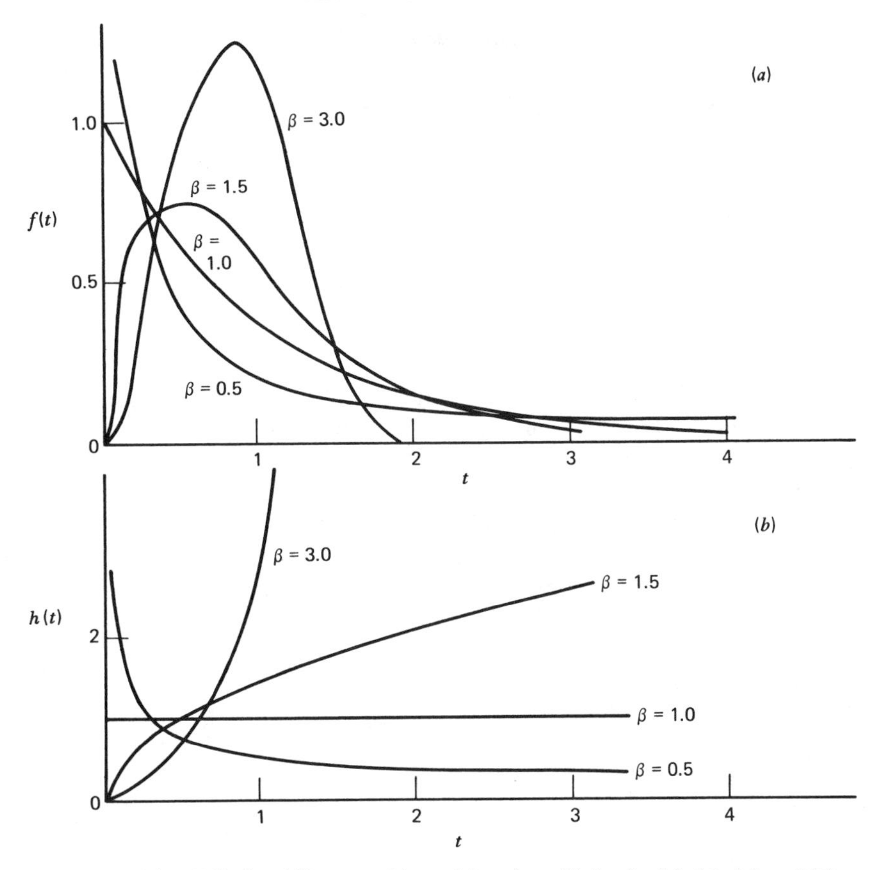

Figure 1.3.2 Weibull p.d.f.'s (a) and hazard functions (b) for $\beta=0.5$, 1.0, 1.5, and 3.0

scale parameter, the effect of different values of λ in Figure 1.3.2 is just to change the scale on the horizontal (t) axis, and not the basic shape of the graph.

The Extreme Value Distribution

It is convenient at this point to introduce a distribution that is closely related to the Weibull distribution. This is the so-called first asymptotic distribution of extreme values, hereafter referred to simply as the extreme value distribution. This distribution is extensively used in a number of areas and sometimes referred to as the Gumbel distribution, after E. J. Gumbel, who had pioneered its use (Gumbel, 1958). Our interest in it is not for its direct use as a lifetime distribution, but rather because of the relationship it bears to the Weibull distribution.

The p.d.f. and survivor function for the extreme value distribution are, respectively,

$$f(x)=b^{-1}\exp\left[\frac{x-u}{b}-\exp\left(\frac{x-u}{b}\right)\right] \qquad -\infty<x<\infty \qquad (1.3.7)$$

$$S(x)=\exp\left[-\exp\left(\frac{x-u}{b}\right)\right] \qquad -\infty<x<\infty \qquad (1.3.8)$$

where $b>0$ and u $(-\infty<u<\infty)$ are parameters. This distribution is directly related to the Weibull distribution by the easily shown fact that if T has a Weibull distribution with p.d.f. (1.3.5), then $X=\log T$ has an extreme value distribution with $b=\beta^{-1}$ and $u=-\log\lambda$. In analyzing data it is often convenient to work with log lifetimes; the extreme value distribution arises when lifetimes are taken to be Weibull distributed.

The extreme value distribution with $u=0$ and $b=1$ is termed the "standard extreme value distribution." A graph of its p.d.f. is given in Figure 1.3.3. Since u is a location and b a scale parameter, values of u and b

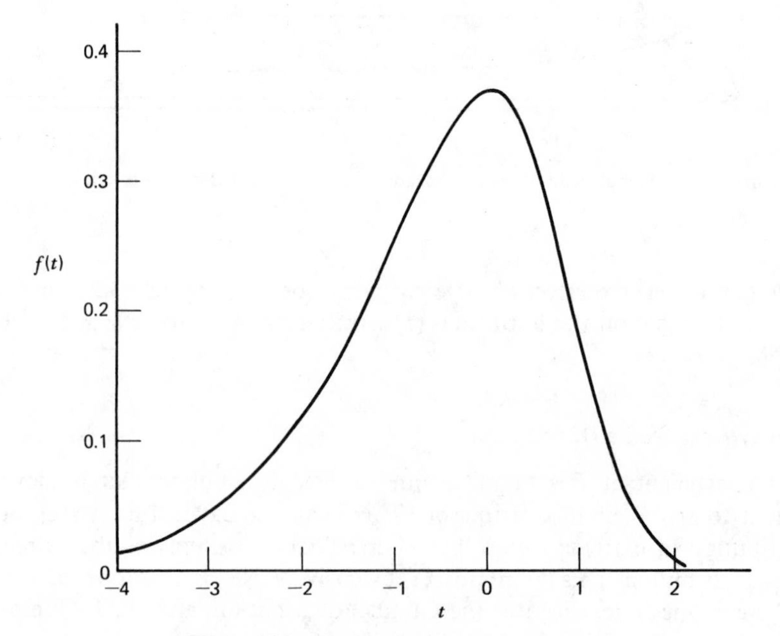

Figure 1.3.3 The standard extreme value p.d.f.

different from 0 and 1 in (1.3.7) do not affect the shape of $f(t)$, but only the location and scale.

To find the moments of the distribution, we will first obtain the moment generating function. For the standard extreme value distribution, this is

$$M(\theta)=\int_{-\infty}^{\infty} e^{\theta x}\exp(x-e^{x})\,dx.$$

Letting $y=e^{x}$, we have

$$M(\theta)=\int_{0}^{\infty} y^{\theta}e^{-y}\,dy$$

$$=\Gamma(1+\theta). \qquad (1.3.9)$$

The mean of the standard extreme value distribution is found from this to be $\Gamma'(1)=-\gamma$, where $\gamma=0.5772\ldots$ is known as Euler's constant; the variance is $\Gamma''(1)-\gamma^{2}=\pi^{2}/6$ (see Appendix B). From this one also has the useful results

$$\int_{-\infty}^{\infty} x\exp(x-e^{x})\,dx=\int_{0}^{\infty}(\log y)e^{-y}\,dy=-\gamma$$

$$\int_{-\infty}^{\infty} x^{2}\exp(x-e^{x})\,dx=\int_{0}^{\infty}(\log y)^{2}e^{-y}\,dy=\frac{\pi^{2}}{6}+\gamma^{2}.$$

The mean and variance of the general distribution (1.3.7) are $u-\gamma b$ and $(\pi^{2}/6)b^{2}$, since $(X-u)/b$ has the standard extreme value distribution. The pth quantile of (1.3.7) is

$$x_{p}=u+b\log[-\log(1-p)]$$

which implies that the location parameter u is the 0.632th quantile.

The statistical analysis of data under a Weibull distribution model is discussed in Chapter 4, and both it and the extreme value distribution are considered further at that point.

1.3.3 The Gamma Distribution

The gamma distribution has a p.d.f. of the form

$$f(t)=\frac{\lambda(\lambda t)^{k-1}e^{-\lambda t}}{\Gamma(k)} \qquad t>0 \qquad (1.3.10)$$

where $k>0$ and $\lambda>0$ are parameters; λ is a scale parameter and k is sometimes called the index or shape parameter. This distribution, like the Weibull distribution, includes the exponential as a special case ($k=1$). The survivor and hazard functions involve the incomplete gamma function [see (B12) of Appendix B]

$$I(k,x)=\frac{1}{\Gamma(k)}\int_0^x u^{k-1}e^{-u}\,du. \tag{1.3.11}$$

Integrating (1.3.10), we find that the survivor function is

$$S(t)=1-I(k,\lambda t).$$

Comments on the calculation of (1.3.11) are given in Appendix B and in Section 5.1. The hazard function is $h(t)=f(t)/S(t)$. It can be shown (see Problem 1.5) to be monotone increasing for $k>1$, with $h(0)=0$ and $\lim_{t\to\infty}h(t)=\lambda$. For $0<k<1$, $h(t)$ is monotone decreasing, with $\lim_{t\to 0^+}h(t)=\infty$ and $\lim_{t\to\infty}h(t)=\lambda$.

The distribution with $\lambda=1$ is called the one-parameter gamma distribution and has p.d.f.

$$\frac{t^{k-1}e^{-t}}{\Gamma(k)} \qquad t>0. \tag{1.3.12}$$

The notation $Y\sim \text{Ga}(k)$ will be used to indicate that a random variable Y has p.d.f. (1.3.12). Note that if T has p.d.f. (1.3.10), then $\lambda T\sim\text{Ga}(k)$. The one-parameter gamma distribution is closely related to the chi square (χ^2) distribution: if $Y\sim\text{Ga}(k)$, then $2Y$ has a χ^2 distribution with $2k$ degrees of freedom, henceforth simply referred to as $\chi^2_{(2k)}$. Figure 1.3.4 shows p.d.f.'s for a few gamma distributions.

The moment generating function of (1.3.12) is

$$\begin{aligned}M(\theta)&=\int_0^\infty \frac{e^{\theta t}t^{k-1}e^{-t}\,dt}{\Gamma(k)}\\ &=(1-\theta)^{-k}\end{aligned}$$

and that of (1.3.10) is $(1-\theta/\lambda)^{-k}$. The moments of the distribution can be readily found from this; for example, $E(T^r)=k(k+1)\cdots(k+r-1)$ for (1.3.12).

The gamma distribution is used as a lifetime model (e.g., Gupta and Groll, 1961), though not nearly as much as the Weibull distribution. This is

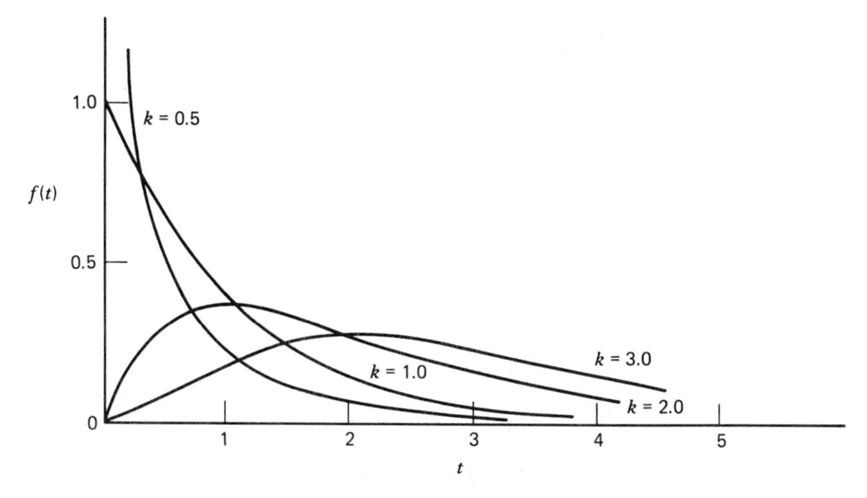

Figure 1.3.4 Gamma p.d.f.'s with $\lambda=1$ and $k=0.5$, 1.0, 2.0, and 3.0

partly because the survivor and hazard functions of the gamma distribution are not expressible in a simple closed form and hence are more difficult to work with than with those of the Weibull distribution. The gamma distribution does fit a wide variety of lifetime data adequately, however, and there are failure process models that lead to it (see Buckland, 1964, Sec. 1.7). The gamma distribution also arises mathematically in some situations in which the exponential distribution is being used, in consequence of the well-known result that sums of independent and identically distributed (i.i.d.) exponential random variables have a gamma distribution. Specifically, if $T_1, \ldots, T_n$ are independent, each with p.d.f. (1.3.2), then $T_1 + \cdots + T_n$ has a gamma distribution with parameters λ and $k=n$. This result is easily established using moment generating functions, since (1.3.2) has moment generating function $M(\theta)=(1-\theta/\lambda)^{-1}$, and thus $T_1 + \cdots + T_n$ has moment generating function $(1-\theta/\lambda)^{-n}$.

The Log-Gamma Distribution

As with the Weibull distribution, it is convenient to note some details about the distribution of a log lifetime when lifetimes follow a gamma distribution. Suppose that T has p.d.f. (1.3.10) or, equivalently, that λT has a one-parameter gamma distribution with p.d.f. (1.3.12). Then $W=\log(\lambda T)=\log\lambda+\log T$ has what is usually termed a "log-gamma distribution" (Bartlett and Kendall, 1946), with p.d.f.

$$\frac{1}{\Gamma(k)}\exp(kw-e^w) \qquad -\infty<w<\infty. \tag{1.3.13}$$

Figure 1.3.5 shows log-gamma distributions for several values of k. It is seen that the distributions are negatively skewed, with skewness decreasing as k increases. The case $k=1$ is the standard extreme value distribution.

The moment generating function for the distribution is

$$\begin{aligned} M(\theta) &= \int_{-\infty}^{\infty} e^{\theta w} \frac{1}{\Gamma(k)} \exp(kw - e^{w})\, dw \\ &= \frac{1}{\Gamma(k)} \int_{0}^{\infty} u^{\theta+k-1} e^{-u}\, du \\ &= \frac{\Gamma(\theta+k)}{\Gamma(k)}. \end{aligned} \tag{1.3.14}$$

From this we find that the mean and variance are

$$E(W) = \frac{d \log \Gamma(k)}{dk} = \psi(k)$$

$$\operatorname{Var}(W) = \frac{d^2 \log \Gamma(k)}{dk^2} = \psi'(k).$$

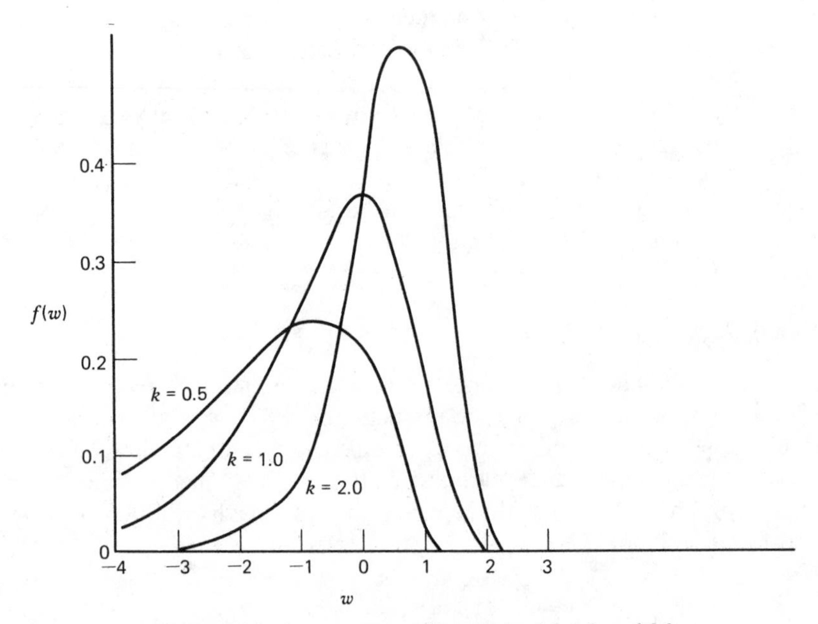

Figure 1.3.5 Log-gamma p.d.f.'s with $k=0.5$, 1.0, and 2.0

The functions $\psi(k)$ and $\psi'(k)$ are called the digamma and trigamma functions, respectively (see Appendix B). These are tabulated, and calculation formulas given, in Abramowitz and Stegun (1965, Ch. 6). For our immediate purposes we note the series expressions

$$\psi(k)=\log k-\frac{1}{2k}-\frac{1}{12k^2}+\frac{1}{120k^4}-\frac{1}{252k^6}+\cdots$$

$$\psi'(k)=\frac{1}{k}+\frac{1}{2k^2}+\frac{1}{6k^3}-\frac{1}{30k^5}+\cdots.$$

For large k the first terms of these series dominate, and it is seen that as $k\to\infty$ the distribution is shifted further and further to the right. As $k\to\infty$ the distribution of the rescaled variate

$$Z=(W-\log k)\sqrt{k} \tag{1.3.15}$$

converges to the standard normal distribution. To see this, we first find from (1.3.13) the p.d.f. of Z, which is

$$g(z;k)=\frac{k^{k-1/2}}{\Gamma(k)}\exp\left(\sqrt{k}z-ke^{z/\sqrt{k}}\right) \qquad -\infty<z<\infty.$$

Expanding $\exp(z/\sqrt{k})$ in a power series and rearranging some terms slightly, we obtain

$$g(z;k)=\frac{k^{k-1/2}e^{-k}}{\Gamma(k)}\exp\left(-\frac{z^2}{2}-\frac{z^3}{6k^{1/2}}-\frac{z^4}{24k}-\cdots\right).$$

Using Stirling's formula (e.g., Abramowitz and Stegun, 1965, Ch. 6), which states that

$$\lim_{k\to\infty}\left(\frac{(2\pi)^{1/2}k^{k-1/2}e^{-k}}{\Gamma(k)}\right)=1$$

we find that as $k\to\infty$, $g(z;k)\to e^{-z^2/2}/(2\pi)^{1/2}$, which is the desired limiting standard normal p.d.f.

The gamma and log-gamma distributions are discussed further in Sections 5.1 and 5.3, where inference procedures for them are presented.

1.3.4 The Log-Normal Distribution

The log-normal distribution, like the Weibull distribution, has been widely used as a lifetime distribution model, in spite of one unattractive feature, which is mentioned below. It has been used in diverse situations, such as the analysis of failure times of electrical insulation (Nelson and Hahn, 1972) and the study of times to the appearance of lung cancer in cigarette smokers (Whittemore and Altschuler, 1976). The distribution is most easily specified by saying that the lifetime T is log-normally distributed if the logarithm $Y=\log T$ of the lifetime is normally distributed, say, with mean μ and variance σ^2. The p.d.f. of Y is therefore

$$\frac{1}{(2\pi)^{1/2}\sigma}\exp\left[-\frac{1}{2}\left(\frac{y-\mu}{\sigma}\right)^2\right] \qquad -\infty<y<\infty$$

and from this the p.d.f. of $T=\exp Y$ is easily found to be

$$f(t)=\frac{1}{(2\pi)^{1/2}\sigma t}\exp\left[-\frac{1}{2}\left(\frac{\log t-\mu}{\sigma}\right)^2\right] \qquad t>0. \qquad (1.3.16)$$

The name "log-normal" is seen to be somewhat of a misnomer, since the log of a log-normal variate actually has a normal distribution.

The survivor and hazard functions for the log-normal distribution involve the standard normal distribution function

$$\Phi(x)=\int_{-\infty}^{x}\frac{1}{(2\pi)^{1/2}}e^{-u^2/2}\,du.$$

The log-normal survivor function is easily seen to be

$$S(t)=1-\Phi\left(\frac{\log t-\mu}{\sigma}\right)$$

and the hazard function is given as $h(t)=f(t)/S(t)$.

The unattractive property of the log-normal distribution alluded to earlier concerns its hazard function. It can be shown (see Problem 1.4) to have the value 0 at $t=0$, increase to a maximum, and then decrease, approaching 0 as $t\to\infty$. Since the hazard function is decreasing for large values of t, the distribution seems implausible as a lifetime model in most situations. Nevertheless, this model is often found suitable for representing lifetimes, especially when very large values of t are not of interest, and it can be derived from fairly plausible assumptions about certain types of failure

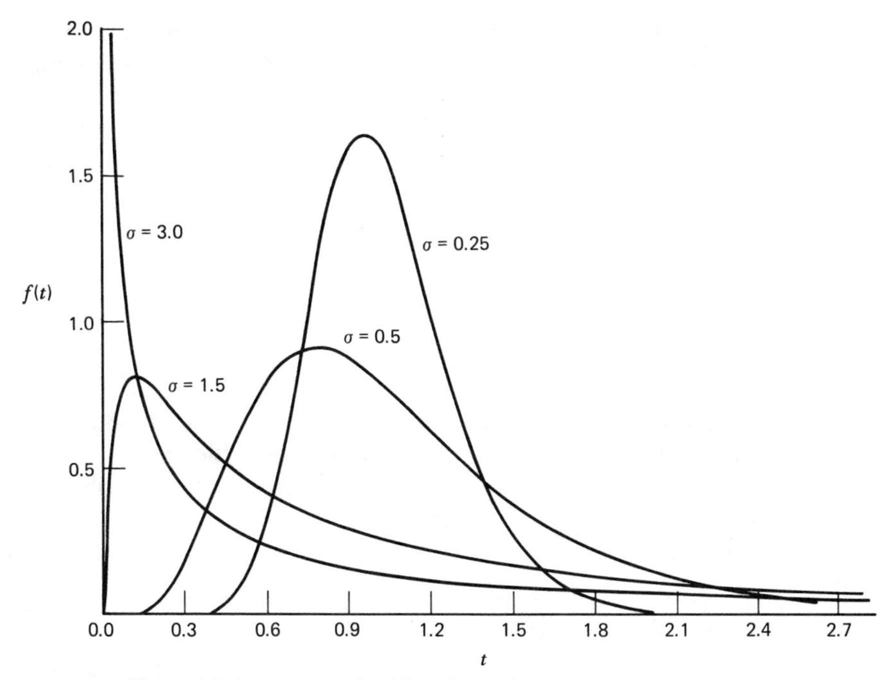

Figure 1.3.6 Log-normal p.d.f.'s with $\mu=0$ and $\sigma=0.25$, 0.5, 1.5, and 3.0

processes. The fact that log lifetimes are normally distributed is sometimes a convenience, though this is less important than in precomputer days.

Figure 1.3.6 shows some log-normal p.d.f.'s for $\mu=0$ and different values of σ. It should be noted that the effect of a nonzero value of μ is just to change the scale on the time axis, and not the basic shapes of the functions portrayed, since μ is a location parameter in the Y $(=\log T)$ form of the model and e^{μ} is a scale parameter in (1.3.16).

Some additional properties of the log-normal distribution are discussed in Problem 1.4., including the fact that the mean and variance are $\exp(\mu+\sigma^2/2)$ and $[\exp(\sigma^2)-1][\exp(2\mu+\sigma^2)]$, respectively. Statistical inference for log-normal distributions is considered in Section 5.2.

1.3.5 The Generalized Gamma Distribution

The generalized gamma distribution is a three-parameter distribution with p.d.f.

$$f(t)=\frac{\lambda\beta}{\Gamma(k)}(\lambda t)^{k\beta-1}\exp\left[-(\lambda t)^{\beta}\right] \qquad t>0 \qquad (1.3.17)$$

where β, λ, and k are all positive. This useful model was introduced by Stacy (1962) and includes as special cases all of the lifetime distributions mentioned in the preceding four sections. The exponential ($\beta=k=1$), Weibull ($k=1$), and gamma ($\beta=1$) distributions are all evident and, in addition, the log-normal distribution appears as a limiting case when $k\to\infty$. This can be seen as follows: from (1.3.17), $(\lambda T)^{\beta}$ has a one-parameter gamma distribution with p.d.f. (1.3.12), and hence

$$W=\log\left[(\lambda T)^{\beta}\right]$$
$$=\beta(\log\lambda+\log T)$$

has a log-gamma distribution with p.d.f. (1.3.13). As shown in Section 1.3.4, $\sqrt{k}(W-\log k)$ then has a standard normal distribution in the limit as $k\to\infty$.

The hazard and survivor functions for the generalized gamma distribution involve the incomplete gamma function (1.3.11). It is readily seen that the survivor function is

$$S(t)=1-I\left[k,(\lambda t)^{\beta}\right]$$

and that the hazard function is $h(t)=f(t)/S(t)$.

The generalized gamma distribution is a flexible three-parameter family of distributions. It is also useful in helping to decide among alternate models such as the Weibull and the log-normal distributions and for examining the effect of departures from an assumed model. Inference procedures for it are given in Section 5.3.

1.3.6 Other Lifetime Distribution Models

The exponential, Weibull, gamma, log-normal, and generalized gamma distributions are the most frequently used parametric lifetime distribution models. Many other models are available, however, and are sometimes used in applications. Although it is not feasible to describe these in detail, we shall list some of them, along with a few references from which more information can be obtained.

1. Distributions in which either $h(t)$ or $\log[h(t)]$ is a lower-order polynomial (e.g., Bain, 1974; Krane, 1963; Gehan and Siddiqui, 1973; Gross and Clark, 1975, Sec. 4.8). Important models of this type are the Gompertz distribution, with $h(t)=\exp(\alpha+\beta t)$, and the Rayleigh, or linear hazard rate, distribution, with $h(t)=a+bt$.

2. The inverse Gaussian distribution (Chhikara and Folks, 1977).
3. The normal distribution $N(\mu, \sigma^2)$ truncated on the left at 0 (Davis, 1952; Barlow and Proschan, 1965).
4. The log-logistic distribution (Kalbfleisch and Prentice, 1980, Sec. 2.2.6).
5. Models with bathtub-shaped hazards. Models capable of giving bathtub and other nonmonotone hazard functions are important, though a practical difficulty is that such models are often harder to handle statistically than the common lifetime distributions. Families of models of this type are discussed by Murthy et al. (1973), Canfield and Borgman (1975), Glaser (1980), Hjorth (1980), and others. Models in (1) can also be of this type when $h(t)$ or $\log[h(t)]$ is a second- or higher-degree polynomial. Glaser and Hjorth give careful discussions of distributions with nonmonotone hazard functions.
6. Discrete distributions. Usually, when a discrete model is used with lifetime data, it is a multinomial distribution and arises because effectively continuous data have been grouped. Methods for multinomial models are discussed in Chapter 2. Very occasionally the situation may call for another discrete distribution, usually over the nonnegative integers. Such situations are best treated individually, but generally one tries to adopt one of the standard discrete distributions (e.g., Johnson and Kotz, 1969).

1.3.7 Mixture Models

Discrete mixture models arise when individuals in a population are each one of k distinct types, with a proportion p_i of the population being of the ith type; the p_i's satisfy $0<p_i<1$ and $\Sigma p_i = 1$. Individuals of type i are assumed to have a lifetime distribution with survivor function $S_i(t)$. An individual randomly selected from this population then has survivor function

$$S(t)=p_1S_1(t)+\cdots+p_kS_k(t). \tag{1.3.18}$$

Models of this kind are termed "discrete mixture models" and are useful in situations where the population is nonhomogeneous but it is not possible to distinguish between individuals of different types. Often the $S_i(t)$'s in (1.3.18) are taken to be from the same parametric family, though this is, of course, unnecessary. The properties of a mixture model are easily derived from the properties of the k distributions, or components, involved in the mixture. Models with k larger than 2 or 3 are rarely used, unless one is in a situation where the $S_i(t)$'s are completely known. Otherwise, the number of unknown parameters is usually large, and estimation of them difficult enough to render the models unattractive. Even models with $k=2$ or 3 are difficult to handle statistically.

Examples of mixture models in a life distribution context are given, for example, by Cox (1959) and Kao (1959). Statistical inference with mixture models is briefly discussed in Section 5.4.

1.3.8 Regression Models

An important way of handling heterogeneity in a population is through the inclusion of concomitant, or regressor, variables in the model. It is very common for data to involve concomitant variables related to lifetime: for example, in a study on survival time for lung cancer patients, factors such as the age and general physical condition of the patient, the type of tumor, the time since diagnosis, and so on may all be relevant (see Example 1.1.8). Similarly, the lifetime of electrical insulation may depend on the voltage the insulation is subjected to while in use (see Example 1.1.5). Regression models, with lifetime as the response variable and the concomitant variables as regressor variables, allow such additional factors to be conveniently incorporated in a statistical analysis.

The models described in this chapter can all be extended to include regressor variables. This can be done in various ways, however, and instead of presenting a thorough discussion of this topic now, we give just a single brief example. Chapters 6 and 7 deal in depth with regression models and regression analysis of lifetime data; Section 6.1 in particular describes the formation of regression models.

Example 1.3.1 Suppose that associated with each individual in a population are a lifetime T and a vector $\mathbf{x}=(x_1,\ldots,x_p)$ of regressor variables, thought to be related to lifetime. Also suppose that given $\mathbf{x}$, the distribution of T is exponential, with survivor function

$$S(t|\mathbf{x})=\exp[-\lambda(\mathbf{x})t]. \tag{1.3.19}$$

Different functional forms for $\lambda(\mathbf{x})$ give different models; one particular form that is used a great deal has $\lambda(\mathbf{x})=\exp(\mathbf{x}\boldsymbol{\beta})$, where $\boldsymbol{\beta}$ is a column vector of regression coefficients (e.g., Breslow, 1974; Prentice, 1973). This and other exponential regression models are discussed in Section 6.2.

1.3.9 Multivariate Lifetime Distributions

Sometimes two or more lifetime variables $T_1,\ldots,T_k$ are of interest simultaneously and a multivariate model is required. For example, a device may have two integral parts, and it may be desired to model the joint life distribution of the parts. A multivariate distribution can be specified in terms of the joint survivor function

$$S(t_1,\ldots,t_k)=\Pr(T_1\geqslant t_1,\ldots,T_k\geqslant t_k). \qquad (1.3.20)$$

In fortunate circumstances $T_1,\ldots,T_k$ can be assumed to be independent, so that (1.3.20) becomes $S_1(t_1)\cdots S_k(t_k)$, where $S_i(t_i)$ is the marginal survivor function for T_i. In this case one is effectively back in the univariate framework.

When $T_1,\ldots,T_k$ cannot be assumed to be independent, the situation is more difficult; in fact, relatively little work has been done on multivariate lifetime distributions. One problem is the scarcity of realistic, tractable multivariate lifetime models. Another is that the hazard function concept is somewhat difficult to extend to the multivariate situation. A third difficulty is that many data that arise in multivariate situations are censored in such a way that one cannot determine whether or not $T_1,\ldots,T_k$ are independent. Because of this there has not been the motivation for developing multivariate lifetime data methods that there might be.

One very important problem of the type just referred to is the so-called "competing risks problem," where it is assumed that individuals can die from any one of k competing causes of death; Example 1.1.9 is an illustration of this. A common approach to this problem is to associate with the ith cause of death a variable T_i that represents the time to death from cause i. The lifetime of an individual is then

$$T=\min(T_1,\ldots,T_k)$$

since the individual is dead as soon as he dies from any particular cause of death. If $T_1,\ldots,T_k$ have the joint survivor function $S(t_1,\ldots,t_k)$, the survivor function of T is

$$\Pr(T\geqslant t)=S(t,\ldots,t).$$

Unfortunately it is not possible on the basis of data on T alone to determine whether the T_i's are independent or not (e.g., Cox, 1959; Tsiatis, 1975).

Further examination of multivariate models is deferred to Chapter 10, where both multivariate distributions and competing risks are discussed.

1.3.10 Remarks on the Choice of a Lifetime Distribution Model

There are obviously many potential life models. In some situations there may be reasons to select a particular family of models: the model may fit data on hand well, past experience may have shown the model to give a good description of lifetime distributions from similar populations, there may be a knowledge of the underlying aging or failure process that suggests the validity of the model, and so on. Sometimes a model is deemed unsuitable because of the form of its hazard function, even though it may fit existing data. The log-normal distribution is a case in point; the fact that it has a decreasing hazard function past a certain point sometimes rules it out as a model, even though it may fit the available data quite well. Two distributions with similar p.d.f.'s can have very different hazard functions, so that specific knowledge or assumptions about the nature of the hazard function can tip the scales in favor of one or another of two competing distributions, even when it is not possible to discriminate between them on the basis of the data on hand.

In situations in which no family of models is singled out as being particularly appropriate, choice of a model is frequently made on the basis of considerations such as (1) the convenience of mathematically handling the model, (2) the statistical methods available in connection with the model, and (3) the degree of complication of the calculations involved in using the model. The early dominance of the exponential distribution as a model was in no small part due to its mathematical simplicity and the availability of good, simple statistical methods for it. Similarly, the emergence of the Weibull distribution as perhaps the most frequently used parametric lifetime distribution was aided by the fact that its p.d.f., survivor function, and hazard function all have simple closed forms and that good statistical methods are available for it.

A point that should be reiterated in connection with most of the commonly used models is that they handle situations that call for a monotone hazard function, but are not capable of giving bathtub or other nonmonotone shapes of hazard functions. A few models capable of nonmonotone hazard functions have been mentioned in Section 1.3.6, and there are many others that could be mentioned. These models generally involve more parameters and are less mathematically and statistically tractable than the common monotone hazard function models, but they are nevertheless important. Some of the problems associated with the use of such models are discussed in Section 5.4.

Three additional points should be mentioned in connection with the choice of a model. First, one test of any model is to see that it fits the available data. Methods of assessing the fit of lifetime distributions are considered in Chapter 9. Second, even though a model may appear acceptable in a given situation, one should be aware of the consequences of departures from the assumed model on any inferences made. This problem is examined in a few places in this book, but deserves more study. Finally, although much use is made in lifetime distribution work of parametric models like those discussed in this section, there are many situations in which it is desirable to avoid strong assumptions about the model. Nonparametric or distribution-free procedures are important in this case. These are discussed in Chapters 2, 7, and 8.

1.4 CENSORING AND STATISTICAL METHODS

1.4.1 Types of Censoring

Lifetime data often come with a feature that creates special problems in the analysis of the data. This feature is known as censoring and, broadly speaking, occurs when exact lifetimes are known for only a portion of the individuals under study; the remainder of the lifetimes are known only to exceed certain values. Censoring arises in various ways; some examples have already been given in Section 1.1. In this section censoring is examined more closely and a number of new ideas are introduced. Specifically, we consider the form of the likelihood function under various types of censored sampling.

Formally, an observation is said to be right censored at L if the exact value of the observation is not known but only that it is greater than or equal to L. Similarly, an observation is said to be left censored at L if it is known only that the observation is less than or equal to L. Right censoring is very common in lifetime data, but left censoring is fairly rare.

For convenience only right censoring is discussed here, though many of the ideas transfer in an obvious way to the case of left censoring. In addition, the term "censoring" will be used, meaning in all instances "right censoring," and when an individual has his lifetime censored at L, we will call L the censoring time for the individual.

To discuss censoring we must consider the way in which data are obtained. In fact, censoring arises for a variety of reasons, and we consequently distinguish among several types of censoring processes in the discussion that follows. The basic problem is to determine the sampling distribution and corresponding likelihood function for a given process and then to determine the properties of statistical methods derived from this. We find that even in relatively simple situations one has to rely heavily on large-sample methods and their asymptotic properties. In complicated situations it may even be difficult to write down a likelihood, let alone determine its asymptotic properties.

For the most part, only continuous models and data are considered in this section. Grouped or discrete data are mentioned briefly in Section 1.4.1d below, and are discussed in more detail in Chapter 2.

1.4.1a *Type II Censoring*

We first describe what has come to be called Type II censoring. A Type II censored sample is one for which only the r smallest observations in a random sample of n items are observed ($1 \leqslant r \leqslant n$). Experiments involving Type II censoring are often used, for example, in life testing; a total of n items is placed on test, but instead of continuing until all n items have failed, the test is terminated at the time of the rth item failure. Such tests can save time and money, since it could take a very long time for all items to fail in some instances. It will be seen that the statistical treatment of Type II censored data is, at least in principle, straightforward.

It should be stressed that with Type II censoring the number of observations r is decided before the data are collected. Formally, the data consist of the r smallest lifetimes $T_{(1)} \leqslant T_{(2)} \leqslant \cdots \leqslant T_{(r)}$ out of a random sample of n lifetimes $T_1, \ldots, T_n$ from the life distribution in question. If $T_1, \ldots, T_n$ are i.i.d. and have a continuous distribution with p.d.f. $f(t)$ and survivor function $S(t)$, it follows from general results on order statistics [see (D2) of Appendix D] that the joint p.d.f. of $T_{(1)}, \ldots, T_{(r)}$ is

$$\frac{n!}{(n-r)!} f(t_{(1)}) \cdots f(t_{(r)}) \left[S(t_{(r)})\right]^{n-r}. \tag{1.4.1}$$

For any given parametric model statistical inference can be based on (1.4.1), which gives the likelihood function and from which one can derive sampling properties of procedures. The analysis of Type II censored data is thus straightforward, at least in principle. Statistical methods for data from various distributions are discussed throughout this book. A few general remarks are made in Section 1.4.2.

Example 1.4.1 Consider a random sample $T_1, \ldots, T_n$ from an exponential distribution with p.d.f. $f(t) = \lambda \exp(-\lambda t)$ and survivor function $S(t) = \exp(-\lambda t)$. Then (1.4.1) becomes

$$\lambda^r \frac{n!}{(n-r)!} \exp\left(-\lambda \sum_{i=1}^{r} t_{(i)}\right) \exp\left[-(n-r)\lambda t_{(r)}\right]$$

$$= \lambda^r \frac{n!}{(n-r)!} \exp\left[-\lambda\left(\sum_{i=1}^{r} t_{(i)} + (n-r) t_{(r)}\right)\right] \tag{1.4.2}$$

which is the joint p.d.f. of the r smallest observations $T_{(1)} \leqslant \cdots \leqslant T_{(r)}$. Statistical inference is very straightforward in this situation, which is discussed in detail in Section 3.1. For example,

$$T=\sum_{i=1}^{r} t_{(i)}+(n-r)t_{(r)}$$

is readily seen to be sufficient for λ; the maximum likelihood estimate (m.l.e.) of λ is $\hat{\lambda}=r/T$, and it can be shown that $2\lambda T\sim\chi^2_{(2r)}$.

Progressive Type II Censoring*

A generalization of Type II censoring is progressive Type II censoring. In this case, the first r_1 failures in a sample of n items are observed; then n_1 of the remaining $n-r_1$ unfailed items are removed from the experiment, leaving $n-r_1-n_1$ items still present. When further r_2 items have failed, n_2 of the still unfailed items are removed, and so on. The experiment terminates after some prearranged series of repetitions of this procedure.

We will obtain the likelihood function in this case, assuming as before that lifetimes are i.i.d. with p.d.f. $f(t)$ and survivor function $S(t)$. For ease of discussion, let us suppose the censoring has just two stages: at the time of the r_1th failure, n_1 of the remaining $n-r_1$ unfailed items are randomly selected and removed. The experiment then terminates when further r_2 items have failed. At this point there will be $n-r_1-n_1-r_2$ items still unfailed and in the experiment. The observations in this case are the r_1 failure times $T_{(1)}\leqslant\cdots\leqslant T_{(r_1)}$ in the first stage of the experiment and the r_2 failure times in the second stage of the experiment, which we will denote by $T^*_{(1)}\leqslant\cdots\leqslant T^*_{(r_2)}$. The experiment is represented in the figure below:

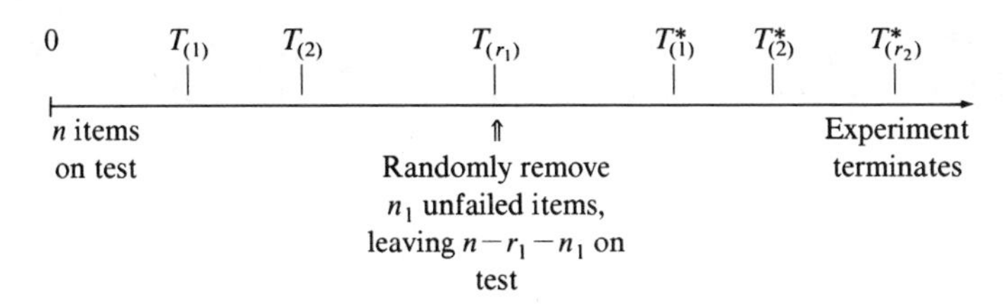

The sampling distribution of the data can be written as

$$g\left(t_{(1)},\ldots,t_{(r_1)},t^*_{(1)},\ldots,t^*_{(r_2)}\right)=g_1\left(t_{(1)},\ldots,t_{(r_1)}\right)g_2\left(t^*_{(1)},\ldots,t^*_{(r_2)}|t_{(1)},\ldots,t_{(r_1)}\right) \tag{1.4.3}$$

where g, g_1, and g_2 represent p.d.f.'s of the variables indicated. The joint p.d.f. $g_1(t_{(1)},\ldots,t_{(r_1)})$ of $T_{(1)},\ldots,T_{(r_1)}$ is given by (1.4.1), with $r=r_1$. To write down the second term in (1.4.3) we observe that given $t_{(1)}, \ldots, t_{(r_1)}$, the lifetimes of the items left in the experiment have a left-truncated distribution with p.d.f. and survivor functions

$$f_1(t) = \frac{f(t)}{S(t_{(r_1)})} \qquad t \geqslant t_{(r_1)}$$

and

$$S_1(t) = \frac{S(t)}{S(t_{(r_1)})} \qquad t \geqslant t_{(r_1)}$$

respectively. $T^*_{(1)}, \ldots, T^*_{(r_2)}$ are the r_2 smallest observations in a random sample of size $n-n_1-r_1$ from this truncated distribution. By (1.4.1), the second term in (1.4.3) is therefore

$$\frac{(n-r_1-n_1)!}{(n-r_1-n_1-r_2)!} f_1(t^*_{(1)}) \cdots f_1(t^*_{(r_2)}) \left[S_1(t^*_{(r_2)})\right]^{n-r_1-n_1-r_2}$$

$$= \frac{(n-r_1-n_1)!}{(n-r_1-n_1-r_2)!} \frac{f(t^*_{(1)}) \cdots f(t^*_{(r_2)})}{\left[S(t_{(r_1)})\right]^{n-r_1-n_1}} \left[S(t^*_{(r_2)})\right]^{n-r_1-n_2-r_2}.$$

Combining the two parts of (1.4.3), we obtain the likelihood function as

$$cf(t_{(1)}) \cdots f(t_{(r_1)}) \left[S(t_{(r_1)})\right]^{n_1} f(t^*_{(1)}) \cdots f(t^*_{(r_2)}) \left[S(t^*_{(r_2)})\right]^{n-r_1-n_1-r_2} \tag{1.4.4}$$

where $c = n!(n-r_1-n_1)!/[(n-r_1)!(n-r_1-n_1-r_2)!]$.

This generalizes in an obvious way to experiments with more than two stages. For any given parametric model (1.4.4) supplies the likelihood function and, in principle, allows the sampling properties of procedures to be calculated, though, of course, these calculations may be complicated.

1.4.1b *Type I Censoring*

Sometimes experiments are run over a fixed time period in such a way that an individual's lifetime will be known exactly only if it is less than some predetermined value. In such situations the data are said to be Type I (or "time") censored. For example, in a life test experiment n items may be placed on test, but a decision made to terminate the test after a time L has elapsed. Lifetimes will then be known exactly only for those items that fail by time L. A more complicated form of Type I censoring exists in Example 1.1.6: a decision had been made to terminate an experiment on August 31, so that lifetimes of certain items were censored. However, each item had its own specific censoring time L_i, since all items were not started on test on the same date. Type I censoring also frequently arises in medical research where, for example, a decision is made to terminate a study at a date on which not all the individuals' lifetimes will be known.

Stated more precisely, a Type I censored sample is one that arises when individuals $1,2,\ldots,n$ are subjected to limited periods of observation $L_1,\ldots,L_n$, so that an individual's lifetime T_i is observed only if $T_i \leqslant L_i$. When all of the L_i's are equal, we sometimes say that the data are singly Type I censored, to distinguish this from the general case. It should be noted that with Type I censoring the number of exact lifetimes observed is random, in contrast to the case of Type II censoring, where it is fixed.

In dealing with Type I censoring, it is convenient to use a different type of notation from before. Suppose that there are n individuals under study and that associated with the ith individual is a lifetime T_i and a fixed censoring time L_i. The T_i's are assumed to be i.i.d. with p.d.f. $f(t)$ and survivor function $S(t)$. The exact lifetime T_i of an individual will be observed only if $T_i \leqslant L_i$. The data from such a setup can be conveniently represented by the n pairs of random variables (t_i, δ_i), where

$$t_i = \min(T_i, L_i) \quad \text{and} \quad \delta_i = \begin{cases} 1 & \text{if } T_i \leqslant L_i \\ 0 & \text{if } T_i > L_i \end{cases}.$$

That is, δ_i indicates whether the lifetime T_i is censored or not, and t_i is equal to T_i if it is observed, and to L_i if it is not.

The joint p.d.f. of t_i and δ_i is

$$f(t_i)^{\delta_i} S(L_i)^{1-\delta_i}.$$

To see this, note that t_i is a mixed random variable with a continuous and a discrete component. For the discrete part we have

$$\begin{aligned} \Pr(t_i = L_i) &= \Pr(\delta_i = 0) \\ &= \Pr(T_i > L_i) = S(L_i). \end{aligned}$$

For values $t_i < L_i$ the continuous p.d.f. is

$$\Pr(t_i|\delta_i=1)=\Pr(t_i|t_i<L_i)$$
$$=\frac{f(t_i)}{1-S(L_i)}$$

where for convenience we have used the notation $\Pr(t_i|\delta_i=1)$ to represent the probability density function of t_i, given that $t_i<L_i$. The distribution of (t_i, δ_i) thus has components

$$\Pr(t_i=L_i, \delta_i=0)=\Pr(\delta_i=0)=S(L_i)$$
$$\Pr(t_i, \delta_i=1)=\Pr(t_i|\delta_i=1)\Pr(\delta_i=1) \qquad t_i<L_i$$
$$=f(t_i).$$

These expressions can be combined into the single expression $\Pr(t_i, \delta_i)= f(t_i)^{\delta_i}S(L_i)^{1-\delta_i}$, and if pairs (t_i, δ_i) are independent, the likelihood function is

$$L=\prod_{i=1}^{n} f(t_i)^{\delta_i}S(L_i)^{1-\delta_i}. \tag{1.4.5}$$

Example 1.4.2 As in Example 1.4.1, suppose that lifetimes are independent and follow an exponential distribution. Then $f(t)=\lambda\exp(-\lambda t)$ and $S(t)=\exp(-\lambda t)$; (1.4.5) becomes

$$L=\prod_{i=1}^{n} (\lambda e^{-\lambda t_i})^{\delta_i} e^{-\lambda L_i(1-\delta_i)}$$
$$=\lambda^r \exp\left(-\lambda \sum_{i=1}^{n} t_i\right)$$

where $r=\Sigma\delta_i$ is the observed number of "deaths," or failures.

The form of the likelihood function (1.4.5) is noteworthy: each observed lifetime contributes a term $f(t_i)$ to the likelihood, and each censoring time contributes a term $S(L_i)$. It can also be noted that although the genesis of (1.4.5) is quite different from that of the likelihood functions (1.4.1) or (1.4.4) obtained for Type II censored sampling, the form of the observed likelihood function is the same in both cases. To see this it is only necessary to note that in both (1.4.1) and (1.4.4) individuals whose lifetimes are observed contribute a term $f(t_i)$ to the observed likelihood function, whereas individuals whose lifetimes are censored contribute a term $S(L_i)$. This form of likelihood function in fact turns out to have wide applicability, as discussed in Section 1.4.1d.

The sampling distribution involved in (1.4.5) is more difficult to deal with than that in (1.4.1), and consequently the statistical analysis of Type I censored data poses slightly more problems than that of Type II censored data. Whereas exact small-sample methods can often be used for Type II censored sampling, it is usually necessary to rely on large-sample methods with Type I censoring. Statistical methods for Type I censored data are discussed throughout this book for a variety of models and situations.

1.4.1c *Random Censoring*

Censoring times are often effectively random. For example, in a medical trial patients may enter the study in a more or less random fashion, according to their time of diagnosis. If the study is terminated at some prearranged date, then censoring times, that is the lengths of time from an individual's entry into the study until the termination of the study, are random. For inference purposes one often works conditionally on the observed censoring times, proceeding as though the censoring were Type I, but the process by which the data were generated needs to be considered in order to justify this. The assumption of a random censoring mechanism is also a useful device for investigating the properties of certain procedures; this approach is used in Section 2.1, for example.

A very simple random censoring process that is often realistic is one in which each individual is assumed to have a lifetime T and a censoring time L, with T and L independent continuous random variables, with survivor functions $S(t)$ and $G(t)$, respectively. Let (T_i, L_i), $i=1,\ldots,n$, be independent and, as in the case of Type I censoring, define $t_i=\min(T_i, L_i)$ and $\delta_i=1$ if $T_i \leqslant L_i$ and $\delta_i=0$ if $T_i > L_i$. The data from observations on n individuals consist of the pairs (t_i, δ_i), $i=1,\ldots,n$. The p.d.f. of (t_i, δ_i) is easily obtained: if $f(t)$ and $g(t)$ are the p.d.f.'s for T_i and L_i, then, using the same notation as in the derivation of (1.4.5), we have

$$\Pr(t_i=t, \delta_i=0)=\Pr(L_i=t, T_i>L_i)$$
$$=g(t)S(t)$$
$$\Pr(t_i=t, \delta_i=1)=\Pr(T_i=t, T_i\leqslant L_i)$$
$$=f(t)G(t).$$

These can be combined into the single expression $\Pr(t_i = t, \delta_i) = [f(t)G(t)]^{\delta_i}[g(t)S(t)]^{1-\delta_i}$, and thus the sampling distribution of (t_i, δ_i), $i=1,\ldots,n$, is

$$\prod_{i=1}^{n} [f(t_i)G(t_i)]^{\delta_i}[g(t_i)S(t_i)]^{1-\delta_i}$$

$$= \left(\prod_{i=1}^{n} G(t_i)^{\delta_i} g(t_i)^{1-\delta_i}\right)\left(\prod_{i=1}^{n} f(t_i)^{\delta_i} S(t_i)^{1-\delta_i}\right).$$

If $G(t)$ and $g(t)$ do not involve any parameters of interest, then the first term can be neglected and the likelihood function taken to be

$$L = \prod_{i=1}^{n} f(t_i)^{\delta_i} S(t_i)^{1-\delta_i}$$

which is of the same form as (1.4.5). The earlier result for Type I censored sampling can in fact be considered as a special case of this if we allow the L_i's to have different, degenerate distributions, each with mass at one fixed point. Another approach that leads directly to this likelihood function is to argue that if $G(t)$ and $g(t)$ do not involve any parameters of interest, then one should condition on the realized censoring times $L_1,\ldots, L_n$ when making inferences about the distribution of T. This leads back to the Type I censoring framework. A point to note is that although it may be desirable to make inferences conditional on the L_i's in any given situation, the properties of procedures averaged over the distribution of the L_i's may be of interest in planning experiments and in assessing methods of estimation and testing.

Although the independent random censorship model is often reasonable, in many situations the censoring process is linked to the failure time process. Suppose, for example, that the termination date for a medical trial is not fixed before the study commences but is chosen later, with the choice influenced by the results of the study up to that time. In such instances it may be difficult to even write down a model that fully represents the process under study. Fortunately, the likelihood function (1.4.5) turns out to be applicable in many such complicated situations. This is discussed in the following section.

1.4.1d *More General Censoring Processes**

Often the censoring–death process is sufficiently complicated to make precise modeling impossible. Nevertheless, it can be shown that under fairly mild conditions the likelihood function (1.4.5) can be used to make inferences about the lifetime distribution under study.

General censoring processes have been studied by several authors, including Cox (1975), Efron (1977), Kalbfleisch and MacKay (1978), and Kalbfleisch and Prentice (1980, Sec. 5.2). The basic idea is to model the censoring–death process as it unfolds in time. We then find that (1.4.5) is valid, provided that a kind of quasi-independence exists between censoring and death.

Suppose that there are n individuals under observation at time 0 and that each of these is followed until he dies or is censored. We consider observations from a continuous lifetime model, but it is convenient to discretize the problem and then obtain results for continuous data by passing to a limit. Thus let the time axis be partitioned into intervals $I_j=[a_{j-1}, a_j)$, $j=1,\ldots, k+1$, where $a_0=0$, $a_{k+1}=\infty$, and $a_k=T$ is an upper limit on observation time. Also let $\Delta_j=a_j-a_{j-1}$; results for continuous data will be obtained by letting $k\to\infty$ and $\Delta_j\to 0$, $j=1,\ldots,k$. The risk set R_j for the interval I_j is defined as the set of individuals known to be alive (i.e., alive and uncensored) at a_{j-1}. Finally, let D_j be the set of individuals who die in I_j and let C_j be the set of individuals who are censored in I_j.

The data consist of $D_1, C_1,\ldots, D_k, C_k$, and in order to develop a likelihood function we will view the process as it evolves in time. The probability distribution of the data can be written as a product of conditional probabilities:

$$\begin{aligned}\Pr(D_1,C_1,\ldots,D_k,C_k)&=\Pr(D_1,C_1)\prod_{j=2}^{k}\Pr(D_j,C_j\mid D_1,C_1,\ldots,D_{j-1},C_{j-1})\\&=\Pr(D_1,C_1)\prod_{j=2}^{k}\Pr(D_j\mid D_1,\ldots,D_{j-1},C_{j-1})Q_j\end{aligned}\tag{1.4.6}$$

where

$$Q_j=\Pr(C_j\mid D_1,\ldots,C_{j-1},D_j)\qquad j=2,\ldots,k.$$

Two assumptions are now made that allow the likelihood function to be specified further:

Assumption 1 The death–censoring mechanisms governing different individuals act independently over I_j, given (C_i, D_i), $i=1,\ldots, j-1$.

Assumption 2 For each individual in R_j, Pr(death, but not censored in $I_j | D_1, C_1, \ldots, D_{j-1}, C_{j-1}) = \Pr(\text{death in } I_j | \text{ survival beyond } I_{j-1})$.

These assumptions are discussed later. Assumption 2 is crucial to what follows and specifies a kind of quasi-independence for lifetimes and censoring times. This assumption would typically be questionable when interval lengths are not small, but in many situations it is appropriate as the Δ_j's approach zero.

Suppose that individual i has survivor function $S_i(t)$. The probability of this individual dying in I_j, given that she survives past I_{j-1}, is then $[S_i(a_{j-1}) - S_i(a_j)]/S_i(a_{j-1})$. Under Assumptions 1 and 2 it follows that for $j = 1, \ldots, k$

$$\Pr(D_j | D_1, \ldots, D_{j-1}, C_{j-1}) = \prod_{i \in D_j} \left(1 - \frac{S_i(a_j)}{S_i(a_{j-1})}\right) \prod_{l \in R_j - D_j} \left(\frac{S_l(a_j)}{S_l(a_{j-1})}\right). \tag{1.4.7}$$

Further discussion will for convenience apply to the case in which all individuals have the same survivor function $S(t)$. Results for the general case in which the $S_i(t)$'s are different follow the same arguments and will be noted later. With all $S_i(t) = S(t)$, (1.4.7) becomes

$$\left(1 - \frac{S(a_j)}{S(a_{j-1})}\right)^{d_j} \left(\frac{S(a_j)}{S(a_{j-1})}\right)^{n_j - d_j}$$

where $d_j = |D_j|$ and $n_j = |R_j|$. From (1.4.6) the likelihood function is therefore

$$\prod_{j=1}^{k} \left(1 - \frac{S(a_j)}{S(a_{j-1})}\right)^{d_j} \left(\frac{S(a_j)}{S(a_{j-1})}\right)^{n_j - d_j} Q$$

where $Q = Q_1 \cdots Q_k$. Letting $c_j = |C_j|$ and noting that $n_{j+1} = n_j - d_j - c_j$, one can rewrite this as

$$\prod_{j=1}^{k} \left[S(a_{j-1}) - S(a_j)\right]^{d_j} \left[S(a_j)\right]^{c_j} Q.$$

If Q does not depend upon any parameters of interest, it can be dropped and the likelihood function taken as

$$\prod_{j=1}^{k}\left[S(a_{j-1})-S(a_j)\right]^{d_j}\left[S(a_j)\right]^{c_j}. \tag{1.4.8}$$

Otherwise (1.4.8) is a partial likelihood (see Appendix E).

For continuous lifetime data the likelihood function is obtained from (1.4.8) by letting $k\to\infty$, $\max(\Delta_j)\to 0$, and $T\to\infty$. Since $S(a_{j-1})-S(a_j)=\Delta_j f(\xi_j)$, where $f(t)=-S'(t)$ and $\xi_j\in[a_{j-1},a_j)$, (1.4.8) can be written as

$$\prod_{j=1}^{k}\left[\Delta_j f(\xi_j)\right]^{d_j}\left[S(a_j)\right]^{c_j}.$$

The limit of the ratio of this to $\Delta_1\Delta_2\cdots\Delta_k$ is the appropriate likelihood function in the continuous case. This gives

$$L=\prod_{i=1}^{n} f(t_i)^{\delta_i}S(L_i)^{1-\delta_i} \tag{1.4.9}$$

where t_i is a lifetime, L_i a censoring time, and $\delta_i=1$ if individual i's lifetime is observed and $\delta_i=0$ if it is censored. This is the same as the likelihood (1.4.5).

For the case in which individuals do not all have the same lifetime distribution, (1.4.9) is replaced by

$$L=\prod_{i=1}^{n} f_i(t_i)^{\delta_i}S_i(L_i)^{1-\delta_i} \tag{1.4.10}$$

where $f_i(t)$ and $S_i(t)$ are the p.d.f. and survivor function for the ith individual. An alternate notation that will be useful later denotes D as the set of individuals for whom lifetimes are observed and C as the set of individuals for whom only censoring times are available. Then L can be rewritten as

$$L=\prod_{i\in D} f_i(t_i)\prod_{i\in C} S_i(L_i). \tag{1.4.11}$$

The likelihood functions (1.4.9) and (1.4.10) are valid under relatively broad conditions, according to the derivation just given. We found earlier that Type I and Type II censoring schemes both gave this form for the likelihood, and now it has been established that more complicated processes do also. In order to comprehend the range of situations for which (1.4.9) and (1.4.10) are appropriate, it is necessary to examine Assumptions 1 and 2 more closely, as well as the step in the derivation where the term $Q_1 \cdots Q_k$ was dropped from the likelihood function. More properly, the assumptions need to be examined in conjunction with the limiting process that produced (1.4.9) and (1.4.10).

When the limit of $Q_1 \cdots Q_k/(\Delta_1 \cdots \Delta_k)$ does not depend upon any parameters of interest, the censoring can be called noninformative. This is the case for Type I and Type II censoring and also holds in many other situations. If the Q_j's do involve parameters of interest, (1.4.9) can still be regarded as a partial likelihood (see Appendix E) and used for inference, though some loss of information may be involved in disregarding the Q_j's. Assumption 1 is straightforward, and reasonable in many situations. Assumption 2 is also often reasonable, at least in the limit as the Δ_j's$\to 0$. Kalbfleisch and MacKay (1978) discuss processes that satisfy Assumption 2 in some detail.

To convince oneself of the validity of (1.4.9) or (1.4.10) for a particular process, it is necessary to check the assumptions. This is often not too difficult; as an example, let us consider the case in which censoring times and lifetimes are continuous independent random variables. It has already been demonstrated in Section 1.4.1 that (1.4.9) is appropriate in this situation, but it is illustrative to view the problem from the approach taken here. Assumption 1 is clearly satisfied with this model. With regard to Assumption 2, suppose that the lifetime and censoring time distributions have hazard functions $h(t)$ and $g(t)$, respectively. Then, if $I_j=[t, t+\delta t)$, say, we have for any individual in R_j

$$\begin{aligned}
&\Pr(\text{death but not censored in } I_j \mid D_1, \ldots, C_{j-1}) \\
&\quad = h(t)\,\delta t[1-g(t)\,\delta t] + o(\delta t) \\
&\quad = h(t)\,\delta t + o(\delta t) \\
&\quad = \Pr(\text{death in } I_j \mid \text{survival to } t) + o(\delta t).
\end{aligned}$$

Thus

$$\lim_{\delta t \to 0} \frac{\Pr(\text{death, but not censored in } I_j \mid D_1, \ldots, C_{j-1})}{\delta t}$$

$$= \lim_{\delta t \to 0} \frac{\Pr(\text{death in } I_j \mid \text{survival to } t)}{\delta t}$$

and Assumption 2 is valid in the limit as $\delta t \to 0$. It can similarly be shown that $Q_j/\delta t \to g(t)$ as $\delta t \to 0$ and therefore does not involve any of the parameters in $h(t)$.

1.4.2 Statistical Inference With Censored Data

The presence of censoring creates special problems for statistical inference, some of which are not completely resolved. With Type II censoring, matters are in principle straightforward: properties of the likelihoods (1.4.1) and (1.4.4) and procedures associated with them can be readily obtained. The books by Sarhan and Greenberg (1962) and David (1970) discuss properties of order statistics and give many procedures for Type II censored data. Exact methods of inference for Type II censored data are discussed throughout this book for a variety of parametric and nonparametric models.

Asymptotic theory associated with (1.4.1) or (1.4.4) can be established under essentially the same conditions as for uncensored samples (e.g., Halperin, 1952; Johnson, 1974). With the likelihood (1.4.1), for example, if $r \to \infty$ and $n \to \infty$ in such a way that $r/n \to p$, then results about the asymptotic normality of the maximum likelihood estimates, the limiting χ^2 distributions of likelihood ratio statistics, and so on are readily established. Standard asymptotic results for maximum likelihood are summarized in Appendix E.

With Type I censored sampling, complicated distributional problems usually make it impossible to work out exact properties of procedures, and there is consequently a heavy reliance on large-sample methods. As for Type II censored sampling, asymptotic results of the usual type can be shown to hold under essentially the same conditions as for complete samples (e.g., Kalbfleisch and Prentice, 1980, Sec. 3.4; Basu and Ghosh, 1980). An added requirement is that the sequence of fixed censoring times $L_1, \ldots, L_n$ obey conditions so that as $n \to \infty$ the expected information in the sample becomes infinite and the information in any finite group of observations does not dominate. A sufficient condition in most instances is that the expected number of observed (i.e., uncensored) lifetimes approach infinity as $n \to \infty$.

The usual large-sample procedures also appear to be valid for the likelihood functions (1.4.9) or (1.4.10) under fairly broad conditions for the lifetime and censoring time processes. Work by Cox (1975), Kalbfleisch and MacKay (1978), and others bears on this, though there are still numerous unanswered questions in this area.

Methods are presented throughout this book for Type I or Type II censored data. When data arise through a more complicated process for which the likelihood (1.4.9) or (1.4.10) is appropriate, they are typically treated as if they were Type I censored. This is sometimes justified by arguing that inferences should be made conditional on observed censoring times. Usually, however, we employ large-sample inference procedures that do not require explicit recognition of the censoring process so that the data are analyzed as if they were Type I censored. It should be remarked that in most cases the observed information, as opposed to the expected information (see Appendix E), is called for in connection with the limiting normal distributions of maximum likelihood estimates. A discussion of this point in sampling from the exponential distribution is given in Section 3.2.3.

1.5 PROBLEMS AND SUPPLEMENTS

1.1 Mean residual lifetime. Let T be a continuous random variable with survivor function $S(t)$. The mean residual life function $m(t)$ is defined as

$$m(t)=E(T-t|T\geqslant t).$$

a. Prove that

$$m(t)=\frac{\int_t^\infty S(x)\,dx}{S(t)}.$$

Also obtain $S(t)$ in terms of $m(t)$, showing that $m(t)$ uniquely defines the distribution of T.

b. Prove that

$$\lim_{t\to\infty} m(t)=\lim_{t\to\infty}\left(-\frac{d}{dt}\log f(t)\right)^{-1}$$

where $f(t)=-S'(t)$ is the p.d.f. of T. Use this to show that for the log-normal distribution $m(t)\to\infty$ as $t\to\infty$.

(Sections 1.2.1, 1.3.4)

1.2 Classifying life distributions. Suppose a continuous lifetime distribution has survivor function $S(t)$, hazard function $h(t)$, cumulative hazard function $H(t)$, and mean residual life function $m(t)$. Consider the following properties that a distribution might have:

I. $h(t)$ is nondecreasing for $t\geqslant 0$. Distributions with this property are often said to have the increasing failure rate (IFR) property.

II. $H(t)/t$ is nondecreasing for $t>0$. Distributions with this property are often said to have the increasing failure rate on the average (IFRA) property.

III. $m(t) \leqslant m(0)$ for all $t \geqslant 0$. Distributions with this property are often said to have the "new better than used" property.

IV. $m(t)$ is a decreasing function for $t \geqslant 0$. This is called the decreasing mean residual life property.

a. Prove that I$\Rightarrow$II$\Rightarrow$III.

b. Prove that I$\Rightarrow$IV$\Rightarrow$III.

(It is sometimes useful to classify distributions according to criteria like these, for example, in applications to system reliability.)

(Section 1.2; Bryson and Siddiqui, 1969; Barlow and Proschan, 1975)

1.3 Distributions with decreasing failure rates. A continuous lifetime distribution is said to have the decreasing failure rate (DFR) property if its hazard function $h(t)$ is nonincreasing for $t \geqslant 0$.

a. Show that $h'(t)<0$ only if $f'(t)<0$ and thus that a necessary condition for a distribution to have a DFR is that its p.d.f. have a unique mode at $t=0$.

b. Prove that a discrete mixture of distributions that all have DFRs has itself a DFR. Show that a discrete mixture of exponential distributions therefore has a decreasing failure rate and also that a mixture of IFR distributions does not necessarily have an IFR.

(Sections 1.2, 1.3.7; Proschan, 1963)

1.4 The log-normal distribution. Consider the log-normal distribution with p.d.f. (1.3.16).

a. Show that the mean and variance of the distribution are

$$E(T)=e^{\mu+\sigma^2/2}$$

$$\operatorname{Var}(T)=\left(e^{\sigma^2}-1\right)\left(e^{2\mu+\sigma^2}\right).$$

b. Show that the log-normal hazard function $h(t)$ has $h(0)=0$, increases to a maximum, then decreases, with $h(t)\to 0$ as $t\to\infty$.

(Section 1.3.4; Watson and Wells 1961; Goldthwaite, 1961; Glaser, 1980).

1.5 The gamma distribution. Consider the gamma distribution with p.d.f. $f(x)$ given by (1.3.10).

a. Show that the hazard function for this distribution is strictly monotone increasing if $k>1$ and strictly monotone decreasing if $k<1$. In both cases show that $\lim_{t\to\infty} h(t)=\lambda$.
b. Show that the gamma distribution mean residual life function $m(t)$ as defined in Problem 1.1 satisfies

$$\lim_{t\to\infty} m(t)=\lambda^{-1}.$$

c. For the case in which the index parameter k is an integer prove by repeated integration by parts that

$$\int_t^{\infty} f(x)\,dx=\sum_{i=0}^{k-1}\frac{e^{-\lambda t}(\lambda t)^i}{i!}.$$

In other words, if T has p.d.f. (1.3.10), then $P(T\geqslant t)=P(Y_{\lambda t}<k)$, where $Y_{\lambda t}$ has a Poisson distribution with mean λt. Note that this result also follows directly from well-known properties of the Poisson process.

(Section 1.3.3)

1.6 The logistic and log-logistic distributions. Lifetime T is said to have a log-logistic distribution if the log lifetime $Y=\log T$ has a logistic distribution with p.d.f.

$$\sigma^{-1}\frac{\exp[(y-\mu)/\sigma]}{\{1+\exp[(y-\mu)/\sigma]\}^2} \qquad -\infty<y<\infty$$

where $-\infty<\mu<\infty$ and $\sigma>0$ are parameters. The logistic distribution closely resembles the normal distribution but is easier to work with when data are censored, since its survivor function has a simple closed form.

a. Show that the moment generating function for $W=(Y-\mu)/\sigma$ is $M(\theta)=E[\exp(\theta W)]=\Gamma(1+\theta)\Gamma(1-\theta)$, and deduce from this that the mean and variance of W are 0 and $\pi^2/3$, respectively. Thus deduce the mean and variance of Y.

b. Show that $T = \exp Y$ has survivor function

$$S(t) = \left[1 + \left(\frac{t}{\alpha}\right)^{\beta}\right]^{-1} \qquad t>0$$

where $\alpha = e^{\mu}$ and $\beta = \sigma^{-1}$.

c. Determine the p.d.f. and hazard function for T. Show that the hazard function is monotone decreasing if $\beta \leqslant 1$ and that it behaves like the log-normal hazard function if $\beta > 1$. That is, for $\beta > 1$, $h(t)$ has $h(0) = 0$, increases to a maximum, then approaches 0 monotonically as $t \to \infty$.

(Section 1.3.6)

1.7 The generalized Pareto distribution. Consider the three-parameter distribution with a hazard function of the form

$$h(t) = \alpha + \frac{\beta}{t+\gamma}.$$

Examine the range of values that α, β, and γ can take. Investigate $h(t)$ and show that it can be monotone increasing or monotone decreasing, according to the values of the parameters. Give the p.d.f. and survivor function for the distribution.

(Davis and Feldstein, 1979)

1.8 Two models capable of bathtub-shaped hazards. Consider the two models that have hazard functions

1. $h(t) = \dfrac{\beta}{t+\gamma} + \delta t$
2. $h(t) = \dfrac{\beta}{\alpha}\left(\dfrac{t}{\alpha}\right)^{\beta-1} \exp\left[\left(\dfrac{t}{\alpha}\right)^{\beta}\right]$

respectively. Examine these models in detail, showing that each is capable of bathtub-shaped hazard functions. Comment on the flexibility of each in allowing a variety of shapes for the hazard function.

(Hjorth, 1980; Smith and Bain, 1975)

1.9 Compare the following asymptotic approximations for the digamma and trigamma functions with the truncated series expressions given in Section 1.3.3, and with exact values (tabulated in Abramowitz and Stegun,

1965, Ch. 6):

$$\psi(k)=\log k-\left(2k-\tfrac{1}{3}+\frac{1}{16k}\right)^{-1}$$

$$\psi'(k)=\left(k-\tfrac{1}{2}+\frac{1}{10k}\right)^{-1}.$$

(Section 1.3.3; Cox and Lewis, 1966, Ch. 2)

1.10 Suppose lifetime T has a gamma distribution (1.3.10) with $k=2$ and $\lambda=0.01$.

a. Determine the mean and variance of $Y=\log T$, which has a log-gamma distribution.
b. Determine the specific extreme value and normal distributions that have the same mean and variance as Y. Graph and compare the p.d.f.'s of the three distributions. Comment on the similarities and dissimilarities in the models, with a view to discriminating among them.
c. Compare in a similar way the p.d.f.'s of $T=\exp Y$ in the three cases.

1.11 Let $X_1, X_2, \ldots$ be i.i.d. random variables with continuous distribution function $F(x)=P(X_i \leqslant x)$ that satisfies the conditions

1. $F(0)=0$.
2. For some $\beta>0, \lim_{t\to 0^+}[F(xt)/F(t)]=x^\beta$, with $x>0$.

The second condition specifies that $F(x)\sim\alpha x^\beta$, where $\alpha>0$, as $x\to 0^+$.

a. Let $Y_n=\min(X_1,\ldots, X_n)$. Determine the survivor function of Y_n and hence the survivor function of $Z_n=n^{1/\beta}Y_n$. Show that as $n\to\infty$ the distribution of Z_n converges to a Weibull distribution.
b. Examine whether or not condition (2) holds when the X_i's have (1) a Weibull distribution, (2) a gamma distribution, and (3) a uniform distribution on $(0, a)$.

(These results are sometimes put forward as motivation for the Weibull model, as, for example, when an individual is assumed to die at the point at which one of many factors reaches a critical level. The approach here is not totally realistic, since the X_i's have been assumed to be i.i.d., but the limiting Weibull form may hold under weaker conditions.)

(Section 1.3.2)

1.12 Let $X_1<X_2<\cdots$ be a sequence of random variables, with X_i representing the amount of degradation in an item at the ith stage in its life. Suppose that the X_i's increase according to a random proportional increments model in which

$$X_{i+1}=(1+\xi_i)X_i \qquad i=1,2,\ldots$$

where the ξ_i's are independent nonnegative random variables. Let $W_n=\log X_n-\log X_1$. Note that

$$W_n=\log(1+\xi_1)+\cdots+\log(1+\xi_n)$$

and prove that if $n\to\infty$, with $\mathrm{Var}(W_n)$ finite, then the distribution of W_n converges to a normal distribution. The distribution of X_n thus converges to a log-normal distribution.

Suppose that failure occurs when W_n exceeds some critical value w. Let N be the smallest integer n such that $W_n>w$. Noting that $\Pr(N>n)=\Pr(W_n\leqslant w)$, find the distribution of N. Suppose that $\log(1+\xi_i)$ has mean μ_i and variance σ_i^2 and that $\mu=\lim_{n\to\infty}(\Sigma_{i=1}^n\mu_i/n)$ and $\sigma^2=\lim_{n\to\infty}(\Sigma_{i=1}^n\sigma_i^2/n)$ exist, and give the form of the distribution of N. (This is essentially the "Birnbaum-Saunders lifetime distribution"; see Mann et al. 1974; Sec. 4.11).

(Section 1.3.4)

1.13 Suppose that a population contains individuals for which lifetimes T are exponentially distributed, but that the hazard function λ varies across individuals. Specifically, suppose that the distribution of T given λ has p.d.f.

$$f(t|\lambda)=\lambda e^{-\lambda t} \qquad t\geqslant 0$$

and that λ itself has a gamma distribution with p.d.f.

$$g(\lambda)=\frac{\lambda^{k-1}e^{-\lambda/\alpha}}{\alpha^k\Gamma(k)} \qquad \lambda>0.$$

a. Find the unconditional p.d.f. and survivor function for T and show that the unconditional hazard function is

$$h(t)=\frac{k\alpha}{1+\alpha t}.$$

Note that this is a special case of the generalized Pareto model of Problem 1.7. Show that $h(t)$ is monotone decreasing.

b. Prove that if the distribution of T, given λ, is exponential and λ has a continuous distribution on $(0,\infty)$, then the hazard function for the marginal distribution of T is monotone decreasing.

c. Prove more generally that if the distribution of T, given λ, has a hazard function $h_\lambda(t)$ that is monotone decreasing for any $\lambda>0$, and λ has a distribution on $(0,\infty)$, then the hazard function for the marginal distribution of T is monotone decreasing. This generalizes results in Problem 1.3.

(Section 1.3; Proschan, 1963; Barlow et al., 1963)

1.14 Discrete models.

a. For the Poisson distribution with probability function

$$\Pr(X=j)=e^{-\lambda}\frac{\lambda^j}{j!} \qquad j=0,1,\ldots$$

show that the hazard function is monotone increasing.

b. For the negative binomial model with probability function

$$\Pr(X=j)=\binom{-\alpha}{j}p^{\alpha}(p-1)^{j} \qquad j=0,1,\ldots$$

where $\alpha>0$ and $0<p<1$, show that the hazard function is monotone decreasing (increasing) if $\alpha<1\,(\alpha>1)$. What happens if $\alpha=1$?

(Section 1.3.6)

1.15 Failure rate in multivariate lifetime distributions. There are various ways in which the hazard function (failure rate) concept can be extended to multivariate distributions. One approach to the idea of increasing hazard functions (Brindley and Thompson, 1972) is as follows: suppose that continuous random variables $T_1,\ldots,T_n$ have the joint survivor function

$$S(t_1,\ldots,t_n)=\Pr(T_1\geqslant t_1,\ldots,T_n\geqslant t_n) \qquad t_i\geqslant 0.$$

Suppose that for any subset $\{i_1,\ldots,i_m\}$ of $\{1,\ldots,n\}$ the joint survivor function $S_{i_1\cdots i_m}(t_{i_1},\ldots,t_{i_m})$ of $T_{i_1},\ldots,T_{i_m}$ is such that

$$\frac{S_{i_1\cdots i_m}(t_{i_1}+x,\ldots,t_{i_m}+x)}{S_{i_1\cdots i_m}(t_{i_1},\ldots,t_{i_m})} \qquad (1.5.1)$$

is monotone decreasing in $t_{i_1},\ldots,t_{i_m}$, for any $x>0$. Then $(T_1,\ldots,T_n)$ is said to have the multivariate increasing failure rate (MIFR) property.

a. For a univariate distribution with survivor function $S(t)$ the MIFR property states that $S(t+x)/S(t)$ is decreasing in t for all fixed $x>0$. Show that this is equivalent to the statement that the hazard function $h(t)=-S'(t)/S(t)$ is monotone increasing; that is, the distribution has an IFR.

b. Prove that $Y=\min(T_1,\ldots,T_n)$ has an IFR if $(T_1,\ldots,T_n)$ has a MIFR.

c. The standard bivariate logistic distribution has distribution function

$$F(y_1,y_2)=(1+e^{-y_1}+e^{-y_2})^{-1} \qquad -\infty<y_1,y_2<\infty.$$

Obtain the joint survivor function for $T_1=\exp Y_1$ and $T_2=\exp Y_2$ and examine (1.5.1) in this case. Does (T_1,T_2) have the MIFR property?

(Section 1.3.9)

1.16 Consider the following two censoring mechanisms.

a. A group of n individuals is observed from time 0; observation stops at the time of the rth death or at time L, whichever occurs first.

b. A group of n individuals is observed from time 0, but each time an individual dies a new individual instantly replaces him in the experiment. The experiment terminates after a preassigned time L has elapsed.

Assuming that lifetimes of individuals are i.i.d. with survivor function $S(t)$ and p.d.f. $f(t)$, give an argument for each of (a) and (b) to show that (1.4.9) is the appropriate likelihood function.

(Section 1.4)

1.17 Left and right censoring. Discuss how (1.4.1), (1.4.5), and (1.4.9) can be generalized in situations where lifetimes can be censored either on the left, on the right, or both.

(Section 1.4)

CHAPTER 2

Life Tables, Graphs, and Related Procedures

2.1 INTRODUCTION

In this chapter we begin our study of statistical methods for lifetime data. Two broad classes of procedures can be distinguished. First, there are procedures based on specific parametric families of distributions. Statistical analysis within such families usually proceeds along familiar lines, though censoring creates some nonstandard problems. Methods associated with the common lifetime distribution models are presented in later chapters, starting in Chapter 3 with the exponential distribution. The second class consists of nonparametric procedures that do not depend on the assumption of a specific family of distributions. Nonparametric and distribution-free methods are discussed in this chapter and in Chapters 7 and 8.

This chapter presents a number of important procedures for describing univariate samples and making inferences about the underlying lifetime distribution. The methods are basically nonparametric but can also be used in conjunction with particular parametric families of distributions. The techniques include the use of relative frequency tables, probability plots, and the empirical distribution function. A reason for discussing these familiar topics at some length is that the presence of censoring creates special problems, with the result that well-known methods for uncensored data have to be modified, or alternate methods found, to handle the censoring.

To illustrate the problems censoring creates let us quickly review one of the simplest procedures in elementary statistics, the formation of a relative frequency table. Suppose we have a complete (i.e., uncensored) sample of n lifetimes from the population under study. Divide the time axis $[0,\infty)$ into $k+1$ intervals $I_j=[a_{j-1}, a_j)$, $j=1,\ldots,k+1$, with $a_0=0$, $a_k=T$, and $a_{k+1}=\infty$, where T is an upper limit of observation. The observed number d_j of lifetimes that lie in the interval I_j is called the frequency for the jth interval, and a frequency table is a list of the intervals and their associated d_js or d_j/ns. An associated relative frequency histogram, consisting of rectangles with bases on $[a_{j-1}, a_j)$ and areas d_j/n $(j=1,\ldots,k)$, is often drawn to portray this. However, when data are censored, it is usually not possible to form a frequency table. If all of the censoring occurs in the last interval $[T, \infty)$, there is no problem, but if censoring times fall into other intervals, we cannot know the exact frequency of deaths associated with each interval, and hence cannot form the frequency table or histogram.

The first topic discussed below, the life table, is essentially a modification of the frequency table to deal with censored data. Life table methods are an important tool for dealing with lifetime data and are given a careful treatment. Next, nonparametric estimation of a survivor function from ungrouped data is considered. This also requires an extension of well-known methods for uncensored data, specifically those involving the empirical distribution and survivor functions. In Section 2.4, plotting techniques are examined; these are useful in suggesting and checking possible models. The last section of this chapter examines simple least squares methods for estimating model parameters. Although this topic is somewhat different from the others in this chapter, the methods described are closely related to graphical procedures discussed here, and it is convenient to consider them at this time.

2.2 LIFE TABLES

The life table is one of the oldest and most widely used methods of portraying lifetime data. The cohort life table discussed here has been employed at least since the beginning of the twentieth century; the so-called population life table, much used by demographers and actuaries, has been around considerably longer. Although life table methods have been used for a long time, the elaboration of their statistical properties has been a much more recent development. Indeed, because of the problems that censoring introduces, a careful investigation of statistical properties is somewhat involved. We shall therefore begin by presenting the practical aspects of the life table and shall then discuss some properties of the procedures.

2.2.1 Standard Life Table Methods: Practice

The life table is primarily a device for portraying the survival experience of a group of individuals, sometimes referred to as a "cohort." In the types of situations discussed in this book the group of individuals is usually assumed to be a random sample from some population, in which case the life table also provides estimates of survival probabilities for the population. The life table is essentially an extension of the relative frequency table to the case of censored data, though with the life table one emphasizes estimation of the conditional probability of death in an interval, given survival to the start of the interval, and the probability of surviving past the end of an interval. If there is no censoring, except possibly in the last interval, it is, of course, easy to estimate these quantities. For example, the proportion of individuals in the sample surviving to time a_j is an estimate of the probability of surviving to a_j. Censoring complicates matters, however, and we therefore give a careful description of the construction of a life table.

As in the previous section, the time axis is divided into $k+1$ intervals $I_j=[a_{j-1}, a_j)$, $j=1,\ldots, k+1$, with $a_0=0$, $a_k=T$, and $a_{k+1}=\infty$, where T is an upper limit on observation. For each member of a random sample of n individuals from some population, suppose that one observes either a lifetime t or a censoring time L. The data are, however, grouped so that it is only known in which intervals particular individuals died or were censored, and not the exact lifetimes and censoring times. The data therefore consist of the numbers of lifetimes and censoring times falling into each of the $k+1$ intervals. In the case of the last interval, I_{k+1}, it can be considered that only lifetimes are in the interval, since all individuals not dead by time T must die sometime in I_{k+1}. We now define the following quantities:

N_j = Number of individuals "at risk" (i.e., alive and not censored) at time a_{j-1}

D_j = Number of "deaths" in (i.e., number of lifetimes observed to fall into) $I_j=[a_{j-1}, a_j)$

W_j = Number of "withdrawals" in (i.e., number of censoring times observed to fall into) $I_j=[a_{j-1}, a_j)$.

The terms "at risk," "deaths," and "withdrawals" are commonly used with life tables, though sometimes other terms are used, such as "number of censoring times" in I_j instead of "number of withdrawals" in I_j. The number of individuals known to be alive at the start of I_j is N_j, and thus $N_1=n$ and

$$N_j=N_{j-1}-D_{j-1}-W_{j-1} \qquad j=2,\ldots,k+1.$$

Let the distribution of lifetimes for the population under study have survivor function $S(t)$, and define the following quantities:

$$\begin{aligned} P_j &= S(a_j) \\ &= \Pr(\text{an individual survives beyond } I_j) \\ p_j &= \Pr(\text{an individual survives beyond } I_j | \text{he survives beyond } I_{j-1}) \\ &= \frac{P_j}{P_{j-1}} \qquad (2.2.1) \\ q_j &= 1 - p_j \\ &= \Pr(\text{an individual dies in } I_j | \text{he survives beyond } I_{j-1}). \end{aligned}$$

In (2.2.1) j ranges over $1,\ldots,k+1$, with P_0 defined to be unity. Note also that $P_{k+1}=0$ and $q_{k+1}=1$. Finally, note that by using the middle expression in (2.2.1), P_j can be written as

$$P_j = p_1 p_2 \cdots p_j \qquad j=1,\ldots,k+1. \qquad (2.2.2)$$

This result, wherein the probability of surviving past I_j is given as the product of conditional probabilities of surviving past intervals up to I_j, given survival to the start of each interval, forms the basis for the approach to life table estimation.

Life table analysis takes as a primary task estimation of the P_j's; these are the probabilities of surviving past the various time points a_j that define the intervals. If the data were uncensored, there would be no problem in doing this: the obvious estimate, and, indeed, the m.l.e. for P_j in this case, is N_{j+1}/n, the proportion of individuals in the sample still alive at time a_j. This will not do if intervals contain withdrawals (i.e., censoring times), however, since then N_{j+1} is not necessarily the number of individuals still alive at time a_j out of the original total of n, but rather the number alive *and* who have not been censored prior to a_j. Since it is very likely that some censored individuals will also still be alive at a_j, N_{j+1}/n will in most instances tend to underestimate P_j. The life table method overcomes this problem.

The idea behind the life table is to employ (2.2.2) in obtaining an estimate of P_j and is based on the observation that even when there is censoring, it is generally possible to give sensible estimates of the p_j's. Life table analysis involves estimating the q_j's and p_j's of (2.2.1), and then via (2.2.2) the P_j's, and displaying the estimates in a format called the life table. The usual procedure is as follows: if a particular interval I_j has no withdrawals in it (i.e., $W_j=0$), then a sensible estimate of q_j is $\hat{q}_j=D_j/N_j$, since q_j is the conditional probability of an individual dying in I_j, given that he is alive at the start of I_j. If, however, the interval has $W_j>0$ withdrawals, D_j/N_j might be expected to underestimate q_j, since it is possible that some of the individuals censored in I_j might have died before the end of I_j, had they not been censored first. It is therefore desirable to make some adjustment for the censored individuals. The most commonly used procedure is to estimate q_j by the so-called standard life table estimate, which is

$$\hat{q}_j=\frac{D_j}{N_j-W_j/2}=\frac{D_j}{N_j'}. \tag{2.2.3}$$

The expression (2.2.3) assumes that $N_j>0$; when $N_j=0$, we define $\hat{q}_j=1$ for reasons of convenience that become apparent later. The denominator $N_j'=N_j-\frac{1}{2}W_j$ can be thought of as an effective number of individuals at risk for the interval I_j; this supposes that, in a sense, a withdrawn individual is at risk for half the interval. This adjustment is arbitrary, but sensible in many situations. Its appropriateness depends on the failure and censoring time process, of course; this point is discussed further in the next section. In some instances other estimates of q_j may be preferable. For example, if all withdrawals in I_j occurred right at the end of I_j, the estimate $\hat{q}_j=D_j/N_j$ would be appropriate, whereas if all withdrawals occurred at the beginning of I_j, $\hat{q}_j=D_j/(N_j-W_j)$ would be appropriate. Still other estimates are useful on certain occasions; we shall return to this point in Section 2.2.3. The present section, however, examines only the standard life table estimate (2.2.3). It is appropriate in a wide range of situations, is particularly easy to use, and is the estimate most employed in practice.

Once estimates $\hat{q}_j$ and $\hat{p}_j=1-\hat{q}_j$ have been calculated, P_j can, by virtue of (2.2.2), be estimated by

$$\hat{P}_j=\hat{p}_1\cdots\hat{p}_j \qquad j=1,\ldots,k+1.$$

The life table itself is a table for displaying the data and the estimates $\hat{q}_j$ and $\hat{P}_j$. The table generally includes columns giving, for each interval, the values of N_j, D_j, W_j, $\hat{q}_j$, and $\hat{P}_j$. Additional columns are sometimes included, giving quantities such as N_j', $\hat{p}_j$, and, occasionally, estimates of other characteristics of the underlying distribution (e.g., Gehan, 1969). The general format is shown below. For the special case in which all W_j's$=0$, $\hat{P}_j$ reduces to the previously mentioned estimate for uncensored data, N_{j+1}/n.

Interval	Number of Deaths	Number of Withdrawals	Number at Risk	N_j'	$\hat{q}_j$	$\hat{p}_j$	$\hat{P}_j$
I_j	D_j	W_j	N_j	$N_j-\frac{1}{2}W_j$	D_j/N_j'	$1-\hat{q}_j$	$\hat{p}_1\cdots\hat{p}_j$

Example 2.2.1 Berkson and Gage (1950) give data describing the survival experience of a group of 374 patients who underwent operations in connection with a type of malignant disease. From these data the life table given in Table 2.2.1 has been formed. Once the intervals I_j have been chosen and D_j, W_j, and N_j determined for each interval, the only computations required in completing the table are the calculation of the $\hat{q}_j$'s, using (2.2.3), and the subsequent calculation of the $\hat{P}_j$'s. The $\hat{P}_j$'s can be conveniently computed recursively, since $\hat{P}_j=\hat{p}_j\hat{P}_{j-1}$.

Variance Estimates

The quantities $\hat{q}_j$, $\hat{p}_j$, and $\hat{P}_j$ are estimates subject to sampling variation, and therefore it is desirable to have some idea of their precision. Under suitable assumptions it is possible to derive estimates of their variances. Details related to this are discussed in Section 2.2.2; here we shall present results, with one or two comments.

The most commonly used variance estimate is one suggested by Greenwood (1926). In this case $\hat{q}_j\hat{p}_j/N_j'$ is taken as an estimate of the

Table 2.2.1 Life Table Computed From Data in Berkson and Gage (1950)

Interval (I_j) in Years	Number of Deaths (D_j)	Number of Withdrawals (W_j)	Number at Risk (N_j)	N_j'	$\hat{q}_j$	$\hat{p}_j$	Estimated Probability of Survival Beyond I_j ($\hat{P}_j$)
[0,1)	90	0	374	374.	0.241	0.759	.759
[1,2)	76	0	284	284.	0.268	0.732	.556
[2,3)	51	0	208	208.	0.245	0.755	.420
[3,4)	25	12	157	151.	0.164	0.834	.350
[4,5)	20	5	120	117.5	0.170	0.830	.291
[5,6)	7	9	95	90.5	0.077	0.923	.268
[6,7)	4	9	79	74.5	0.054	0.946	.254
[7,8)	1	3	66	64.5	0.016	0.984	.250
[8,9)	3	5	62	59.5	0.050	0.950	.237
[9,10)	2	5	54	51.5	0.039	0.961	.228
[10,∞)	47	0	47	47.	1.000	0.000	0

variance of $\hat{q}_j$ (or $\hat{p}_j$), and an approximation to the variance of $\hat{P}_j = \hat{p}_1 \cdots \hat{p}_j$ is then derived using a standard asymptotic approximation (see Section 2.2.2) to give

$$\widehat{\operatorname{Var}}(\hat{P}_j) = \hat{P}_j^2 \sum_{i=1}^{j} \frac{\hat{q}_i}{N_i' \hat{p}_i}. \tag{2.2.4}$$

This estimate is reasonable provided that $E(N_j')$ is not too small, though when there is a lot of censoring, k should also not be too small. It sometimes tends to underestimate the variance of $\hat{P}_j$ for intervals in the right-hand tail of the life distribution, essentially when $E(N_j')$ is quite small. However, in such instances the distribution of $\hat{P}_j$ is typically highly skewed, and its variance is not a particularly good indication of estimation precision anyway. Confidence intervals for the P_j's are examined in Section 8.1.1, and a more thorough discussion of precision is given there.

It can be noted that (2.2.4) gives the usual estimate for the variance of $\hat{P}_j$ in the case in which there is no censoring. To see this, note that when there are no withdrawals, $N_i = n\hat{P}_{i-1}$ and thus $N_i \hat{p}_i = n\hat{p}_i \hat{P}_{i-1} = n\hat{P}_i$. Therefore (2.2.4) equals

$$\begin{aligned} \hat{P}_j^2 \sum_{i=1}^{j} \frac{\hat{q}_i}{n\hat{P}_i} &= \frac{\hat{P}_j^2}{n} \sum_{i=1}^{j} \frac{1-\hat{p}_i}{\hat{P}_i} \\ &= \frac{\hat{P}_j^2}{n} \sum_{i=1}^{j} \left(\frac{1}{\hat{P}_i} - \frac{1}{\hat{P}_{i-1}} \right) \\ &= \frac{\hat{P}_j(1-\hat{P}_j)}{n} \end{aligned} \tag{2.2.5}$$

which is the usual estimate of $\operatorname{Var}(\hat{P}_j) = P_j(1-P_j)/n$.

A few general remarks about life tables are in order. When lifetime data are grouped, life tables give a concise picture of the survival experience of the individuals in the sample and also supply nonparametric estimates of survival probabilities. Even when lifetimes and censoring times are known exactly, life table methods are useful in summarizing large bodies of data, though if precise estimation of the underlying survivor function is important, methods described in Section 2.3 are preferable. The life table has traditionally been widely used in medical areas, for example, in connection with follow-up studies on individuals suffering from chronic diseases. Such studies frequently involve fairly large numbers of individuals, and persons often become "lost to follow-up" sometime during the course of the study, or are still alive when the data are collected. In either case their lifetime is censored. Life tables are often used to estimate survival probabilities in such instances and are an effective way of presenting the data. In situations like that described in Example 2.2.1, estimates of quantities such as 5-year survival probabilities are often wanted. In the example this would be $\hat{P}_5 = .291$; the estimated variance of $\hat{P}_5$ given by (2.2.4) is $.024^2$.

In forming a life table, there is no need to make the intervals of equal length, though it may be convenient to do so. The number of intervals used will depend on the amount of data available and on the aims of the analysis. Certain statistical properties of the estimates are enhanced when the number of intervals is fairly large, as described in the next section. On the other hand, if an easily comprehended summary of the data is wanted, it may be sensible not to have too many intervals, though it is generally desirable to have at least 8 or 10.

It should be remembered that for life table methods (and other methods of analysis throughout this book) to be appropriate it is necessary for individuals who are censored to have lifetimes from the same distribution as those not censored. If, as sometimes happens in medical studies, this is not the case, then the methods given here may give misleading results. Finally, the appropriateness of the standard life table estimate (2.2.3) or variance formula (2.2.4) depends on certain assumptions about the censoring pattern in the population being approximately satisfied. This is discussed in the next section, but for practical purposes it can be said that the methods described above are satisfactory in a broad range of situations.

2.2.2 Standard Life Table Methods: Theory*

Life table methods have been used for many years, but the thorough study of their statistical properties is a much more recent development and is still not complete. Many expressions, such as Greenwood's formula (2.2.4), were originally derived on intuitive grounds or as relatively *ad hoc* modifications to procedures that would apply if there were no censoring and their behaviors in realistic situations involving censoring are not totally known. To better understand the basis and validity of life table methods, let us examine their properties in a few situations.

The Case of No Withdrawals

The no withdrawals case is straightforward, but it is convenient to consider it first, since exact results can be obtained and because several formulas used when there are withdrawals are closely related to these results. When there is no possible censoring, the numbers of deaths $D_1, \ldots, D_k$ in intervals

$I_1,\ldots,I_k$ follow a multinomial distribution with probability function

$$\Pr(D_1,\ldots,D_k)=\frac{n!}{D_1!\cdots D_k!D_{k+1}!}\prod_{j=1}^{k+1}\pi_j^{D_j} \tag{2.2.6}$$

where

$$D_1+\cdots+D_{k+1}=n \quad \text{and} \quad \pi_1+\cdots+\pi_{k+1}=1$$

and

$$\pi_j=P_{j-1}-P_j=p_1\cdots p_{j-1}q_j$$

is the unconditional probability of an individual dying in I_j. The likelihood function arising from (2.2.6) is therefore proportional to

$$\prod_{j=1}^{k+1}(p_1\cdots p_{j-1}q_j)^{D_j}=\prod_{j=1}^{k+1}q_j^{D_j}p_j^{N_j-D_j}$$

where $N_j=n-D_1-\cdots-D_{j-1}$. This is maximized at $\hat{q}_j=1-\hat{p}_j=D_j/N_j$ $(j=1,\ldots,k+1)$, as long as $N_j>0$ for all j. If $N_j=0$ for some j, then the likelihood function does not involve that q_j or p_j and no m.l.e of q_j is obtained. In such cases we define $\hat{q}_j=1$, for reasons of convenience made clear later.

The m.l.e. of P_j is

$$\hat{P}_j=\hat{p}_1\cdots\hat{p}_j$$

$$=\frac{N_{j+1}}{n} \tag{2.2.7}$$

where we note that $N_{j+1}=N_j-D_j$. Here we make use of the fact that $\hat{p}_j=0$ if $N_j=0$. The distribution of N_{j+1} is binomial, with

$$\Pr(N_{j+1}=x)=\binom{n}{x}P_j^x(1-P_j)^{n-x}. \tag{2.2.8}$$

The distribution of $\hat{P}_j$ is thus known exactly. The mean and variance are respectively

$$E(\hat{P}_j)=P_j, \qquad \mathrm{Var}(\hat{P}_j)=\frac{P_j(1-P_j)}{n}. \tag{2.2.9}$$

The covariance between $\hat{P}_j$ and $\hat{P}_l$ is also needed in some instances. Supposing $j<l\leqslant k$, we have

$$\begin{aligned}
\operatorname{Cov}(\hat{P}_j, \hat{P}_l) &= \frac{1}{n^2}\operatorname{Cov}(N_{j+1}, N_{l+1}) \\
&= \frac{1}{n^2}\operatorname{Cov}(n-N_{j+1}, n-N_{l+1}) \\
&= \frac{1}{n^2}\operatorname{Cov}(D_1+\cdots+D_j, D_1+\cdots+D_l) \\
&= \frac{1}{n^2}\operatorname{Var}(D_1+\cdots+D_j) \\
&\quad + \frac{1}{n^2}\operatorname{Cov}(D_1+\cdots+D_j, D_{j+1}+\cdots+D_l) \\
&= \frac{1}{n^2}nP_j(1-P_j) + \frac{1}{n^2}\sum_{i=1}^{j}\sum_{s=j+1}^{l}\operatorname{Cov}(D_i, D_s).
\end{aligned}$$

Since $\operatorname{Cov}(D_i, D_s) = -n\pi_i\pi_s\ (i\neq s)$, this equals

$$\frac{P_j(1-P_j)}{n} - \frac{1}{n}\sum_{i=1}^{j}\pi_i\sum_{s=j+1}^{l}\pi_s = \frac{P_j(1-P_j)}{n} - \frac{(1-P_j)(P_j-P_l)}{n}$$

and therefore

$$\operatorname{Cov}(\hat{P}_j, \hat{P}_l) = \frac{(1-P_j)P_l}{n} \qquad j<l. \tag{2.2.10}$$

Let us also determine the first two moments of the $\hat{q}_j$'s, since these will be of interest when we take up the censored case. To do this assume that $N_j>0$ and note that in this case the distribution of D_j, given N_j, is binomial, with parameters N_j and q_j. Using conditional expectation, we obtain the mean and variance of $\hat{q}_j = D_j/N_j$ as follows:

$$\begin{aligned}
E(\hat{q}_j) &= E_{N_j}\left[E(\hat{q}_j|N_j)\right] \\
&= E_{N_j}(q_j) = q_j
\end{aligned} \tag{2.2.11}$$

$$\begin{aligned}
\operatorname{Var}(\hat{q}_j) &= E_{N_j}\left[\operatorname{Var}(\hat{q}_j|N_j)\right] + \operatorname{Var}_{N_j}\left[E(\hat{q}_j|N_j)\right] \\
&= E_{N_j}\left(\frac{p_jq_j}{N_j}\right) + \operatorname{Var}_{N_j}(q_j) = p_jq_jE\left(\frac{1}{N_j}\right).
\end{aligned} \tag{2.2.12}$$

These expressions are not quite correct unconditionally, since when $N_j=0$, $\hat{q}_j$ was defined for convenience to be unity. The amount by which they are in error is small, however, as long as the probability that $N_j=0$ is small, as it will generally be. If the results are viewed as conditional on N_j being positive, they are exact.

Note that for $i<j$, and the same comment about N_j being positive,

$$E[(\hat{q}_j-q_j)|\hat{q}_i]=E\left[\left(\frac{D_j}{N_j}-q_j\right)\middle|\frac{D_i}{N_i}\right]$$

$$=E_{N_j}\left[E_{D_j}\left(\frac{D_j}{N_j}-q_j\right)\middle|\frac{D_i}{N_i},N_j\right]$$

$$=E_{N_j}(0|\hat{q}_i)=0.$$

Thus for $i<j$

$$\operatorname{Cov}(\hat{q}_i,\hat{q}_j)=E_{\hat{q}_i}[(\hat{q}_i-q_i)E(\hat{q}_j-q_j|\hat{q}_i)]=0. \tag{2.2.13}$$

Although $\hat{q}_i$ and $\hat{q}_j$ have zero covariance, they are not independent, as is easily shown by example (see Problem 2.4). It should be noted that since $\hat{p}_j=1-\hat{q}_j$, the variances and covariances of the $\hat{p}_j$'s are also given by (2.2.12) and (2.2.13) and that $E(\hat{p}_j)=1-E(\hat{q}_j)=p_j$.

The uncensored case is thus straightforward. Exact means, variances, and covariances of estimates are available, and we can, in addition, obtain confidence intervals on $P_j=S(a_j)$, using standard theory for the binomial distribution. We now turn to the censored case, where matters are more difficult.

The General Case Involving Withdrawals

The properties of life table methods when withdrawals occur will depend on factors such as the mechanism that leads to withdrawals. On the other hand, results like (2.2.4) are derived in a relatively *ad hoc* manner, without specific reference to any particular form of withdrawal mechanism, so it is legitimate to wonder about their validity. Before examining this problem, let us consider the origins of the standard life table estimate (2.2.3) and the variance estimate (2.2.4). The estimate (2.2.3) derives from an *ad hoc* adjustment to the estimate in the no withdrawals case, as discussed earlier. The genesis of Greenwood's variance formula (2.2.4) is similarly tied to the no withdrawals case. This derives from the first-order approximation to the variance of $P_j=p_1\cdots p_j$, obtained from formula (C2) in Appendix C, which gives

$$\operatorname{Var}(\hat{P}_j) \doteq \sum_{i=1}^{j} \sum_{l=1}^{j} \frac{\partial P_j}{\partial p_i} \frac{\partial P_j}{\partial p_l} \operatorname{Cov}(\hat{p}_i, \hat{p}_l). \tag{2.2.14}$$

We obtain Greenwood's formula by assuming, as in the no withdrawals case, that $\operatorname{Cov}(\hat{p}_i, \hat{p}_l)=0$ $(i \neq l)$ and by estimating $\operatorname{Var}(\hat{p}_i)=\operatorname{Cov}(\hat{p}_i, \hat{p}_i)$ by

$$\widehat{\operatorname{Var}}(\hat{p}_i) = \frac{\hat{p}_i \hat{q}_i}{N_i'}. \tag{2.2.15}$$

This gives

$$\widehat{\operatorname{Var}}(\hat{P}_j) = \hat{P}_j^2 \sum_{i=1}^{j} \frac{\hat{q}_i}{\hat{p}_i N_i'}$$

which is (2.2.4).

There are thus several approximations involved in (2.2.4), not all of which are easily assessed. The approximation (2.2.14) is reasonable if n is sufficiently large, but the approximations $\operatorname{Cov}(\hat{p}_i, \hat{p}_l)=0$ $(i \neq l)$ and (2.2.15), both of which are written down by analogy with the no withdrawals case, are more questionable. Their adequacy will actually depend on the censoring mechanism and lifetime distribution in the problem under study. Note, incidentally, that although (2.2.4) gives the correct estimated variance of $\hat{P}_j$ in the no withdrawals case, this is because the two approximations (2.2.14) and (2.2.15) effectively cancel each other out. The variance of $\hat{p}_i$ in the no withdrawals case is actually $p_i q_i E(1/N_i)$, as given in (2.2.12), and when $E(N_i)$ is small, $p_i q_i / N_i$ will tend to underestimate it.

In addition to questions about the adequacy of the variance estimate (2.2.4), there is the question of bias in $\hat{q}_j$, $\hat{p}_j$, and $\hat{P}_j$. In the no withdrawals case these are unbiased estimates of q_j, p_j, and P_j, respectively, but this will almost certainly not be true when there is censoring.

A problem in examining the properties of the life table estimates is in specifying a realistic model for censoring. The properties of the estimates in a given situation depend on the particular censoring mechanism in operation. It seems plausible that if censoring times tend to be reasonably well distributed across intervals that are not too wide, and if they are independent of lifetimes in some sense, then the standard life table estimate (2.2.3) should behave reasonably. A really thorough investigation of this is not easy, however. One useful exercise is to examine the properties of (2.2.3) and (2.2.4) under a model in which censoring times are assumed to be i.i.d.

random variables, independent of lifetimes. Breslow and Crowley (1974) have done this, and some of their results are sketched here. After looking at random censorship, we shall briefly consider other situations.

Asymptotic Properties of Estimates Under a Random Censorship Model

Suppose that lifetimes $T_1,\ldots,T_n$ of n individuals under study are independent and identically distributed with survivor function $S(t)$. The primary aim of the life table methods is to estimate $S(a_j), j=1,\ldots,k$, where intervals $I_j=[a_{j-1},a_j)$ are defined as before. Suppose that a censoring time L_i is also associated with each individual and that the L_i's are independent and identically distributed, independent of the lifetimes, with survivor function $G(t)$. The data for the life table consist of the numbers of lifetimes (deaths) and censoring times (withdrawals) in the $k+1$ intervals. This can be represented by the vector $\mathbf{D}=(D_1,W_1,\ldots,D_k,W_k,N_{k+1})'$, where D_j and W_j represent the number of deaths and withdrawals, respectively, in I_j, and $N_{k+1}=n-D_1-W_1-\cdots-D_k-W_k$ is the number of individuals surviving past time $a_k=T$. With an obvious notation, $\mathbf{D}$ has a multinomial distribution with parameters n and $\boldsymbol{\pi}=(\pi_1^D,\pi_2^W,\ldots,\pi_k^D,\pi_k^W,\pi_{k+1}^N)'$, where $\pi_1^D+\pi_1^W+\cdots+\pi_{k+1}^N=1$. The probabilities in $\boldsymbol{\pi}$ can be determined in terms of $S(t)$ and $G(t)$. For example, π_1^D is the unconditional probability of an individual being observed to die in I_1, that is,

$$\begin{aligned}\pi_1^D&=\Pr(\text{an individual dies in } I_1 \text{ and is observed to do so})\\&=\Pr(T_i\leqslant a_1, T_i\leqslant L_i)\\&=\int_0^{a_1}G(x)|dS(x)|.\end{aligned}$$

For convenience, integrals here and in a few other places are considered Riemann–Stieltjes integrals (see Appendix A). In general, we find that for $j=1,\ldots,k$

$$\pi_j^D=\int_{a_{j-1}}^{a_j}G(x)|dS(x)|$$

$$\pi_j^W=\int_{a_{j-1}}^{a_j}S(x)|dG(x)|.$$

Since $\mathbf{D}$ is multinomial, it follows that as $n\to\infty$, the distribution of $n^{-1/2}(\mathbf{D}-n\boldsymbol{\pi})$ converges to a multivariate normal distribution with mean $\mathbf{0}$ and covariance matrix $\Sigma=\text{diag}(\pi_1^D,\pi_1^W,\ldots,\pi_{k+1}^N)-\boldsymbol{\pi}\boldsymbol{\pi}'$ (e.g., Bishop et al.,

1975, p. 470). The standard life table estimates $\hat{q}_j = D_j/(N_j - \frac{1}{2}W_j)$ are smooth functions of $D_1, W_1, \ldots, N_{k+1}$, and hence the distribution of $\sqrt{n}(\hat{\mathbf{q}} - \mathbf{q}^*)$ also converges to a multivariate normal distribution with mean $\mathbf{0}$ and covariance matrix Σ_q, say, where $\hat{\mathbf{q}} = (\hat{q}_1, \ldots, \hat{q}_k)$ and $\mathbf{q}^* = (q_1^*, \ldots, q_k^*)$ is the quantity for which $\hat{\mathbf{q}}$ is a consistent estimate. Since

$$\hat{q}_j = \frac{D_j/n}{N_j/n - W_j/2n}$$

it follows that

$$q_j^* = \frac{\pi_j^D}{\pi_j^N - \pi_j^W/2}$$

where $\pi_j^N = E(N_j/n) = G(a_{j-1})S(a_{j-1})$. Thus

$$q_j^* = \left(\int_{a_{j-1}}^{a_j} G(x)|dS(x)| \right) \Big/ \left(G(a_{j-1})S(a_{j-1}) - \frac{1}{2}\int_{a_{j-1}}^{a_j} S(x)|dG(x)| \right). \tag{2.2.16}$$

In general q_j^* will not equal

$$q_j = \frac{S(a_{j-1}) - S(a_j)}{S(a_{j-1})}.$$

(Note, however, that if no censoring can occur until after time a_j, (2.2.16) gives $q_1^* = q_1, \ldots, q_j^* = q_j$, as indeed it must.) Since $q_j^* \neq q_j$ in general, the standard life table estimate (2.2.3) is not a consistent estimate of q_j and, also, $\hat{P}_j = \hat{p}_1 \cdots \hat{p}_j$ is not a consistent estimate of $P_j = p_1 \cdots p_j$. Therefore an important practical question is whether the asymptotic bias in the estimates is sufficiently small to render this inconsistency relatively harmless. It appears that this is in fact the case in many situations, as we shall see momentarily.

The entries in the asymptotic covariance matrix Σ_q of $\hat{\mathbf{q}}$ can be determined by a straightforward but tedious application of the asymptotic formula for the covariance of two functions (see Appendix C). Using this (see Problem 2.5), we find that Σ_q is a diagonal matrix, and hence $\hat{q}_j$ and $\hat{q}_l$ $(j \neq l)$ are asymptotically uncorrelated, just as in the no withdrawals case, where they also happen to be uncorrelated in finite samples. The asymptotic variance of $\sqrt{n}\,(\hat{q}_j - q_j^*)$ turns out to be

$$\operatorname{Asvar}\left[\sqrt{n}\,(\hat{q}_j - q_j^*)\right] = \frac{q_j^* - q_j^{*2}\left[\left(\pi_j^N - \pi_j^W/4\right)/\left(\pi_j^N - \pi_j^W/2\right)\right]}{\pi_j^N - \pi_j^W/2}. \tag{2.2.17}$$

The usual life table estimate of $\operatorname{Var}(\hat{q}_j)$ given by (2.2.15) is

$$\widehat{\operatorname{Var}}(\hat{q}_j) = \frac{\hat{q}_j - \hat{q}_j^2}{N_j'}. \tag{2.2.18}$$

If q_j and q_j^* are not too different, this tends to overestimate the true variance somewhat, since N_j'/n converges in probability to the denominator of (2.2.17), and the term in square brackets in (2.2.17) is less than unity. If q_j is small, the second terms in (2.2.17) and (2.2.18) are small relative to the first and the agreement between the two formulas is improved.

The limiting distribution of the $\sqrt{n}\,(\hat{P}_j - P_j)$'s is multivariate normal, with means, variances, and covariances that can be determined by the usual methods. Application of the standard first-order variance approximation (see Appendix C) to $\hat{P}_j = \hat{p}_1 \cdots \hat{p}_j$ gives formulas for the asymptotic variances and covariances of the $\sqrt{n}\,\hat{P}_j$'s. This yields for $j \leqslant l$

$$\begin{aligned}
\operatorname{Cov}\left(\sqrt{n}\,\hat{P}_j, \sqrt{n}\,\hat{P}_l\right) &\doteq \sum_{i=1}^{j} \sum_{S=1}^{l} \left.\frac{\partial \hat{P}_j}{\partial \hat{p}_i} \frac{\partial \hat{P}_l}{\partial \hat{p}_s}\right|_{\mathbf{p}^*} \operatorname{Cov}\left(\sqrt{n}\,\hat{p}_i, \sqrt{n}\,\hat{p}_s\right) \\
&= \sum_{i=1}^{j} \sum_{s=1}^{l} \frac{P_j^*}{p_i^*} \frac{P_l^*}{p_s^*} \operatorname{Cov}\left(\sqrt{n}\,\hat{p}_i, \sqrt{n}\,\hat{p}_s\right) \\
&= P_j^* P_l^* \sum_{i=1}^{j} \frac{\operatorname{Var}\left(\sqrt{n}\,\hat{p}_i\right)}{p_i^{*2}}.
\end{aligned}$$

Let $\hat{\mathbf{P}} = (\hat{P}_1, \ldots, \hat{P}_k)$ and $\mathbf{P}^* = (P_1^*, \ldots, P_k^*)$, where $P_j^* = p_1^* \cdots p_j^*$; the limiting distribution of $\sqrt{n}\,(\hat{\mathbf{P}} - \mathbf{P}^*)$ is multivariate normal with mean $\mathbf{0}$ and a covariance matrix whose (j, l) term is, for $j \leqslant l$,

$$P_j^* P_l^* \sum_{i=1}^{j} \frac{\operatorname{Var}\left[\sqrt{n}\,(\hat{q}_i - q_i^*)\right]}{(1 - q_i^*)^2} \tag{2.2.19}$$

Putting $j=l$, using (2.2.18), and replacing P_i^* and q_i^* with $\hat{P}_i$ and $\hat{q}_i$, we get from (2.2.19) the approximation

$$\operatorname{Var}(\hat{P}_j) \doteq \hat{P}_j^2 \sum_{i=1}^{j} \frac{\hat{q}_i - \hat{q}_i^2}{(1-\hat{q}_i)^2 N_i'}$$

$$= \hat{P}_j^2 \sum_{i=1}^{j} \frac{\hat{q}_i}{N_i' \hat{p}_i}$$

which is Greenwood's formula (2.2.4). In view of the earlier comment about (2.2.18) slightly overestimating (2.2.17), Greenwood's formula will in sufficiently large samples tend to slightly overestimate $\operatorname{Var}(\hat{P}_j)$.

Provided that the asymptotic biases $q_j^* - q_j$ in the $\hat{q}_j$'s are not too large, the standard life table methods should therefore be satisfactory, at least for reasonably large sample sizes. The asymptotic bias can be determined exactly by using (2.2.16) for any particular model in which a lifetime and a censoring time distribution are specified. There has been only a little investigation of this, but the results appear encouraging. Crowley (1970) investigated $q_j^* - q_j$ and the associated bias $P_k^* - P_k$ for cases in which lifetimes were exponential or uniform and censoring times were uniformly distributed over the period of observation. He found asymptotic bias to be small, provided that k was as large as 10 or so. A few results of a similar nature were also reported by Littell (1952). Others, such as Kuzma (1967) and Drolette (1975), have investigated just the bias in the estimates $\hat{q}_j$ under assumptions about the censoring times and lifetimes in I_j, conditional on being alive and uncensored at a_{j-1}, and have found it to be small if q_j is not too large and censoring is not too heavy.

Broadly speaking, the behavior of the life table estimates of Section 2.2.1 is acceptable under random independent censorship provided that censoring is fairly evenly distributed across individual intervals and not too heavy, intervals are not too wide, and sample sizes not too small. It is nevertheless wise to remember that the properties of (2.2.3) and (2.2.4) depend on the censoring and lifetime distributions at hand, that estimates of survival probabilities will be slightly biased, and that the adequacy of the variance estimate (2.2.4) is not fully known unless censoring is very light. Some further discussion on this can be found in Section 8.1.1, where confidence interval estimation of P_j is discussed.

Other Situations

There has been relatively little investigation of life table estimates under other models, neither have the properties for moderate-size samples been very thoroughly examined. In many situations the random censorship model is a reasonable approximation to the actual state of affairs, and the guidelines given should not be misleading, at least for fairly large samples. These results can also be extended to situations in which the censoring times are fixed and, if censoring is reasonably evenly distributed across individual intervals, the comments concerning the adequacy of (2.2.3) and (2.2.4) still apply. More generally, the procedures should be satisfactory under somewhat broader conditions concerning independence of the lifetime and censoring processes, such as those discussed in Section 1.4.1d.

Two additional points should be mentioned. First, it is sometimes desirable to use other life table estimates than (2.2.3), primarily when it appears that use of (2.2.3) may incur nonnegligible bias. Trivial examples of this, such as when all withdrawals occur at the beginnings or ends of intervals, were mentioned in Section 2.2.1. Formal approaches like those in the following section are occasionally useful on such occasions, but often *ad hoc* modifications to the standard estimates will suffice. A second important point is that if exact lifetimes and censoring times are observed, consistent estimates of survival probabilities can be obtained if we let the lengths of intervals in the life table approach zero while the number of intervals becomes arbitrarily large. This procedure is discussed in Section 2.3.

2.2.3 Other Approaches to Life Table Estimation*

Although the methods presented in Section 2.2.1 are adequate for most practical purposes, several other approaches to life table estimation might be mentioned. Most alternate approaches are based on specific assumptions about the lifetime and censoring processes in the population. These methods are not very widely used and are not discussed in detail here. On the other hand, it is worth examining another approach, at least briefly, to see how procedures can be developed from a more formal model. We shall therefore give a short presentation of the life table methods of Chiang (1960a, b, 1968). These are perhaps the most frequently referenced life table procedures, after the standard ones of the previous two sections.

Chiang's Method

Chiang's methods require a knowledge of the censoring times of all individuals under study, even those who are observed to die. Supposing this information to be available, we can partition N_j and D_j into

$$N_j = N_{j1} + N_{j2} \qquad D_j = D_{j1} + D_{j2},$$

where

N_{j1} = Number of individuals alive at a_{j-1} and not due to be censored in I_j

N_{j2} = Number of individuals alive at a_{j-1} and due to be censored in I_j

D_{j1} = Number of deaths in I_j among the N_{j1} individuals not due to be censored

D_{j2} = Number of deaths in I_j among the N_{j2} individuals due to be censored.

Note that by knowing the actual censoring times of all individuals, we can determine which individuals are in the "due to be censored" group and those in the "not due to be censored" group for each interval. When all individuals have the same lifetime distribution, it follows that given N_{j1}, D_{j1} is binomial with parameters N_{j1} and q_j. One could estimate q_j just as D_{j1}/N_{j1}; this is sometimes called the "reduced sample estimate of q_j" and it is an unbiased estimate. However, unless censoring is very light, this estimate neglects a good deal of information and is therefore not often used. To utilize information related to the individuals due to be censored in I_j we need to consider the distribution of D_{j2}. In general, this cannot be done without knowledge of the censoring time and lifetime distributions. Chiang introduces the additional assumptions that, conditional on survival to the start of I_j, censoring times for individuals due for withdrawal in I_j are uniformly distributed over I_j and that the lifetimes of these individuals are exponentially distributed. Under these assumptions it can be shown (see Problem 2.7) that given N_{j2}, D_{j2} has a binomial distribution with parameters N_{j2} and $q_j^* = 1 + q_j/\log(1-q_j)$. Combining the two binomial likelihood functions for D_{j1} and D_{j2}, we get a likelihood function for q_j,

$$L(q_j) = \binom{N_{j1}}{D_{j1}} q_j^{D_{j1}}(1-q_j)^{N_{j1}-D_{j1}} \binom{N_{j1}}{D_{j2}} q_j^{*D_{j2}}(1-q_j^*)^{N_{j2}-D_{j2}}$$

which can be maximized to obtain the m.l.e. of q_j. This has to be done numerically, however, and Chiang employs an approximation that gives a closed form estimate. This is based on the observation that if q_j is small ($q_j \leqslant 0.30$ is adequate), the approximation $q_j^* = 1 + q_j/\log(1-q_j) \doteq 1-(1-q_j)^{1/2}$ is very good. Substituting $1-(1-q_j)^{1/2}$ for q_j^* in $L(q_j)$ gives a function that is maximized for

$$\hat{q}_j = 1 - \left(\frac{-0.5D_{j2} + \left[0.25D_{j2}^2 + 4(N_{j1} + 0.5N_{j2})(N_{j1} - D_{j1} + 0.5W_j)\right]^{1/2}}{2(N_{j1} + 0.5N_{j2})} \right)^2 \tag{2.2.20}$$

where $W_j = N_{j2} - D_{j2}$. In addition, standard maximum likelihood large-sample theory applied to the approximation to $L(q_j)$ gives an approximate variance for $\hat{q}_j$ of

$$\left(\frac{M_j}{p_j q_j} + \pi_j \right)^{-1} \tag{2.2.21}$$

where $M_j = N_{j1} + N_{j2}(1 + p_j^{1/2})^{-1}$ and

$$\pi_j = \frac{N_{j2}(1 - p_j)^{1/2}}{4(1 + p_j^{1/2})p_j^{3/2}}.$$

Other assumptions about lifetime and censoring time distributions would lead to other estimates of q_j and other variance estimates. Elveback (1958), for example, derives an estimate of q_j in a manner similar to Chiang's by assuming that, given an individual dies in I_j, the time of death is uniformly distributed, as are censoring times within intervals. This leads to a likelihood function $L(q_j)$ of the same form as Chiang's, except that $q_j^* = q_j/2$. Elveback's estimate and an estimate of its variance are discussed in Problem 2.8.

Discussion

Estimates like Chiang's and Elveback's are derived on slightly more formal grounds than the life table estimate (2.2.3), though there is still much that is arbitrary in their development. It is not clear, for example, what the underlying failure–censoring mechanism must be in order for the assumptions made about conditional probabilities of death to be valid. In addition, the "likelihood" function formed by taking the product of the $L(q_j)$'s is not a likelihood function in the usual strict sense. The properties of the estimates $P_j = p_1 \cdots p_j$ are consequently not evident, nor are the properties of variance estimates for P_j obtained by using (2.2.21) in conjunction with the large-sample approximation (2.2.14). Two other points are also important. One is that the Chiang, Elveback, and similar methods require knowledge of censoring times for all individuals under study, and this is often unavailable. A second point is that unless censoring is quite heavy, or intervals are rather wide, there is usually very little difference between estimates such as Chiang's and Elveback's and the standard estimates (e.g., Kuzma, 1967; Drolette, 1975; Elandt-Johnson, 1977; Johnson, 1977). The practical consequences of this are that in most situations one should use the standard estimate (2.2.3). When there is some doubt about the adequacy of this estimate, other procedures can be considered, though often a relatively *ad hoc* modification to the standard method will be as reasonable as some more involved procedure.

2.3 NONPARAMETRIC ESTIMATION OF THE SURVIVOR FUNCTION

2.3.1 The Product-Limit Estimate

A useful way of portraying ungrouped univariate survival data is to compute and graph the empirical survivor function or, equivalently, the empirical distribution function. This also provides a nonparametric estimate of the survivor or distribution function for the life distribution under study. If there are no censored observations in a sample of size n, the empirical survivor function (ESF) is defined as

$$\hat{S}(t)=\frac{\text{Number of observations}\geqslant t}{n} \qquad t\geqslant 0. \tag{2.3.1}$$

This is a step function that decreases by $1/n$ just after each observed lifetime if all observations are distinct. More generally, if there are d lifetimes equal to t, the ESF drops by d/n just past t.

When dealing with censored data, some modification of (2.3.1) is necessary, since the number of lifetimes greater than or equal to t will not generally be known exactly. The modification of (2.3.1) described here has come to be called the "product-limit" (PL) estimate of the survivor function or, sometimes, the Kaplan–Meier estimate, from the authors who first discussed its properties (Kaplan and Meier, 1958). The estimate is defined as follows: suppose that there are observations on n individuals and that there are k ($k\leqslant n$) distinct times $t_1<t_2<\cdots<t_k$ at which deaths occur. The possibility of there being more than one death at t_j is allowed, and we let d_j represent the number of deaths at t_j. In addition to the lifetimes $t_1,\ldots,t_k$, there are also censoring times L_i for individuals whose lifetimes are not

observed. The product-limit estimate of $S(t)$ is defined as

$$\hat{S}(t)=\prod_{j:t_j<t}\frac{n_j-d_j}{n_j} \tag{2.3.2}$$

where n_j is the number of individuals at risk at t_j, that is, the number of individuals alive and uncensored just prior to t_j. If a censoring time L_i and a lifetime t_j are recorded as equal, we adopt the convention that censoring times are adjusted an infinitesimal amount to the right so that L_i is considered to be infinitesimally larger than t_j. In other words, any individuals with censoring times recorded as equal to t_j are included in the set of n_j individuals at risk at t_j, as are individuals who die at t_j. This convention is sensible, since an individual censored at time L almost certainly survives past L. Another point about (2.3.2) concerns situations in which the largest observed time in the sample is a censoring time rather than a lifetime. In this case the PL estimate is taken as being defined only up to this last observation. The reason for this is explained later.

The motivation for (2.3.2) is essentially the same as that for the survival probability estimates $\hat{P}_j=\hat{p}_1\cdots\hat{p}_j$ in the life tables of the previous section. That is, the estimate $\hat{S}(t)$ is built up as a product, and each term in the product can be thought of as an estimate of the conditional probability of surviving past time t_j, given survival till just prior to t_j. The estimate (2.3.2) is in fact a limiting case of the standard life table procedure, obtained when the number of intervals in the life table becomes infinite and the lengths of all intervals except the last approach zero. This connection is explored below. It will be noted when there is no censoring, $n_1=n$ and $n_j=n_{j-1}-d_{j-1}$ ($j=2,\ldots,k$), and (2.3.2) reduces to the ordinary ESF (2.3.1). In both the censored and uncensored cases $\hat{S}(t)$ is a step function that equals 1 at $t=0$ and drops by a factor $(n_j-d_j)/n_j$ immediately after each lifetime t_j. The estimate does not change at censoring times L_i; the effect of the censoring times is, however, felt in the values of n_j and hence in the sizes of the steps in $\hat{S}(t)$.

Before we examine the PL estimate and its properties further, let us consider it in an example.

Example 2.3.1 (Example 1.1.7 revisited) Example 1.1.7 gave remission times for two groups of leukemia patients, one given the drug 6-MP and the other a placebo. Table 2.3.1 outlines the calculation of the PL estimates of the survivor functions for the two groups, and Figure 2.3.1 shows these on a graph. The PL estimate has jumps just after each observed lifetime, so in the table we have employed the convenient notation $\hat{S}(t_j+0)=\lim_{x\to 0^+}\hat{S}(t_j+x)$. The PL estimate is easily calculated recursively, since $\hat{S}(t_1+0)=(n_1-$

Table 2.3.1 Computation of Two PL Estimates

Placebo				Drug 6-MP			
t_j	n_j	d_j	$\hat{S}(t_j+0)$	t_j	n_j	d_j	$\hat{S}(t_j+0)$
6	21	3	.857	1	21	2	.905
7	17	1	.807	2	19	2	.810
10	15	1	.753	3	17	1	.762
13	12	1	.690	4	16	2	.667
16	11	1	.627	5	14	2	.571
22	7	1	.538	8	12	4	.381
23	6	1	.448	11	8	2	.286
				12	6	2	.190
				15	4	1	.143
				17	3	1	.095
				22	2	1	.048
				23	1	1	.0

$d_1)/n_1$ and

$$\hat{S}(t_j+0)=\hat{S}(t_{j-1}+0)\frac{n_j-d_j}{n_j} \qquad j=2,\ldots,k.$$

It will be noted that the PL estimate for the drug 6-MP group is defined only up to $t=35$, since the last observed time for that sample is a censoring time, $L=35$.

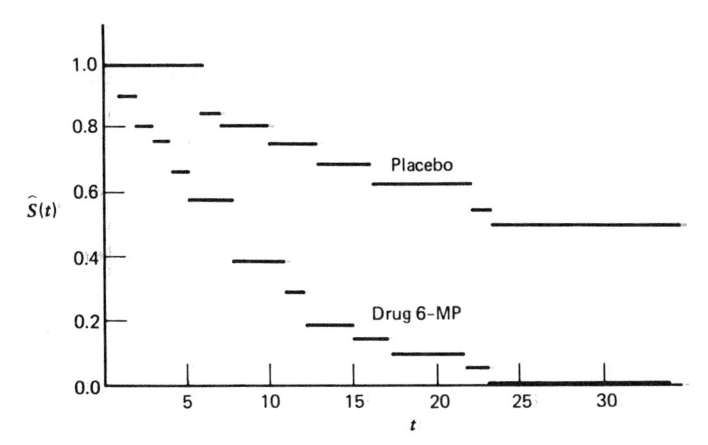

Figure 2.3.1 Product-limit estimates from two samples (Example 2.3.1)

The graph is a very useful representation of the survival experience of the two groups and suggests a clear superiority of the drug 6-MP over the placebo in prolonging survival. Formal methods of testing and estimating differences in two or more life distributions are discussed in later chapters.

Variance Estimation

To effectively assess results when using PL estimates it is desirable to have an estimate of the variance of $\hat{S}(t)$. Proceeding along lines described below, it is possible to obtain the estimate

$$\widehat{\mathrm{Var}}\left[\hat{S}(t)\right]=\hat{S}(t)^2 \sum_{j:t_j<t} \frac{d_j}{n_j\left(n_j-d_j\right)}. \tag{2.3.3}$$

The similarity of (2.3.3) to Greenwood's formula (2.2.4) is apparent and will be discussed later. It is easily shown, as with the life table, that when there is no censoring, (2.3.3) reduces to the usual variance estimate $\hat{S}(t)[1-\hat{S}(t)]/n$. As an illustration of (2.3.3), we find an estimate of the variance of $\hat{S}(16)$ for the drug 6-MP group in Example 2.3.1 to be

$$\widehat{\mathrm{Var}}\left[\hat{S}(16)\right]=0.690^2\left(\frac{3}{21(18)}+\frac{1}{17(16)}+\frac{1}{15(14)}+\frac{1}{12(11)}\right).$$

$$=0.011403$$

which gives an estimated standard deviation of 0.107.

The PL Estimate as an M.L.E.

The PL estimate (2.3.2) possesses a number of important properties, a main one being that $\hat{S}(t)$ is a consistent estimate of $S(t)$ under quite general conditions. Some of these properties are examined in the next section, but first a derivation of the PL estimate as a kind of nonparametric m.l.e. of $S(t)$ will be sketched. This derivation was originally given by Kaplan and Meier (1958), and although it is not particularly helpful for investigating properties of $\hat{S}(t)$, it provides additional motivation for the definition of the PL estimate and will be useful later.

Recall that in the definition of $\hat{S}(t)$ there are assumed to be k distinct lifetimes $t_1<\cdots<t_k$, with d_j deaths at t_j and n_j individuals at risk at t_j. There are, in addition, some censoring times; suppose that in the interval $[t_{j-1}, t_j)$ there are λ_j observed censoring times L_i^j $(i=1,\ldots,\lambda_j)$. By letting $t_0=0$, $t_{k+1}=\infty$, and j range over $1,\ldots,k+1$, we include censoring times

that occur prior to the first or after the last observed lifetime. We continue to use the convention set out earlier, that censoring times are adjusted an infinitesimal amount to the right so that any censoring at t_j is considered to have occurred just after t_j. The underlying survivor function for lifetimes is $S(t)$, and the probability of an individual dying at t_j is thus $S(t_j)-S(t_j+0)$, where we recall that $S(t)$ is a nonincreasing left-continuous function. Under assumptions regarding censoring of the type discussed in Section 1.4.1, the observed likelihood function is of the form

$$L=\prod_{j=1}^{k}\left[\left(\prod_{i=1}^{\lambda_j} S(L_i^j)\right)\left[S(t_j)-S(t_j+0)\right]^{d_j}\right]\prod_{i=1}^{\lambda_{k+1}} S(L_i^{k+1}). \quad (2.3.4)$$

In seeking to maximize L with respect to $S(t)$, we first observe that $\hat{S}(t)$ must be discontinuous at the t_j's, or else $L=0$. Further, the $\hat{S}(L_i^j)$'s should be as large as possible, in accordance with the restrictions on $S(t)$. This implies that

$$\hat{S}(t_1)=\hat{S}(L_i^1)=1 \qquad i=1,\ldots,\lambda_1$$

$$\hat{S}(L_i^{j+1})=\hat{S}(t_j+0)=\hat{S}(t_{j+1}) \qquad j=1,\ldots,k \quad \text{and} \quad i=1,\ldots,\lambda_{j+1}.$$

Writing $S(t_j+0)=P_j$ $(j=1,\ldots,k)$ and defining $P_0=1$, we see from (2.3.4) that it is only necessary to maximize

$$L_1=\prod_{j=1}^{k}(P_{j-1}-P_j)^{d_j}P_j^{\lambda_{j+1}} \qquad (2.3.5)$$

with respect to $P_1,\ldots,P_k$. Letting $p_j=P_j/P_{j-1}$ and $q_j=1-p_j$ $(j=1,\ldots,k)$, we have

$$L_1=\prod_{j=1}^{k}(p_1\cdots p_{j-1}q_j)^{d_j}(p_1\cdots p_j)^{\lambda_{j+1}}=\prod_{j=1}^{k}q_j^{d_j}p_j^{n_j-d_j}$$

which is maximized for $\hat{p}_j=(n_j-d_j)/n_j$. Hence L is maximized with $S(t)$ such that $\hat{S}(t_j+0)=\hat{S}(t_{j-1}+0)(n_j-d_j)/n_j$, which yields

$$\hat{S}(t)=\prod_{j:t_j<t}\frac{n_j-d_j}{n_j}.$$

Note that if the last observation is a censoring time L^*, in which case $\lambda_{k+1}>0$, then $\hat{S}(L^*)=\hat{S}(t_k+0)$, but $\hat{S}(t)$ is undefined for $t>L^*$. This is because L does not depend on $S(t)$ for $t>L^*$, whereas $\hat{S}(L^*)>0$; thus $S(t)$ is arbitrary here, except, of course, that it must be nonincreasing and lie between $\hat{S}(L^*)$ and 0.

This derivation ignores technical difficulties that arise in attempting to precisely specify L and its parameter space. These can be circumvented, and a more rigorous discussion of the PL estimate and its properties given, by discretizing the time axis and then passing to a limit by letting the number of intervals go to infinity. This approach is briefly discussed in the next section. Johansen (1978) provides a more rigorous discussion of the PL estimate as a maximum likelihood estimate.

2.3.2 Properties of the Product-Limit Estimate*

The product-limit estimate possesses several desirable large-sample properties, a main one being that $\hat{S}(t)$ is a consistent estimate of $S(t)$, under suitable assumptions about censoring. A thorough study of the properties of the PL estimate is rather involved; work in this area has been done by Breslow and Crowley (1974), Meier (1975), Efron (1967), Kaplan and Meier (1958), Peterson (1977), Johansen (1978), and others. We shall merely outline a few pertinent results and refer the reader to these papers for more details. The important points are that $\hat{S}(t)$ is a consistent estimate of $S(t)$ under quite broad conditions and that (2.2.3) is a valid asymptotic variance estimate.

A rigorous approach can be taken by discretizing the time axis and then passing to a limit. Suppose, as in the life table work of Section 2.2, that the time axis is partitioned into intervals $I_j=[a_{j-1}, a_j)$, $j=1,\ldots,k+1$, with $a_0=0$, $a_k=T$, and $a_{k+1}=\infty$. Once again T is an upper limit on the observation time; asymptotic results about $\hat{S}(t)$ will refer to the interval $[0,T)$. The PL estimate (2.3.2) can be viewed as the left-continuous limit that the standard life table estimates give in estimating the $S(a_j)$'s when $k\to\infty$ while $\max(|a_j-a_{j-1}|)\to 0$. Here we suppose, as before, that censoring times are never equal to lifetimes.

To study properties of $\hat{S}(t)$ it is necessary to make specific assumptions about censoring. For example, Breslow and Crowley (1974) adopt the independent random censorship model employed in Section 2.2.2 and Meier (1975) assumes that censoring times are fixed but that the sequence of censoring times has certain properties as the sample size $n\to\infty$. In a careful treatment of this problem it is necessary to verify the conditions under which the two limiting operations $n\to\infty$ and $k\to\infty$ can be interchanged. Breslow and Crowley (1974) and Meier (1975) give rigorous discussions of this.

Breslow and Crowley establish the following result under the random censorship model introduced in Section 2.2.2, in which censoring times $L_1, \ldots, L_n$ for individuals are assumed to be i.i.d. random variables, independent of lifetimes of the individuals, with survivor function $G(t)$.

THEOREM 2.3.1 *Let $T<\infty$ satisfy $S(T)>0$ and suppose that $S(x)$ and $G(x)$ are continuous. Then the random function*

$$\sqrt{n}\left[\hat{S}(x)-S(x)\right] \qquad 0<x<T$$

converges weakly to a mean zero Gaussian process with covariance function

$$S(x)S(y)\int_0^x \frac{|dS(u)|}{S^2(u)G(u)} \qquad 0<x\leqslant y<T. \tag{2.3.6}$$

This theorem implies that under the stated conditions $\hat{S}(t)$ is a consistent estimate of $S(t)$. In addition, (2.3.6) suggests that in large samples the variance of $\hat{S}(t)$ can be approximated as

$$\operatorname{Var}\left[\hat{S}(t)\right] \doteq \frac{1}{n} S(t)^2 \int_0^t \frac{|dS(u)|}{S(u)S^0(u)} \tag{2.3.7}$$

where $S^0(u)=S(u)G(u)$. Similar results have been obtained by Meier (1975), under a slightly different model. Meier treats censoring times L_i as fixed but having certain properties as $n\to\infty$ and proves that $\hat{S}(t)$ is consistent and asymptotically normal. He obtains an expression for variance analogous to (2.3.7).

An estimate of (2.3.7) can be obtained by replacing $S(x)$ and $S^0(x)$ with estimates. A sensible procedure would be to replace $S(x)$ with the PL estimate $\hat{S}(x)$ and to replace $S^0(x)$, which is the expected proportion of individuals alive and at risk at x, by $\hat{S}^0(x)=n_x/n$, where n_x is the number of individuals at risk at x. There is a technical difficulty involved, however, in that the denominator of the integral in (2.3.7) is then a step function with jumps at the same points as the step function $\hat{S}(x)$ involved in the numerator, and, strictly speaking, the Riemann–Stieltjes integral is not defined in this case (see Appendix A). If we improvise and replace $S(x)$ in the denominator of (2.3.7) with $\hat{S}(x+0)$, (2.3.7) becomes

$$\frac{1}{n}\hat{S}(t)^2 \sum_{j:t_j<t} \frac{\hat{S}_{j-1}-\hat{S}_j}{\hat{S}_j(n_j/n)}$$

where $\hat{S}_j = \prod_{i=1}^{j}(n_i - d_i)/n_i$. This reduces to

$$\widehat{\mathrm{Var}}[\hat{S}(t)] = \hat{S}(t)^2 \sum_{j:t_j<t} \frac{d_j}{n_j(n_j-d_j)} \tag{2.3.8}$$

which is the commonly used estimate of $\mathrm{Var}[\hat{S}(t)]$, given earlier as (2.3.3). We might note that if one inserts $\hat{S}(x-0)$ for $S(x)$ in the denominator of (2.3.7), the variance estimate

$$\hat{S}(t)^2 \sum_{j:t_j<t} d_j/n_j^2$$

is obtained.

The consistency and asymptotic normality of $\hat{S}(t)$ can be proved under weaker conditions than those required for Theorem 2.3.1. The deeper result of the theorem, that $\sqrt{n}[\hat{S}(t)-S(t)]$, $0<t<T$, converges to a Gaussian process, is useful in certain contexts, however. One such situation is connected with estimation of the mean lifetime, which can be represented as $\mu = \int_0^\infty S(x)\,dx$ or, more generally, the mean lifetime restricted to T, defined as

$$\mu_T = \int_0^T S(x)\,dx. \tag{2.3.9}$$

This can be estimated by

$$\hat{\mu}_T = \int_0^T \hat{S}(x)\,dx$$

where $\hat{S}(x)$ is the PL estimate. Theorem 2.3.1 then tells us that $\sqrt{n}(\hat{\mu}_T - \mu_T)$ is asymptotically normal, with variance given by

$$\int_0^T \frac{A(u)^2 |dS(u)|}{S(u)S^0(u)} \tag{2.3.10}$$

where $A(u) = \int_u^\infty S(x)\,dx$. The derivation of this result and an estimate of variance based on it are discussed in Problem 2.10.

We have only mentioned properties of the PL estimate under two simple censorship models. The estimate and the associated variance estimate (2.3.3) presumably possess desired asymptotic properties under a fairly broad range of conditions, the essential requirements being the usual ones of quasi-independence of censoring and lifetimes and the assurance that the censoring process does not effectively censor all individuals prior to T. Further study of $\hat{S}(t)$ and its distribution in moderate-size samples is needed. Some results bearing on this are discussed in Section 8.1.1, where confidence interval estimation of $S(t)$ is considered.

2.3.3 The Empirical Cumulative Hazard Function

The cumulative hazard function is defined as $H(t) = -\log S(t)$, so a natural estimate of it is

$$\hat{H}(t) = -\log \hat{S}(t)$$

where $\hat{S}(t)$ is the PL estimate of the survivor function. An alternate estimate of $H(t)$ is

$$\tilde{H}(t) = \sum_{j: t_j < t} \frac{d_j}{n_j} \tag{2.3.11}$$

which is sometimes called the empirical (cumulative) hazard function. This estimate has been discussed by several authors, including Altschuler (1970), Breslow and Crowley (1974), Nelson (1972b), and Efron (1977). In the continuous case, where $H(t) = \int_0^t h(u)\,du$, it can be motivated by thinking of d_j/n_j as a contribution to the hazard function $h(t)$ at t_j. In addition, when the data are Type II censored, it can be shown (see Section 2.4.2) that $E[H(t_{(j)})] = \tilde{H}(t_{(j)}+0)$, where $t_{(j)}$ is the jth observed lifetime.

For continuous models $\hat{H}(t)$ and $\tilde{H}(t)$ are asymptotically equivalent and do not differ greatly in most situations, except for large values of t. Since $\hat{H}(t) = -\log \hat{S}(t)$

$$\hat{H}(t) = -\sum_{j: t_j < t} \log\left(1 - \frac{d_j}{n_j}\right)$$

$$= \sum_{j: t_j < t} \left(\frac{d_j}{n_j} + \frac{d_j^2}{2n_j^2} + \cdots\right) \tag{2.3.12}$$

and thus $\tilde{H}(t)$ is a first-order approximation to $\hat{H}(t)$. Breslow and Crowley (1974) explicitly employ the asymptotic equivalence of $\hat{H}(t)$ and $\tilde{H}(t)$ in proving Theorem 2.3.1.

Because of the simple relationship of $H(t)$ to $S(t)$, a lengthy discussion of $H(t)$ and its estimates is not required, but one or two points can be made.

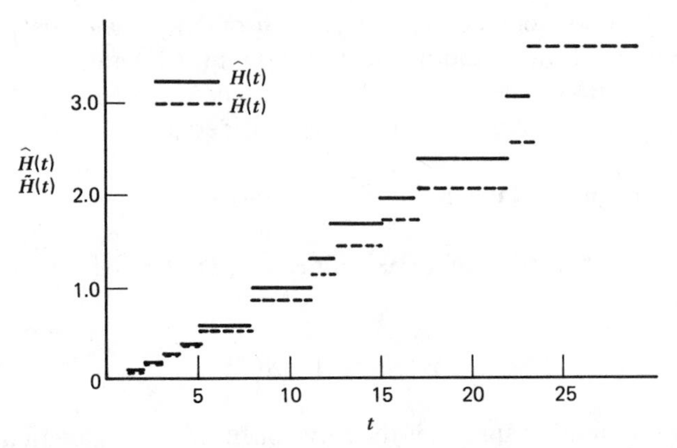

Figure 2.3.2 Hazard function estimates for placebo data (Example 2.3.2)

First, plots of $\hat{H}(t)$ or $\tilde{H}(t)$ are often very useful. For example, it is much easier to assess from such plots whether a life distribution might have a constant, decreasing, or increasing hazard function than from plots of $\hat{S}(t)$. This is because $H(t)$ will be a linear function in t if $h(t)$ is constant, a convex function with increasing first derivative if $h(t)$ is increasing, and a convex function with decreasing first derivative if $h(t)$ is decreasing. Second, there are no theoretical grounds for preferring either $\hat{H}(t)$ or $\tilde{H}(t)$, one to the other in general; $\tilde{H}(t)$ is slightly easier to compute, but this is not an overriding factor. In the same vein $\tilde{S}(t)=\exp[-\tilde{H}(t)]$ can be used as an estimate of $S(t)$, though $\hat{S}(t)$ is usually preferred, partly because of its close connection to life table methods.

Example 2.3.2 (Example 2.3.1 revisited) Figure 2.3.2 shows graphs of $\hat{H}(t)$ and $\tilde{H}(t)$ for the placebo group in Example 2.3.1. The values $\hat{H}(t_j+0)$ and $\tilde{H}(t_j+0)$ are easily calculated from the quantities given in Table 2.3.1 and are shown in Table 2.3.2.

As would be expected from (2.3.12), agreement between the two estimates is good for smaller values of t, but poorer as t becomes larger. Both graphs suggest the possibility of an exponential lifetime distribution, since the plots lie near a straight line, though since there are relatively few observations, too much should not be inferred from the graph. For an idea of the sampling variation in $\hat{H}(t)$ or $\tilde{H}(t)$ the estimate

$$\widehat{\mathrm{Var}}\left[\hat{H}(t)\,\mathrm{or}\,\tilde{H}(t)\right]=\frac{\widehat{\mathrm{Var}}\left[\hat{S}(t)\right]}{\hat{S}(t)^2}$$

Table 2.3.2 Calculation of Hazard Function Estimates

t_j	$\hat{H}(t_j+0)$	$\tilde{H}(t_j+0)$	t_j	$\hat{H}(t_j+0)$	$\tilde{H}(t_j+0)$
1	0.100	0.095	11	1.252	1.110
2	0.211	0.201	12	1.661	1.444
3	0.272	0.259	15	1.945	1.694
4	0.405	0.384	17	2.354	2.027
5	0.560	0.527	22	3.037	2.527
8	0.965	0.860	23	∞	3.527

can be used, where $\widehat{\text{Var}}[\hat{S}(t)]$ is the estimate (2.3.3). This is obtained from the first-order approximation for variance of a function [(C4) of Appendix C] applied to $\hat{H}(t)=-\log\hat{S}(t)$, but applies to $\tilde{H}(t)$ as well, since $\tilde{H}(t)$ and $\hat{H}(t)$ are asymptotically equivalent.

2.4 PLOTTING PROCEDURES

2.4.1 Plots Involving Estimated Survivor or Hazard Functions

Plots of estimated survivor or cumulative hazard functions provide useful pictures of univariate lifetime data, as well as information on the underlying life distribution. They can be used for informal checks on the appropriateness of a model and for obtaining parameter estimates within a model.

Plots of $\hat{H}(t)=-\log\hat{S}(t)$, or of $\tilde{H}(t)$ given by (2.3.11), have already been suggested for help in assessing the shape of the hazard function in a distribution. Similar plots can often be used to help assess whether a specific parametric family of models is reasonable. The basic idea is to make plots that should be roughly linear if the proposed family of models is appropriate, since departures from linearity can be readily appreciated by eye. Suppose, for example, that the possibility of an underlying exponential distribution is being considered. The survivor function for the exponential distribution satisfies

$$\log S(t)=-\lambda t.$$

Therefore, if $\log\hat{S}(t)$ is plotted against t, the resultant graph should be roughly linear and pass through the origin, if an exponential model is appropriate. Alternately, $\tilde{H}(t)$ could be plotted against t, with the same features expected. Such plots also provide an estimate of λ when an exponential model appears reasonable. This is obtained by fitting a straight line to the plot and estimating λ as the slope of the line.

For the Weibull distribution $S(t)$ is of the form $\exp[-(\lambda t)^\beta]$, and

$$\log[-\log S(t)] = \beta \log t + \beta \log \lambda.$$

Thus a plot of $\log[-\log \hat{S}(t)]$ versus $\log t$ should be approximately linear if a Weibull model is appropriate. In addition, when the plot is approximately linear, one can obtain graphical estimates of λ and β by fitting a straight line to the plot and calculating the slope and intercept. In particular, if $\log[-\log \hat{S}(t)]$ is treated as the ordinate (y variate) and $\log t$ as the abscissa (x variate), the slope of the line is an estimate of β and the x-intercept is an estimate of $-\log \lambda$.

The types of procedure just described can be used for models in which some transform of a lifetime T, say $X = g(T)$, has a location-scale parameter distribution. In this case X has a survivor function of the form (assuming X an increasing function of T)

$$\Pr(X \geqslant x) = S_1\left(\frac{x-\mu}{\sigma}\right)$$

$$= S(t) = \Pr(T \geqslant t)$$

where $t = g^{-1}(x)$. Then $S_1^{-1}[S(t)] = (x-\mu)/\sigma$ is a linear function of $x = g(t)$, and a plot of $S_1^{-1}[\hat{S}(t)]$ versus $g(t)$ should be roughly linear if the family of models being considered is reasonable. The Weibull and exponential distributions both fall into this category: for the Weibull distribution $X = \log T$ has an extreme value distribution (see Section 1.3.2), with $S_1(z) = \exp(-e^z)$. This leads to the plot of $\log[-\log \hat{S}(t)]$ versus $\log t$ suggested here. Two approaches are possible for the exponential distribution. Since the exponential distribution is a special case of the Weibull distribution, with $\beta = 1$, one could plot $\log[-\log \hat{S}(t)]$ versus $\log t$. Alternately, one can plot $\log \hat{S}(t)$ versus t, since λ is a scale parameter in the exponential model.

Similar checks can be made with other distributions. For example, one can check for a log-normal distribution for T by plotting $Q^{-1}[\hat{S}(t)]$ or $\Phi^{-1}[\hat{S}(t)]$ versus $\log t$, where

$$Q(x) = 1 - \Phi(x) = \int_x^\infty \frac{1}{(2\pi)^{1/2}} e^{-z^2/2}\, dz$$

is the standard normal survivor function. Sometimes a little ingenuity may help to find a suitable plot. For example, the linear hazard rate distribution with hazard function $h(t) = \alpha + \beta t$ has cumulative hazard function $H(t) = \alpha t + \beta t^2/2$. Thus $t^{-1}H(t) = \alpha + \beta t/2$ is a linear function of t, and approximately linear plots of $t^{-1}\hat{H}(t)$ or $t^{-1}\tilde{H}(t)$ versus t should result if the model is reasonable.

Example 2.4.1 Pike (1966) gives results of a laboratory experiment concerning vaginal cancer in female rats. In one experiment 19 rats were painted with the carcinogen DMBA, and the number of days T until the appearance of a carcinoma was the variable of interest. At the time the data were collected only 17 out of the 19 rats had developed a carcinoma, so that two of the times below (marked *) are censoring times. The times were

$$143, 164, 188, 188, 190, 192, 206, 209, 213, 216,$$
$$220, 227, 230, 234, 246, 265, 304, 216^*, 244^*.$$

The PL estimate $\hat{S}(t)$ or the empirical hazard function $\tilde{H}(t)$ is readily calculated using (2.3.2) or (2.3.11), respectively. A Weibull model has been frequently found adequate in this kind of problem, and thus we consider a plot of $\log[-\log \hat{S}(t)]$ versus $\log t$. Figure 2.4.1*a* shows this to be roughly linear, suggesting that a Weibull model could be reasonable. There is also some reason to believe that there may be a threshold value $\mu>0$ before

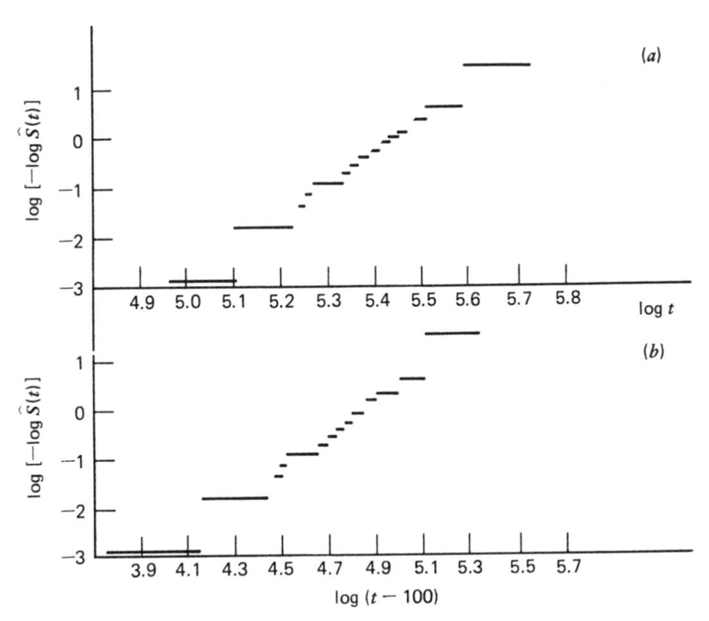

Figure 2.4.1 $\text{Log}[-\log \hat{S}(t)]$ plots for rat carcinogenesis data (Example 2.4.1)

which no carcinoma can appear and that a more reasonable model would be to treat $T-\mu$ as having a Weibull distribution. Figure 2.4.1b shows a plot of $\log[-\log \hat{S}(t)]$ versus $\log(t-100)$, with the value $\mu=100$ arbitrarily selected for illustration. This plot is also approximately linear, though it is not possible to say on this basis that the model with $\mu=100$ is necessarily more plausible than that with $\mu=0$. Similar pictures are obtained for other μ values in the range 0 to 120 or so. It thus appears that a Weibull model, perhaps with a nonzero threshold parameter μ, is not unreasonable, though there is evidently little information about the value of μ to be gleaned from the plots.

Graphical estimates of parameters can be obtained from the plots. For example, in a Weibull model with $\mu=100$, $S(t)=\exp[-\lambda(t-100)^{\beta}]$, so that $\log[-\log S(t)]=\beta\log\lambda+\beta\log(t-100)$. If a straight line is drawn through the plot in Figure 2.4.1b, β can therefore be estimated as the slope of the line and λ estimated from the intercept. A line was drawn in Figure 2.4.1b by eye; the slope of the line was approximately 3.8, giving $\tilde{\beta}=3.8$ as an estimate of β. To estimate λ one can use the fact that when $\log[-\log \hat{S}(t)]=0$, $\beta\log\lambda+\beta\log(t-100)=0$, which gives $\log\lambda=-\log(t-100)$. The fitted straight line crossed the line $\log[-\log \hat{S}(t)]=0$ at about $\log(t-100)=4.87$, which gives $\lambda=\exp(-4.87)=0.0077$.

This example is reexamined in Example 4.4.1 of Chapter 4 and, by comparison, the m.l.e.'s of β and λ are found there to be 3.38 and 0.0076, so the graphical estimates obtained here are quite good.

2.4.2 Probability and Hazard Plots

When the data are uncensored or Type II censored, it is customary to use so-called probability plots, rather than the type discussed in the previous section. These are similar to the plots of the preceding section, except that, instead of the entire estimated survivor function, a single point is plotted for each lifetime.

Probability Plots

Probability plots in their most common form are used with location–scale parameter models. Suppose that X is a random variable with distribution function of the form $F[(x-\mu)/\sigma]$, where σ is a scale parameter and μ a location parameter ($\sigma>0, -\infty<\mu<\infty$). Let $x_{(1)}<x_{(2)}<\cdots<x_{(n)}$ be the ordered observations in a random sample of size n from the distribution of X. A probability plot is a plot of the $x_{(i)}$'s against quantities $m_i=F^{-1}(a_i)$, where a_i is a fixed estimate of $F[(x_{(i)}-\mu)/\sigma]$. Since $F^{-1}\{F[(x_{(i)}-\mu)/\sigma]\}=(x_{(i)}-\mu)/\sigma$, if the stated model is reasonable the plot of the points

$(x_{(i)}, m_i)$ should be roughly linear. In fact, the points should lie fairly near the line $x=\mu+\sigma m$, and thus estimates of μ and σ can be obtained from the plot.

The a_i's are sometimes referred to as the plotting positions. Several choices for the a_i's are used in practice, but the two most popular are $a_i=(i-0.5)/n$ and $a_i=i/(n+1)$. The former is motivated by the fact that the empirical distribution function changes from $(i-1)/n$ to i/n at $x_{(i)}$, so that one can think of $x_{(i)}$ as corresponding to something between the $(i-1)/n$ and i/n quantiles. Taking $(i-0.5)/n$, which is midway between these two values, and then equating $x_{(i)}$ and the $(i-0.5)/n$ quantile of the distribution, we get $F[(x_{(i)}-\mu)/\sigma]=(i-0.5)/n=a_i$. The second choice mentioned, $a_i=i/(n+1)$, is similarly plausible and has the added feature that $E\{F[(x_{(i)}-\mu)/\sigma]\}=i/(n+1)$ (see Appendix D). Still another choice is $a_i=E\{[(x_{(i)}-\mu)/\sigma]\}$, provided that these quantities are available for the distribution in question. For the extreme value distribution these are given by White (1969), and for the normal distribution they are given by Sarhan and Greenberg (1962), among others. For most purposes it does not matter much which a_i's are used, though when the objective is a precise estimation of μ and σ, the choice of a_i's can make a difference (e.g., Barnett, 1975).

To facilitate probability plots, special "probability" graph papers are available for the more common distributions. These graph papers have a scale based on values of $F^{-1}(a)$ but labeled with an a-scale, so that to effectively plot the points $[x_{(i)}, F^{-1}(a_i)]$, one need only plot the points $(x_{(i)}, a_i)$. This saves the trouble of computing $F^{-1}(a)$. Probability papers for the extreme value and normal distributions are particularly useful in life distribution work.

The exponential distribution should be mentioned as a special case. The exponential distribution can be treated as a Weibull distribution with $\beta=1$, and probability plots based on the extreme value distribution can be used to check the model. A simpler procedure, however, is to plot ordered lifetimes $t_{(i)}$ against, say, $\log[1-(i-0.5)/n]$ or $\log[1-i/(n+1)]$. Since $F(t)=1-\exp(-\lambda t)$, this should result in a plot that is approximately linear with slope $-\lambda$. An alternate approach is to plot the points $(t_{(i)}, \alpha_i)$, where

$$\alpha_i=\sum_{j=1}^{i}(n-j+1)^{-1}.$$

It is shown in Chapter 3 (see Lemma 3.4.1) that $\alpha_i=E(\lambda t_{(i)})$, so this plot should be approximately linear with slope λ.

An example will help clarify these ideas and illustrate the use of probability graph paper.

Example 2.4.2 Mann and Fertig (1973) give failure times of airplane components subjected to a life test. The data are Type II censored: 13 components were placed on test, but the test was terminated at the time of the tenth failure. Failure times $t_{(i)}$ (in hours) of the 10 components that failed were 0.22, 0.50, 0.88, 1.00, 1.32, 1.33, 1.54, 1.76, 2.50, 3.00.

The Weibull distribution has often been found a suitable model in such situations, and as an informal check on this we can construct a Weibull probability plot. If T has a Weibull distribution, then (see Section 1.3.2) $X=\log T$ has an extreme value distribution with a distribution function of the form $F[(x-\mu)/\sigma]$, where

$$F(z)=1-\exp[-\exp(z)] \qquad -\infty<z<\infty$$

and

$$F^{-1}(a)=\log[-\log(1-a)] \qquad 0<a<1.$$

A probability plot of a sample $x_1,\ldots,x_n$ from an extreme value distribution consists of plotting $x_{(i)}$ versus, say, $m_i=F^{-1}[(i-0.5)/n]=\log\{-\log[1-(i-0.5)/n]\}$, for $i=1,\ldots,n$. If the data are Type II censored and only $x_{(1)},\ldots,x_{(r)}$ $(r<n)$ are available, the probability plot can still be made but will include points only for the first r ordered observations.

To obtain the desired probability plot we thus compute log lifetimes $x_{(i)}=\log t_{(i)}$ from the data and plot the 10 points $(x_{(i)}, m_i)$, $i=1,\ldots, 10$. If extreme value probability paper is used, this job is easier, since the vertical scale on the paper is set up so that the points can be located as having coordinates $[x_{(i)},(i-0.5)/13]$. Weibull probability paper takes this one step further and gives on the horizontal scale values of both x and $t=\exp x$. This means that one does not need to calculate $x_{(i)}=\log t_{(i)}$ explicitly, but can use the t-scale on the graph and plot the points as $[t_{(i)},(i-0.5)/13]$, $i=1,\ldots, 10$.

Figure 2.4.2 shows the resulting plot on Weibull probability paper. The points in the plot fall roughly on a straight line and give no obvious evidence that the data are not from a Weibull distribution. Graphical estimates of μ and σ, and of the related parameters $\beta=1/\sigma$ and $\lambda=\exp(-\mu)$ in the Weibull survivor function $S(t)=\exp[-(\lambda t)^{\beta}]$, can be obtained. For the extreme value distribution the relationship between X and its distribution function becomes

$$\begin{aligned} x &= \mu+\sigma F^{-1}(a) \\ &= \mu+\sigma\log[-\log(1-a)] \\ &= \mu+\sigma Y. \end{aligned}$$

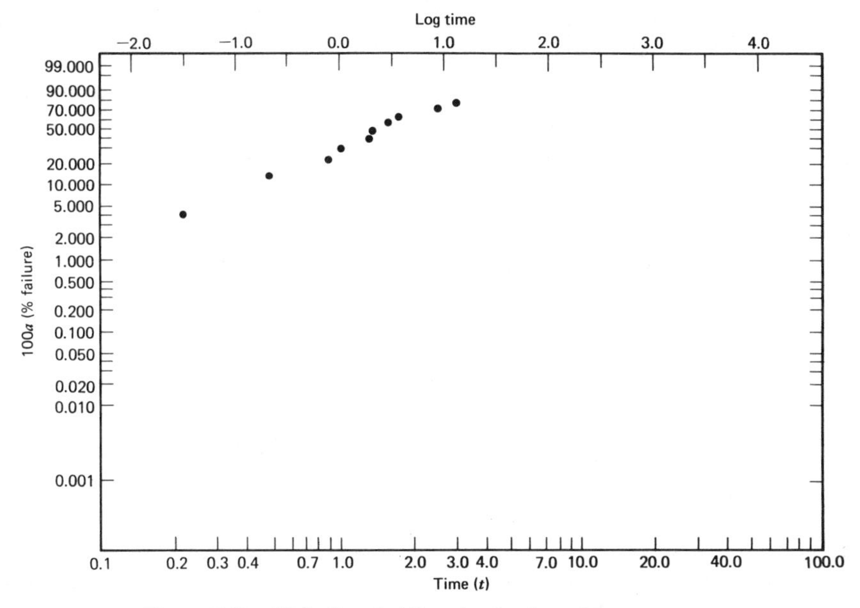

Figure 2.4.2 Weibull probability plot for data of Example 2.4.2

Rewriting this as $Y=(1/\sigma)x-(\mu/\sigma)$, we see that if a straight line is fitted through the points in the plot, then the slope of the line provides an estimate of $1/\sigma$ and that the line $Y=0$ gives the x-intercept as an estimate of μ. A straight line was drawn by eye on Figure 2.4.2, and the slope was approximately 1.45, giving $\tilde{\sigma}=1.45^{-1}=0.69$. In addition, the value $x=0.77$ corresponded to the value $Y=0$ and gave the estimate $\tilde{\mu}=0.77$. These can be converted into estimates for the Weibull parameters:

$$\tilde{\beta}=\frac{1}{\tilde{\sigma}}=1.45 \quad \text{and} \quad \tilde{\lambda}=\exp(-\tilde{\mu})=0.46.$$

Probability plots have been described in terms of the distribution function $F(t)$ rather than the survivor function $S(t)$, since this is the conventional approach, but it is clear that $S(t)$ could replace $F(t)$ in the discussion, with one or two obvious modifications. In addition, probability plots are primarily intended for complete or Type II censored samples, but point plots are sometimes used with Type I censored data. Probability paper is useful for making either these or complete plots of $\hat{S}(t)$. With Type I censored data jumps in $\hat{S}(t)$ occur at points $[t_j, \hat{S}(t_j)]$, where t_j is an observed lifetime. The a-scale on the probability paper is used for $\hat{S}(t)$ values, and if the entire estimated survivor function is plotted on the paper, the result should be a step function that lies roughly along a straight line, if the proposed model is appropriate. It is usually preferable to plot $\hat{S}(t)$ in its entirety rather than single points from it. If just points are plotted, the points $\{t_j, [\hat{S}(t_j)+\hat{S}(t_j+0)]/2\}$ give a better impression than $[t_j, \hat{S}(t_j)]$ or $[t_j, \hat{S}(t_j+0)]$ of the entire function $\hat{S}(t)$.

Hazard Plots

Another frequently used plot is the so-called hazard plot (Nelson, 1972b). This is essentially the same as a probability plot or a point plot obtained from $\hat{S}(t)$, except that instead of being based on the PL estimate, the plots are based on the empirical hazard function (2.3.11),

$$\tilde{H}(t) = \sum_{j:t_j<t} \frac{d_j}{n_j}.$$

The usual procedure is either to plot $\tilde{H}(t)$ or a transform of it in full or to base a point plot on $\tilde{H}(t)$. When the data are Type II censored, the observed lifetimes $t_{(1)} < \cdots < t_{(k)}$ are the first k lifetimes in a sample of size n, and the number of individuals at risk just prior to $t_{(j)}$ is $n_j = n-j+1$. This gives

$$\tilde{H}\left(t_{(i)}+0\right) = \sum_{j=1}^{i} (n-j+1)^{-1} \qquad i=1,\ldots,k \tag{2.4.1}$$

in this case.

Plots involving $\tilde{H}(t)$ are used in exactly the same way as plots involving $\hat{S}(t)$, the connection being obvious in view of the fact that $H(t) = -\log S(t)$. For the Weibull distribution, for example, $S(t) = \exp[-(\lambda t)^{\beta}]$ and $\log[-\log S(t)] = \log H(t) = \beta \log t + \beta \log \lambda$. Thus a plot of $\log \tilde{H}(t)$ versus $\log t$ should be roughly linear. Point plots are often based on (2.4.1), and when the data are Type II censored, these hazard plots are analogous to probability plots. The most commonly used procedure is to plot the points

$$\left\{g_1\left(t_{(i)}\right), g_2\left[\tilde{H}\left(t_{(i)}+0\right)\right]\right\} \qquad i=1,\ldots,k \tag{2.4.2}$$

where $g_2[H(t)]$ is a transform of $H(t)$, which is linear in $g_1(t)$. The plot should be roughly linear if the proposed model is appropriate.

The points (2.4.2) are usually preferred to other choices such as $\{g_1(t_{(i)}), g_2[\tilde{H}(t_{(i)})]\}$, $i=1,\ldots,k$. One argument in support of (2.4.2) is that with Type II censoring it can be shown that

$$E\left[H\left(t_{(i)}+0\right)\right] = \tilde{H}\left(t_{(i)}+0\right) \qquad i=1,\ldots,n.$$

To prove this suppose that T is a random variable with survivor function $S(t)$. As is well known, the random variable $U=S(T)$ has a uniform distribution on $(0,1)$, and hence $W=-\log U=H(T)$ has an exponential distribution with p.d.f. $f(w)=\exp(-w), w\geqslant 0$. Therefore, if $T_{(1)}<\cdots<T_{(n)}$ are the ordered observations in a random sample of size n, the random variable $W_{(i)}=H(T_{(i)})$ is the ith ordered observation in a random sample of size n from the standard exponential distribution. It is easily shown (see Lemma 3.4.1) that

$$E(W_{(i)})=\sum_{j=1}^{i}(n-j+1)^{-1}$$

and thus the stated result follows from (2.4.1). For the exponential distribution this leads to the plots of $t_{(i)}$ versus $E(W_{(i)})$ mentioned earlier.

It is straightforward but slightly more tedious to show that if the data are progressively Type II censored, the result $E[H(t_{(i)}+0)]=\tilde{H}(t_{(i)}+0)$ still holds, where $t_{(i)}$ represents the ith smallest observed lifetime. Hazard plots involving the points (2.4.2) are also frequently used when the data are Type I censored, though in this case the property just stated no longer holds.

Hazard plots can be made on ordinary probability paper, but since $H(t)=-\log S(t)$, values of $\tilde{H}(t)$ need to be converted to values $\tilde{S}(t)=\exp[-\tilde{H}(t)]$ to correspond to the a-scale on the paper. To enable hazard plots to be made with as few calculations as possible, commercial hazard plotting papers have been prepared for certain common distributions. The scales on the paper are set up so that the points $[t, H(t)]$ give a straight line, and thus if the assumed model is appropriate, a plot of the points $[t_{(i)}, \tilde{H}(t_{(i)}+0)]$ should be roughly linear.

Example 2.4.3 (Example 2.4.1 revisited) Example 2.4.1 presented a sample with 19 observations from a laboratory carcinogenesis experiment, two of which were censored. There we calculated the PL estimate $\hat{S}(t)$ and graphed $\log[-\log\hat{S}(t)]$ versus $\log t$, and also $\log[-\log\hat{S}(t)]$ versus $\log(t-100)$, in checking to see whether a Weibull model appeared reasonable. One could alternately calculate and graph the empirical hazard function $\tilde{H}(t)$, either in its entirety or as a hazard plot.

Successive values $\tilde{H}(t_j+0)$ are easily calculated recursively, since $\tilde{H}(t_j+0)=\tilde{H}(t_{j-1}+0)+d_j/n_j$. The first few values are shown in Table 2.3.3. The values of $\log\tilde{H}(t_j+0)$ and $\log(t_j-100)$ are also shown. Figure 2.4.3 shows a hazard plot of the points $[\log(t_j-100), \log\tilde{H}(t_j+0)]$ for the 16 distinct observed lifetimes. This plot is appropriate for examining whether $T-100$ might have a Weibull distribution, since in this case $\log H(t)=\beta\log\lambda+$

Table 2.3.3 Calculation of Hazard Function Estimate

t_j	n_j	d_j	$\tilde{H}(t_j+0)$	$\log \tilde{H}(t_j+0)$	$\log(t_j-100)$
143	19	1	0.053	−2.94	3.76
164	18	1	0.108	−2.23	4.16
188	17	2	0.226	−1.49	4.48
190	15	1	0.293	−1.23	4.50
192	14	1	0.364	−1.01	4.52

$\beta\log(t-100)$. Alternately, we could plot the entire step function $\log \tilde{H}(t)$ versus $\log(t-100)$. The points in the plot lie reasonably close to a straight line, though there is a hint of a systematic departure from linearity in the very slight S shape of the plot.

As in Example 2.4.1, graphical estimates of the parameters can be obtained. A straight line was drawn by eye in Figure 2.4.3, and by using the same approach as in Example 2.4.1, we obtain estimates $\tilde{\beta}=3.2$ and $\tilde{\lambda}=\exp(-4.87)=0.0077$ from the slope and from the intersection of the line with the line $\log \tilde{H}(t)=0$. These agree well with the estimates obtained earlier.

Discussion

When plots are used informally, it does not make too much difference exactly what plotting positions are chosen and whether plots are based on $\hat{S}(t)$ or $\tilde{H}(t)$. One should be careful not to read too much into a plot, however, especially one based on relatively few observations. It is desirable

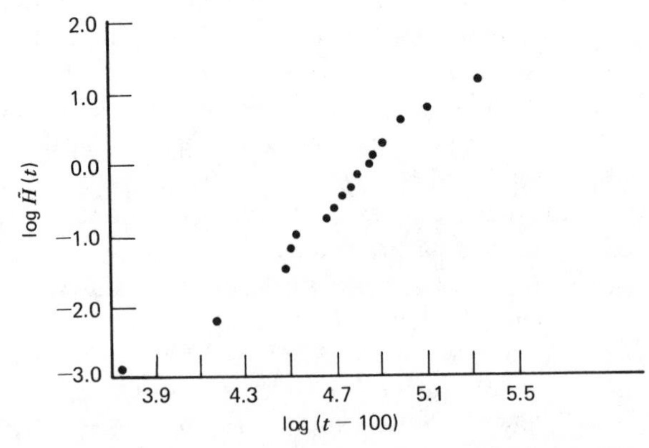

Figure 2.4.3 Hazard plot for rat carcinogenesis data (Example 2.4.3)

to have an idea of the variability inherent in a plotting procedure: one should be aware of approximate variances of the quantities plotted, dependence among the points in the plot, and so on. Differences in precision for estimates in the tails, as opposed to the center of the distribution, should be noted. Cox (1978) provides an illuminating discussion of graphical procedures and their properties.

Plots provide methods of displaying information and suggesting and checking models and give crude parameter estimates. They are not meant as a substitute for formal test and estimation procedures. A thorough statistical analysis will typically involve a combination of graphical and other informal techniques and more formal methods. Tests and estimation under various models are developed in detail in other chapters.

2.5 LEAST SQUARES ESTIMATION OF PARAMETERS

This chapter deals primarily with nonparametric methods of estimation and with simple methods of portraying and examining lifetime data. Parameter estimation within specific families of models has not yet been considered formally, though probability plots can be used to provide graphical estimates in certain situations. Parameter estimation is treated at length in subsequent chapters, but before concluding this chapter we will mention simple least squares procedures for estimating parameters. This is closely related in certain cases to the plotting methods described here and can be used to provide quick parameter estimates in many models.

In the examples of the previous section parameters were estimated from a probability plot by fitting a straight line through the points by eye, but it is clear that the line could have been determined by least squares or some other formal method. A similar idea can be used more generally to give parameter estimates in certain instances. We consider models in which the unknown parameters can be linearly related to some transform of the survivor function. To be specific, suppose that

$$g[S(t)] = \alpha_1 l_1(t) + \cdots + \alpha_m l_m(t) \tag{2.5.1}$$

where the $l_i(t)$'s are known and form a linearly independent set of functions of t and $\alpha_1, \ldots, \alpha_m$ are unknown parameters. Many distributions can be put in this form. For example, for the Weibull distribution

$$\begin{aligned} \log[-\log S(t)] &= \beta \log \lambda + \beta \log t \\ &= \alpha_1 + \alpha_2 \log t \end{aligned}$$

with $\alpha_1 = \beta \log \lambda$ and $\alpha_2 = \beta$. Another family with this property is the family of distributions with polynomial hazard functions $h(t) = \beta_0 + \beta_1 t + \cdots + \beta_{m-1} t^{m-1}$ (see Section 1.3.6). In this case the cumulative hazard function is also a polynomial,

$$\begin{aligned} H(t) &= -\log S(t) \\ &= \alpha_1 t + \alpha_2 t^2 + \cdots + \alpha_m t^m \end{aligned}$$

where $\alpha_i = \beta_{i-1}/i$ $(i = 1, \ldots, m)$.

With models for which a relationship of the form (2.5.1) holds, a form of least squares can be used to estimate $\alpha_1, \ldots, \alpha_m$. Suppose that t_i $(i = 1, \ldots, k)$ are the observed lifetimes in a censored sample, let $\tilde{S}(t_i)$ be some convenient estimate of $S(t_i)$ and let $Y_i = g[\tilde{S}(t_i)]$. Then $\alpha_1, \ldots, \alpha_m$ can be estimated by minimizing

$$Q = \sum_{i=1}^{k} \left[Y_i - \alpha_1 l_1(t_i) - \cdots - \alpha_m l_m(t_i) \right]^2$$

provided that $k \geqslant m$. This procedure is crude but is often useful as a simple way of obtaining estimates that can, for example, be used as initial estimates in procedures for obtaining m.l.e.'s. Various refinements, including the use of a weighted least squares criterion, are possible but will not be discussed here.

When the data are Type II censored and the t_i's are the k smallest observations $t_{(1)} < t_{(2)} < \cdots < t_{(k)}$ in a sample of size n, a convenient procedure is to estimate $S(t_{(i)})$ by either $1 - i/(n+1)$ or $1 - (i - 0.5)/n$. For motivation of these choices, recall the earlier discussion of probability plotting positions. With Type I censored data $S(t_i)$ can be estimated various ways: either $\hat{S}(t_i)$ or $\hat{S}(t_i + 0)$ is convenient, where $\hat{S}(t)$ is the PL estimate, though the choice $\frac{1}{2}[\hat{S}(t_i) + \hat{S}(t_i + 0)]$ is perhaps preferable to either of these. Another possibility is to estimate $S(t_i)$ using the empirical hazard function $\tilde{H}(t)$, for example, as $\exp[-\tilde{H}(t_i + 0)]$.

When the data are Type II censored and the lifetime model can be transformed to a location–scale parameter model for $X = g(T)$, say, minimizing Q is equivalent to fitting a straight line by least squares to the points in a probability plot, assuming that $S(x_{(i)})$ is estimated by the a_i's used in forming the probability plot. Refinement of the least squares procedure to take into account the variances and covariances of the $X_{(i)}$'s is possible and produces best linear unbiased estimates of the location parameter μ and scale parameter σ (e.g., Sarhan and Greenberg, 1962). This will not be explored here, since the aim is simply to describe a quick method of estimation. Best linear unbiased estimation is discussed for the extreme value and normal distributions in Chapters 4 and 5, respectively.

Least squares methods can also be applied when the data are grouped. Discussion of this is deferred to Section 5.5, where the problem of estimating parameters from grouped data is examined. The brief treatment of the present section concludes with an example.

Example 2.5.1 (Example 2.4.2 revisited) Consider the estimation of parameters in the Weibull distribution, assumed to be an appropriate model for the data of Example 2.4.2. The Weibull survivor function $S(t)=\exp[-(\lambda t)^{\beta}]$ satisfies

$$\log[-\log S(t)] = \beta\log\lambda + \beta\log t$$

$$= \alpha_1 + \alpha_2 \log t.$$

Define

$$Y_i = \log\left[-\log\left(1 - \frac{i-0.5}{n}\right)\right]$$

where $1-(i-0.5)/n$ estimates $S(t_{(i)})$. The data in question are Type II censored, with $k=10$ out of a total of $n=13$ lifetimes observed. Letting $x_{(i)}=\log t_{(i)}$, we need to minimize

$$Q = \sum_{i=1}^{10} \left(Y_i - \alpha_1 - \alpha_2 x_{(i)}\right)^2$$

in order to obtain $\hat{\alpha}_1$ and $\hat{\alpha}_2$. As is well known, $\hat{\alpha}_1$ and $\hat{\alpha}_2$ are given by

$$\hat{\alpha}_2 = \left(\sum_{i=1}^{10} (y_i - \bar{y})(x_{(i)} - \bar{x})\right) \Big/ \left(\sum_{i=1}^{10} (x_{(i)} - \bar{x})^2\right)$$

$$\hat{\alpha}_1 = \bar{y} - \hat{\alpha}_2 \bar{x}$$

where $\bar{y}=\Sigma y_i/10$ and $\bar{x}=\Sigma x_{(i)}/10$. We find that $\hat{\alpha}_2=1.3833$ and $\hat{\alpha}_1=-1.1360$, which gives $\tilde{\beta}=1.38$ and $\tilde{\lambda}=0.44$. These can be compared with the estimates obtained in Example 2.4.2 by fitting a line through a probability plot by eye, which were $\tilde{\beta}=1.45$ and $\tilde{\lambda}=0.46$. They can also be compared with the m.l.e.'s of β and λ, which are found in Example 4.1.1 of Chapter 4 to be $\hat{\beta}=1.416$ and $\hat{\lambda}=0.440$. The simple least squares procedure produces estimates that are close to the m.l.e.'s in this instance.

2.6 PROBLEMS AND SUPPLEMENTS

2.1 Sometimes it is desired to estimate the hazard function from grouped (i.e., life table) data. For the jth interval $I_j=[a_{j-1}, a_j)$ in a life table define $h_j=a_j-a_{j-1}$ and let $t_{mj}=a_{j-1}+h_j/2$ be the interval midpoint. Two estimates that have been suggested for $h(t_{mj})$ are

$$\hat{h}_1(t_{mj})=\frac{2\hat{q}_j}{h_j(1+\hat{p}_j)} \qquad \text{(Kimball, 1960; Gehan, 1969)}$$

$$\hat{h}_2(t_{mj})=\frac{-\log \hat{p}_j}{h_j} \qquad \text{(Sacher, 1956).}$$

a. Can you motivate these choices of estimate? Compare the estimates by expanding them in powers of $\hat{q}_j$.
b. Derive variance estimates for the two estimates. What use (if any) would you be inclined to make of these?

(Section 2.2)

2.2 Consider the properties of the standard life table estimates under the random censorship model of Section 1.4.1c. Show that if the censoring times have a continuous survivor function $G(x)$ on $[0, T]$, then the standard life table method gives a consistent estimate of $S(a_i), i=1,\ldots,k-1$, for any choice of interval endpoints if and only if

$$S(x)=[1+c-cG(x)]^{-1/2}$$

for some constant $c>0$. Examine the form of $S(x)$ for the cases in which

1. Censoring times are uniformly distributed on $(0, T)$.
2. Censoring times have an exponential distribution truncated at T.

(Section 2.2.2; Breslow and Crowley, 1974)

2.3 The data below are survival times for a group of 121 breast cancer patients treated over the period 1929 to 1938, quoted in Boag (1949). Times are given in months and starred observations are censoring times.

0.3	0.3*	4.0*	5.0	5.6	6.2	6.3	6.6	6.8
7.4*	7.5	8.4	8.4	10.3	11.0	11.8	12.2	12.3
13.5	14.4	14.4	14.8	15.5*	15.7	16.2	16.3	16.5
16.8	17.2	17.3	17.5	17.9	19.8	20.4	20.9	21.0
21.0	21.1	23.0	23.4*	23.6	24.0	24.0	27.9	28.2
29.1	30	31	31	32	35	35	37*	37*
37*	38	38*	38*	39*	39*	40	40*	40*
41	41	41*	42	43*	43*	43*	44	45*
45*	46*	46*	47*	48	49*	51	51	51*
52	54	55*	56	57*	58*	59*	60	60*
60*	61*	62*	65*	65*	67*	67*	68*	69*
78	80	83*	88*	89	90	93*	96*	103*
105*	109*	109*	111*	115*	117*	125*	126	127*
129*	129*	139*	154*					

a. Calculate the product-limit estimate of the survivor function. Estimate 1- and 5-year survival probabilities and give an approximate variance for these estimates. Do any parametric models suggest themselves for these data?

b. Group the data into a life table with 1-year intervals. Compare the 1- and 5-year survival probability estimates with those obtained in part (a).

c. In the data given by Boag the individuals with censored survival times are actually known to fall into one of three groups:

   (i) Individuals free from signs or symptoms of breast cancer but who died from some other cause.

   (ii) Individuals free from signs or symptoms of breast cancer and still alive at the time the data were collected.

   (iii) Individuals still alive at the time the data were collected but who were suffering a persistence or recurrence of the cancer that was unlikely to yield to further treatment.

   How would you take this information into account in analyzing the data?

(Sections 2.2, 2.3)

2.4 Consider the life table in the case in which there is no censoring, and show by counterexample that although $\mathrm{Cov}(\hat{p}_i, \hat{p}_j)=0$, $\hat{p}_i$ and $\hat{p}_j$ are not in general independent. (Hint: consider $\hat{p}_1$ and $\hat{p}_2$.)

(Section 2.2.2)

2.5 Using (C2) and (C5) of Appendix C, derive expression (2.2.17) for the asymptotic variance of the $\sqrt{n}\hat{q}_j$'s and show that their asymptotic covariances are zero.

(Section 2.2.2)

2.6 Examine the asymptotic bias in the standard life table estimates when the lifetime distribution is exponential and the censoring time distribution is (1) exponential, (2) uniform over some interval $[a, b)$.

(Section 2.2.2)

2.7 ***Chiang's life table estimates.*** Under the assumptions stated in Section 2.2.3, namely, that lifetimes of individuals follow an exponential distribution and that censoring times for individuals due for withdrawal in interval I_j have a uniform distribution over I_j, show that given N_{j2}, D_{j2} has a binomial distribution with parameters N_{j2} and $q_j^* = 1 + q_j/\log(1-q_j)$. Show that when the approximation $q_j^* = 1-(1-q_j)^{1/2}$ is substituted into the likelihood function, the likelihood is maximized by (2.2.20).

(Section 2.2.3; Chiang, 1960a, b, 1968)

2.8 ***Elveback's life table estimates.*** Consider the same situation as in Section 2.2.3, where Chiang's life table estimates were derived. Suppose that, conditional on survival till the start of I_j, censoring times for individuals due for withdrawal in I_j are uniformly distributed over I_j and that the lifetimes of these individuals are also uniformly distributed over I_j. Show that the conditional probability of an individual who is alive but due for withdrawal in I_j being observed to die in I_j is then $q_j^* = q_j/2$. Proceeding in a manner similar to that used in Section 2.2.3, show that maximization of the joint likelihood arising from the distribution of D_{j1} and D_{j2} leads to the estimate

$$\hat{q}_j = 1 - \frac{W_j - D_{j1} - D_{j2} + \left\{\left[2(N_{j1}+N_{j2}) - W_j + D_{j1} + D_{j2}\right]^2 - 8(D_{j1}+D_{j2})(N_{j1}+N_{j2})\right\}^{1/2}}{2(N_{j1}+N_{j2})}.$$

Also use this likelihood to obtain as an asymptotic expression for the variance of $\hat{q}_j$

$$\text{Asvar}(\hat{q}_j) = \frac{q_j(1-q_j)(2-q_j)}{(N_{j1}+N_{j2})\left[2-q_j-N_{j1}/(N_{j1}+N_{j2})\right]}.$$

(Section 2.2.3; Elveback, 1958)

Table 2.6.1 Life Table With Information on Withdrawals

		Individuals Not Due for Withdrawal in I_j		Individuals Due for Withdrawal in I_j		
Interval I_j (in years)	Alive at Start of Interval (N_j)	N_{j1}	D_{j1}	N_{j2}	D_{j2}	W_j
[0,1)	356	356	60	0	0	0
[1,2)	296	296	48	0	0	0
[2,3)	248	248	30	0	0	0
[3,4)	218	178	23	40	5	35
[4,5)	155	100	13	55	6	49
[5,6)	87	41	7	46	5	41

2.9 For the life table shown in Table 2.6.1 compare Chiang's, Elveback's, and the standard life table estimation methods. In particular, calculate estimates of the 5-year survival probability under each model, along with an estimate of variance. Comment on similarities and/or discrepancies in the results of the different methods.

(Section 2.2)

2.10 Consider the mean life restricted to T, defined in Section 2.3.2 as

$$\mu_T = \int_0^T S(x)\,dx$$

and the associated estimate

$$\hat{\mu}_T = \int_0^T \hat{S}(x)\,dx.$$

a. Show that $\mu_T = E[\min(t, T)]$, where t represents lifetime and T is a fixed (censoring) time.

b. Using the result of Theorem 2.3.1, derive the asymptotic variance of $\sqrt{n}\hat{\mu}_T$ given by (2.3.10). Note that asymptotically

$$\operatorname{Var}(\sqrt{n}\hat{\mu}_T) = \int_0^T \int_0^T \operatorname{Cov}\left[\sqrt{n}\hat{S}(x), \sqrt{n}\hat{S}(y)\right] dx\,dy.$$

c. Use the result of part (b) to motivate the variance estimate

$$\widehat{\text{Var}}(\hat{\mu}_T) = \sum_{j:\,t_j \leqslant T} \frac{\hat{A}_j^2 d_j}{n_j(n_j - d_j)}$$

where

$$\hat{A}_j = (t_{j+1} - t_j)\hat{S}(t_j) + (t_{j+2} - t_{j+1})\hat{S}(t_{j+1}) + \cdots + (T - t_m)\hat{S}(t_m)$$

with t_m being the largest observed lifetime less than or equal to T.

d. In the special case in which there is no censoring possible, let $T \to \infty$ and show that $\hat{\mu} = \hat{\mu}_\infty$ reduces to $\bar{t} = \Sigma t_i / n$ and that (2.3.10) reduces to $\sigma^2 = \text{Var}(t_i)$, where it is assumed that $\text{Var}(t_i)$ exists.

(Section 2.3; Kaplan and Meier, 1958)

2.11 The data below show survival times (in months) of patients with Hodgkin's disease who were treated with nitrogen mustards (Bartolucci and Dickey, 1977). Group A patients received little or no prior therapy, whereas Group B patients received heavy prior therapy. Starred observations are censoring times.

Group A	1.25, 1.41, 4.98, 5.25, 5.38, 6.92, 8.89, 10.98, 11.18, 13.11, 13.21, 16.33, 19.77, 21.08, 21.84*, 22.07, 31.38*, 32.62*, 37.18*, 42.92
Group B	1.05, 2.92, 3.61, 4.20, 4.49, 6.72, 7.31, 9.08, 9.11, 14.49*, 16.85, 18.82*, 26.59*, 30.26*, 41.34*.

a. Obtain and compare PL estimates for the two groups. Does there appear to be a difference in the 1-year survival probability for the two types of patients?

b. Do any parametric models whereby one might compare the two distributions suggest themselves?

c. Use (hazard) plots of the empirical hazard function $\hat{H}(t)$ to examine and compare the two life distributions.

(Sections 2.3, 2.4)

2.12 Suppose the hazard function $h(t)$ for a life distribution is a step function

$$h(t) = h_i \quad \text{for} \quad a_{i-1} \leqslant t < a_i \qquad i = 1, \ldots, k+1$$

where $a_0 = 0$ and $a_{k+1} = \infty$.

a. Determine the survivor function $S(t)$ and the p.d.f. $f(t)$ for the distribution.

b. Describe how you would estimate $h(t)$, and thus $S(t)$, given data consisting of both lifetimes and censoring times. What happens to this estimate as k increases and the intervals $[a_{i-1}, a_i)$ become small?

(Section 2.3)

2.13 Let $X_{(i)}$ be the ith order statistic in a random sample of size n from a continuous distribution with p.d.f. $f(x)$ and let x_p denote the pth quantile of the distribution $(0<p<1)$. Let $n \to \infty$ and $i \to \infty$ in such a way that $i/n \to p$. Show that in large samples $X_{(i)}$ can be considered to be approximately normally distributed with mean x_p and variance

$$p(1-p) \Big/ \left[nf(x_p)^2 \right].$$

What are the implications of this with regard to probability plots?

(Section 2.4.2; Cramer, 1946; Appendix D)

2.14 Consider the data given in Example 1.1.5 of Chapter 1 concerning the failure times of electrical insulation specimens subjected to a constant voltage stress. Make Weibull probability plots of the data for the experiments run at 30 and 32 kilovolts, respectively.

a. Does the suggestion of a Weibull failure time distribution for each situation with the shape parameters, but not the scale parameters, having the same value in the two cases seem plausible?

b. Compute estimates of parameters in the Weibull models using least squares and compare the estimated survivor functions from these with the empirical survivor function (PL estimate).

(Sections 2.4, 2.5)

CHAPTER 3

Inference Procedures for Exponential Distributions

The exponential distribution occupies an important position in lifetime distribution work, and an entire chapter will be devoted to a discussion of statistical procedures for it. (The Weibull distribution will be given similarly favored treatment in Chapter 4.) Historically the exponential distribution was the first lifetime model for which statistical methods were extensively developed. Early work by Sukhatme (1937) and later work by Epstein and Sobel (1953, 1954, 1955) and Epstein (1954, 1960a) gave numerous results and popularized the exponential as a lifetime distribution, especially in the area of industrial life testing. Many authors have contributed to the statistical methodology of the distribution. The lengthy bibliographies of Mendenhall (1958), Govindarajulu (1964), and Johnson and Kotz (1970, Ch. 18) give some idea of the very large number of papers in this area, even up to 1970.

This chapter covers basic inference procedures for the exponential distribution. Estimation and significance tests for single samples are considered first, with separate treatments for Type II and Type I censoring. The comparison of distributions is the topic of Section 3.3. In Section 3.4 we examine life test acceptance procedures and discuss the planning of life test experiments. Although the results in these sections are specific to the exponential distribution, the discussion illustrates several general points about inference with censored data in parametric models. Methods for the two-parameter exponential distribution are considered in Section 3.5, where a threshold parameter is included in the model. Section 3.6 concludes with some cautionary remarks about the sensitivity of these procedures to departures from the exponential model.

3.1 SINGLE SAMPLES: COMPLETE OR TYPE II CENSORED DATA

The exponential model will be written with its p.d.f. in the form

$$f(t;\theta)=\theta^{-1}e^{-t/\theta} \qquad t\geqslant 0 \tag{3.1.1}$$

in which case θ is the mean of the distribution and $\lambda=\theta^{-1}$ is the hazard function.

With complete (i.e., uncensored) samples inference procedures are simple and well known. Although these can be given as a special case of the results for Type II censored data, let us review them briefly by way of introduction. If $t_1,\ldots,t_n$ is a random sample from (3.1.1), then the likelihood function based on this is

$$L(\theta)=\prod_{i=1}^{n} f(t_i;\theta)=\frac{1}{\theta^n}\exp\left(-\sum_{i=1}^{n}\frac{t_i}{\theta}\right). \tag{3.1.2}$$

(For notational convenience we will not adhere completely to the habit of representing random variables by capital letters and the realized values of the random variables by lower case letters. No ambiguity should result from sometimes using lower case letters to represent random variables as well as the realized values of the random variables.) The m.l.e. of θ, obtained by maximizing (3.1.2), is easily found to be $\hat{\theta}=T/n$, where $T=\Sigma t_i$. It is also easily seen that T is sufficient for θ, and since the t_i/θ's are independent standard exponential variates, T/θ has a one-parameter gamma distribution with index parameter n. Equivalently, $2T/\theta\sim\chi^2_{(2n)}$ (see Section 1.3.4).

Similar results hold for Type II censored sampling. Suppose that only the first r observations $t_{(1)}<t_{(2)}<\cdots<t_{(r)}$ are available in a total sample of size n. From (1.4.1) the joint p.d.f. of $t_{(1)},\ldots,t_{(r)}$ is

$$\frac{n!}{(n-r)!}\left(\prod_{i=1}^{r}\frac{1}{\theta}e^{-t_{(i)}/\theta}\right)\left(e^{-t_{(r)}/\theta}\right)^{n-r}$$

$$=\frac{n!}{(n-r)!}\frac{1}{\theta^r}\exp\left[-\left(\sum_{i=1}^{r}t_{(i)}+(n-r)t_{(r)}\right)\Big/\theta\right]. \tag{3.1.3}$$

If we let

$$T=\sum_{i=1}^{r}t_{(i)}+(n-r)t_{(r)}$$

the likelihood function can be taken to be

$$L(\theta)=\frac{1}{\theta^r}e^{-T/\theta} \tag{3.1.4}$$

dropping the constant term $n!/(n-r)!$ in (3.1.3). Clearly T is sufficient for θ and the m.l.e. is $\hat{\theta}=T/r$. Note that the definition of T agrees with the earlier definition $T=\Sigma t_i$ in the case in which the data are uncensored.

Remark T is sometimes referred to as the "total observed lifetime," or the "total time on test," since it is the total of the observed lifetimes for all n individuals. This statistic is central to inference for the exponential distribution.

The distribution of T is easily found: make the change of variables

$$W_1=nt_{(1)}$$

$$W_i=(n-i+1)\left(t_{(i)}-t_{(i-1)}\right) \qquad i=2,\ldots,r. \tag{3.1.5}$$

Since

$$T=\sum_{i=1}^{r} t_{(i)}+(n-r)t_{(r)}=\sum_{i=1}^{r} W_i$$

and the Jacobian is

$$\frac{\partial(W_1,\ldots,W_r)}{\partial\left(t_{(1)},\ldots,t_{(r)}\right)}=\frac{n!}{(n-r)!}$$

the joint p.d.f. of $W_1,\ldots,W_r$ is found from (3.1.3) to be

$$\frac{1}{\theta^r}\exp\left(-\sum_{i=1}^{r}\frac{w_i}{\theta}\right) \qquad w_i>0.$$

We have proved the following result:

THEOREM 3.1.1 *Let $t_{(1)},\ldots,t_{(r)}$ be the first r ordered observations of a random sample of size n from the exponential distribution* (3.1.1). *Then the quantities $W_1,\ldots,W_r$ given by* (3.1.5) *are independent and identically distributed, also with p.d.f.* (3.1.1).

Since $T=\Sigma_{i=1}^{r} W_i$, we also immediately have

Corollary 3.1.1 Under the conditions of Theorem 3.1.1,

$$T=\sum_{i=1}^{r} t_{(i)}+(n-r)t_{(r)}$$

has a distribution given by $2T/\theta \sim \chi^2_{(2r)}$.

3.1.1 Tests and Confidence Intervals

Tests and confidence intervals for θ are easily obtained using the pivotal quantity $2T/\theta$. For example, to obtain an equitailed, two-sided $1-\alpha$ confidence interval for θ, we take

$$\Pr\left(\chi^2_{(2r),\alpha/2} \leqslant \frac{2T}{\theta} \leqslant \chi^2_{(2r),1-\alpha/2}\right)=1-\alpha,$$

where $\chi^2_{(2r),p}$ is the pth quantile of $\chi^2_{(2r)}$. Then

$$\frac{2T}{\chi^2_{(2r),1-\alpha/2}} \leqslant \theta \leqslant \frac{2T}{\chi^2_{(2r),\alpha/2}}$$

is the $1-\alpha$ confidence interval for θ.

Example 3.1.1 The first 8 observations in a random sample of 12 lifetimes from an assumed exponential distribution are, in hours

$$31, 58, 157, 185, 300, 470, 497, 673.$$

Hence $n=12$, $r=8$, and $T=5063$. The m.l.e. for θ is $\hat{\theta}=5063/8=632.9$ hours. To obtain, for example, a two-sided .95 confidence interval for θ, we find by using tables of the χ^2 distribution that $\Pr(6.91 \leqslant 2T/\theta \leqslant 28.8)=\Pr(6.91 \leqslant \chi^2_{(16)} \leqslant 28.8)=.95$, which gives $(2T/28.8, 2T/6.91)$ as a .95 confidence interval for θ. For the sample observed, $T=5063$, and the realized .95 confidence interval for θ is therefore $(351.6, 1465.4)$.

Tests or interval estimates for other characteristics of the distribution are similarly found, since these are simple functions of θ. In particular, confidence intervals and tests are readily obtained for (1) the constant hazard rate $\lambda=1/\theta$, (2) the survivor (i.e., reliability) function at time t_0, given by $S(t_0)=\exp(-t_0/\theta)$, and (3) the pth quantile of the distribution, given by

$t_p=\theta[-\log(1-p)]$. Specifically, if

$$A(T)\leqslant\theta\leqslant B(T)$$

is an α confidence interval for θ, then

1. $1/B(T)\leqslant\lambda\leqslant 1/A(T)$ is an α confidence interval for $\lambda=\theta^{-1}$.
2. $\exp[-t_0/A(T)]\leqslant S(t_0)\leqslant\exp[-t_0/B(T)]$ is an α confidence interval for $S(t_0)$.
3. $[-\log(1-p)]A(T)\leqslant t_p\leqslant[-\log(1-p)]B(T)$ is an α confidence interval for t_p.

Example 3.1.1 (continued) Confidence intervals for the survivor function or selected quantiles of the distribution are often needed in life testing work. (Confidence intervals for quantiles are sometimes called tolerance intervals for the distribution.) For example, suppose one wants a .95 lower confidence limit on $S(500)$, the probability of an item surviving past 500 hours. To obtain this we first find a lower .95 confidence limit on θ. Since $\Pr(\chi^2_{(16)}\leqslant 26.3)=\Pr(2T/\theta\leqslant 26.3)=.95$, a .95 lower confidence limit for θ is $2T/26.3$, which, for the given data, becomes 385.0 hours. Hence the lower .95 confidence limit for $S(500)$ is $\exp(-500/385)=0.273$.

Similarly, a .95 lower confidence limit on the .10 quantile $t_{.10}=-\log(.90)=0.105\theta$ is $0.105(385.0)=40.6$ hours. Note, incidentally, that to say that the .10 quantile exceeds 40.6 hours is equivalent to saying that the proportion of the population that lives past 40.6 hours is at least 90 percent, or that $S(40.6)\geqslant.90$. This complementary relationship between the quantiles and survivor function of a distribution is used later in a number of places.

Before considering Type I censored data we make two additional remarks. First, graphical methods are useful with the exponential distribution. This is discussed in Sections 2.4.1 and 2.4.2, and we do not return to this topic here. Second, formal hypothesis testing and acceptance sampling plans for the exponential distribution are examined in Section 3.4. Tests in the present section and in Section 3.2 and 3.3 are viewed as significance tests, with no discussion of a test's operating characteristic. In situations where optimal tests exist, however, these are essentially the ones given.

Example 3.1.1 (continued) Suppose that it has been desired to test the hypothesis that $\theta=1000$ hours, with alternative values of θ less than 1000 hours in mind. Formally, we wish to test $H_0:\theta=1000$ versus $H_1:\theta<1000$. In this case small values of T or, equivalently, $\hat{\theta}$ provide evidence against

H_0. For the data given, $T = 5063$ and the observed significance level (p value) for H_0 versus H_1 is $\Pr(T \leqslant 5063;\ \theta = 1000) = \Pr[2T/1000 \leqslant 2(5063)/1000] = \Pr(\chi^2_{(16)} \leqslant 10.126) = .15$. This gives no reason to doubt the hypothesized value of 1000 hours.

3.2 SINGLE SAMPLES: TYPE I CENSORED DATA

3.2.1 Inferences About θ

We shall employ the notation introduced in Section 1.4. Suppose there is a random sample of n individuals with lifetimes $T_1, \ldots, T_n$, but that associated with each individual is also a fixed censoring time $L_i > 0$. We observe T_i only if $T_i \leqslant L_i$ and the data therefore consist of pairs

$$(t_i, \delta_i) \qquad i = 1, \ldots, n$$

where

$$t_i = \min(T_i, L_i) \quad \text{and} \quad \delta_i = \begin{cases} 1 & \text{if} \quad t_i = T_i \\ 0 & \text{if} \quad t_i = L_i \end{cases}.$$

The general form of the likelihood function for Type I censored data was given in Section 1.4.1 as (1.4.5), and for the exponential model (3.1.1) this becomes

$$\begin{aligned} L(\theta) &= \prod_{i=1}^{n} \left(\frac{1}{\theta} e^{-T_i \delta_i / \theta} \right) e^{-L_i(1-\delta_i)/\theta} \\ &= \frac{1}{\theta^r} \exp\left(-\sum_{i=1}^{n} \frac{t_i}{\theta} \right) \end{aligned} \tag{3.2.1}$$

where $r = \sum \delta_i$ is the observed number of lifetimes. Note that

$$T = \sum_{i=1}^{n} t_i = \sum_{i \in D} T_i + \sum_{i \in C} L_i$$

is the total observed lifetime for the n individuals, where D and C denote the sets of individuals for whom lifetimes are observed and censored, respectively.

The form of the likelihood function is identical to that in the case of Type II censored sampling, though the sampling properties are different. Here the statistic (r, T) is minimally sufficient for θ, and the m.l.e., found by maximizing (3.2.1), is, assuming $r>0$,

$$\hat{\theta}=\frac{T}{r}.$$

If $r=0$, the likelihood function is monotone increasing, approaching unity as $\theta\to\infty$, and so does not possess a finite maximum. It is tacitly assumed henceforth that $r>0$ and the probability of r being 0 is ignored in a few subsequent calculations. For n sufficiently large this probability is negligible.

The m.l.e. is, as in the case of Type II censoring, the sum of observed lifetimes and censoring times, divided by the number of observed lifetimes. However, r is random here, and exact small-sample test and interval estimation procedures are not easily obtainable, as they were for Type II censoring. We therefore consider methods based on large-sample properties of maximum likelihood.

The first and second derivatives of the log likelihood function are, from (3.2.1),

$$\frac{d\log L}{d\theta}=\frac{-r}{\theta}+\frac{1}{\theta^2}\sum_{i=1}^{n} t_i \tag{3.2.2}$$

$$\frac{d^2\log L}{d\theta^2}=\frac{r}{\theta^2}-\frac{2}{\theta^3}\sum_{i=1}^{n} t_i. \tag{3.2.3}$$

To obtain the Fisher (or "expected") information we need the expectation of $-d^2\log L/d\theta^2$. To get this, note that $\Pr(\delta_i=0)=\exp(-L_i/\theta)=1-\Pr(\delta_i=1)$. Also $E(t_i|\delta_i=0)=L_i$ and

$$\begin{aligned}E(t_i|\delta_i=1)&=E(T_i|T_i\leqslant L_i)\\&=\int_0^{L_i}x\frac{\theta^{-1}e^{-x/\theta}}{1-e^{-L_i/\theta}}dx\\&=\theta-\frac{L_ie^{-L_i/\theta}}{1-e^{-L_i/\theta}}.\end{aligned}$$

Thus

$$\begin{aligned}E(t_i)&=E(t_i|\delta_i=0)\Pr(\delta_i=0)+E(t_i|\delta_i=1)\Pr(\delta_i=1)\\&=L_ie^{-L_i/\theta}+\left(\theta-\frac{L_ie^{-L_i/\theta}}{1-e^{-L_i/\theta}}\right)(1-e^{-L_i/\theta})\\&=\theta(1-e^{-L_i/\theta}).\end{aligned}$$

Since

$$E(r)=\sum E(\delta_i)=\sum(1-e^{-L_i/\theta})=Q$$

say, the expected information in the sample is

$$I(\theta)=E\left(\frac{-d^2\log L}{d\theta^2}\right)$$

$$=\frac{1}{\theta^2}\sum_{i=1}^{n}(1-e^{-L_i/\theta})=\frac{Q}{\theta^2}. \tag{3.2.4}$$

Any of several maximum likelihood large-sample procedures might be used to make inferences about θ (see Appendix E). One possibility is to employ the asymptotic normal approximation

$$\frac{\hat{\theta}-\theta}{I(\hat{\theta})^{-1/2}}\sim N(0,1) \tag{3.2.5}$$

to obtain confidence intervals or significance tests for θ. For example, $\hat{\theta}\pm 1.96I(\hat{\theta})^{-1/2}$ would then be an approximate .95 confidence interval for θ. It should be noted that in order to calculate $I(\hat{\theta})$ one requires the censoring times L_i of all individuals in the sample. Often these are not all available, since the potential censoring times of individuals observed to die may be unknown. As an alternative to (3.2.5), one can employ the asymptotically equivalent approximation

$$\frac{\hat{\theta}-\theta}{I_0^{-1/2}}\sim N(0,1) \tag{3.2.6}$$

where $I_0=(-d^2\log L/d\theta^2)_{\theta=\hat{\theta}}$ is the observed information. Note that (3.2.2) and (3.2.3) imply that

$$I_0=\frac{r}{\hat{\theta}^2}.$$

One qualification about approximations like (3.2.5) and (3.2.6) is that they may not be very good in small- or moderate-size samples. It can in fact be shown that these approximations are rather poor unless the sample size is fairly large. Fortunately, there are alternate approximate procedures that can be recommended, even with small samples. We will mention three such procedures:

1. Sprott (1973) and others have indicated that the distribution of $\hat{\phi}=\hat{\theta}^{-1/3}$ in small samples is much more closely approximated by a normal distribution than is the distribution of $\hat{\theta}$. The distribution of $\hat{\phi}$ is approximately normal, with mean $\phi=\theta^{-1/3}$ and variance $(d\phi/d\theta)^2\,\mathrm{Asvar}(\hat{\theta})=\phi^2/9Q$, where $\mathrm{Asvar}(\hat{\theta})=\theta^2/Q$. The procedure analogous to (3.2.5) is thus to use the normal approximation

$$\frac{\hat{\phi}-\phi}{(\hat{\phi}^2/9\hat{Q})^{1/2}} \sim N(0,1) \tag{3.2.7}$$

where $\hat{Q}=\Sigma(1-e^{-L_i/\hat{\theta}})$. Alternatively, the approximation analogous to (3.2.6) can be used; this is

$$\frac{\hat{\phi}-\phi}{(\hat{\phi}^2/9r)^{1/2}} \sim N(0,1). \tag{3.2.8}$$

These approximations can be used to obtain confidence intervals or tests for ϕ, which are readily converted to confidence intervals or tests for θ.

2. Likelihood ratio methods can be used for either tests or interval estimation. Under the hypothesis $H_0: \theta=\theta_0$ the likelihood ratio statistic

$$\Lambda=-2\log\left(\frac{L(\theta_0)}{L(\hat{\theta})}\right) \tag{3.2.9}$$

where $L(\theta)$ is given by (3.2.1), is approximately distributed as $\chi^2_{(1)}$. Significance tests can be carried out by treating Λ as $\chi^2_{(1)}$, with large values of Λ indicating evidence against H_0. Confidence intervals for θ can be obtained by inverting this test: to get an α confidence interval one needs to find the set of values θ_0 for which Λ is $\leqslant \chi^2_{(1),\alpha}$. It turns out, as in many situations, that the χ^2 approximation to Λ is quite good, even for small samples, and thus the confidence intervals obtained have close to the desired coverage probabilities.

3. A final suggestion, often attributed to Cox (1953), is to treat $2r\hat{\theta}/\theta$ as being approximately $\chi^2_{(2r+1)}$. This approximation appears to give satisfactory tests and confidence intervals.

Confidence intervals for $S(t)$ or t_p are readily found from confidence intervals for θ, as described in Section 3.1. The use of the various methods is illustrated in the following example.

Example 3.2.1 (Example 1.1.6 revisited) The data given in Example 1.1.6 concerned the lifetimes of 10 pieces of equipment. The data were as follows:

Item number	1	2	3	4	5	6	7	8	9	10
T_i	2	—	51	—	33	27	14	24	4	—
L_i	81	72	70	60	41	31	31	30	29	21

The lifetimes of items 2, 4, and 10 are censored. It will be noted that the effective censoring times are known for all 10 items. The data give $r=7$ and $\Sigma t_i=308$, so that $\hat{\theta}=308/7=44.0$. Using (3.2.4), we find $I(\hat{\theta})=6.15/44^2$ and $I(\hat{\theta})^{-1/2}=17.7$. If we use (3.2.5) and treat $(\hat{\theta}-\theta)/17.7$ as $N(0,1)$, a two-sided .95 confidence interval for θ is obtained as $9.3\leqslant\theta\leqslant78.7$ (days). Alternately, (3.2.6) gives $(\hat{\theta}-\theta)/16.6\sim N(0,1)$, which gives the two-sided .95 confidence interval $11.5\leqslant\theta\leqslant76.5$.

Methods (1), (2), and (3) also yield approximate confidence intervals, as follows:

1. We have $\hat{\phi}=\hat{\theta}^{-1/3}=0.2833$, and since $\hat{Q}=\Sigma(1-e^{-L_i/\hat{\theta}})=6.15$, we find $\hat{\phi}^2/9\hat{Q}=0.0014425$; (3.2.7) then gives an approximate .95 confidence interval $0.2088\leqslant\phi\leqslant0.3577$. Since $\phi=\theta^{-1/3}$, this converts into an approximate .95 confidence interval for θ of $21.8\leqslant\theta\leqslant109.9$.
2. The likelihood ratio statistic is, from (3.2.1) and (3.2.9),

$$\Lambda=-2r\log\left(\frac{\hat{\theta}}{\theta}\right)+2r\left(\frac{\hat{\theta}}{\theta}-1\right).$$

   Since $\Pr(\chi^2_{(1)}\leqslant3.84)=.95$, a .95 confidence interval for θ is found as the set of all θ values giving $\Lambda\leqslant3.84$. By computing Λ for several values of θ and then using trial and error to locate the exact values of θ that make $\Lambda=3.84$, we find the approximate .95 confidence interval to be $22.8\leqslant\theta\leqslant102.5$.
3. With $r=7$, $14\hat{\theta}/\theta$ is treated as approximately $\chi^2_{(15)}$. Since $\Pr(6.26\leqslant\chi^2_{(15)}\leqslant27.5)=.95$, this gives $22.4\leqslant\theta\leqslant98.4$ as an approximate .95 confidence interval for θ.

It is observed that the approximate methods (1), (2), and (3) give results that are in broad agreement but that the large-sample procedures based directly on $\hat{\theta}$ give somewhat different answers. Some guidelines on the adequacy of the various approximations are desirable. This is briefly examined in the next section; it appears that for practical purposes all of (1), (2), and (3) can be recommended, even for quite small sample sizes. The method one uses is a matter of choice though (1) and (3) are slightly simpler than (2).

The methods described here are all based on large-sample approximations. The exact distribution of $\hat{\theta}$ has been obtained by Bartholomew (1963), but except when the L_i's are equal, it is too complicated to be of use. When the L_i's are equal, the distribution is still extremely complicated, though Barlow et al. (1968) have presented a computer program that uses this to obtain confidence intervals. Since methods (1) through (3) are adequate even for small sample sizes, there is little need for this.

3.2.2 Adequacy of the Approximate Methods*

To indicate the extent to which the large-sample methods of the previous section give appropriate coverage probabilities in small samples, some results of a small simulation study are presented in Table 3.2.1. Results are shown for situations in which the normal approximation (3.2.5) and the methods (1), (2), and (3) of the preceding section were used to obtain lower α confidence limits on θ. Two sample sizes, $n=10$ and 20, and three single Type I censoring patterns, are represented. In each case all individuals had the same censoring time L, with L selected in the three censoring patterns to give values $Q'=\exp(-L/\theta)$ of .10, .25, and .50. Here Q' is the effective censoring fraction, since any observation has a probability Q' of being censored.

Table 3.2.1 Proportion of the Time (Out of 2000 Trials) That Approximate One-Sided .75, .90, .95, .99 Confidence Intervals Contained θ

		$Q'=.10$				$Q'=.25$				$Q'=.50$			
n	Method	.75	.90	.95	.99	.75	.90	.95	.99	.75	.90	.95	.99
10	(1)	.766	.913	.960	.991	.756	.908	.951	.989	.746	.893	.950	.990
	(2)	.767	.910	.960	.991	.759	.905	.944	.989	.743	.891	.945	.990
	(3)	.778	.923	.962	.993	.775	.913	.956	.990	.779	.913	.945	.990
	$\hat{\theta}$	.798	.928	1.000	1.000	.808	.996	1.000	1.000	.825	1.000	1.000	1.000
20	(1)	.751	.908	.951	.992	.742	.903	.951	.989	.734	.892	.949	.992
	(2)	.750	.907	.950	.990	.737	.901	.949	.989	.734	.893	.949	.991
	(3)	.763	.914	.954	.992	.754	.912	.953	.989	.752	.905	.956	.992
	$\hat{\theta}$	.777	.952	.993	1.000	.774	.961	.995	1.000	.768	.981	1.000	1.000

There were 2000 samples generated for each (Q', n) combination. The results are valid whatever the value of θ, since censoring times were chosen to be fixed multiples of θ. The table shows the observed coverage proportions for confidence intervals with nominal coverage probabilities .75, .90, .95, and .99. It is clear that methods (1), (2), and (3) all give close to the desired coverage probabilities. On the other hand, the normal approximation (3.2.5) for $\hat{\theta}$ tends to produce confidence bounds that are too low, giving coverage probabilities that are higher than they should be. Similar behavior is noted for the method based on (3.2.6), though results for it are not given here.

The approximate methods (1), (2), and (3) are adequate, even for small sample sizes, in the situations represented in the table. Simulations for problems in which censoring times L_i are unequal gave similarly good results. When the methods are used to obtain upper .25, .10, .05, and .01 confidence limits, coverage probabilities are not quite as close to the nominal ones as for the lower confidence limits, but they are still broadly acceptable. In many applications lower confidence limits are called for, however, so it is more important that these be satisfactory.

3.2.3 Discussion

The methods in this section apply to Type I censored data but can be used more generally for situations in which the likelihood (1.4.9) is valid. Small-sample properties of procedures will depend on the censoring–lifetime mechanism present, but the procedures are used in essentially the same way as when the censoring is Type I. The observed information I_0 should be employed in normal approximations for $\hat{\theta}$ or other quantities, since the expected information $I(\theta)$ given in Section 3.2.1 is based on the assumption of Type I censoring. Throughout this book methods are presented for Type I censored data with the understanding that they can be used in other situations where (1.4.9) is valid.

3.3 COMPARISON OF EXPONENTIAL DISTRIBUTIONS

The comparison of two or more life distributions is often important in statistical analyses of lifetime data. When the distributions are one-parameter exponential distributions, this amounts to a comparison of their means. This topic is considered here, once again with separate treatments for Type II and Type I censored data.

3.3.1 Type II Censored Data

Suppose that there are two exponential distributions with means θ_1 and θ_2 and that independent Type II censored samples from the two distributions, based on n_1 and n_2 individuals, respectively, have observations

$$t_{1(1)} < \cdots < t_{1(r_1)} \qquad (r_1 \leqslant n_1)$$

$$t_{2(1)} < \cdots < t_{2(r_2)} \qquad (r_2 \leqslant n_2).$$

Letting

$$T_i = \sum_{j=1}^{r_i} t_{i(j)} + (n_i - r_i) t_{i(r_i)} \qquad i = 1,2$$

we have from Corollary 3.1.1 that $2T_1/\theta_1 \sim \chi^2_{(2r_1)}$ and $2T_2/\theta_2 \sim \chi^2_{(2r_2)}$, with T_1 and T_2 independent. Thus

$$U = \frac{2T_1/2r_1\theta_1}{2T_2/2r_2\theta_2} = \frac{\hat{\theta}_1\theta_2}{\hat{\theta}_2\theta_1} \tag{3.3.1}$$

has a F distribution with $(2r_1, 2r_2)$ degrees of freedom. Tests of equality of θ_1 and θ_2 are hence easily carried out using tables of the F distribution. Confidence intervals for θ_2/θ_1, which is the ratio of the means and of the hazard functions for the two distributions, are also easily obtained.

Example 3.3.1 Samples censored at the tenth out of 15 observations, from each of two exponential distributions, gave $T_1 = 700$ and $T_2 = 840$, so that $\hat{\theta}_1 = 70$ and $\hat{\theta}_2 = 84$. Under $H_0: \theta_1 = \theta_2$, $\hat{\theta}_1/\hat{\theta}_2$ has by (3.3.1) a $F_{(20,20)}$ distribution. The .025 and .975 quantiles of $F_{(20,20)}$ are 0.406 and 2.46, respectively; the observed value $\hat{\theta}_1/\hat{\theta}_2 = 70/84 = 0.83$ clearly provides no evidence against H_0. To obtain, say, a two-sided .95 confidence interval for θ_2/θ_1, we use the pivotal quantity U of (3.3.1). Since $\Pr(0.406 \leqslant U \leqslant 2.46) =$ .95, $(0.406\hat{\theta}_2/\hat{\theta}_1, 2.46\hat{\theta}_2/\hat{\theta}_1)$ is a .95 confidence interval for θ_2/θ_1. Here the data give $\hat{\theta}_1 = 70$ and $\hat{\theta}_2 = 84$, and the realized confidence interval is (0.49, 2.95).

Comparison of m Exponential Means

The standard m-sample problem is to test for the equality of the distributions, that is, to test

$$H_0: \theta_1 = \theta_2 = \cdots = \theta_m.$$

One way to test H_0 is with a likelihood ratio test. The combined likelihood function from m independent samples is, from (3.1.4),

$$L(\theta_1,\dots,\theta_m)=\prod_{i=1}^{m}\frac{1}{\theta_i^{r_i}}e^{-T_i/\theta_i} \tag{3.3.2}$$

where the sample $\{t_{i(j)};\ j=1,\dots,n_i\}$ from the ith population is censored at the r_ith out of n_i observations and

$$T_i=\sum_{j=1}^{r_i}t_{i(j)}+(n_i-r_i)t_{i(r_i)} \qquad i=1,\dots,m.$$

Under H_0 the θ_i's are equal and the m.l.e. of θ_i, obtained by maximizing (3.3.2), is $\tilde{\theta}=\Sigma T_i/\Sigma r_i$. The m.l.e.'s under the full unrestricted model (with the alternative hypothesis that $\theta_i>0, i=1,\dots,m$) are $\hat{\theta}_i=T_i/r_i(i=1,\dots,m)$. The likelihood ratio statistic for testing H_0 against this alternative is thus

$$\Lambda=-2\log\left(\frac{L(\tilde{\theta},\dots,\tilde{\theta}\,)}{L(\hat{\theta}_1,\dots,\hat{\theta}_m)}\right)$$

$$=\left(2\sum_{i=1}^{m}r_i\right)\log\tilde{\theta}-2\sum_{i=1}^{m}r_i\log\hat{\theta}_i. \tag{3.3.3}$$

Asymptotically Λ has a $\chi^2_{(m-1)}$ distribution, and if the r_i's are not too small, it is suitable to treat Λ as being $\chi^2_{(m-1)}$. Large values of Λ give evidence against H_0.

An adjustment that improves the χ^2 approximation to Λ in small- or moderate-size samples is obtained by noting that (3.3.3) is just Bartlett's (1937) test statistic, developed for testing the equality of m normal variances. Bartlett's statistic can be written in the form

$$\Lambda'=v\log\sum_{i=1}^{m}\frac{S_i^2}{v}-\sum_{i=1}^{m}v_i\log\left(\frac{S_i^2}{v_i}\right)$$

where $S_i^2\sim\chi^2_{(v_i)}$ and are independent and $v=\Sigma v_i$. Since $S_i^2/2$ has a one-parameter gamma distribution $\mathrm{Ga}(v_i/2)$, the statistics Λ' and Λ are identical when $S_i^2/2$ is replaced with $\hat{\theta}_i$, and $v_i/2$ with r_i. Bartlett showed (see Problem 3.6) that instead of taking Λ' to be $\chi^2_{(m-1)}$, a better approximation is to take $\Lambda_1 = C\Lambda'$ to be $\chi^2_{(m-1)}$, where

$$C^{-1} = 1 + \frac{1}{3(m-1)}\left(\sum_{i=1}^{m} v_i^{-1} - v^{-1}\right).$$

Since Bartlett's $v_i/2$ is equivalent to our r_i, the appropriate procedure here is to define

$$C^{-1} = 1 + \frac{1}{6(m-1)}\left(\sum_{i=1}^{m} r_i^{-1} - r^{-1}\right) \tag{3.3.4}$$

where $r = \Sigma r_i$, and to use the approximation

$$C\Lambda \sim \chi^2_{(m-1)}.$$

Bartlett's approximation is good unless the r_i's are smaller than about 4, or m is large; these are both uncommon in practice. Glaser (1976a, b) and Chao and Glaser (1978) give exact expressions for the p.d.f. of Λ, which, though complicated, can be used with a computer to evaluate exact percentage points. Glaser (1976b) and Dyer and Keating (1980) give tables of exact percentage points for cases in which the r_i's are equal. In addition, Pearson and Hartley (1966, pp. 63–66) tabulate an improved approximation to the distribution of Λ, developed by Hartley. One of these alternatives can be taken up in the infrequent situations in which the r_i's are too small to use (3.3.4).

Example 3.3.2 As a numerical illustration, suppose that samples from four exponential distributions were censored at the seventh out of 10 observations, with maximum likelihood estimates $\hat{\theta}_1 = 106$, $\theta_2 = 80$, $\hat{\theta}_3 = 140$, and $\hat{\theta}_4 = 158$. In order to test whether $\theta_1 = \theta_2 = \theta_3 = \theta_4$ let us consider the likelihood ratio statistic (3.3.3). With $r_1 = r_2 = r_3 = r_4 = 7$ and $\tilde{\theta} = \Sigma r_i \hat{\theta}_i / 28 = 121$, the observed value of Λ is calculated to be 1.87. Using the approximation $\Lambda \sim \chi^2_{(3)}$ gives a significance level of about .60, which provides no evidence against the hypothesis of equal means. The correction factor (3.3.4) is found to be $C = 0.971$ and clearly does not alter any conclusions we might make.

To illustrate the adequacy of the χ^2 approximation in this situation, Table 3.3.1 gives upper quantiles of Λ determined by using the χ^2 approximations, with and without the correction factor, and exact quantiles evaluated using Table 1 in Glaser (1976b). Those wishing to refer to Glaser's table should note that he tabulates percentage points of $\exp(-\Lambda/2)$ rather than of Λ.

Table 3.3.1 Approximate and Exact Quantiles of Λ for $k=4; r_1=r_2=r_3=r_4=7$

Quantile	.90	.95	.99
$\Lambda \sim \chi^2_{(3)}$	6.251	7.815	11.341
$C\Lambda \sim \chi^2_{(3)}$	6.437	8.049	11.681
Exact	6.438	8.044	11.669

The excellence of the χ^2 approximation, especially when the correction factor (3.3.4) is used, is evident.

In the case $m=2$, the likelihood ratio test yields a F test based on (3.3.1), since when $m=2$, $\tilde{\theta}=(r_1\hat{\theta}_1+r_2\hat{\theta}_2)/(r_1+r_2)$ and the likelihood ratio reduces to

$$\frac{L(\tilde{\theta},\tilde{\theta})}{L(\hat{\theta}_1,\hat{\theta}_2)}=\left(\frac{\hat{\theta}_1}{\hat{\theta}_2}\right)^{r_1}\Bigg/\left(\frac{r_1}{r_1+r_2}\frac{\hat{\theta}_1}{\hat{\theta}_2}+\frac{r_2}{r_1+r_2}\right)^{r_1+r_2}.$$

This is a function of $U=\hat{\theta}_1/\hat{\theta}_2$. In this case, of course, one does not need an approximation for the distribution of the test statistic, since it is known exactly.

As a final remark we mention that other tests of equality of $\theta_1,\dots,\theta_m$ may also be of interest, depending on the types of departure from equality envisaged. This is not discussed here, but note that alternate tests to Bartlett's for comparing normal variances (e.g., Pearson and Hartley, 1966, pp. 66–67), can be adopted for comparing exponential means.

3.3.2 Type I Censored Data

When the data are Type I censored, the likelihood ratio test can be used to test for equality of $\theta_1,\dots,\theta_m$. In view of the similarity of the likelihood functions for Types I and II censoring, the likelihood ratio statistic is again given by (3.3.3), with r_i now being the number of observed lifetimes in the ith sample. That is, the joint likelihood function of m independent Type I censored samples is, by (3.2.1),

$$L(\theta_1,\dots,\theta_m)=\prod_{i=1}^{m}\frac{1}{\theta_i^{r_i}}e^{-T_i/\theta_i} \tag{3.3.5}$$

where $T_i=\sum_{j=1}^{n_i}t_{ij}$ is the total observed lifetime for the n_i individuals in the

ith sample. Under $H_0: \theta_1 = \cdots = \theta_m$ the m.l.e. of θ_i is

$$\tilde{\theta} = \sum_{i=1}^{m} T_i \Big/ \sum_{i=1}^{m} r_i$$

and the unrestricted m.l.e.'s are $\hat{\theta}_i = T_i / r_i$. The likelihood ratio statistic for testing H_0 against the alternative that the θ_i's are not all equal is

$$\Lambda = \left(2 \sum_{i=1}^{m} r_i\right) \log \tilde{\theta} - 2 \sum_{i=1}^{m} r_i \log \hat{\theta}_i. \qquad (3.3.6)$$

The distribution of Λ under H_0 is asymptotically $\chi^2_{(m-1)}$, under mild assumptions about censoring times, and unless the r_i's are quite small, it is satisfactory to treat Λ as $\chi^2_{(m-1)}$. Note, however, that $2T_i/\theta_i$ does not have a χ^2 distribution, so that the results concerning the exact and approximate distribution of Bartlett's test statistic (3.3.3), discussed in the previous section, are not applicable.

Confidence Intervals for θ_1/θ_2

When $m=2$, we may wish to obtain confidence intervals for θ_1/θ_2. One way to do this is to invert the likelihood ratio test of $H_0: \theta_1 = a\theta_2$. The m.l.e.'s of θ_1 and θ_2 under the restriction $\theta_1 = a\theta_2$ are readily found from maximizing (3.3.5) to be given by

$$\tilde{\theta}_1 = a\tilde{\theta}_2$$
$$= \frac{T_1 + aT_2}{r_1 + r_2}.$$

The likelihood ratio statistic for testing $H_0: \theta_1 = a\theta_2$ versus $H_1: \theta_1 \neq a\theta_2$ is then

$$\Lambda_1 = -2\log\left(\frac{L(\tilde{\theta}_1, \tilde{\theta}_2)}{L(\hat{\theta}_1, \hat{\theta}_2)}\right)$$

$$= 2r_1 \log\left(\frac{\tilde{\theta}_1}{\hat{\theta}_1}\right) + 2r_2 \log\left(\frac{\tilde{\theta}_2}{\hat{\theta}_2}\right). \qquad (3.3.7)$$

An α confidence interval for θ_1/θ_2 consists of the set of values a for which H_0 is not rejected at the $1-\alpha$ level of significance. If we use the approximation $\Lambda_1 \sim \chi^2_{(1)}$, this entails finding all values of a for which $\Lambda_1 \leqslant \chi^2_{(1),\alpha}$.

A second, somewhat simpler, approach is to make use of approximation (3) of Section 3.2.1, which treats $2r_i\hat{\theta}_i/\theta_i$ as being $\chi^2_{(2r_i+1)}$. This gives the approximation

$$U_1 = \frac{(1+0.5/r_1)}{(1+0.5/r_2)} \frac{\theta_2}{\theta_1} \frac{\hat{\theta}_1}{\hat{\theta}_2} \sim F_{(2r_1+1,2r_2+1)}. \tag{3.3.8}$$

U_1 can be used to obtain approximate tests and confidence intervals for θ_1/θ_2. Note, incidentally, that by using the approximations $2r_i\hat{\theta}_i/\theta_i \sim \chi^2_{(2r_i+1)}$ in the general m-sample case, and using the analogy with Bartlett's test, a test of $\theta_1 = \cdots = \theta_m$ slightly different than that represented by (3.3.6) could be developed. Regal (1980) investigates a statistic similar to U_1 in a few situations.

Example 3.3.3 In a clinical trial to compare the duration of remission achieved by two drugs used in the treatment of leukemia, two groups of 20 patients each were used. The data were Type I censored, with $r_1 = r_2 = 10$ observed remission times in each group, and the data from each group appeared consonant with an exponential model. Sums of remission and censoring times for the two groups were $T_1 = 700$ and $T_2 = 540$ weeks, giving $\hat{\theta}_1 = 70$ and $\hat{\theta}_2 = 54$ weeks, respectively.

Let us obtain (approximate) .95 confidence intervals for θ_1/θ_2, the ratio of the mean duration times. To use the F approximation (3.3.8) we find from F tables that $\Pr(0.42 \leqslant F_{(21,21)} \leqslant 2.41) = .95$, and this, along with (3.3.8), gives a confidence interval $0.32 \leqslant \theta_2/\theta_1 \leqslant 1.80$. Alternately, we could use the likelihood ratio statistic (3.3.7), with $\tilde{\theta}_1 = a\tilde{\theta}_2 = (T_1 + aT_2)/20$, to get a confidence interval for $a = \theta_1/\theta_2$. With $T_1 = 700$ and $T_2 = 540$, (3.3.7) becomes

$$\Lambda_1 = 20\log(0.50 + 0.386a) + 20\log\left(\frac{0.648}{a} + 0.50\right).$$

To obtain the .95 confidence interval we need to find all a values that make $\Lambda_1 \leqslant \chi^2_{(1),.95} = 3.841$. It is readily found by trial and error that the set of values is $0.51 \leqslant a \leqslant 3.15$, which gives a confidence interval of $0.32 \leqslant \theta_2/\theta_1 \leqslant 1.96$ for $\theta_2/\theta_1 = 1/a$. This is in good agreement with the first method, considering the small sample sizes.

3.4 EXPERIMENTAL PLANS AND LIFE TEST PROCEDURES

In some situations, physical constraints related to the problem under study, or a lack of prior knowledge about the problem, can make precise planning of an investigation difficult. In well-controlled situations, on the other hand, experiments can often be planned to satisfy defined objectives. We consider in this section problems involved in planning an experiment under an assumed exponential model. Much of the discussion concerns life test procedures, for several reasons. One is that life test plans with stated economic objectives are important in many areas and widely used. A second reason is that many of the considerations involved with them are relevant in planning any lifetime distribution investigation. Finally, by examining different experimental plans for the relatively simple exponential model, we gain insight into the difficulties of designing plans for other distributions.

We begin by outlining the ideas involved with a life test acceptance procedure. By such a procedure, we mean an experimental plan for testing lifetimes of items in a population, along with a formal rule for deciding whether to accept or reject a particular hypothesis about the underlying lifetime distribution. For the case in which the underlying distribution is exponential such plans have been studied by Epstein and Sobel (1953, 1955), Epstein (1954, 1960a), and many others. Widely used life test acceptance plans as those in MIL-STD-781C (1977) have been based on this work. Only problems associated with a single population of lifetimes are discussed, though plans for comparing two or more distributions are also important. In addition, the treatment of acceptance sampling given here is elementary; for a thorough discussion of this area see, for example, Wetherill (1977).

The standard framework for studying accept–reject rules involves the Neyman–Pearson theory of hypothesis testing and decision theory. Any hypothesis concerning the one-parameter exponential distribution can be given in terms of the mean θ for the distribution. The most common life testing problem, and the only one we discuss in any detail, involves testing a specific value θ_0 of θ against values less than θ_0. For example, a consumer may want the mean lifetime of a particular type of item (or equivalently, some other characteristic of the item's life distribution) to be satisfactorily high. With this in mind, a plan is set up whereby one can test that the mean lifetime is θ_0, against the alternative that it is less than θ_0. We therefore consider testing

$$H_0: \theta = \theta_0 \quad \text{vs.} \quad H_1: \theta < \theta_0. \tag{3.4.1}$$

Life test plans are generally designed so that the size and power of the test at some particular value $\theta_1 < \theta_0$ are specified. Recall that the size of the test is defined as

$$\alpha = \Pr(\text{reject } H_0;\, \theta = \theta_0)$$

and the power function, defined for $\theta_1 < \theta_0$, is given by

$$P(\theta_1) = \Pr(\text{reject } H_0;\ \theta = \theta_1).$$

(Here "$;\theta=\theta_0$," for example, means "when the true value of θ is θ_0.")

Some of the problems in selecting a life test plan will now be examined by considering plans based on Type II censored sampling.

3.4.1 Type II Censored (Nonreplacement) Life Test Plans

Consider the problem of testing the hypothesis (3.4.1) on the basis of a Type II censored sample containing the r smallest lifetimes $t_{(1)} < \cdots < t_{(r)}$ in a total sample of size n. The likelihood function for this situation has been given in Section 3.1.1, and it is easily seen from general results on hypothesis testing (e.g., Lehmann, 1959; also see Epstein and Sobel, 1953) that for a given r and n, a size α uniformly most powerful test of H_0 versus H_1 exists and has acceptance rule of the form

$$\text{Accept } H_0 \text{ if } \hat{\theta} > C_\alpha = \frac{\theta_0 \chi^2_{(2r),\alpha}}{2r} \tag{3.4.2}$$

where

$$\hat{\theta} = \left[\sum t_{(i)} + (n-r)t_{(r)}\right]/r.$$

For any positive integer r one can get a size α test. If we also require the power of the test at $\theta = \theta_1$ to be $1-\beta$ ($1-\beta$ is used to represent the power at θ_1, to conform with usual acceptance sampling notation, where β is often called the producers risk), then we need

$$\begin{aligned} P(\theta_1) &= \Pr\left(\hat{\theta} \leqslant C_\alpha;\ \theta = \theta_1\right) \\ &= 1-\beta. \end{aligned}$$

But if $\theta = \theta_1$, then, by Corollary 3.1.1, $2r\hat{\theta}/\theta_1 \sim \chi^2_{(2r)}$, and so

$$\begin{aligned} P(\theta_1) &= \Pr\left(\frac{2r\hat{\theta}}{\theta_1} \leqslant \frac{2rC_\alpha}{\theta_1}\right) \\ &= \Pr\left(\chi^2_{(2r)} \leqslant \frac{2rC_\alpha}{\theta_1}\right). \end{aligned} \tag{3.4.3}$$

Thus $\chi^2_{(2r),1-\beta}=2rC_\alpha/\theta_1$ or, since $C_\alpha=\theta_0\chi^2_{(2r),\alpha}/2r$,

$$\frac{\chi^2_{(2r),\alpha}}{\chi^2_{(2r),1-\beta}}=\frac{\theta_1}{\theta_0}. \tag{3.4.4}$$

Hence, to make $P(\theta_1)$ equal to $1-\beta$, we must choose r such that (3.4.4) is satisfied. There will not generally be an integral r value that exactly satisfies (3.4.4). However, it can be seen that for $\alpha < .5$ and $\beta < .5$ the quotient on the left-hand side of (3.4.4) is an increasing function of r and approaches unity from below as $r\to\infty$. Since $\theta_1/\theta_0<1$, there is a smallest value r_0 of r such that the left-hand side of (3.4.4) is $\geqslant\theta_1/\theta_0$, and then for any $r\geqslant r_0$ $P(\theta_1)\geqslant 1-\beta$. The choice $r=r_0$ therefore gives a test with the desired size and (approximately) the desired power at $\theta=\theta_1$. The larger $1-\beta$ is, the larger r_0 will be. The entire power function for the test can be calculated from (3.4.3). Epstein (1960a) provides a table of r_0 for selected values of θ_1/θ_0, α, and β.

It will be observed that no particular value of n is indicated by the above arguments. Two tests with the same value of r but different values of n have identical power functions. However, although n does not enter into the power calculations, it is an important factor in the test plan, since the larger n is, the less the time generally required to complete the test. One aspect of this is given in the following result.

Lemma 3.4.1. Let $t_{(r)}$ be the rth smallest observation in a random sample of size n from the exponential distribution (3.1.1) with mean θ. Then

$$E(t_{(r)})=\theta\sum_{i=1}^{r}\frac{1}{n-i+1}. \tag{3.4.5}$$

Proof By Theorem 3.1.1, $W_1=nt_{(1)}$ and $W_i=(n-i+1)(t_{(i)}-t_{(i-1)})$, $i=2,\ldots,r$, are independent random variables all having the same exponential distribution as the original observations. Thus $E(W_i)=\theta$, $i=1,\ldots,r$. But $t_{(r)}$ can be written as

$$t_{(r)}=\frac{W_1}{n}+\frac{W_2}{n-1}+\cdots+\frac{W_r}{n-r+1}$$

and hence the stated result follows. ■

Pearson and Hartley (1972, Table 19) give a table of $E(t_{(r)})/\theta$. For fixed r (3.4.5) indicates the extent to which increasing n decreases the expected time required to complete the test. Of course, a larger n involves more testing, and if testing is expensive it may be necessary to keep n fairly small. Another relevant point, discussed later, is that highly censored tests are more affected by departures from the assumed exponential model than are tests with light censoring. These factors have to be balanced against the potential time savings a larger n promises.

Example 3.4.1 A particular electrical device has a lifetime distribution adequately modeled by an exponential distribution. In setting up a screening procedure for consignments of these devices, it is decided to institute a Type II censored life test plan, with $\theta_0 = 1000$ hours, $\theta_1 = 400$ hours, $\alpha = .05$, and $\beta = .10$. In other words, the test is to have only a 5% chance of rejecting a distribution with mean 1000 hours, but a 90% chance of rejecting one with mean 400 hours.

The smallest integer r such that the left side of (3.4.4) exceeds $\theta_1/\theta_0 = 0.4$ is $r = 11$, which gives $\chi^2_{(22),.05}/\chi^2_{(22),.90} = 12.338/30.813 = 0.4004$. Then (3.4.2) gives $C_{.05} = 1000(12.338)/22 = 561$. The plan therefore stipulates that we use a Type II censored life test with $r = 11$ and reject H_0 if $\hat{\theta} \leqslant 561$.

To show the effect of n, the total number of items on test, we can use (3.4.5) to calculate expected durations of the test (or see Pearson and Hartley, 1972, Table 19). One finds, for example, that for $n = 11$, 13, 15, and 20, $E(t_{(11)}) = 3.02\theta$, 1.68θ, 1.23θ, and 0.77θ, respectively. A decision as to how large n should be can be based on considerations involving the costs of testing, the amount of time available for the test, the possibility of departures from the exponential model, and so on.

3.4.2 Some Other Life Test Plans

There are many ways to run a life test experiment. Other possibilities include plans with Type I censoring, a mixture of Type I and Type II censoring, or a sequential procedure. In addition, tests can sometimes be run with replacement, whereby items that fail are immediately replaced by new items, so that there are always n items on test. Still another possibility is to use partial replacement, replacing only a portion of the failed items. A few plans are given below.

3.4.2a *Type II Censoring With Replacement*

Sometimes it is feasible to replace failed items immediately, with the result that n items are continually on test. If the test is terminated at the time T_r, of the rth item failure, we say that there is Type II censoring with replacement. This censoring mechanism satisfies conditions laid down in

Section 1.4.1, so that the likelihood function is, from (1.4.9),

$$L(\theta)=\frac{1}{\theta^r}e^{-\Sigma t_i/\theta} \tag{3.4.6}$$

where Σt_i is the total observed lifetime, or the "total time on test." Since there are n items on test at all times and the test terminates at time T_r, Σt_i must equal nT_r, and T_r is sufficient for θ.

An alternate, more direct, derivation of $L(\theta)$ is to note that the observed failure times in the experiment are the times of occurrence of events in a Poisson process with intensity n/θ (e.g., Cox and Lewis, 1966, Ch. 2). Because T_r is the time to the occurrence of the rth event, $nT_r/\theta \sim \text{Ga}(r)$ or, equivalently, $2nT_r/\theta = 2r\hat{\theta}/\theta \sim \chi^2_{(2r)}$. Since the same result, namely, $2r\hat{\theta}/\theta \sim \chi^2_{(2r)}$, holds as in the case of Type II censoring without replacement, formulas given in Section 3.4.1 can be used again here to obtain accept – reject rules. In addition, confidence intervals for θ or functions of θ are easily obtained along exactly the same lines as in Section 3.1. Note, of course, that the expected duration of a life test with replacement of items is substantially less than that of one without replacement. The expected duration of the replacement test is $E(T_r)=r\theta/n$, which can be compared with the nonreplacement result (3.4.5). For example, for $r=11$ and $n=11$, 13, 15, and 20, as in Example 3.4.1, mean duration times are θ, 0.85θ, 0.73θ, and 0.55θ, respectively, compared with the corresponding nonreplacement times of 3.02θ, 1.68θ, 1.23θ, and 0.77θ.

3.4.2b *Type I Censoring With Replacement*

If failed items are replaced immediately, so that n items are always on test, and if testing terminates at some prespecified time L_0, we say that there is Type I censoring with replacement. Once again, the likelihood is of the form (1.4.9), which gives

$$L(\theta)=\frac{1}{\theta^r}e^{-\Sigma t_i/\theta}$$

where r is the observed number of failures and Σt_i is the total time on test. Since $\Sigma t_i = nL_0$ and L_0 is fixed, r is sufficient for θ. The distribution of r is a Poisson distribution with mean nL_0/θ, since r is the number of events in the interval $(0, L_0)$ in a Poisson process with intensity n/θ. Confidence intervals for θ can be obtained using standard methods for the Poisson distribution (e.g., Cox and Lewis, 1966, Ch. 2).

It is interesting that in setting up an accept–reject rule with this type of life test one is once again led to the same formula, (3.4.4), as for the two types of plans previously considered. To see this, first note that tests of $H_0: \theta=\theta_0$ versus $H_1: \theta<\theta_0$ will have acceptance regions for H_0 of the form $r<r_0$. We now use the fact (see Problem 1.5) that if N_t is the number of events in time t in a Poisson process with intensity λ, then $\Pr(N_t \leqslant k-1) = \Pr(\chi^2_{(2k)} > 2\lambda t)$. Hence

$$\begin{aligned} 1-\alpha &= \Pr(\text{accept } H_0; \theta=\theta_0) \\ &= \Pr(r<r_0; \theta=\theta_0) \\ &= \Pr\left(\chi^2_{(2r_0)} \geqslant 2nL_0/\theta_0\right). \end{aligned} \tag{3.4.7}$$

Also,

$$\begin{aligned} \beta &= \Pr(\text{accept } H_0; \theta=\theta_1) \\ &= \Pr(r<r_0; \theta=\theta_1) \\ &= \Pr\left(\chi^2_{(2r_0)} \geqslant 2nL_0/\theta_1\right). \end{aligned} \tag{3.4.8}$$

From (3.4.7) one must have $\chi^2_{(2r_0),\alpha} = 2nL_0/\theta_0$, and from (3.4.8), $\chi^2_{(2r_0),1-\beta} = 2nL_0/\theta_1$. Hence r_0 must be such that

$$\frac{\chi^2_{(2r_0),\alpha}}{\chi^2_{(2r_0),1-\beta}} = \frac{\theta_1}{\theta_0} \tag{3.4.9}$$

which is the same as (3.4.4). Note that nL_0 must be chosen so that (3.4.7) and (3.4.8) are satisfied. Actually, since it is generally not possible to choose r_0 to satisfy (3.4.9) exactly, both (3.4.7) and (3.4.8) cannot be satisfied exactly. For example, if (3.4.7) is satisfied with nL_0 chosen such that $nL_0 = \frac{1}{2}\theta_0\chi^2_{(2r_0),\alpha}$, the test will have size α but $P(\theta_1)$ will only approximately equal $1-\beta$. Finally, trade-offs between n, the number of items on test, and L_0, the length of the test, can be made. Since a particular value of nL_0 is required, n can be increased at the expense of decreasing L_0, and vice-versa.

3.4.2c *Type I Censoring Without Replacement*

Type I censoring without replacement has been thoroughly discussed from the point of view of significance tests and confidence interval estimation in

Section 3.2.1. Recall that in this case there is no one-dimensional sufficient statistic for θ, the likelihood function being, when all items have the same censoring time L_0,

$$L(\theta)=\frac{1}{\theta^r}\exp\left(-\sum_{i=1}^{r}\frac{t_{(i)}}{\theta}-\frac{(n-r)L_0}{\theta}\right).$$

Accept–reject rules can be based on the excellent approximation (1) of Section 3.2.1, in which it is assumed that $\hat{\phi}\sim N[\phi, v(\phi)]$, where $\phi=\theta^{-1/3}$, $\hat{\phi}=\hat{\theta}^{-1/3}$, and $v(\phi)=\phi^2/9n[1-\exp(-L_0/\theta)]$. Acceptance regions for $H_0: \theta=\theta_0$ versus $H_1: \theta<\theta_0$ are of the form $\hat{\phi}\leqslant C$, and the normal approximation to the distribution of $\hat{\phi}$ can be used to determine C and n, for given L_0, to give specified size and power. For a test with size α and power $1-\beta$ at $\theta=\theta_1$, C and n must satisfy the two equations

$$3\sqrt{n}\left(1-e^{-L_0/\theta_0}\right)^{1/2}\left(\frac{C}{\phi_0}-1\right)=N_{1-\alpha}$$

$$3\sqrt{n}\left(1-e^{-L_0/\theta_1}\right)^{1/2}\left(\frac{C}{\phi_1}-1\right)=N_{\beta}$$

where N_p is the pth quantile of the standard normal distribution.

An alternate method of constructing acceptance plans in this situation is to base these on just r, which has a binomial distribution (e.g., Epstein, 1960a; Mann et al., 1974, p. 312). This method is simple and convenient, but ignores information contained in the observed failure times.

3.4.2d *Sequential Plans*

If the primary purpose of an experiment is to provide an accept–reject decision for H_0 versus H_1, then a sequential procedure can often be valuable. Discussion of sequential methods is beyond the scope of this book, but the basic idea is that the life test is more or less continuously monitored, so that the decisions to accept or reject H_0 can be made as soon as there is sufficient evidence to reach such a decision.

Epstein and Sobel (1955) present a test based on Wald's sequential probability ratio test, in which the decision made at time t essentially depends on the inequality

$$B<\left(\frac{\theta_0}{\theta_1}\right)^{r(t)}\exp\left[\left(\theta_1^{-1}-\theta_0^{-1}\right)T(t)\right]<A \qquad (3.4.10)$$

where $r(t)$ is the number of failures observed by time t and $T(t)$ is the total time on test up to time t, that is, the total lifetime lived by all items, failed and unfailed, up to time t. At time t experimentation continues as long as (3.4.10) is satisfied; on the other hand, if the function in the middle of (3.4.10) is $\leqslant B$, H_0 is rejected, and if it is $\geqslant A$, H_0 is accepted. A slight modification consists of truncating the tests to avoid very long test times. The constants B and A are selected to give the test desired power $1-\beta$ at $\theta=\theta_1$ and size α; it turns out that to a close approximation $A=(1-\beta)/\alpha$ and $B=\beta/(1-\alpha)$. Epstein and Sobel (1955) and Epstein (1960a) give approximate formulas for calculating the power function and other characteristics of this test when testing is with or without replacement. Woodall and Kurkjian (1962), Aroian (1976), Bryant and Schmee (1979), and Kao et al. (1979) contain additional results and many references to other work. Tests of this type are tabled in MIL-STD 781C (1977).

The main advantage of a sequential plan is that the time needed to reach a decision about H_0 versus H_1 can be substantially reduced from that required by a similar nonsequential plan. If, however, one is not just interested in a decision rule, but also in estimation, sequential procedures create complications, though it is possible to obtain conservative confidence limits from them (e.g., Bryant and Schmee, 1979). Another qualification of the sequential tests is that their properties depend rather heavily on the exponentiality of the underlying lifetime distribution, and so the effect of possible departures from the model needs to be carefully considered.

3.4.3 General Remarks on Experimental Plans

Several life test plans have been examined for situations where the lifetime distribution is assumed to be exponential. If one's aim is to devise an efficient accept–reject procedure, then considerations of size, power, and economic constraints on the amount and type of testing feasible are important factors in the choice of a plan. Both sequential and nonsequential plans of the types discussed have been tabulated and are widely used (e.g., MIL-STD-781C, 1977). Although the emphasis here has been on acceptance procedures, similar power considerations are relevant when planning other types of trials or experiments. In addition, estimation has taken a back seat in our discussion; for estimation, plans can be chosen to give satisfactorily short or selective confidence intervals for θ, satisfactorily small variance for $\hat{\theta}$, or, indeed, to satisfy any particular objective of interest. MacKay (1979) discusses experiments whose object is estimation and shows that in one sense Type II censored plans are preferable to Type I censored plans. Another problem that has not been discussed here is the design of experiments for comparing two or more exponential distributions. Much work has been done in this area, especially in the context of medical trials for the comparison of treatments. Some of this work, and related references, can be found in Armitage (1975), Breslow and Haug (1972), and Louis (1977).

It should be remembered that this section has dealt with experimental plans under an assumed exponential model, with the focus being on power considerations. In practice, the planning of an experiment will involve many other important factors that may dominate questions of pure statistical efficiency. A good discussion of the design of clinical trials, for example, is given by Peto et al. (1976). Finally, the properties of the plans considered are sensitive to departures from the exponential distribution. This, of course, has important practical ramifications and is discussed in Section 3.6.

3.5 THE TWO-PARAMETER EXPONENTIAL DISTRIBUTION

The two-parameter exponential distribution has p.d.f.

$$f(t;\mu,\theta)=\frac{1}{\theta}e^{-(t-\mu)/\theta} \qquad t\geqslant\mu \tag{3.5.1}$$

where $\mu\geqslant 0$ is a threshold, or "guarantee time," parameter. This model is employed in situations where it is thought that death cannot occur before some particular time μ. If μ is known, statistical analysis can be carried out as for the one-parameter distribution (3.1.1), since $t-\mu$ has a one-parameter exponential distribution. Methods will be presented for the case in which both parameters in (3.5.1) are unknown, but first we make one or two general remarks about models with threshold parameters.

The best way to handle an unknown threshold parameter μ is often to use methods that assume μ to be known, but to take care to assess what values of μ are plausible and to repeat the analysis for a number of different μ values. One reason for this is that models with unknown threshold parameters pose some rather special statistical problems. Also, the data often contain little information about threshold parameters, and this can create numerical problems, for example, in computing m.l.e.'s. For the exponential distribution a threshold parameter is more easily handled than for other common lifetime distributions, however (e.g., Epstein and Sobel, 1954), and a brief treatment of this follows.

3.5.1 Type II Censored Data

When the data are Type II censored, exact procedures are available. From (3.5.1) and (1.4.1), the joint p.d.f. of the r smallest observations $t_{(1)}<\cdots<t_{(r)}$

in a random sample of size n from (3.5.1) is

$$\frac{n!}{(n-r)!}\frac{1}{\theta^r}\exp\left(-\frac{1}{\theta}\sum_{i=1}^{r}\left(t_{(i)}-\mu\right)-\frac{n-r}{\theta}\left(t_{(r)}-\mu\right)\right) \qquad t_{(i)}\geqslant\mu. \tag{3.5.2}$$

To obtain the m.l.e.'s $\hat{\mu}$ and $\hat{\theta}$ note that for any $\theta>0$ (3.5.2) decreases as μ decreases; also note that, regardless of the value of θ, $t_{(1)}$ is the largest value μ can take on in (3.5.2), since $\mu\leqslant t_{(1)}<\cdots<t_{(r)}$. Hence

$$\hat{\mu}=t_{(1)}.$$

We then find that (3.5.2) with $\mu=t_{(1)}$ is maximized for θ by

$$\hat{\theta}=\left(\sum_{i=1}^{r}t_{(i)}+(n-r)t_{(r)}-nt_{(1)}\right)\Big/r.$$

It is readily seen that $\hat{\mu}$ and $\hat{\theta}$ are jointly sufficient for μ and θ. The distributions of $\hat{\mu}$ and $\hat{\theta}$ are given in the following theorem.

THEOREM 3.5.1 *Let $\hat{\mu}$ and $\hat{\theta}$ be the maximum likelihood estimators of μ and θ, based on a Type II censored sample from* (3.5.1), *consisting of the r smallest observations in a random sample of n. Then $\hat{\mu}$ and $\hat{\theta}$ are independent, and $2n(\hat{\mu}-\mu)/\theta$ and $2r\hat{\theta}/\theta$ are distributed as $\chi^2_{(2)}$ and $\chi^2_{(2r-2)}$, respectively.*

Proof The random variables $t_{(1)}-\mu,\ldots,t_{(r)}-\mu$ are the first r ordered observations in a sample of size n from the one-parameter exponential distribution (3.1.1). By Theorem 3.1.1, the quantities

$$W_1=n\left(t_{(1)}-\mu\right)$$

$$W_i=(n-i+1)\left(t_{(i)}-t_{(i-1)}\right) \qquad i=2,\ldots,r$$

are independent and have one-parameter exponential distributions (3.1.1). Hence $2n(\hat{\mu}-\mu)/\theta=2W_1/\theta\sim\chi^2_{(2)}$ and

$$\frac{2r\hat{\theta}}{\theta}=2\sum_{i=2}^{r}\frac{W_i}{\theta}\sim\chi^2_{(2r-2)}$$

as stated. ■

Remark The results in the proof of this theorem imply that given $t_{(1)}$, the variables $t_{(i)}-t_{(1)}$, $i=2,\ldots,n$, are distributed as the order statistics in a random sample of size $n-1$ from the one-parameter exponential distribution with mean θ. This means that one can assess the adequacy of the two-parameter model by basing probability plots on the $(t_{(i)}-t_{(1)})$'s.

Confidence Intervals for θ or μ

Confidence intervals for θ or μ are easily obtained from the results of Theorem 3.5.1. Confidence intervals for θ can be derived through the pivotal quantity

$$\frac{2r\hat{\theta}}{\theta} \sim \chi^2_{(2r-2)}.$$

To get confidence intervals for θ we consider the pivotal

$$\frac{n(r-1)(\hat{\mu}-\mu)}{r\hat{\theta}} \sim F_{(2,2r-2)}. \tag{3.5.3}$$

That this has a F distribution follows directly from the results of Theorem 3.5.1. Exact quantiles of the $F_{(2,2r-2)}$ distribution can be given in closed form, so that it is not necessary to refer to tables in using (3.5.3). Specifically, the pth quantile of $F_{(2,2r-2)}$ is

$$F_{(2,2r-2),p} = (r-1)\left[(1-p)^{-1/(r-1)}-1\right]. \tag{3.5.4}$$

Confidence Intervals for t_p or S(t)

Confidence intervals for quantiles or the survivor function of (3.5.1) can be obtained, but with more difficulty. The pth quantile of (3.5.1) is $t_p = \mu+\theta[-\log(1-p)]$. Confidence intervals for t_p can be based on the pivotal quantity

$$Z_p = \frac{\hat{\mu}-t_p}{\hat{\theta}}.$$

If $z_{p,\gamma}$ is the γth quantile of Z_p, then

$$\begin{aligned}\Pr\left(Z_p \leqslant z_{p,\gamma}\right) &= \gamma \\ &= \Pr\left(t_p \geqslant \hat{\mu}-z_{p,\gamma}\hat{\theta}\right)\end{aligned} \tag{3.5.5}$$

and $\hat{\mu}-z_{p,\gamma}\hat{\theta}$ is a lower γ confidence limit for t_p. Confidence intervals for the survivor function $S(t_0)=\exp[-(t_0-\mu)/\theta]$ can also be obtained through (3.5.5). To see this, note that

$$\Pr(t_p \geq t_0)=\Pr[S(t_0)\geq 1-p].$$

To get a lower γ confidence limit for $S(t_0)$, for given t_0, we merely need to determine p such that $\Pr(t_p \geq t_0)=\gamma$. For any observed set of data, the realized confidence limit $1-p$ is found by determining p such that the lower confidence limit $\hat{\mu}-z_{p,\gamma}\hat{\theta}$ in (3.5.5) equals t_0. That is, one must find p such that $z_{p,\gamma}$ in (3.5.5) satisfies $z_{p,\gamma}=-(t_0-\hat{\mu})/\hat{\theta}$.

It is relatively straightfoward, though tedious, to determine the quantiles $z_{p,\gamma}$ used in (3.5.5). This has been discussed by Engelhardt and Bain (1978b), Guenther et al. (1976), and Dunsmore (1978). Grubbs (1971) and Pierce (1973) discuss the complementary problem of the direct calculation of confidence limits for $S(t)$. The results of Engelhardt and Bain are the most convenient to use and are presented below. First, however, we mention that some authors, among them Guenther et al. and Engelhardt and Bain, allow the threshold parameter to range over the entire real line, whereas (5.3.1) makes the restriction $\mu \geq 0$. Although there are certain mathematical advantages to leaving μ completely arbitrary, μ is naturally nonnegative when dealing with lifetimes. The results of Guenther et al. and Engelhardt and Bain occasionally yield negative confidence limits for t_p, as does the pivotal (3.5.3) in giving confidence limits for μ. When μ is restricted to be ≥ 0, the appropriate procedure is to replace any negative confidence limits with 0. Cox and Hinkley (1974, pp. 224–226) provide some general discussion on this point.

Let us return to (3.5.5). Temporarily dropping the subscripts on $z_{p,\gamma}$ and letting $q=-\log(1-p)$, we can rewrite this as

$$\begin{aligned}\Pr(Z_p \leq z) &= \Pr(\hat{\mu}-\mu-q\theta \leq z\hat{\theta}) \\ &= \Pr\left[\frac{\hat{\mu}-\mu}{\theta}-z\left(\frac{\hat{\theta}}{\theta}\right)\leq q\right]=\gamma. \end{aligned} \tag{3.5.6}$$

By Theorem 3.5.1, $W=2n(\hat{\mu}-\mu)/\theta \sim \chi^2_{(2)}$, $T=r\hat{\theta}/\theta \sim \chi^2_{(2r-2)}$, and W and T are independent. Hence (3.5.6) reduces to the problem of finding the value of $k=-z/r$ such that

$$\Pr\left(\frac{W}{2n}+kT \leq q\right)=\gamma \tag{3.5.7}$$

for a given r, n, q, and γ. Having found this, the desired lower γ confidence bound on t_p is, from (3.5.5), $\hat{\mu}+rk\hat{\theta}$.

There are various ways of calculating k. For cases in which $k \leqslant 0$, Engelhardt and Bain (1978a) show that it can be given explicitly. They show that $k \leqslant 0$ if and only if $(1-p)^n/(1-\gamma) \geqslant 1$ and that in this case

$$k=\frac{1}{n}\left[1-\left(\frac{(1-p)^n}{1-\gamma}\right)^{1/(r-1)}\right]. \tag{3.5.8}$$

This result is also given by Dunsmore (1978). Equating $-z$ to $(t_0-\hat{\mu})/\hat{\theta}$ and solving for $1-p$, we get a lower γ confidence bound on $S(t_0)$ as

$$(1-\gamma)^{1/n}\left(1-\frac{n(t_0-\hat{\mu})}{r\hat{\theta}}\right)^{(r-1)/n}. \tag{3.5.9}$$

For the case in which $(1-p)^n/(1-\gamma)<1$ it is not possible to obtain an explicit expression for k. Exact values of k can be computed with some effort, but this seems unnecessary in view of very good approximations given by Engelhardt and Bain. They show that

$$k \doteq \frac{1}{n}\left[-m(p)-N_\gamma\left(\frac{m^2(p)}{r}+\frac{1}{r^2}\right)^{1/2}\right] \tag{3.5.10}$$

where $m(p)=[1+n\log(1-p)]/(r-\frac{5}{2})$ and N_γ is the γth quantile of the standard normal distribution. The approximate lower γ confidence bound on $S(t_0)$ obtained from this is

$$\exp\left[-\frac{1}{n}+\frac{r}{n}\frac{r-5/2}{a}\left(Y-\frac{N_\gamma}{r}(rY^2+a)^{1/2}\right)\right] \tag{3.5.11}$$

where $Y=n(\hat{\mu}-t_0)/\hat{\theta}$ and $a=r^2(1-N_\gamma^2/r)$. These approximations, which are discussed further by Engelhardt and Bain, are sufficiently accurate for virtually all practical purposes.

Example 3.5.1 Let us consider an example discussed by Grubbs (1971), Engelhardt and Bain (1978a), and others. The data are mileages for 19 military personnel carriers that failed in service and appear consonant with an exponential model. There is no censoring, and the mileages are 162, 200, 271, 320, 393, 508, 539, 629, 706, 777, 884, 1008, 1101, 1182, 1463, 1603, 1984, 2355, and 2880. Hence $n=r=19$, $\hat{\mu}=t_{(1)}=162$, and $\hat{\theta}=(\Sigma t_i-19t_{(1)})/19=15{,}869/19=835.2$. To obtain confidence intervals for θ and μ, respectively, we use the results of Theorem 3.5.1 and (3.5.3), which tell us that $38\hat{\theta}/\theta \sim \chi^2_{(36)}$ and $18(\hat{\mu}-\mu)/\hat{\theta} \sim F_{(2,36)}$. From χ^2 and F tables we find that $\Pr(21.38 \leqslant \chi^2_{(36)} \leqslant 54.40) = .95$ and $\Pr(F_{(2,36)} \leqslant 3.254) = .95$, which leads to confidence intervals $583.4 \leqslant \theta \leqslant 1484$ and $11.0 \leqslant \mu (\leqslant 162)$. Tests of the hypothesis $\mu = 0$ are of interest. In this case the observed F value for $H_0: \mu = 0$ is 3.491, which gives a significance level between .02 and .03. There is thus some evidence against the value $\mu = 0$.

Let us also calculate a lower .90 confidence limit for the .10 quantile of the distribution. Since $\gamma = .90$ and $p = .10$, we find that $(1-p)^n/(1-\gamma) = .90^{19}/.90 < 1$, so approximation (3.5.10) must be used. Then $m(.9) = [1 + 19\log(.9)]/16.5 = -0.0607$ and $N_{.90} = 1.281$, so

$$k = \frac{1}{19}\left\{0.0607 - 1.281\left[\frac{0.0607^{\,2}}{19} + \left(\frac{1}{19}\right)^2\right]^{1/2}\right\}$$

$$= 0.0069.$$

The desired lower .90 confidence limit for $t_{.10}$ is therefore $\hat{\mu} + 0.0069(19\hat{\theta}) = 271.5$.

3.5.2 Type I Censored Data

If the data are Type I censored, matters are more difficult. Suppose, as usual, that individual i in a sample of size n has lifetime T_i and a fixed censoring time L_i and that $t_i = \min(T_i, L_i)$. Let D and C denote the individuals for which $t_i = T_i$ and $t_i = L_i$, respectively. One point to note here is that if a particular value of μ is under consideration, then if $L_i < \mu$, the information $T_i \geqslant L_i$ is superceded by the information, implicit in the model, that $T_i \geqslant \mu$. With this observation, the likelihood function is, by (1.4.5),

$$L(\theta, \mu) = \frac{1}{\theta^r} \exp\left(-\sum_{i \in D} \frac{T_i - \mu}{\theta} - \sum_{i \in C} \frac{\max(L_i, \mu) - \mu}{\theta}\right) \qquad T_{(1)} \geqslant \mu$$

where r is the observed number of lifetimes and $T_{(1)} = \min_{i \in D} T_i$ is the smallest observed lifetime. As in Section 3.2, it will be tacitly assumed that $r > 0$; when $r = 0$, the likelihood function does not possess a finite maximum. For $r \geqslant 0$, $L(\theta, \mu)$ is maximized by

$$\hat{\mu} = T_{(1)} \qquad \hat{\theta} = \frac{1}{r}\left(\sum_{i \in D} (T_i - \hat{\mu}) + \sum_{i \in C} \max(L_i - \hat{\mu}, 0)\right). \tag{3.5.12}$$

This follows by the same argument used in the preceding section for Type II censored sampling.

The exact distribution of $\hat{\mu}$ and $\hat{\theta}$ is too complicated to be used for inference purposes. Wright et al. (1978) present test and interval estimation procedures for the special case of singly Type I censored data, based on certain conditional distributions. Likelihood ratio or other approximate methods can also be used for inference about μ and θ. We will take this approach, since it is straightfoward and applies to arbitrarily censored data.

Tests and interval estimates for θ are readily obtained. The likelihood ratio statistic for testing $H_0: \theta=\theta_0$ versus $H_1: \theta\neq\theta_0$ is

$$\Lambda=-2\log\left(\frac{L(\theta_0,\hat{\mu})}{L(\hat{\theta},\hat{\mu})}\right)$$

$$=2r\left[\hat{\theta}-1-\log\left(\frac{\hat{\theta}}{\theta_0}\right)\right].$$

In large samples the distribution of Λ is approximately $\chi^2_{(1)}$, under H_0. An alternate approach to inference about θ is to use an approximation like (3) in Section 3.2, which treats $2r\hat{\theta}/\theta$ as being distributed as $\chi^2_{(2r+1)}$. This approximation was derived for the one-parameter exponential model; if one works conditionally on $T_{(1)}$ here, the approximation $2r\hat{\theta}/\theta\sim\chi^2_{(2r-1)}$ suggests itself.

Inference about μ is also straightfoward, though since μ is a threshold parameter, the asymptotic results concerning the likelihood ratio are slightly different. The likelihood ratio statistic for testing $H_0: \mu=\mu_0$ versus $H_1: \mu\neq\mu_0$ is

$$\Lambda=-2\log\left(\frac{L(\tilde{\theta},\mu_0)}{L(\hat{\theta},\hat{\mu})}\right)$$

where $\tilde{\theta}=\tilde{\theta}(\mu_0)$ is the m.l.e. of θ when $\mu=\mu_0$. This is

$$\tilde{\theta}(\mu_0)=\frac{1}{r}\left(\sum_{i\in D}(T_i-\mu_0)+\sum_{i\in C}\max(L_i-\mu_0,0)\right).$$

It is easily seen that $\Lambda=-2r\log(\hat{\theta}/\tilde{\theta})$. Now, only values $\mu_0\leqslant\hat{\mu}$ need be considered, since μ_0 must be $\leqslant T_{(1)}$, and if all L_i $(i\in C)$ exceed $T_{(1)}$, then Λ can be further rewritten as

$$\Lambda=2r\log\left(1+\frac{n(\hat{\mu}-\mu_0)}{r\hat{\theta}}\right). \tag{3.5.13}$$

It turns out that in large samples Λ is approximately $\chi^2_{(2)}$, instead of the $\chi^2_{(1)}$ that one has in regular problems. The $\chi^2_{(2)}$ limiting distribution for Λ arises because μ is a threshold parameter (e.g., see Hogg, 1956).

Instead of treating Λ as being $\chi^2_{(2)}$, we could derive a F approximation that parallels the exact F distribution used for inference about μ in the case of Type II censoring. If the L_i's are sufficiently large so that $\Pr[T_{(1)} \leq L_{(1)}]$ is essentially unity, then, by Theorem 3.5.1, $2n(\hat{\mu}-\mu_0)/\theta \sim \chi^2_{(2)}$ and is independent of $\hat{\theta}$, to a very close approximation. Combining this with the approximation $2r\hat{\theta}/\theta \sim \chi^2_{(2r-1)}$ suggested above, we get the approximation

$$\frac{(2r-1)n(\hat{\mu}-\mu_0)}{2r\hat{\theta}} \sim F_{(2,2r-1)}. \qquad (3.5.14)$$

This can be used to obtain approximate tests and confidence intervals for μ.

Confidence intervals for quantiles or the survivor function will not be discussed explicitly. The only approach that appears feasible at present is to use the methods given in Section 3.5.1 for the case of Type II censoring, replacing the value of r there with the value $r+\frac{1}{2}$. The arguments leading to (3.5.14) suggest that this should be a reasonable method of obtaining approximate confidence intervals.

Discussion

Much work has been published on the two-parameter exponential distribution; some references are given in Johnson and Kotz (1970, Ch. 18). Little work has been directed at Type I censored data, however, so that questions remain concerning the adequacy of approximate methods in small samples. Only one-sample problems have been discussed here; m-sample tests ($m \geq 2$) are considered by many authors, including Epstein and Tsao (1953), Kumar and Patel (1971), and Kendall and Stuart (1967, Probs. 24.11 through 24.13). A few results in this area are considered in Problem 3.10.

3.6 CAUTIONARY REMARKS

The exponential distributions has been demonstrated to provide good approximations to many lifetime distributions, and inference procedures based on the model are widely used. A serious drawback of the exponential distribution, however, is that procedures based on it tend to be highly nonrobust. That is, fairly small departures from the assumed model can alter the sampling properties of a procedure considerably. Inferences should be based on the exponential model only when observed data are consonant with the model, of course, but departures from the model that are difficult to detect with small- or moderate-size samples can have a substantial effect.

Nonrobustness is particularly critical in relation to life test acceptance plans, as several studies have shown. Zelen and Dannemiller (1961) and Harter and Moore (1976) have examined the behavior of the plans in Section 3.4 when the underlying life distribution is a Weibull rather than an exponential distribution. They find that the power functions for the tests are grossly distorted, even by fairly moderate departures from the exponential model. A further point of interest is that procedures that are the most efficient (statistically or economically) under an exponential model generally turn out to be the least robust. Thus, though increasing the total number of units on test can decrease test time (see Section 3.4.1), inferences based on highly censored plans are more critically affected by model departures than are those based on plans with lighter censoring. A third point is that sequential plans are more vulnerable to robustness problems than nonsequential ones. Other evidence on robustness under the exponential model is given by Fryer and Holt (1970, 1976) and MacKay (1979).

As a general rule, careful thought should be given to the possibility and likely effects of model inadequacy when using the exponential distribution. Methods for detecting departures from the exponential distribution and assessing the effect of departures on inferences are hence important. Tests of fit are discussed in Chapter 9. A convenient way of assessing the effect of model departures is to embed the exponential distribution in a more general model that reflects the types of departure envisaged. Numerous models considered in this book include the exponential distribution as a special case and they can be used for this purpose.

3.7 PROBLEMS AND SUPPLEMENTS

3.1 The following data are times $t_1, \ldots, t_n$ between successive failures of air conditioning equipment in a Boeing 720 airplane (Proschan, 1963): 74, 57, 48, 29, 502, 12, 70, 21, 29, 386, 59, 27, 153, 26, and 326. Assuming that the data have arisen from an exponential distribution with mean θ, compare confidence intervals for θ obtained by using the methods below, with the exact intervals obtained by using the fact that $2\Sigma t_i/\theta \sim \chi^2_{(2n)}$: (1) $\sqrt{n}(\hat{\theta}-\theta)/\hat{\theta} \sim N(0,1)$, (2) $3\sqrt{n}\hat{\phi}(\hat{\phi}-\phi) \sim N(0,1)$, where $\phi=\theta^{-1/3}$, (3) the likelihood ratio statistic.

(Sections 3.1, 3.2)

3.2 Methods based on the number of failure times. Consider Type I censored data from the exponential distribution. In some instances inferences can be conveniently based on r, the observed number of failures among n individuals, though there will inevitably be some loss of information in neglecting the actual failure times. Suppose that the data are singly Type I censored, with all individuals having the same fixed censoring time L. In this case r has a binomial distribution, with parameters n and $p=1-\exp(-L/\theta)$.

a. Compare the precision with which θ is estimated in large samples when only r is used with that of the procedure of Section 3.2.1, which uses both r and Σt_i. In what situations is the loss of information entailed by using r the smallest?

b. Suppose that it is wished to estimate $S(t_0)$, where $0<t_0<L$. Compare the two approaches of part (a) with the nonparametric approach based on the distribution of m, where m is the number of observations less than or equal to t_0. What advantage does the latter method possess?

c. In an experiment 20 items are placed on test, with the test being terminated after 150 hours: 15 items fail, at times 3, 19, 23, 26, 27, 37, 38, 41, 45, 58, 84, 90, 99, 109, and 138. Obtained two-sided .90 confidence intervals for (1) θ, (2) $S(100)$, based on the methods of part (a) for (1) and parts (a) and (b) for (2).

(Section 3.2)

3.3 The effect of grouping. Lifetime data are often grouped or rounded off to some degree. Suppose that lifetimes from an exponential distribution that are recorded as t actually lie in the interval $(t-h/2, t+h/2)$; censoring times recorded as t will be assumed to be exactly equal to t. Consider a censored sample of n observations involving r lifetimes and $n-r$ censoring times; this leads, in the continuous case, to a likelihood of the form

$$L(\theta)=\theta^{-r}\exp\left(-\sum_{i=1}^{n}\frac{t_i}{\theta}\right)$$

as discussed in Problem 3.9.

a. Show that the likelihood function in the present case of grouped data is

$$L_1(\theta)=\left(e^{h/2\theta}-e^{-h/2\theta}\right)^r\exp\left(-\sum_{i=1}^{n}\frac{t_i}{\theta}\right).$$

Obtain the exact m.l.e. by maximizing $L_1(\theta)$ and compare it with the m.l.e. obtained from $L(\theta)$.

b. Show that the expected information based on $L_1(\theta)$ is

$$I_1(\theta)=\frac{E(r)}{\theta^2}g\left(\frac{h}{\theta}\right)$$

where $g(a)=a^2e^{-a}/(1+e^{-a})^2$. Examine the loss of information entailed by grouping.

c. Examine the effect of rounding off observations to the nearest week on $\hat{\theta}$ and $L(\theta)$ for the two sets of leukemia remission time data given in Example 1.1.7, assuming one-parameter exponential lifetime distributions in each case. Compute two-sided .90 confidence intervals for θ, using the two likelihoods.

(Section 3.2)

3.4 Consider the likelihood function $L(\theta)$ obtained in the case of Type I censored sampling from the exponential distribution and let $\phi=g(\theta)$ be an arbitrary one-to-one transformation. Show that $(\partial^3\log L/\partial\phi^3)_{\hat{\phi}}=0$ if and only if $\phi\propto\theta^{-1/3}$. (The likelihood function for ϕ thus looks more "normal" than that for θ and suggest that treating $\hat{\phi}$ as normally distributed is preferable to treating $\hat{\theta}$ as normally distributed in obtaining confidence intervals.)

(Section 3.2.2; Anscombe, 1964; Sprott, 1973)

3.5 The following data are remission times, in weeks, for a group of 30 leukemia patients in a certain type of therapy; starred observations are censoring times: 1, 1, 2, 4, 4, 6, 6, 6, 7, 8, 9, 9, 10, 12, 13, 14, 18, 19, 24, 26, 29, 31*, 42, 45*, 50*, 57, 60, 71*, 85*, 91.

a. Estimate the mean remission time, (1) using the nonparametric method described in Problem 2.10, and (2) assuming that the underlying distribution of remission times is exponential. Obtain and compare confidence intervals for the mean using the two methods.

b. Similarly compare estimates of $S(26)$, the probability of a remission lasting more than 26 weeks, using the nonparametric PL estimate and the exponential model, respectively.

(Sections 3.2, 2.3)

3.6 Consider the likelihood ratio statistic (3.3.3) for testing the equality of m exponential means. Note that under $H_0:\theta_1=\cdots=\theta_m$ the quantities $\log\hat{\theta}$ and $\log\hat{\theta}_i$ $(i=1,\ldots,m)$ have log-gamma distributions (see Section 1.3.3) and show that $E(\Lambda)=m-1+0(r_i^{-1})$ but that $E(C\Lambda)=m-1+0(r_i^{-3})$, where C is the correction factor defined in Equation 3.3.4.

(Section 3.3; Bartlett, 1937)

3.7 Suppose that an acceptance plan is desired that under the one-parameter exponential model will accept $H_0: \theta=1000$ with probability .90 when $\theta=1000$ hours and probability .05 when $\theta=300$ hours. Obtain Type II censored plans, both with and without replacement, and sketch the power functions for the plans.

(Section 3.4)

3.8 Suppose that in running a life test experiment the following costs are incurred:

C_1 = cost of running the experiment for a unit of time

C_2 = cost of placing an item on test.

a. Suppose that a Type II censored (nonreplacement) life test plan with a given value of r has been selected. Determine the value of n^* of n, the total number of items on test, that minimizes the expected cost of the experiment (this will depend on r, C_1, C_2, and θ). Show that n^* can be expressed in terms of r and $C_1\theta/C_2$ and examine n^* numerically for the life test discussed in Example 3.4.1.
b. For Type II censored life test plans with a fixed value of r, compare sampling with and without replacement of items under this cost structure.

(Section 3.4)

3.9 Consider general censoring schemes for which the likelihood (1.4.9) is valid. For the exponential model this likelihood function is

$$L(\theta)=\frac{1}{\theta^r}e^{-T/\theta}$$

where $T=\Sigma t_i$ is the sum of observed lifetimes and censoring times and r is the observed number of lifetimes. By noting that $E(d\log L/d\theta)=0$ under suitable conditions, prove that the expected information based on $L(\theta)$ is $I(\theta)=E(r)/\theta^2$, so that $I(\theta)$ is completely determined by the expected number of failures. Use this to obtain the information in the cases of Type

II and Type I censoring and for an experiment in which observation ceases either at time t_0 or at the time of rth failure, whichever occurs first. Compare the expected duration of experiments of these three types, all having the same number of items, n, on test, and the same value $E(r)$.

(Section 3.4; MacKay, 1979)

3.10 Comparison of two-parameter exponential distributions. Consider $m \geqslant 2$ exponential distributions (3.5.1), with parameters (μ_i, θ_i), $i=1,\ldots,m$. Given Type II censored samples consisting of the first r_i out of n_i observations from the ith distribution $(i=1,\ldots,m)$,

a. Describe how one could test the equality of $\theta_1,\ldots,\theta_m$, with no assumptions about $\mu_1,\ldots,\mu_m$, using a likelihood ratio procedure. Discuss how to obtain significance levels in the cases (1) $m=2$ and (2) $m>2$.

b. Consider $H_0: \mu_1 = \cdots = \mu_m$ in the case in which the θ_i's are equal. Obtain the likelihood ratio statistic for testing H_0 versus H_1: "μ_i's are not all equal" and determine its exact distribution.

c. Develop a test of H_0 in the case in which the θ_i's are not necessarily equal. Discuss the distribution of the test statistic in the cases (1) $m=2$ and (2) $m>2$.

(Section 3.5)

3.11 The data below represent failure times, in minutes, for two types of electrical insulation in an experiment in which the insulation was subjected to a continuously increasing voltage stress.

Type A	219.3	79.4	86.0	150.2	21.7	18.5
	121.9	40.5	147.1	35.1	42.3	48.7
Type B	21.8	70.7	24.4	138.6	151.9	75.3
	12.3	95.5	98.1	43.2	28.6	46.9

Examine graphically whether the two sets of data might be considered to be random samples from different two-parameter exponential distributions. (See the Remark following Theorem 3.5.1). If this appears reasonable, compare the two distributions and, in particular, test that they have the same threshold parameter value.

(Section 3.5)

3.12 Robustness

a. Let $t_1, \ldots, t_n$ be a complete random sample from an exponential distribution with mean θ. Consider life test plans that test $H_0: \theta = 1000$ versus $H_1: \theta < 1000$ and have size 0.10. Graph the power functions of the tests for sample sizes $n = 10$ and $n = 20$.

b. Suppose that $t_1, \ldots, t_n$ actually come from a Weibull distribution with p.d.f. $(\beta/\alpha)(t/\alpha)^{\beta-1}\exp[-(t/\alpha)^{\beta}]$, $t > 0$, where $\beta = 1.5$ and $\alpha = \theta/\Gamma(1+1/1.5)$; this distribution also has mean θ. It can be shown that the distribution of $\Sigma t_i/\alpha$ is well approximated by a χ^2 distribution,

$$\sum_{i=1}^{n} \frac{t_i}{\alpha} \sim c\chi^2_{(b)}$$

where c and b are selected so that $c\chi^2_{(b)}$ has the same mean and variance as $\Sigma t_i/\alpha$. Show that this yields the values $c = [\Gamma(1+2/1.5) - \Gamma(1+1/1.5)^2]/2\Gamma(1+1/1.5)$ and $b = nc^{-1}\Gamma(1+1/1.5)$.

c. Use the χ^2 approximation of part (b) to examine the power function of the tests in part (a) when the underlying distribution is a Weibull distribution with $\beta = 1.5$ rather than an exponential distribution.

(Section 3.6)

3.13 Let $t_{(1)} \leqslant \cdots \leqslant t_{(r)}$ be the r smallest observations in a random sample of size n from (3.1.1) and define the W_i's as in (3.1.5). Let

$$T_j = W_1 + \cdots + W_j \qquad j = 1, \ldots, r$$

and show that given T_r, the T_j/T_r's $(j = 1, \ldots, r-1)$ are distributed like the ordered observations in a random sample of size $r-1$ from the uniform distribution on $(0,1)$. Discuss how this might be used in assessing the adequacy of the model (3.1.1).

3.14 Predicting the duration of a life test. Sometimes it is desired to predict the total duration of a life test on the basis of early results in the test. Suppose, for example, that a test is to terminate at the time $t_{(r)}$ of the rth failure. If the sth failure has just occurred $(1 \leqslant s < r)$, we can predict $t_{(r)}$.

a. If the data came from a one-parameter exponential distribution with mean θ, prove that $t_{(r)} - t_{(s)}$ and

$$T_s = \sum_{i=1}^{s} t_{(i)} + (n-s)t_{(s)}$$

are independent, and that $U=(t_{(r)}-t_{(s)})/T_s$ is pivotal, with distribution function

$$\Pr(U\leqslant t)=1-\frac{(n-s)!}{(r-s-1)!(n-r)!}$$

$$\cdot\sum_{i=0}^{r-s-1}\binom{r-s-1}{i}(-1)^i\Big/(n-r+i+1)[1+(n-r+i+1)t]^s.$$

Show how U can be used to obtain prediction intervals for $t_{(r)}$, based on $t_{(1)},\ldots,t_{(s)}$.

b. Show that $Q=(t_{(r)}-t_{(s)})/\theta$ can be expressed as a linear combination of independent $\chi^2_{(1)}$ random variables and that Q has mean and variance

$$m=\sum_{i=s+1}^{r}\frac{1}{n-i+1}\qquad v=\sum_{i=s+1}^{r}\frac{1}{(n-i+1)^2}$$

respectively. It turns out that the distribution of Q is well approximated by the distribution of $c\chi^2_{(d)}$, where c and d are selected so that the mean and variance of $c\chi^2_{(d)}$ equal the mean and variance of Q. Show that this leads to the approximation

$$\frac{2m}{v}Q\sim\chi^2_{(2m^2/v)}.$$

c. Using the χ^2 approximation for Q and the fact that $2T_s/\theta\sim\chi^2_{(2s)}$ and is independent of Q, one can obtain the approximation

$$\frac{s}{m}U\sim F_{(2m^2/v,2s)}.$$

This is simpler to work with than the exact distribution of U, given in (a). Check the adequacy of the F approximation for the case $n=15$, $r=10$, $s=5$.

(Lawless, 1971; Mann et al., 1974)

CHAPTER 4

Inference Procedures for Weibull and Extreme Value Distributions

This chapter is devoted to a discussion of statistical procedures for the Weibull distribution. The Weibull p.d.f. is

$$\frac{\beta}{\alpha}\left(\frac{t}{\alpha}\right)^{\beta-1}\exp\left[-\left(\frac{t}{\alpha}\right)^{\beta}\right] \qquad t\geqslant 0 \qquad (4.0.1)$$

where $\beta>0$ and $\alpha>0$ are parameters sometimes referred to as the shape and scale parameters of the distribution, respectively. In place of the Weibull distribution, it is often more convenient to work with the equivalent extreme value distribution with p.d.f.

$$f(x;u,b)=\frac{1}{b}e^{(x-u)/b}\exp\left(-e^{(x-u)/b}\right) \qquad -\infty<x<\infty \qquad (4.0.2)$$

where u $(-\infty<u<\infty)$ and b $(b>0)$ are parameters. As noted in Section 1.3.2, if T has p.d.f. (4.0.1), then $X=\log T$ has p.d.f. (4.0.2), with $u=\log\alpha$ and $b=\beta^{-1}$. The main convenience in working with the extreme value distribution stems from the fact that u and b are location and scale parameters. Any results derived in terms of one distribution are easily transfered to the other.

The Weibull distribution is a particularly important life distribution, and a large body of literature on statistical methods has evolved for it. One reason that so many papers have been written on the Weibull distribution concerns its statistical properties. There is, in general, no two-dimensional sufficient statistic for β and α of (4.0.1) or, equivalently, u and b of (4.0.2), and the possibilities for producing estimators are many. In addition, the distributions of most estimators and other statistics associated with the Weibull and extreme value distributions are mathematically intractable.

This has led to extensive work in producing tables for carrying out inferences and in the development of approximations to distributions of certain types of estimators. Fortunately, good statistical procedures that are relatively easy to use are now available, thanks largely to the existence of high-speed computers.

Problems involving Type II censored data are discussed first, followed by inference with Type I censored data, the comparison of Weibull distributions, and other topics.

4.1 SINGLE SAMPLES: COMPLETE OR TYPE II CENSORED DATA

For complete or Type II censored data the statistical theory behind various methods is fairly straightforward, though there can be mathematical or computational complications. Suppose that $t_1 \leqslant t_2 \leqslant \cdots \leqslant t_r$ are the r smallest observations in a random sample of n from the Weibull distribution (4.0.1) or, equivalently, that $x_1 \leqslant \cdots \leqslant x_r$, where $x_i = \log t_i$, are the r smallest observations in a sample of size n from (4.0.2). (For convenience we shall omit brackets on the subscripts i in $t_{(i)}$ or $x_{(i)}$; so unless otherwise specified, in this section x_i represents the ith smallest observation.) The joint p.d.f. of $x_1, \ldots, x_r$ is, by (1.4.1),

$$\frac{n!}{(n-r)!}\left(\prod_{i=1}^{r} \frac{1}{b} e^{(x_i-u)/b} \exp\left(-e^{(x_i-u)/b}\right)\right)\left[\exp\left(-e^{(x_r-u)/b}\right)\right]^{n-r}. \quad (4.1.1)$$

This defines the likelihood function for u and b, and it can be seen that $(x_1, \ldots, x_r)$ is the minimal sufficient statistic for the sample.

Point estimation for u and b will be discussed first, and then tests and interval estimation. Derivations and results are given mainly in terms of the extreme value distribution.

4.1.1 Point Estimation

Maximum Likelihood Estimates

From (4.1.1) the likelihood function can be taken as

$$L(u,b) = \frac{1}{b^r} \exp\left(\sum_{i=1}^{r} \frac{x_i - u}{b} - \sum_{i=1}^{r}{}^{*} \exp \frac{x_i - u}{b}\right)$$

where we introduce the useful notation

$$\sum_{i=1}^{r}{}^{*} w_i = \sum_{i=1}^{r} w_i + (n-r) w_r$$

for any sequence $w_1, \ldots, w_r$. The log likelihood function is

$$\log L(u,b) = -r\log b + \sum_{i=1}^{r} \frac{x_i - u}{b} - \sum_{i=1}^{r}{}^{*} \exp\left(\frac{x_i - u}{b}\right) \tag{4.1.2}$$

and thus

$$\frac{\partial \log L}{\partial u} = -\frac{r}{b} + \frac{1}{b} \sum_{i=1}^{r}{}^{*} \exp\left(\frac{x_i - u}{b}\right) \tag{4.1.3}$$

$$\frac{\partial \log L}{\partial b} = -\frac{r}{b} - \frac{1}{b} \sum_{i=1}^{r} \frac{x_i - u}{b} + \frac{1}{b} \sum_{i=1}^{r}{}^{*} \frac{x_i - u}{b} \exp\left(\frac{x_i - u}{b}\right). \tag{4.1.4}$$

The m.l.e.'s $\hat{u}$ and $\hat{b}$ can be obtained by simultaneously solving $\partial \log L/\partial u = 0$ and $\partial \log L/\partial b = 0$. One convenient way to do this is to note that setting (4.1.3) equal to 0 gives

$$e^{\hat{u}} = \left[\frac{1}{r} \sum_{i=1}^{r}{}^{*} \exp\left(\frac{x_i}{\hat{b}}\right)\right]^{\hat{b}}. \tag{4.1.5}$$

Substituting this into $\partial \log L/\partial b = 0$, we get

$$\sum_{i=1}^{r}{}^{*} x_i \exp\left(\frac{x_i}{\hat{b}}\right) \Big/ \sum_{i=1}^{r}{}^{*} \exp\left(\frac{x_i}{\hat{b}}\right) - \hat{b} - \frac{1}{r} \sum_{i=1}^{r} x_i = 0. \tag{4.1.6}$$

To find $\hat{u}$ and $\hat{b}$ one can therefore determine $\hat{b}$ as the solution to (4.1.6) and then obtain $\hat{u}$ from (4.1.5). Since (4.1.6) cannot be solved analytically for $\hat{b}$, some numerical method must be employed. The solution of (4.1.6) with an iterative procedure such as Newton's method (see Appendix F) poses no problems.

The m.l.e.'s of the Weibull parameters β and α are $\hat{\alpha} = \exp \hat{u}$ and $\hat{\beta} = \hat{b}^{-1}$. If desired, the maximum likelihood equations (4.1.5) and (4.1.6) can be written in Weibull form and solved directly for $\hat{\alpha}$ and $\hat{\beta}$ from the start, though there is no particular convenience in doing this. The equations are

$$\hat{\alpha} = \left(\frac{1}{r} \sum_{i=1}^{r}{}^{*} t_i^{\hat{\beta}}\right)^{1/\hat{\beta}} \tag{4.1.7}$$

$$\sum_{i=1}^{r}{}^{*} t_i^{\hat{\beta}} \log t_i \Big/ \sum_{i=1}^{r}{}^{*} t_i^{\hat{\beta}} - \frac{1}{\hat{\beta}} - \frac{1}{r} \sum_{i=1}^{r} \log t_i = 0. \tag{4.1.8}$$

Linear Estimates

Maximum likelihood estimation of Weibull or extreme value distribution parameters is straightforward, though a computer will usually be needed to solve (4.1.6) or (4.1.8). Sometimes it is convenient to have estimators that do not require a computer for their calculation. In this respect linear estimators of u and b are useful: these are estimators of the form

$$\tilde{u} = \sum_{i=1}^{r} a_i(n,r)x_i$$

$$\tilde{b} = \sum_{i=1}^{r} c_i(n,r)x_i \tag{4.1.9}$$

where the a_i's and c_i's are constant coefficients that do, however, depend on r and n. Given the a_i's and c_i's, linear estimates are obviously easily calculated. However, for most of the better linear estimators there are no simple formulas giving the a_i's and c_i's in terms of r and n, and thus tables of coefficient values must be constructed. Best linear unbiased estimators (b.l.u.e.'s), that is, estimators with minimum variance in the class of linear unbiased estimators, can be obtained from general results on the linear estimation of location and scale parameters (Lloyd, 1952; Kendall and Stuart, 1967, pp. 87–91; see also Problem 4.1) Lieblein and Zelen (1956) and others have derived such estimators for the extreme value distribution. Mann (1967a), on the other hand, considers best linear invariant estimators (b.l.i.e.'s) of u and b. These estimators possess minimum mean-square error in the class of linear estimators for which (mean-square error)$/b^2$ is invariant under location and scale transformations of the x_i's. For both types of estimators determination of the a_i's and c_i's in (4.1.9) requires the evaluation of means, variances, and covariances of order statistics from the standard extreme value distribution (see Problem 4.1). Mann et al. (1974, Ch. 5) give a good discussion of linear estimators for the extreme value distribution, and Sarhan and Greenberg (1962) and David (1970), for example, discuss general aspects of linear estimation.

For small- to moderate-size samples the b.l.i.e.'s are probably the most convenient. Tables of $a_i(n,r)$ and $c_i(n,r)$ for these estimators are readily accessible: Mann (1967a) and Mann et al. (1974, pp. 194–207) give tables for samples with $2 \leq r \leq n \leq 13$, and Mann (1967b) gives tables covering $2 \leq r \leq n \leq 25$. The b.l.i.e.'s are comparable to the m.l.e.'s as point estimators of u and b. Both are asymptotically fully efficient and they have similar properties in small- to moderate-size samples. The b.l.i.e.'s are more easily calculated, though they can at present be used only for samples with $n \leq 25$, since necessary tables do not exist beyond this range. At any rate, ease of calculation is not an overriding factor. In fact, because maximum likelihood estimation also applies to Type I censored data, whereas linear estimation in general does not, anyone who routinely uses the Weibull model in situations where data are sometimes Type I censored will require a computer program for maximum likelihood estimation. Whether the m.l.e. or a linear estimator is used with Type II censored data is largely a matter of preference.

Many other linear estimators are suggested in the literature. D'Agostino (1971) and others discuss approximations to b.l.u.e.'s and b.l.i.e.'s, and many authors have provided linear estimators that require fewer tables than b.l.u.e.'s or b.l.i.e.'s. Mann et al. (1974, Ch. 5), Mann and Fertig (1977) and Engelhardt and Bain (1977a) survey the best of these. Two such estimators are discussed in Section 4.1.2d, where they are used to obtain confidence intervals for u and b. Finally, Thomas and Wilson (1972) discuss linear estimation with progressively Type II censored samples.

We may sometimes want quick or very easily calculated estimates of u and b. For example, in order to calculate the m.l.e.'s $\hat{u}$ and $\hat{b}$ an initial guess at the value of $\hat{b}$ is required. One can use a simple linear estimate, such as that discussed in Section 4.1.2d, or a least squares estimate for this. Another possibility is to employ graphical estimates obtained from probability plots. A general description of these has been given in Section 2.4.2, but let us briefly review the process for the extreme value distribution. If a probability plot is constructed as in Section 2.4.2, estimates of u and b are obtained by fitting a straight line through the points in the plot. This can be done with some form of least squares (see Section 2.5; White, 1969), but it will usually be satisfactory to fit the line by eye. Since the extreme value survivor function satisfies $\log[-\log S(x)]=(x-u)/b$, the fitted line will have equation

$$
\begin{aligned}
Y &= \log[-\log S(x)] \\
&= (x-u_1)/b_1.
\end{aligned}
$$

The slope b_1^{-1} of this line estimates b^{-1} and the x-intercept u_1 estimates u. When extreme value probability paper with an $S(x)$ scale on it is used, the x-intercept is given by the intersection of the fitted line and the line $Y=0$, which corresponds to $S(x)=e^{-1}=0.368$.

Graphical and other methods of estimation are illustrated in the following example.

Example 4.1.1 (Example 2.4.2 revisited) Mann and Fertig (1973) give failure times of airplane components for a life test in which 13 components were placed on test, with the test terminating at the time of the tenth failure. Failure times in hours were 0.22, 0.50, 0.88, 1.00, 1.32, 1.33, 1.54, 1.76, 2.50, and 3.00.

Let us use the various estimation procedures described above, assuming these data arose from a Weibull distribution. To calculate linear estimates we transform the data to extreme value form: the logs of the 10 observations are -1.541, -0.693, -0.128, 0, 0.278, 0.285, 0.432, 0.565, 0.916, and 1.099. Using Table 5.3 in Mann et al. (1974) to obtain the necessary weights, we can calculate the b.l.i.e.'s, which are

$$\tilde{u} = -0.002927(-1.541) + \cdots + 0.615348(1.099) = 0.873$$

$$\tilde{b} = -0.083170(-1.541) + \cdots + 0.528441(1.099) = 0.715.$$

Graphical estimation for this example has been described previously in Example 2.4.2, where the estimates $\tilde{u} = 0.77$ and $\tilde{b} = 0.69$ were obtained from a line drawn by eye through a probability plot.

To determine the m.l.e.'s we first solve (4.1.16) iteratively for $\hat{b}$. With the graphical estimate 0.69 as an initial guess, Newton's method, for example, gives the m.l.e. in a couple of iterations as $\hat{b} = 0.706$. Then (4.1.5) gives $\hat{u} = 0.821$.

Estimates of other characteristics of the lifetime distribution are obtained directly from the parameter estimates. For example, if $\tilde{u}$ and $\tilde{b}$ estimate u and b, then the pth quantile $x_p = u + b\log[-\log(1-p)]$ is estimated by $\tilde{x}_p = \tilde{u} + \tilde{b}\log[-\log(1-p)]$. If m.l.e.'s of u and b are used in this way, the resulting estimators are of course m.l.e.'s of the quantities they are estimating. B.l.i.e.'s and b.l.u.e.'s of u and b yield b.l.i.e.'s and b.l.u.e.'s, respectively, of x_p, but if used to obtain estimates, for example, of $S(x_0) = \exp(-e^{(x_0-u)/b})$, the resulting estimators are obviously no longer linear nor do they possess certain other properties of the original estimators.

4.1.2 Confidence Intervals and Tests

General methods of obtaining confidence intervals for location and scale parameters are well known, though for distributions other than the normal distribution, mathematical and computational intractability is usually a problem. These problems are discussed in Appendix G, but for ease of exposition they are reviewed in Section 4.1.2a before being applied to the extreme value distribution. Tests of significance are not discussed explicitly but are easily obtained from the pivotal quantities used to construct confidence intervals. Several methods of getting confidence intervals for u,

b, and x_p are discussed in Sections 4.1.2b, c, and d. Estimation of the survivor function is discussed in Section 4.1.2e.

4.1.2a *Confidence Intervals for Location and Scale Parameters*

Consider a family of distributions with location parameter u ($-\infty<u<\infty$) and scale parameter b ($b>0$), with p.d.f. of the form

$$f(x; u, b)=\frac{1}{b}g\left(\frac{x-u}{b}\right) \qquad -\infty<x<\infty \tag{4.1.10}$$

and survivor function $G[(x-u)/b]$, where

$$G(y)=\int_y^\infty g(z)\,dz.$$

Suppose that $x_1\leqslant \cdots \leqslant x_r$ is a Type II censored sample consisting of the r smallest observations in a total sample of size n ($r\leqslant n$) from (4.1.10). Let $\tilde{u}=\tilde{u}(x_1,\ldots,x_r)$ and $\tilde{b}=\tilde{b}(x_1,\ldots,x_r)$ be estimators of u and b possessing the following invariance properties:

$$\tilde{u}(dx_1+c,\ldots,dx_r+c)=d\tilde{u}(x_1,\ldots,x_r)+c \tag{4.1.11}$$

$$\tilde{b}(dx_1+c,\ldots,dx_r+c)=d\tilde{b}(x_1,\ldots,x_r) \tag{4.1.12}$$

for any real constants c ($-\infty<c<\infty$) and d ($d>0$). Such estimators are termed "equivariant." The requirements (4.1.11) and (4.1.12) are simply that location and scale changes on the data should induce the same location and scale change in $\tilde{u}$ and the same scale change in $\tilde{b}$. These are natural requirements for estimators of location and scale parameters, and all of the point estimators discussed in Section 4.1.1 can be shown to satisfy (4.1.11) and (4.1.12) (see Problem 4.2).

The following theorem is proved in Appendix G (see Theorem G2):

THEOREM 4.1.1 *If $\tilde{u}$ and $\tilde{b}$ are equivariant estimators of u and b, based on a Type II censored sample $x_1\leqslant\cdots\leqslant x_r$ from* (4.1.10), *then*

(i) $Z_1=(\tilde{u}-u)/\tilde{b}$, $Z_2=\tilde{b}/b$, *and* $Z_3=(\tilde{u}-u)/b$ *are pivotal (parameter-free) quantities.*

(ii) *The quantities $a_i=(x_i-\tilde{u})/\tilde{b}$ form a set of ancillary statistics (i.e., statistics whose distribution does not depend on u or b), of which only $r-2$ are functionally independent.*

The pivotals Z_1 and Z_2, based on a particular pair of equivariant estimators, can be used to construct confidence intervals for u and b. For example, if l_1 and l_2 are such that $\Pr(l_1 \leq Z_1 \leq l_2)=\gamma$, then $(\tilde{u}-l_2\tilde{b}, \tilde{u}-l_1\tilde{b})$ is a γ confidence interval for u. Similarly, if $\Pr(l_1 \leq Z_2 \leq l_2)=\gamma$, then $(\tilde{b}/l_2, \tilde{b}/l_1)$ is a γ confidence interval for b. Confidence intervals for quantiles of the distribution can also be obtained. The pth quantile of (4.1.10) is $x_p=u+w_p b$, where w_p satisfies $G(w_p)=1-p$. The pivotal quantity

$$Z_p = \frac{(\tilde{u}-u)-w_p b}{\tilde{b}}$$

$$= \frac{\tilde{u}-x_p}{\tilde{b}} \tag{4.1.13}$$

can be used to get confidence intervals for x_p, since $\Pr(l_1 \leq Z_p \leq l_2)=\gamma$ implies $\Pr(\tilde{u}-l_2\tilde{b} \leq x_p \leq \tilde{u}-l_1\tilde{b})$, giving $(\tilde{u}-l_2\tilde{b}, \tilde{u}-l_1\tilde{b})$ as a γ confidence interval for x_p. That Z_p is pivotal follows from the fact that $Z_p=Z_1-w_p Z_2^{-1}$.

Although confidence intervals for u, b, or x_p can in principle be obtained from the pivotals Z_1, Z_2, and Z_p, a practical difficulty is that their distributions can be very complicated. The following theorem is proved in Appendix G (see Theorem G3):

THEOREM 4.1.2 *Let $\tilde{u}$ and $\tilde{b}$ be equivariant estimators of u and b under the conditions of Theorem* 4.1.1. *Then the joint p.d.f. of $Z_1, Z_2, a_1, \ldots, a_{r-2}$ is of the form*

$$k(\mathbf{a}, r, n) z_2^{r-1}\left(\prod_{i=1}^{r} g(a_i z_2 + z_1 z_2)\right)\left[G(a_r z_2 + z_1 z_2)\right]^{n-r} \tag{4.1.14}$$

where $k(\mathbf{a}, r, n)$ is a function of $a_1, \ldots, a_{r-2}$, r, and n only. The conditional p.d.f. of (Z_1, Z_2) given $\mathbf{a}=(a_1, \ldots, a_r)$ is also of the form (4.1.14).

Expression (4.1.14) shows the form of the joint p.d.f. of $Z_1, Z_2, a_1, \ldots, a_{r-2}$ for an arbitrary model. Unfortunately, except for the case of uncensored samples from the normal distribution, it turns out that the manner in which the a_i's enter $k(\mathbf{a}, r, n)$ and the form of the rest of (4.1.14) make it impossible to integrate out $a_1, \ldots, a_{r-2}$, so one cannot obtain the distributions of Z_1 and Z_2 analytically. The consequences of this in the case of the extreme value distribution are discussed in Section 4.1.2c.

Another point needs to be mentioned. Except for uncensored data from the normal distribution, no pair of estimates $\tilde{u}$ and $\tilde{b}$ can be sufficient for u

and b in a model of the form (4.1.10). It has already been remarked in Section 4.1.1 that this is the case for the extreme value distribution. There is an $(r-2)$-dimensional ancillary statistic, however, and, strictly speaking, one should make inferences conditional on the observed value of $\mathbf{a}$ (see Appendix G). That is, confidence intervals should be obtained from the conditional distributions of Z_1, Z_2, or Z_p, given $\mathbf{a}$. To get confidence intervals for u, for example, we need to find l_1 and l_2 such that

$$\Pr(l_1 \leqslant Z_1 \leqslant l_2 | \mathbf{a}) = \gamma. \tag{4.1.15}$$

Though this might seem to be an additional complication in the inference problem, it turns out to be very much easier to calculate probabilities like (4.1.15) than unconditional probabilities such as $\Pr(l_1 \leqslant Z_1 \leqslant l_2)$.

The conditional method of obtaining confidence intervals is discussed in Section 4.1.2b. Other approaches are given in Sections 4.1.2c and d. Conditional methods in general location–scale parameter models are discussed further in Appendix G and references cited therein. Lawless (1978) reviews this area, with special reference to the extreme value distribution.

For the extreme value distribution the conditional method has the advantage of being able to obtain confidence intervals in any given situation, though a computer is required to do this. Simpler methods are available for certain problems, and these are presented in Sections 4.1.2c and d. Section 4.1.2f contains some remarks on the choice of method.

4.1.2b *Confidence Intervals Obtained by the Conditional Method*

Let $\tilde{u}$ and $\tilde{b}$ be equivariant estimators of u and b, based on a Type II censored sample. The extreme value distribution (4.0.2) has p.d.f. of the form (4.1.10), with $g(y) = \exp(y - e^y)$ and $G(y) = \exp(-e^y)$. It follows from Theorem 4.1.2 that the p.d.f. of $Z_1 = (\tilde{u} - u)/\tilde{b}$ and $Z_2 = \tilde{b}/b$, given $\mathbf{a}$, is of the form

$$k'(\mathbf{a}, r, n) z_2^{r-1} \exp\left(\sum_{i=1}^{r} (a_i z_2 + z_1 z_2) - \sum_{i=1}^{r}{}^{*} e^{a_i z_2 + z_1 z_2} \right) \tag{4.1.16}$$

where we once again employ the notation

$$\sum_{i=1}^{r}{}^{*} w_i = \sum_{i=1}^{r} w_i + (n-r) w_r$$

introduced in Section 4.1.1. From (4.1.16) we can derive the marginal distributions for each of Z_1, Z_p, and Z_2, conditional on $\mathbf{a}$. The results in the following theorem are given by Lawless (1972, 1975) (see also Lawless, 1978).

THEOREM 4.1.3 *Let* $Z_p=(\tilde{u}-x_p)/\tilde{b}=(\tilde{u}-u-w_p b)/\tilde{b}$ *and* $Z_2=\tilde{b}/b$, *where* $w_p=\log[-\log(1-p)]$, $\tilde{u}$ *and* $\tilde{b}$ *are equivariant estimators of u and b based on a Type II censored sample* $x_1\leqslant\cdots\leqslant x_r$ *from the extreme value distribution* (4.0.2), *and* $a_i=(x_i-\tilde{u})/\tilde{b}$, $i=1,\ldots,r$. *Then*

(i) *The conditional p.d.f. of* Z_2, *given* **a**, *is of the form*

$$h_2(z|\mathbf{a})=\frac{k(\mathbf{a},r,n)z^{r-2}\exp\left((z-1)\sum_{i=1}^{r}a_i\right)}{\left(\frac{1}{r}\sum_{i=1}^{r}{}^{*}e^{a_i z}\right)^{r}}\qquad z\geqslant 0\qquad(4.1.17)$$

(ii) *The conditional distribution function* (*d.f.*) *of* Z_p, *given* **a**, *is*

$$\Pr(Z_p\leqslant t|\mathbf{a})=\int_0^{\infty}h_2(z|\mathbf{a})I\left(r,e^{w_p+tz}\sum_{i=1}^{r}{}^{*}e^{a_i z}\right)dz\qquad(4.1.18)$$

where $I(r,s)$ *is the incomplete gamma function* (*B*12). *The d.f. of* Z_1, *given* **a**, *is given by* (4.1.18), *with* $w_p=0$.

Proof

(i) To obtain (4.1.17) we integrate z_1 out of (4.1.16). This takes the form

$$h_2(z_2|\mathbf{a})=k'(\mathbf{a},r,n)z_2^{r-1}\int_{-\infty}^{\infty}\exp\left(z_2\sum_{i=1}^{r}a_i+rz_1z_2-e^{z_1z_2}\sum_{i=1}^{r}{}^{*}e^{a_iz_2}\right)dz_1.$$

Letting

$$y=\left(\sum_{i=1}^{r}{}^{*}e^{a_iz_2}\right)e^{z_1z_2}$$

one finds

$$h_2(z_2|\mathbf{a})=\left[k'(\mathbf{a},r,n)\exp\left(\sum_{i=1}^{r}a_iz_2\right)z_2^{r-2}\Big/\left(\sum_{i=1}^{r}{}^{*}e^{a_iz_2}\right)^{r}\right]\Gamma(r).$$

This is essentially (4.1.17), where for later numerical convenience several terms are grouped together to form the new constant

$$k(\mathbf{a}, r, n) = \frac{k'(\mathbf{a}, r, n)\Gamma(r)\exp\left(\sum a_i\right)}{r^r}.$$

(ii) It is not possible to integrate z_2 out of (4.1.16) analytically, so we work with the d.f. of Z_p, given **a**. The joint p.d.f. of Z_2 and $Z_p = Z_1 - w_p Z_2^{-1}$, given **a**, is easily found from (4.1.16) to be

$$h(z_p, z_2|\mathbf{a}) = k'(\mathbf{a}, r, n) z_2^{r-1} \exp\left(\sum_{i=1}^{r} (a_i z_2 + z_2 z_p + w_p) - \sum_{i=1}^{r}{}^{*} \exp(a_i z_2 + z_2 z_p + w_p) \right)$$

where $z_2 > 0$, $-\infty < z_p < \infty$. The d.f. of Z_p, given **a**, is

$$\Pr(Z_p \leqslant t|\mathbf{a}) = \int_0^\infty \int_{-\infty}^{t} h(z_p, z_2|\mathbf{a})\, dz_p\, dz_2.$$

Making the change of variables

$$y = e^{z_p z_2} \sum_{i=1}^{r}{}^{*} e^{a_i z_2 + w_p} \qquad z_2 = z_2$$

we get

$$\Pr(Z_p \leqslant t|\mathbf{a}) = \int_0^\infty \left[\frac{k'(\mathbf{a}, r, n) z_2^{r-2} \exp\left(\sum_{i=1}^{r} a_i z_2 + r w_p\right)}{\left(\sum_{i=1}^{*r} e^{a_i z_2 + w_p}\right)^r} \times \int_0^{t^*} y^{r-1} e^{-y}\, dy \right] dz_2$$

where

$$t^* = \left(\sum{}^{*} e^{a_i z_2} \right) \exp(t z_2 + w_p).$$

After a little rearrangement this gives (4.1.18). ■

To construct confidence intervals for u, b, or x_p one needs to obtain percentage points for Z_1, Z_2, or Z_p. These are readily found from the results of Theorem 4.1.3, though numerical integration is necessary to integrate (4.1.17) and to evaluate (4.1.18). This is described in Example 4.1.2. It can be shown (see Appendix G) that different equivariant estimators yield the same confidence intervals for a given sample and therefore it is immaterial what estimators are used to form the pivotals and **a**. Since the computer is required for numerical integration, one might as well obtain and use the m.l.e.'s, though any equivariant estimators will suffice. Note that although $h(z_2|\mathbf{a})$ involves an unknown constant $k(\mathbf{a}, r, n)$, this can be evaluated by using the fact that $h_2(z_2|\mathbf{a})$ must integrate to unity. The mechanics of the method will be made clear in the following example.

Example 4.1.2 These data, with $r=28$ and $n=40$, are given by Lawless (1975, p. 258). The observations x_i in extreme value form are -2.982, -2.849, -2.546, -2.350, -1.983, -1.492, -1.443, -1.394, -1.386, -1.269, -1.195, -1.174, -0.845, -0.620, -0.576, -0.548, -0.247, -0.195, -0.056, -0.013, 0.006, 0.033, 0.037, 0.046, 0.084, 0.221, 0.245, and 0.296.

We shall work with pivotal quantities and ancillaries based on the m.l.e.'s. The m.l.e.'s turn out to be $\hat{u}=0.1563$ and $\hat{b}=0.9104$, and then the ancillaries are defined as $a_i=(x_i-0.1563)/0.9104$, $i=1,\ldots,28$. Before calculating some confidence intervals let us consider the type of calculations required. Once the a_i's are obtained all but the constant $k(\mathbf{a}, r, n)$ in (4.1.17) is known. This can be evaluated from the fact that

$$\int_0^{\infty} h_2(z|\mathbf{a})\,dz=1$$

which implies that

$$k(\mathbf{a}, r, n)=\left[\int_0^{\infty} z^{r-2}\exp\left((z-1)\sum_{i=1}^{r} a_i\right)\Big/\left(\frac{1}{r}\sum_{i=1}^{r}{}^{*} e^{a_i z}\right)^{r} dz\right]^{-1}. \tag{4.1.19}$$

The integrand in (4.1.19) is well behaved and the integral is easily numerically evaluated. A simple procedure, especially if one is using an interactive computing facility, is to examine the integrand and determine a range $d_1 \leqslant z \leqslant d_2$ so that all but a negligible amount of the area lies between d_1 and d_2. Then a simple numerical procedure such as Simpson's rule can be used to calculate the integral. When the m.l.e.'s are employed, as here, it will hardly ever be necessary to consider the integrand outside of the range 0 to 10.

Having obtained $k(\mathbf{a}, r, n)$, we can easily determine percentage points for Z_2, using

$$\Pr(Z_2 \leq l|\mathbf{a}) = \int_0^l h_2(z|\mathbf{a})\, dz. \tag{4.1.20}$$

Exact percentage points $l = z_{2,\gamma}$ making (4.1.20) equal to γ can be obtained iteratively.

Getting percentage points for Z_p involves similar computations with (4.1.18). This integral behaves in much the same way as integrals (4.1.20). Since $I(r,s) \leq 1$, the integrand in (4.1.18) is in fact always less than or equal to the integrand in (4.1.20). The calculation of the incomplete gamma function $I(r,s)$ is discussed in Appendix B. Finally, note that if a significance test for some specified value of u or b is wanted, one needs to calculate only a single probability for Z_2 or Z_p.

Let us now return to our numerical example. It is quickly seen that in this case the area under the integrand in (4.1.19) outside of the range (0,3) is negligible. By numerical integration, the total area under the integrand is found to be 0.4355, and hence $k(\mathbf{a}, r, n) = 2.2961$. We now have the complete p.d.f. $h_2(z|\mathbf{a})$, and can calculate any desired probabilities for Z_2, Z_1, or Z_p. Suppose, for example, we want a two-sided .90 confidence interval for b. Integrating $h_2(z|\mathbf{a})$ numerically, we determine that $P(Z_2 \leq 0.713|\mathbf{a}) = .05$ and $\Pr(Z_2 \leq 1.257|\mathbf{a}) = .95$. Thus

$$\Pr\left(0.713 \leq \hat{b}/b \leq 1.257|\mathbf{a}\right) = .90$$

and this yields $0.724 \leq b \leq 1.277$ as a .90 confidence interval for b from the observed value $\hat{b} = 0.9104$.

Suppose we also want a lower .95 confidence limit for $x_{.10}$. From (4.1.18) we find that

$$\Pr(Z_{.10} \leq 3.153|\mathbf{a}) = .95$$

where $Z_{.10} = (\hat{u} - x_{.10})/\hat{b}$. Thus $x_{.10} \geq \hat{u} - 3.153\hat{b}$ is the desired confidence interval, which gives $x_{.10} \geq -2.714$ as the realized interval.

An advantage of the conditional method is that it can be used to obtain confidence intervals for u, b, or x_p, whatever the size of the Type II censored sample. It can also be used to obtain confidence intervals for other quantities such as the survivor function (see Section 4.1.2e) and can be used with Type II progressively censored data (see Problem 4.4). The approach is straightforward, but since it requires numerical integration, it must be done on the computer. Calculations like those required here are very conveniently carried out on an interactive computing facility. Lawless (1978) has some additional advice on this and references a FORTRAN program that calculates confidence intervals with very little input from the user.

In some situations methods that require less calculation than the conditional approach are available. These are discussed in the next two sections.

4.1.2c *Other Methods of Obtaining Exact Confidence Intervals*

Instead of basing confidence intervals on the conditional distributions of pivotals, given **a**, one can base them on unconditional distributions of the pivotals. Whereas with the conditional method it is immaterial what equivariant estimators are used to form pivotals, this is not the case for the unconditional approach, however. It is preferable to use estimators with good properties, that is, those with sampling distributions concentrated as closely as possible around the parameters.

Several methods have been suggested for estimating the extreme value parameters u and b. Mann (1968a) and Mann and Fertig (1973) consider pivotals based on the b.l.i.e.'s of u and b, and Thoman et al. (1969) and McCool (1970) propose pivotals based on the m.l.e.'s. A difficulty is that the distributions of Z_1, Z_2, and Z_p are in both cases mathematically intractable, and it is impossible to obtain by analytical means exact percentage points for them. It is possible, however, to produce very close estimates of percentage points by Monte Carlo methods, and tables of (approximate) percentage points have been constructed in this way. The basic idea is as follows: since Z_1, Z_2, and Z_p are pivotals, their distributions (for given values of r and n) are the same whatever the values of u and b in the underlying extreme value distribution. When $u=0$ and $b=1$, the pivotals become

$$Z_1=\frac{\tilde{u}}{\tilde{b}}, \qquad Z_2=\tilde{b}, \qquad Z_p=\frac{\tilde{u}-\log[-\log(1-p)]}{\tilde{b}}.$$

The distribution of Z_1, for example, can thus be estimated by generating samples from the standard extreme value distribution ($u=0, b=1$) on the computer, computing $\tilde{u}$ and $\tilde{b}$ and obtaining values of $Z_1=\tilde{u}/\tilde{b}$. By generating many samples (as many as 40,000 were used in some of the situations mentioned here) we can obtain a very good estimate of the distribution of Z_1. The distributions of Z_2 and Z_p can be similarly estimated.

The following tables of percentage points have been produced by Monte Carlo methods and are available for pivotals based on b.l.i.e.'s or m.l.e.'s.

1. Pivotals based on b.l.i.e.'s: Mann et al. (1971) give tables of percentage points for Z_1, Z_2, and Z_p ($p=.01,.05,.10$), for samples with $3 \leqslant r \leqslant n \leqslant 25$. These tables are also given in Mann and Fertig (1973) and Mann et al. (1974, Ch. 5), but only for $3 \leqslant r \leqslant n \leqslant 13$, and $p=.05$ and $.10$ for Z_p.
2. Pivotals based on m.l.e.'s: Thoman et al. (1969) give tables that yield percentage points for Z_1 and Z_2 for complete samples up to size $n=120$. Thoman et al. (1970) provide tables that give confidence intervals on the survivor function for complete samples up to size 100. For censored samples Billmann et al. (1972) present similar tables; these are quite sparse, covering only samples with $n=40$, 60, 80, 100, or 120 and $r=0.75n$ or $0.50n$, but additional values can be obtained fairly accurately by interpolation. Finally, McCool (1970, 1974) gives tables covering 17 combinations of r and n for Z_1, Z_2, and Z_p ($p=.01,.10,.50$). The (r, n) combinations covered are $n=5$ $(r=3,5)$, $n=10$ $(r=3,5,10)$, $n=15$ $(r=5,10,15)$, $n=20$ $(r=5,10,15,20)$, and $n=30$ $(r=5,10,15,20,30)$.

Tables like these are very expensive to produce, since each combination of values of r and n must be treated separately. Consequently, there are no tables covering all of the sample sizes likely to arise in practice. In addition, the tables take up a fair amount of space and are not collected in one spot. Nevertheless, they are very convenient, and anyone who uses the Weibull distribution a lot would do well to acquire copies of them.

A pair of numerical examples will illustrate the use of the tables.

Example 4.1.3 (Example 4.1.1 revisited) The data given in Example 4.4.1 on failure times of aircraft components involved a Type II censored sample with $n=13$ and $r=10$, in which log lifetimes were assumed to have extreme value distributions. Suppose that we want two-sided .90 confidence intervals for u and b and a .95 lower confidence limit for $x_{.10}$.

Of the tables listed, only those given by Mann and Fertig cover samples with $n=13$ and $r=10$. These are based on the b.l.i.e.'s $\tilde{u}$ and $\tilde{b}$ of u and b. From Tables 5.7, 5.8, and 5.10 in Mann et al. (1974) we find that (1) $\Pr(0.52 \leqslant Z_2 \leqslant 1.40)=.90$, (2) $\Pr(-0.72 \leqslant Z_1 \leqslant 0.58)=.90$, (3) $\Pr(z_{.10} \leqslant 4.37) =.95$, where $Z_1=(\tilde{u}-u)/\tilde{b}$, $Z_2=\tilde{b}/b$, and $Z_{.10}=(\tilde{u}-x_{.10})/\tilde{b}$. These give the confidence intervals (1) $\tilde{b}/1.40 \leqslant b \leqslant \tilde{b}/0.52$, (2) $\tilde{u}-0.58\tilde{b} \leqslant u \leqslant \tilde{u}+0.72\tilde{b}$, (3) $\tilde{u}-4.37\tilde{b} \leqslant x_{.10}$. With the observed values $\tilde{u}=0.873$ and $\tilde{b}=0.715$, these become $0.511 \leqslant b \leqslant 1.375$, $0.458 \leqslant u \leqslant 1.388$ and $-2.252 \leqslant x_{.10}$, respectively.

These confidence intervals can be transformed to confidence intervals for the corresponding Weibull parameters or quantiles. For example, the .10 quantile of the Weibull distribution or lifetimes is $t_{.10} = \exp x_{.10}$, so that the lower .95 confidence limit on $t_{.10}$ is $\exp(-2.252) = 0.105$.

Example 4.1.4 As a second example, consider the following Type II censored sample from the extreme value distribution, with $n=20$ and $r=10$: -3.57, -2.55, -2.02, -1.66, -1.36, -1.15, -0.95, -0.77, -0.61, and -0.45. In this case, to obtain confidence intervals we could use either the tables based on m.l.e.'s, given by McCool (1974), or those based on the b.l.i.e.'s, given by Mann et al. (1971). Let us calculate estimates based on each of these, and for further comparison let us also get confidence intervals by the conditional method of Section 4.1.2b. From the given data the m.l.e.'s are found to be $\hat{u} = -0.112$ and $\hat{b} = 0.907$. The b.l.i.e.'s, on the other hand, are $\tilde{u} = -0.048$ and $\tilde{b} = 0.915$. Using tables of percentage points for Z_1, Z_2, and $Z_{.10}$ given in Mann et al. (1971) and McCool (1974) for the case $n=20$ and $r=10$, we obtain the two-sided .90 confidence intervals on b, u, and $x_{.10}$ given in Table 4.1.1. Similar confidence intervals obtained by the conditional method are also shown. Confidence limits are given to only two decimal places for comparison purposes, since the degree of accuracy in the tables does not support a more precise comparison. Clearly there is no practical difference in the results of the three methods in this example.

In most situations there will be very little difference between results obtained from the unconditional distributions of pivotals based on the b.l.i.e.'s and those obtained from the unconditional distributions of pivotals based on the m.l.e.'s, and also between these and results obtained by the conditional method. Differences can occur in small or heavily censored samples, but for almost all practical purposes the methods are roughly equivalent. Section 4.1.2f contains a few comments on the pros and cons of the methods.

4.1.2d *Approximations to Distributions of Pivotals*

Because it is unfeasible to obtain percentage points of Z_1, Z_2, and Z_p by Monte Carlo methods for all cases of practical importance, work has gone

Table 4.1.1 Two-Sided .90 Confidence Intervals Obtained by Three Methods

B.l.i.e. Pivotals	M.l.e. Pivotals	Conditional Method
$0.64 \leqslant b \leqslant 1.82$	$0.64 \leqslant b \leqslant 1.81$	$0.64 \leqslant b \leqslant 1.82$
$-0.51 \leqslant u \leqslant 0.89$	$-0.51 \leqslant u \leqslant 0.90$	$-0.51 \leqslant u \leqslant 0.89$
$-3.76 \leqslant x_{.10} \leqslant -1.51$	$-3.74 \leqslant x_{.10} \leqslant -1.49$	$-3.76 \leqslant x_{.10} \leqslant -1.51$

into finding approximations to these distributions. Some simple approximations with applications here and beyond the present topics of discussion are now considered.

A χ^2 Approximation for the Distribution of $\hat{b}/b$

A very simple approximation has been found for the distribution of $Z_2=\hat{b}/b$, where $\hat{b}$ is the m.l.e. of b. The approximation involves the χ^2 distribution and is adequate for virtually all practical situations. In addition to providing a simple method of obtaining confidence intervals for b, the approximation is useful for comparing two or more Weibull or extreme value distributions, as discussed in Section 4.3.1.

The approximation, which was developed empirically, is of the form

$$g\left(\frac{\hat{b}}{b}\right)\sim\chi^2_{(h)} \tag{4.1.21}$$

where $\hat{b}$ is the m.l.e. of b from a Type II censored sample containing the first r out of n observations in a random sample and $g=g(r,n)$ and $h=h(r,n)$ are constants. McCool (1975b), Lawless and Mann (1976), and others have suggested approximating the distribution of $\hat{b}/b$ in this way. The usual procedure is to choose g and h so that the mean and variance of the two sides of (4.1.21) agree. This gives the two equations

$$gE\left(\frac{\hat{b}}{b}\right)=h$$

$$g^2\mathrm{Var}\left(\frac{\hat{b}}{b}\right)=2h$$

from which we obtain $g=2/\mathrm{Var}(\hat{b}/b)$ and $h=gE(\hat{b}/b)$. A difficulty is that $E(\hat{b}/b)$ and $\mathrm{Var}(\hat{b}/b)$ are not known exactly for all (r,n). However, Harter and Moore (1968) have estimated these by simulation for censored samples with $n=10$ or 20, and McCool (1975b) and Lawless and Mann (1976) discuss other methods of estimating $E(\hat{b}/b)$ and $\mathrm{Var}(\hat{b}/b)$.

Table 4.1.2 was prepared by combining results given by Harter and Moore (1968), McCool (1975b), and Lawless and Mann (1976). It gives values of $h(r,n)$ for various (r,n) combinations; $g(r,n)$ is given by $g(r,n)=h(r,n)+2$. Comparison of percentage points of $\hat{b}/b$ obtained by using (4.1.21) with these values of g and h, with the essentially exact percentage points given by McCool (1974, 1975b) and Billmann et al. (1972), shows that the approximation is adequate for virtually all situations. In general, the approximation improves as n or r/n increases and is exact in the limit as $n\to\infty$ for fixed r/n.

Table 4.1.2 Values $h(r, n)$ for Use in Approximation (4.1.21)[a]

		n							
		5	10	20	40	60	80	100	∞
	.1	—	—	2.0	6.0	10.0	14.1	18.1	$0.205n$
	.2	—	2.0	6.2	14.6	23.0	31.5	39.9	$0.420n$
	.3	—	4.3	10.9	24.0	37.0	50.1	63.2	$0.652n$
	.4	2.2	6.7	15.8	33.8	51.8	69.9	87.9	$0.899n$
r/n	.5	3.5	9.1	20.7	44.0	67.3	90.6	113.9	$1.165n$
	.6	4.7	11.4	25.8	54.7	83.5	112.3	141.1	$1.457n$
	.7	6.0	14.8	32.6	68.1	103.8	139.5	175.0	$1.782n$
	.8	7.8	18.5	40.0	83.3	126.4	169.5	212.5	$2.155n$
	.9	10.3	23.0	49.0	100.9	153.0	204.9	256.9	$2.607n$
	1.0	12.9	29.3	62.4	128.2	194.8	257.6	325.5	$3.290n$

[a] $g(r, n) = h(r, n) + 2$.

For values of n and r/n not included in Table 4.1.2, values of h can be obtained by linear interpolation. For samples with r bigger than 10 or so the approximation is very accurate, slightly more so in the lower tail of the distribution than in the upper tail. An example of the accuracy of the approximation is given below. Note that in (4.1.21) the χ^2 degrees of freedom parameter h will be nonintegral. For small values of h probabilities and percentage points for $\chi^2_{(h)}$ can be either calculated directly by employing the relationship between the χ^2 cumulative distribution function and the incomplete gamma function (see Appendix B), or determined by interpolating in χ^2 tables for integral degrees of freedom (e.g., Pearson and Hartley, 1966; Mardia and Zemroch, 1978). For h larger than about 10, very accurate approximations to χ^2 probabilities and quantiles can be obtained from the Wilson–Hilferty transformations [see (B14) and (B15) of Appendix B]. For probabilities, the approximation is

$$\left(\frac{9h}{2}\right)^{1/2}\left[\left(\frac{\chi^2_{(h)}}{h}\right)^{1/3}+\frac{2}{9h}-1\right]\sim N(0,1) \tag{4.1.22}$$

and for quantiles of $\chi^2_{(h)}$

$$\chi^2_{(h),p} \doteq h\left[1-\frac{2}{9h}+N_p\left(\frac{2}{9h}\right)^{1/2}\right]^3 \tag{4.1.23}$$

where N_p is the pth quantile of the standard normal distribution. These approximations allow easy and accurate calculations of probabilities and quantiles for most of the situations covered in Table 4.1.2.

Example 4.1.5 As an illustration of the χ^2 approximation, consider a Type II censored sample with $n=20$ and $r=10$. With $n=20$ and $r/n=.5$, one obtains the values $h=20.7$ and $g=22.7$ from Table 4.1.2. The χ^2 approximation (4.1.21) is then

$$22.7\frac{\hat{b}}{b} \sim \chi^2_{(20.7)}.$$

Table 4.1.3 gives some percentage points of the distribution of $\hat{b}/b$ that were calculated using this approximation, along with essentially exact percentage points for $\hat{b}/b$ available, in this case, from the tables of McCool (1975b). Except for a fairly small deviation in the 99% points, the approximation is extremely accurate.

Approximations for Pivotals Based on Linear Estimators

Several approximations have been developed for the distributions of pivotals based on linear estimators, two of which will be discussed here. These are very useful, especially in conjunction with two simple linear estimators of u and b, which are introduced below. The approximations apply to unbiased linear estimators of u and b; this is not restrictive, since other estimators such as the b.l.i.e.'s can be expressed in terms of unbiased estimators.

Let $\tilde{u}$ and $\tilde{b}$ be unbiased linear estimators of u and b, based on a Type II censored sample involving the r smallest observations in a random sample of n. Then $E(\tilde{u})=u$ and $E(\tilde{b})=b$; in addition, let

$$A(r,n)=\frac{\operatorname{Var}(\tilde{u})}{b^2} \qquad B(r,n)=\frac{\operatorname{Cov}(\tilde{u},\tilde{b})}{b^2} \qquad C(r,n)=\frac{\operatorname{Var}(\tilde{b})}{b^2}.$$

The values of $A(r,n)$, $B(r,n)$, and $C(r,n)$ for any particular estimator can, in principle, be readily obtained, since these will be functions of the means,

Table 4.1.3 Exact and Approximate Percentage Points of $\hat{b}/b$

	Percent						
	1	5	10	50	90	95	99
Exact	0.38	0.50	0.58	0.88	1.29	1.42	1.67
Approximate (4.1.21)	0.38	0.50	0.57	0.88	1.29	1.42	1.70

variances, and covariances of the order statistics from a standard extreme value distribution. On a practical level there may be difficulty, however, because these quantities have not all been tabulated for samples with $n>25$.

The two approximations, which match moments to approximate the distributions of pivotals with χ^2 and F distributions, are as follows:

1. The distribution of $Z_2=\tilde{b}/b$ can often be well approximated by a χ^2 distribution in the same way the distribution of $\hat{b}/b$ was approximated in (4.1.21). Engelhardt (1975), Lawless and Mann (1976), and others use this approach. Here $E(\tilde{b}/b)=1$, $\text{Var}(\tilde{b}/b)=C(r,n)$, and the approximation is

$$g\left(\frac{\tilde{b}}{b}\right)\sim\chi^2_{(h)} \tag{4.1.24}$$

where $g=g(r,n)$ and $h=h(r,n)$ depend on r and n. Matching the first two moments of the two sides of (4.1.24), we get

$$g(r,n)=h(r,n)=\frac{2}{C(r,n)}.$$

2. Mann et al. (1974, p. 241) and Mann (1977) derive a F approximation to the distribution of Z_p. Let $x_p=u+b\log[-\log(1-p)]$ be the pth quantile of the extreme value distribution and let $x^*=\tilde{u}-B(r,n)\tilde{b}/C(r,n)$. The approximation is then

$$(x^*-x_p)\Big/\tilde{b}\left(-\frac{B(r,n)}{C(r,n)}-w_p\right)\sim F_{(d_1,d_2)} \tag{4.1.25}$$

where $w_p=\log[-\log(1-p)]$

$$d_1=2\left(-\frac{B(r,n)}{C(r,n)}-w_p\right)^2\Big/\left(A(r,n)-\frac{B(r,n)^2}{C(r,n)}\right)$$

$$d_2=\frac{2}{C(r,n)}.$$

This approximation, when used with the best linear unbiased estimators (b.l.u.e.'s), is satisfactory only for certain situations, though most situations of practical importance are covered. Details are given in Mann (1977), but, broadly speaking, the approximation is suitable for $p\leqslant.25$, provided that

$d_2 \geqslant 5.5$ and $2d_2 < d_1 < 18d_2 - 50$. Note that d_1 and d_2 in (4.1.25) will typically be nonintegral values. Determination of F probabilities under these circumstances is discussed in Appendix B, Section 4.3, and in Example 4.1.6.

These approximations can be used with the b.l.u.e.'s, though there is no need to do this, since the Mann-Fertig (1973) tables cover the distributions of pivotals based on the b.l.i.e.'s, which can be related to the b.l.u.e.'s, up to samples with $n=25$. For samples with $n>25$ not all of $A(r, n)$, $B(r, n)$, and $C(r, n)$ have been tabulated, so the approximations are unuseable. Because of this there has been a great deal of interest in finding simple linear estimators with good properties. References to this work are given by Mann and Fertig (1977) and Engelhardt and Bain (1977a). A pair of estimators with very good statistical properties is presented here. When used with the approximations (4.1.24) and (4.1.25), they provide simple methods of constructing tests and confidence intervals.

Simple Linear Estimators

The estimators and results considered here were developed by Bain (1972), Engelhardt and Bain (1973, 1974), Engelhardt (1975), and Mann and Fertig (1975a). The estimators take one of two forms, depending on whether the sample is Type II censored ($r<n$) or complete. They are as follows:

1. For censored samples ($r<n$)

$$\tilde{b} = \sum_{i=1}^{r} \frac{|x_i - x_r|}{nk(r, n)}$$

$$\tilde{u} = x_r - w(r, n)\tilde{b} \tag{4.1.26}$$

where $x_1 \leqslant \cdots \leqslant x_r$ is a Type II censored extreme value sample, $k(r, n)$ is a constant chosen so that $E(\tilde{b}) = b$, and $w(r, n)$ is the expected value of the rth smallest observation in a sample of n from the standard extreme value distribution. Table 4.1.4, taken from Engelhardt (1975), provides coefficients that can be used to calculate $k(r, n)$ and $w(r, n)$. The values obtained using Table 4.1.4 are actually approximate, but are only in error by one unit in the fourth decimal place at most.

Table 4.1.5 gives variances and covariance $A(r, n)$, $B(r,n)$, and $C(r, n)$ for (4.1.26). These are required for use with the approximations (4.1.24) and (4.1.25) and have been obtained from Engelhardt and Bain (1973, 1974) and Engelhardt (1975).

Table 4.1.4 Coefficients Used in Calculating Factors for Simple Linear Estimators[a, b]

	r/n								
	.1	.2	.3	.4	.5	.6	.7	.8	.9
k_0	0.10265	0.21129	0.32723	0.45234	0.58937	0.74274	0.92026	1.1382	1.14436
k_1	−1.0271	−1.0622	−1.1060	−1.1634	−1.2415	−1.3540	−1.8313	−1.8567	−2.6929
k_2	0.0	0.030	0.054	0.089	0.145	0.242	0.433	0.906	2.796
w_0	−2.2504	−1.4999	−1.0309	−0.67173	−0.36651	−0.08742	0.18563	0.47589	0.83403
w_1	−5.5743	−3.0740	−2.2859	−1.9301	−1.7619	−1.7114	−1.7727	−2.0110	−2.773
w_2	−7.848	−1.886	−0.767	−0.335	−0.091	0.111	0.369	0.891	2.825

[a]Adapted, with permission, from Engelhardt (1975).
[b]Coefficients are given by

$$k(r,n)=k_0+\frac{k_1}{n}+\frac{k_2}{n^2}$$

$$w(r,n)=w_0+\frac{w_1}{n}+\frac{w_2}{n^2}.$$

Linear interpolation can be used to find $k(r,n)$ and $w(r,n)$ for r/n values not represented in the table.

Table 4.1.5 Variances and Covariances of the Simple Linear Estimators ($r<n$)[a]

		r/n								
	n	.1	.2	.3	.4	.5	.6	.7	.8	.9
$n\,\mathrm{Var}(\tilde{u}/b)$	10		39.04	12.052	5.609	3.233	2.172	1.650	1.384	1.255
$=nA(r,n)$	20	140.7	23.96	9.136	4.666	2.850	2.000	1.570	1.350	1.248
	30	100.4	20.96	8.416	4.410	2.743	1.949	1.546	1.339	1.248
	40	87.06	19.68	8.088	4.292	2.692	1.925	1.534	1.335	1.249
	50	80.39	18.97	7.901	4.223	2.662	1.911	1.528	1.332	1.249
	60	76.40	18.52	7.781	4.179	2.643	1.902	1.524	1.331	1.249
	∞	60.53	16.50	7.219	3.967	2.550	1.859	1.503	1.323	1.251
$n\,\mathrm{Cov}(\tilde{u}/b, \tilde{b}/b)$	10		17.58	6.109	2.868	1.474	0.7502	0.3344	0.0826	−0.0694
$=nB(r,n)$	20	49.91	10.75	4.505	2.254	1.184	0.5975	0.2500	0.0373	−0.0856
	30	35.98	9.397	4.107	2.089	1.102	0.5533	0.2253	0.0245	−0.0883
	40	31.36	8.819	3.927	2.012	1.064	0.5323	0.2136	0.0185	−0.0891
	50	29.06	8.499	3.825	1.967	1.041	0.5200	0.2068	0.0150	−0.0894
	60	27.68	8.296	3.750	1.938	1.026	0.5120	0.2023	0.0127	−0.0895
	∞	22.19	7.383	3.450	1.801	1.008	0.4734	0.1807	0.0019	−0.0891
$n\,\mathrm{Var}(\tilde{b}/b)$	10		9.488	4.609	2.979	2.161	1.667	1.336	1.096	0.9197
$=nC(r,n)$	20	19.49	6.324	3.686	2.552	1.920	1.515	1.234	1.028	0.8784
	30	14.62	5.691	3.455	2.436	1.851	1.471	1.204	1.008	0.8683
	40	13.00	5.420	3.350	2.382	1.819	1.450	1.189	0.9981	0.8641
	50	12.18	5.269	3.290	2.350	1.800	1.437	1.181	0.9925	0.8619
	60	11.70	5.173	3.251	2.330	1.787	1.429	1.175	0.9888	0.8605
	∞	9.746	4.742	3.070	2.232	1.728	1.390	1.148	0.9610	0.8549

[a]Adapted, with permission, from Engelhardt and Bain (1973, 1974) and Engelhardt (1975).

2. For complete samples

$$\tilde{b}=\left(-\sum_{i=1}^{s} x_i+\frac{s}{n-s}\sum_{i=s+1}^{n} x_i\right)\Big/ nk_n$$

$$\tilde{u}=\bar{x}+0.5772\tilde{b} \qquad (4.1.27)$$

where $x_1 \leqslant \cdots \leqslant x_n$ are the ordered observations in an extreme value sample, $s=[0.84n]$ is the largest integer not exceeding $0.84n$, k_n is an unbiasing constant, and $\bar{x}$ is the mean of $x_1,\ldots,x_n$. Table 4.1.6, taken from Engelhardt and Bain (1977a), gives values of k_n, variances, and covariances for $\tilde{u}$ and $\tilde{b}$.

The estimators in (4.1.26) and (4.1.27) have been shown to possess good properties, and though they are slightly inferior to the b.l.i.e.'s or m.l.e.'s, their simplicity makes them very useful, especially for samples with $n \geqslant 25$. In addition, their distributions are well approximated by (4.1.24) and (4.1.25) in the same situations the distributions of the b.l.u.e.'s are, and they can be used in pivotals to obtain tests and confidence intervals for u, b, or x_p.

An example will illustrate the use of the simple linear estimators.

Example 4.1.6 (Example 4.1.2 revisited) Recall the Type II censored data of Example 4.1.2, for which $n=40$ and $r=28$. For these data $\Sigma|x_i-x_{28}|=32.483$ and $x_{28}=0.296$. Using Table 4.1.4, we find $k(28,40)=0.8822$ and $w(28,40)=0.1415$. Thus $\tilde{b}=32.483/0.8822(40)=0.921$ and $\tilde{u}=0.296-0.1415(0.921)=0.166$.

As an illustration of the χ^2 and F approximations (4.1.24) and (4.1.25), suppose first that we want a two-sided .90 confidence interval for b. From Table 4.1.5 we find that $40A(28,40)=40\,\mathrm{Var}(\tilde{u}/b)=1.534$, $40B(28,40)=0.2136$ and $40C(28,40)=1.1894$. The χ^2 approximation (4.1.24) uses degrees of freedom $h=2/C(28,40)=67.3$ and is $67.3\tilde{b}/b \sim \chi^2_{(67.3)}$. From (4.1.23) we can find that $\Pr(0.734 \leqslant \chi^2_{(67.3)}/67.3 \leqslant 1.30)=.90$, and this gives $0.708 \leqslant b \leqslant 1.255$ as the .90 confidence interval for b. This is close to the confidence interval obtained by the conditional method in Example 4.1.2, which was $0.724 \leqslant b \leqslant 1.277$.

To obtain a confidence interval for the quantile $x_{.10}$, say, we consider (4.1.25). With $w_p=\log[-\log(.90)]=-2.25$, we calculate $d_2=67.3$ and

$$d_1=\frac{80[(-0.2136/1.1894)+2.25]^2}{1.534-0.2186^2/1.1894}=229.3.$$

Table 4.1.6 Constants k_n, Variances, and Covariances for Simple Linear Estimators for Complete Samples[a]

n	k_n	$n\text{Var}(\tilde{b}/b)$	$n\text{Cov}(\tilde{u}/b, \tilde{b}/b)$	$n\text{Var}(\tilde{u}/b)$
2	0.6931	1.4237	0.1286	1.3191
3	0.9808	1.0428	−0.0621	1.2258
4	1.1507	0.9198	−0.1151	1.2056
5	1.2674	0.8623	−0.1357	1.2009
6	1.3545	0.8312	−0.1442	1.2015
7	1.1828	0.7958	−0.1872	1.1637
8	1.2547	0.7700	−0.1938	1.1647
9	1.3141	0.7526	−0.1970	1.1668
10	1.3644	0.7405	−0.1981	1.1695
11	1.4079	0.7321	−0.1980	1.1725
12	1.4461	0.7263	−0.1970	1.1755
13	1.3332	0.7202	−0.2148	1.1570
14	1.3686	0.7129	−0.2148	1.1594
15	1.4004	0.7072	−0.2144	1.1618
16	1.4293	0.7029	−0.2135	1.1643
17	1.4556	0.6996	−0.2123	1.1668
18	1.4799	0.6972	−0.2109	1.1692
19	1.3960	0.6954	−0.2224	1.1565
20	1.4192	0.6919	−0.2216	1.1585
21	1.4408	0.6891	−0.2207	1.1606
22	1.4609	0.6869	−0.2196	1.1626
23	1.4797	0.6852	−0.2184	1.1645
24	1.4975	0.6838	−0.2171	1.1665
25	1.5142	0.6828	−0.2157	1.1684
26	1.4479	0.6811	−0.2248	1.1584
27	1.4642	0.6795	−0.2239	1.1601
28	1.4796	0.6781	−0.2228	1.1618
29	1.4943	0.6770	−0.2217	1.1634
30	1.5083	0.6761	−0.2206	1.1650
31	1.5216	0.6755	−0.2194	1.1666
32	1.4665	0.6746	−0.2267	1.1585
33	1.4795	0.6735	−0.2258	1.1599
34	1.4920	0.6725	−0.2248	1.1613
35	1.5040	0.6718	−0.2238	1.1627
36	1.5156	0.6712	−0.2228	1.1641
37	1.5266	0.6707	−0.2218	1.1655
38	1.4795	0.6702	−0.2279	1.1586
39	1.4904	0.6694	−0.2270	1.1598
40	1.5009	0.6687	−0.2262	1.1611
41	1.5110	0.6682	−0.2253	1.1623
42	1.5208	0.6677	−0.2244	1.1635

Table 4.1.6 Continued

n	k_n	$n\text{Var}(\tilde{b}/b)$	$n\text{Cov}(\tilde{u}/b, \tilde{b}/b)$	$n\text{Var}(\tilde{u}/b)$
43	1.5303	0.6674	−0.2234	1.1646
44	1.4891	0.6671	−0.2287	1.1587
45	1.4984	0.6664	−0.2279	1.1598
46	1.5075	0.6659	−0.2271	1.1609
47	1.5163	0.6655	−0.2263	1.1619
48	1.5248	0.6651	−0.2255	1.1630
49	1.5331	0.6649	−0.2247	1.1640
50	1.5411	0.6647	−0.2238	1.1651
51	1.5046	0.6642	−0.2285	1.1598
52	1.5126	0.6638	−0.2278	1.1608
53	1.5204	0.6635	−0.2271	1.1617
54	1.5279	0.6632	−0.2264	1.1627
55	1.5352	0.6630	−0.2256	1.1636
56	1.5424	0.6628	−0.2249	1.1645
57	1.5096	0.6625	−0.2291	1.1598
58	1.5167	0.6621	−0.2284	1.1607
59	1.5236	0.6619	−0.2277	1.1615
60	1.5304	0.6616	−0.2271	1.1624
∞	1.5692	0.6484	−0.2309	1.1624

[a]Reprinted, with permission, from Engelhardt and Bain (1977a).

The approximation is then

$$\frac{x^* - x_{.10}}{\tilde{b}(2.070)} \sim F_{(229.3,67.3)}$$

where $x^* = \tilde{u} - B(28,40)\tilde{b}/C(28,40)$. Let us obtain a lower .95 confidence limit on $x_{.10}$. Using interpolation in tables of the F distribution (see Section 4.3 or Appendix B), we find that $\Pr(F_{(229.3,67.3)} \leq 1.38) = .95$. Hence $x^* - 1.38(2.070)\tilde{b} = -2.630$ is the desired lower confidence limit on $x_{.10}$. This is not substantially different from the limit -2.714 obtained in Example 4.1.2.

4.1.2e *Estimation of the Survivor (Reliability) Function*

We now consider confidence limits for the extreme value survivor function $S(x_0) = \exp(-e^{(x_0 - u)/b})$ for a given x_0. Confidence limits for $S(x_0)$ can in fact be obtained from the pivotal Z_p used to get confidence limits for quantiles because of the relationship between the survivor function and the quantiles of a distribution, mentioned earlier in Section 3.5.1. Specifically, suppose that $l(\mathbf{x})$ is a γ lower confidence limit on x_p based on data $\mathbf{x}$; that is, $\Pr[l(\mathbf{x}) \leqslant x_p] = \gamma$. This is true if and only if $\Pr[S(l(\mathbf{x})) \geqslant 1-p] = \gamma$. Thus, if one determines $p = p(\mathbf{x})$ such that $l(\mathbf{x}) = x_0$, then $1-p$ will be a γ lower confidence limit for $S(x_0)$.

To obtain a γ lower confidence limit on $S(x_0)$ with the pivotal $Z_p = (\tilde{u} - x_p)/\tilde{b}$, one can proceed as follows: since a γ lower confidence limit $l(\mathbf{x})$ on x_p is of the form $l(\mathbf{x}) = \tilde{u} - \tilde{b}z_{p,\gamma}$, where $z_{p,\gamma}$ is the γth quantile of Z_p, we merely need to determine p such that

$$-z_{p,\gamma} = \frac{x_0 - \tilde{u}}{\tilde{b}}. \tag{4.1.28}$$

The γ lower confidence limit on $S(x_0)$ is then $1-p$. With the conditional method of Section 4.1.2b, it is easy to find the required p. For unconditional methods, however, existing tabulations for Z_p based on the b.l.i.e.'s or m.l.e.'s cover only the cases $p = .01, .05$, and $.10$, and thus they are of little use in obtaining confidence limits on $S(x_0)$. For the case of the m.l.e.'s a few tables have been constructed that give confidence limits on $S(x_0)$ directly. One given by Thoman et al. (1970) covers complete samples only, and tables by Billmann et al. (1972) cover censored samples, but only for $n = 40, 60, 80, 100$, and 120, and $r = 0.50n$ or $0.75n$. In addition, Smith (1977) gives a few tables that can be used with certain simple linear estimators, and Engelhardt and Bain (1977a) discuss a log χ^2 approximation that can be used with linear estimators to get confidence intervals on $S(x_0)$ directly.

Example 4.1.7 (Example 4.1.2 revisited) Consider again the data given in Example 4.1.2, involving a Type II censored extreme value sample with $n = 40$ and $r = 28$. Suppose that a .95 lower confidence limit on $S(-1.0) = \exp(-e^{(-1-u)/b})$ is wanted, that is, a limit $1 - p(\mathbf{x})$ such that $\Pr[S(-1) \geqslant 1 - p(\mathbf{x})] = .95$. According to (4.1.28), we need to determine p such that

$$-z_{p,.95} = \frac{-1-\hat{u}}{\hat{b}} = -1.2071$$

recalling that the m.l.e.'s for u and b are $\hat{u} = 0.1563$ and $\hat{b} = 0.9104$. That is, we must find p such that $\Pr(\hat{Z}_p \leqslant -1.27) = .95$, using (4.1.18). Noting that $\hat{S}(-1.0) = \exp(-e^{(-1-\hat{u})/\hat{b}}) = .755$, and starting with $1-p$ in the neighbourhood of 0.70, we find after a few iterations with (4.1.18) that the required value of $1-p$ is .647. Hence the desired .95 confidence interval for $S(-1.0)$ is $S(-1.0) \geqslant .647$.

Alternately, Table 6 in Billmann et al. (1972) could be used to get the lower confidence limit. Although this table covers only the cases $r=20$, $n=40$ and $r=30$, $n=40$, it can be observed that both these cases lead to the same confidence limits, and thus so will $r=28$ and $n=40$, when $\hat{S}(t)=.755$. Using linear interpolation in the table gives the desired .95 lower confidence limit as .647; the conditional and unconditional procedures thus give exactly the same result in this case.

4.1.2f *Comments on Choice of Methods*

There is a huge literature dealing with inferences for Type II censored extreme value or Weibull distributed samples. This rather lengthy section has covered only the better of the available procedures, but even so, one is still left in many situations with a choice of methods.

If one is willing to write a few computer programs, the conditional methods of Section 4.1.2b are perhaps most convenient. They apply without modification to any size sample and can be used to get confidence limits on parameters, arbitrary quantiles, or the survivor function. In addition, the methods extend easily to handle progressively Type II censored data. The exact unconditional methods of Section 4.1.2c, on the other hand, require fairly extensive tables, and the tables at present exist only for certain ranges of sample size and for certain problems. Nevertheless, if one is willing to collect the necessary tables, these procedures are easy to use.

The approximate methods discussed in Section 4.1.2d are very useful alternatives to the exact methods, especially for situations not covered by the tables discussed in Section 4.1.2c. In most situations little is lost by using the approximate procedures, as long as the sample size is sufficiently large for the approximations to be accurate. As noted in Section 4.1.2d, this covers most situations of practical importance.

Two other points can be mentioned. First, from a strictly technical point of view there is some reason to prefer the conditional approach of Section 4.1.2b. For the extreme value distribution, however, the different methods discussed here tend to give results that are in close agreement, except perhaps for very small sample sizes. The choice among methods should consequently be based mainly on convenience. It should also be remarked that with quite large samples, procedures based on large-sample properties of m.l.e.'s and likelihood ratio tests can be used. These have not been discussed here, since there is no real need for them. However, large-sample methods are discussed in Section 4.2, and although that section deals with Type I censored data, the procedures given there apply to the case of Type II censoring as well.

4.2 SINGLE SAMPLES: TYPE I CENSORED DATA

With Type I censored data from the extreme value or Weibull distribution, there is generally no alternative but to use large-sample procedures based on maximum likelihood. We shall examine methods based on normal approximations to the distributions of the m.l.e.'s and likelihood ratio methods.

4.2.1 Maximum Likelihood Estimation

We continue to discuss results in terms of the extreme value distribution, though we shall work with the Weibull distribution when it is convenient to do so. Consider the usual situation leading to a Type I censored sample, where T_i represents the lifetime and L_i the fixed censoring time of the ith individual in a random sample of n individuals. One observes only $t_i = \min(T_i, L_i)$ and whether the observation is a lifetime or a censoring time. The T_i's are assumed to have a Weibull distribution or, equivalently, $X_i = \log T_i$ has an extreme value distribution with parameters u and b. Let $\eta_i = \log L_i$, $x_i = \log t_i$, and $\delta_i = 1$ or 0, according to whether $t_i = T_i$ or $t_i = L_i$, respectively; the likelihood function is then, by (1.4.9),

$$L(u, b) = \prod_{i=1}^{n} \left[\frac{1}{b}\exp\left(\frac{x_i - u}{b} - e^{(x_i - u)/b}\right)\right]^{\delta_i} \left[\exp\left(-e^{(\eta_i - u)/b}\right)\right]^{1-\delta_i}.$$

Letting $r = \Sigma\delta_i$ denote the number of observed lifetimes and D denote the set consisting of those individuals for which $\delta_i = 1$ (i.e., individuals whose lifetimes are uncensored), we have

$$\log L(u, b) = -r\log b + \sum_{i\in D} \frac{x_i - u}{b} - \sum_{i=1}^{n} \exp\left(\frac{x_i - u}{b}\right). \tag{4.2.1}$$

Note that $\log L(u, b)$ is of exactly the same form as in the case of Type II censoring [see (4.1.2)], since in the latter case x_r is the censoring time for the $n - r$ individuals whose lifetimes are not observed. The maximum likelihood equations are consequently of the same form as (4.1.5) and (4.1.6) and can be written as (provided that $r > 0$)

$$\sum_{i=1}^{n} x_i \exp\left(\frac{x_i}{\hat{b}}\right) \Big/ \sum_{i=1}^{n} \exp\left(\frac{x_i}{\hat{b}}\right) - \hat{b} - \frac{1}{r}\sum_{i\in D} x_i = 0 \tag{4.2.2}$$

$$e^{\hat{u}} = \left(\frac{1}{r}\sum_{i=1}^{n} e^{x_i/\hat{b}}\right)^{\hat{b}}. \tag{4.2.3}$$

Equation (4.2.2) can be solved iteratively for $\hat{b}$, then $\hat{u}$ calculated from (4.2.3).

Equivalently, in terms of the Weibull distribution (4.0.1) and its parameters, the log likelihood and maximum likelihood equations are respectively

$$\log L(\alpha,\beta)=r\log\beta-r\beta\log\alpha+(\beta-1)\sum_{i\in D}\log t_i-\sum_{i=1}^{n}\left(\frac{t_i}{\alpha}\right)^{\beta} \tag{4.2.4}$$

and

$$\sum_{i=1}^{n} t_i^{\hat{\beta}}\log t_i\Big/\sum_{i=1}^{n} t_i^{\hat{\beta}}-\frac{1}{\hat{\beta}}-\frac{1}{r}\sum_{i\in D}\log t_i=0 \tag{4.2.5}$$

$$\hat{\alpha}=\left(\frac{1}{r}\sum_{i=1}^{n} t_i^{\hat{\beta}}\right)^{1/\hat{\beta}}. \tag{4.2.6}$$

Rough estimates of u and b can be obtained by plotting the estimated (product-limit) survivor function $\hat{S}(x)$ against x, as described in Section 2.4. If $\log[-\log\hat{S}(x)]$ is plotted against x, then b and u can be estimated from the slope and intercept of a straight line that the plot should approximate if the extreme value model is appropriate. This can be used for an initial guess at the m.l.e. $\hat{b}$. As an alternate first guess at $\hat{b}$, the value of unity is often satisfactory.

4.2.2 Large-Sample Procedures Based on the M.l.e.'s

The asymptotic normal distribution for the m.l.e.'s is obtained in the usual way. The contribution to the log likelihood from a single observation can be written as

$$\log L_i=\delta_i\left[\frac{x_i-u}{b}-\log b-\exp\left(\frac{x_i-u}{b}\right)\right]+(1-\delta_i)\left[-\exp\left(\frac{\eta_i-u}{b}\right)\right].$$

For ease of notation, set $z_i=(x_i-u)/b$ and $c_i=(\eta_i-u)/b$; we now obtain the following derivatives of the log likelihood:

$$\frac{\partial\log L_i}{\partial u}=\delta_i\left(-\frac{1}{b}+\frac{1}{b}\exp z_i\right)+(1-\delta_i)\left(\frac{1}{b}\exp c_i\right)$$

$$\frac{\partial\log L_i}{\partial b}=\delta_i\left(-\frac{1}{b}-\frac{z_i}{b}+\frac{z_i}{b}\exp z_i\right)+(1-\delta_i)\left(\frac{c_i}{b}\exp c_i\right)$$

$$\frac{\partial^2 \log L_i}{\partial u^2} = -\delta_i\left(\frac{1}{b^2}\exp z_i\right) - (1-\delta_i)\left(\frac{1}{b^2}\exp c_i\right) \tag{4.2.7}$$

$$\frac{\partial^2 \log L_i}{\partial b^2} = \delta_i\left(\frac{1}{b^2} + \frac{2z_i}{b^2} - \frac{2z_i}{b^2}e^{z_i} - \frac{z_i^2}{b^2}e^{z_i}\right) + (1-\delta_i)\left(-\frac{2c_i}{b^2}e^{c_i} - \frac{c_i^2}{b^2}e^{c_i}\right)$$

$$\frac{\partial^2 \log L_i}{\partial u\,\partial b} = \delta_i\left(\frac{1}{b^2} - \frac{1}{b^2}e^{z_i} - \frac{z_i}{b^2}e^{z_i}\right) + (1-\delta_i)\left(-\frac{1}{b^2}e^{c_i} - \frac{c_i}{b^2}e^{c_i}\right).$$

To obtain the Fisher information matrix **I** we require the expectations of minus the second derivatives of $\log L$. These can be found by noting that given $\delta_i = 1$, z_i has a standard extreme value distribution truncated at c_i, with p.d.f.

$$g(w) = \frac{e^w \exp(-e^w)}{1-\exp(-e^{c_i})} \qquad -\infty < w \leqslant c_i.$$

Also noting that $\Pr(\delta_i = 1) = 1 - \exp(-e^{c_i}) = 1 - \Pr(\delta_i = 0)$, we obtain the following expectations:

$$I_{uu,i} = E\left(\frac{-\partial^2 \log L_i}{\partial u^2}\right) = \frac{1}{b^2}\left(\int_{-\infty}^{c} e^{2z}e^{-e^z}dz + e^c e^{-e^c}\right) = \frac{1}{b^2}\left(1 - e^{-e^c} + e^c e^{-e^c}\right)$$

$$I_{bb,i} = E\left(\frac{-\partial^2 \log L_i}{\partial b^2}\right) = \frac{1}{b^2}\left(\int_{-\infty}^{c}(1+z^2e^z)e^z e^{-e^z}dz + c^2 e^c e^{-e^c}\right) \tag{4.2.8}$$

$$I_{ub,i} = E\left(\frac{-\partial^2 \log L}{\partial u\,\partial b}\right) = \frac{1}{b^2}\left(\int_{-\infty}^{c} z e^{2z} e^{-e^z}dz + c e^c e^{-e^c}\right).$$

The Fisher (or expected) information matrix **I** is then

$$\mathbf{I}(u,b) = \begin{pmatrix} \sum_{i=1}^n I_{uu,i} & \sum_{i=1}^n I_{ub,i} \\ \sum_{i=1}^n I_{ub,i} & \sum_{i=1}^n I_{bb,i} \end{pmatrix} \tag{4.2.9}$$

and the usual large-sample normal approximation to the joint distribution of $\hat{u}$ and $\hat{b}$ is to treat $(\hat{u}, \hat{b})$ as being approximately bivariate normal with mean (u, b) and covariance matrix $\mathbf{I}^{-1}$. In practice, we usually estimate $\mathbf{I}(u,b)^{-1}$ by $\mathbf{I}(\hat{u},\hat{b})^{-1}$. The integrals involved in the calculation of (4.2.9) can be related to incomplete gamma, digamma, and trigamma functions but need for the most part to be evaluated numerically. Although Harter and Moore (1968) and Meeker and Nelson (1975, 1977) give tabulations that can in some instances be used to obtain $\mathbf{I}(\hat{u}, \hat{b})$, it is tiresome to evaluate (4.2.9). In addition, log censoring times η_i will often not be known for all individuals in the sample.

A simpler and equally valid procedure is to use the approximation

$$(\hat{u}, \hat{b}) \sim N_2[(u, b), \mathbf{I}_0^{-1}]$$

where $\mathbf{I}_0$ is the observed information matrix

$$\mathbf{I}_0 = \begin{pmatrix} -\partial^2 \log L/\partial u^2 & -\partial^2 \log L/\partial u\,\partial b \\ -\partial^2 \log L/\partial u\,\partial b & -\partial^2 \log L/\partial b^2 \end{pmatrix}_{(\hat{u}, \hat{b})}.$$

The form of $\mathbf{I}_0$ is simple; with $\hat{z}_i = (x_i - \hat{u})/\hat{b}$, (4.2.7) gives

$$\left.\frac{-\partial^2 \log L}{\partial u^2}\right|_{(\hat{u}, \hat{b})} = \frac{1}{\hat{b}^2} \sum_{i=1}^{n} \exp \hat{z}_i$$

$$\left.\frac{-\partial^2 \log L}{\partial b^2}\right|_{(\hat{u}, \hat{b})} = \frac{1}{\hat{b}^2} \left(-r - 2 \sum_{i \in D} \hat{z}_i + 2 \sum_{i=1}^{n} \hat{z}_i \exp \hat{z}_i + \sum_{i=1}^{n} \hat{z}_i^2 \exp \hat{z}_i \right)$$

$$\left.\frac{-\partial^2 \log L}{\partial u\,\partial b}\right|_{(\hat{u}, \hat{b})} = \frac{1}{\hat{b}^2} \left(-r + \sum_{i=1}^{n} \exp \hat{z}_i + \sum_{i=1}^{n} \hat{z}_i \exp \hat{z}_i \right).$$

These can be simplified even more by noting from the maximum likelihood equations $\partial \log L/\partial u = 0$ and $\partial \log L/\partial b = 0$ that $\Sigma \exp \hat{z}_i = r$ and $\Sigma \hat{z}_i + \Sigma \hat{z}_i^2 \exp \hat{z}_i = r$. Using these we get

$$\mathbf{I}_0 = \frac{1}{\hat{b}^2} \begin{pmatrix} r & \Sigma_{i=1}^{n} \hat{z}_i \exp \hat{z}_i \\ \Sigma_{i=1}^{n} \hat{z}_i \exp \hat{z}_i & r + \Sigma_{i=1}^{n} \hat{z}_i^2 \exp \hat{z}_i \end{pmatrix}. \tag{4.2.10}$$

Approximate confidence intervals for u and b are found by taking $(\hat{u}, \hat{b})$ to be bivariate normally distributed with mean (u, b) and covariance matrix $\mathbf{I}_0^{-1}$. Concerning confidence intervals for $x_p = u + w_p b$, the normal approximation for $(\hat{u}, \hat{b})$ implies that $\hat{x}_p = \hat{u} + w_p \hat{b}$ is approximately normal, with mean x_p and variance

$$\text{Asvar}(\hat{x}_p) = \text{Asvar}(\hat{u}) + 2w_p \text{Ascov}(\hat{u}, \hat{b}) + w_p^2\, Asvar(\hat{b})$$

where we write "Asvar($\hat{u}$)" and so on to denote variances of the approximating normal distributions for $\hat{u}$ and so on. An example of the use of these methods is deferred to Section 4.2.3.

It should be remarked that although these procedures are adequate for quite large samples, the approximations on which they are based are rather poor for small- and moderate-size samples (e.g., Billmann et al., 1972; Lawless, 1975). One way of alleviating this problem would be to find transformations of the m.l.e.'s, whose distributions are close to normal even in small samples. For example, treating $\log \hat{b}$ as approximately normal is somewhat preferable to treating $\hat{b}$ as approximately normal. However, other such transformations, especially one that would apply to $\hat{x}_p$, are difficult to find. A second approach is to use likelihood ratio procedures. The χ^2 distributions that approximate the distributions of likelihood ratio test statistics are often found to be adequate, even for quite small sample sizes. These procedures will be discussed now.

4.2.3 Likelihood Ratio Methods

We shall work in this section with the Weibull form of the model, since various expressions can be slightly more compactly expressed in this form. The procedures do not, of course, depend on the parameterization we choose to work with. The log likelihood, given by (4.2.4), is

$$\log L(\alpha,\beta)=r\log\beta-r\beta\log\alpha+(\beta-1)\sum_{i\in D}\log t_i-\sum_{i=1}^{n}\left(\frac{t_i}{\alpha}\right)^{\beta}.$$

Procedures for obtaining tests or confidence intervals for β, α, and $t_p=\alpha[-\log(1-p)]^{1/\beta}$ will be described in turn.

Tests or Confidence Intervals for β

Consider the null hypothesis $H_0:\beta=\beta_0$, with the alternative being $H_1:\beta\neq\beta_0$. For a likelihood ratio test of H_0 versus H_1 one needs the m.l.e.'s of α and β under H_0. The m.l.e. of β is $\tilde{\beta}=\beta_0$, and the m.l.e. $\tilde{\alpha}$ of α is found by maximizing $\log L(\alpha,\beta_0)$ with respect to α. This gives

$$\tilde{\alpha}=\left(\sum_{i=1}^{n}t_i^{\beta_0}\Big/r\right)^{1/\beta_0}.$$

The likelihood ratio statistic for testing H_0 versus H_1 is then

$$\Lambda=-2\log L(\tilde{\alpha},\beta_0)+2\log L(\hat{\alpha},\hat{\beta})$$

where $\hat{\alpha}$ and $\hat{\beta}$ are the m.l.e.'s from the unrestricted model, given as the solutions to (4.2.5) and (4.2.6). Large values of Λ provide evidence against H_0. Approximate significance levels can be obtained by using the fact that in large samples Λ is approximately $\chi^2_{(1)}$, under H_0. To obtain an approximate γ confidence interval for β, we find the set of values β_0 for which H_0 is not rejected at the $1-\gamma$ level of significance, that is, the set of values β_0 such that $\Lambda \leq \chi^2_{(1),\gamma}$.

Tests or Confidence Intervals for α

These can be given as a special case of tests or confidence intervals for t_p immediately below, but for convenience the results will be given separately. To test $H_0: \alpha=\alpha_0$ versus $H_1: \alpha \neq \alpha_0$ requires maximizing $\log L(\alpha, \beta)$ under the restriction $\alpha=\alpha_0$. Setting $\partial \log L(\alpha_0, \beta)/\partial\beta = 0$, we get the equation

$$\frac{r}{\beta} - r\log\alpha_0 + \sum_{i\in D}\log t_i - \sum_{i=1}^{n}\left(\frac{t_i}{\alpha_0}\right)^{\beta}\log\left(\frac{t_i}{\alpha_0}\right) = 0. \qquad (4.2.11)$$

This equation can be solved iteratively to give $\tilde{\beta}$. The likelihood ratio statistic for testing H_0 versus H_1 is

$$\Lambda = -2\log(\alpha_0, \tilde{\beta}) + 2\log L(\hat{\alpha}, \hat{\beta}).$$

Large values of Λ give evidence against H_0, and if samples are not too small, Λ can be treated as being $\chi^2_{(1)}$, under H_0. An approximate γ confidence interval for α is obtained by finding the set of all values α_0 such that $\Lambda \leq \chi^2_{(1),\gamma}$.

Tests or Confidence Intervals for t_p

Consider the hypothesis $H_0: t_p = Q_0$ versus $H_1: t_p \neq Q_0$. That is, H_0, specifies that $\alpha[-\log(1-p)]^{1/\beta} = Q_0$ or that $\alpha = Q_0/[-\log(1-p)]^{1/\beta}$. To find the m.l.e.'s $\tilde{\alpha}$ and $\tilde{\beta}$ of α and β under H_0, we can set $\alpha = Q_0/[-\log(1-p)]^{1/\beta}$ in $\log L(\alpha, \beta)$ and differentiate with respect to β. Setting this equal to zero, we get the equation

$$\frac{r}{\beta} - r\log Q_0 + \sum_{i\in D}\log t_i + \log(1-p)\sum_{i=1}^{n}\left(\frac{t_i}{Q_0}\right)^{\beta}\log\left(\frac{t_i}{Q_0}\right) = 0. \qquad (4.2.12)$$

This equation can be solved iteratively to give $\tilde{\beta}$ and then $\tilde{\alpha} = Q_0/[-\log(1-p)]^{1/\tilde{\beta}}$. The likelihood ratio statistic for testing H_0 versus H_1 is

$$\Lambda = -2\log L(\tilde{\alpha}, \tilde{\beta}) + 2\log L(\hat{\alpha}, \hat{\beta}).$$

Large values of Λ are evidence against H_0, and approximate significance levels can be calculated by treating Λ as $\chi^2_{(1)}$, under H_0. A γ confidence interval for t_p is found by determining the set of values Q_0 for which $\Lambda \leqslant \chi^2_{(1),\gamma}$.

Likelihood ratio tests about α, β, or t_p are straightforward, though in either of the last two sections it is necessary to obtain estimates under H_0 iteratively. To get confidence intervals it is also necessary to find the set of values satisfying $\Lambda \leqslant \chi^2_{(1),\gamma}$ iteratively. This is easily done with the aid of a computer. Tests and confidence limits on the survivor function can also be found by the third procedure. Suppose that a confidence interval or test for $S(t_0)=\exp[-(t_0/\alpha)^\beta]$ is wanted, where t_0 is given. Consider the hypothesis $H_0: S(t_0)=S_0$ versus $H_1: S(t_0)\neq S_0$. Since $S(t_0)=S_0$ implies that

$$\alpha=\frac{t_0}{(-\log S_0)^{1/\beta}},$$

to maximize $\log L(\alpha,\beta)$ under H_0 we can use (4.2.12) to obtain $\tilde{\beta}$, replacing Q_0 with t_0 and $1-p$ with S_0. An approximate γ confidence interval for $S(t_0)$ consists of the set of values S_0 such that $\Lambda=-2\log L(\tilde{\alpha},\tilde{\beta})+2\log L(\hat{\alpha},\hat{\beta}) \leqslant \chi^2_{(1),\gamma}$, where $\tilde{\alpha}=t_0/(-\log S_0)^{1/\tilde{\beta}}$.

Example 4.2.1 (Example 2.3.1 revisited) Let us reconsider the leukemia remission time data first discussed in Example 1.1.7. The data were remission times, in weeks, for two groups of patients, one given a treatment (drug 6-MP) and the other a placebo. In the first group, 12 times out of 21 were censored, and in the second, 0 out of 21 were censored. The times were as follows, with asterisks denoting censoring times:

Drug 6-MP	6, 6, 6, 6*, 7, 9*, 10, 10*, 11*, 13, 16, 17*, 19*, 20*, 22, 23, 25*, 32*, 32*, 34*, 35*
Placebo	1, 1, 2, 2, 3, 4, 4, 5, 5, 8, 8, 8, 8, 11, 11, 12, 12, 15, 17, 22, 23

Product-limit estimates of the survivor function for the two distributions were calculated in Example 2.3.1, and subsequently we noted in Example 2.3.2 that exponential models might be reasonable. As a generalization of this, results obtained by treating the data as samples from two separate Weibull distributions are given below. Grouping is ignored, although with the many ties present, a check on this should be made (see Section 5.5). M.l.e.'s of α and β and approximate confidence intervals for β, $t_{.50}$, and S(10), calculated using the likelihood ratio procedures of this section, are given for each group. Since there is no censoring for the Placebo group, the methods of Section 4.1 could actually have been used there, but this was not done. Two-sample tests could also be carried out; these are discussed in Section 4.3. For these data the results that would be obtained from a two-sample test are, however, quite clear.

	Drug 6-MP	Placebo
	$\tilde{\beta}=1.35 \quad \tilde{\alpha}=33.77$	$\tilde{\beta}=1.37 \quad \tilde{\alpha}=9.482$
.95 confidence interval for β	$0.72 \leqslant \beta \leqslant 2.21$	$0.95 \leqslant \beta \leqslant 1.88$
.95 confidence interval for $t_{.50}$	$16.2 \leqslant t_{.50} \leqslant 51.6$	$4.75 \leqslant t_{.50} \leqslant 10.3$
.95 confidence interval for S(10)	$.637 \leqslant S(10) \leqslant .931$	$.197 \leqslant S(10) \leqslant .513$

Confidence intervals are obtained in each case by finding the set of parameter values such that $\Lambda \leqslant \chi^2_{(1),.95} = 3.84$. Tests of the hypothesis $\beta=1$ are also of interest, since $\beta=1$ gives the exponential distribution. There is some evidence, though not strong, against this hypothesis in both groups. It is clear that there is no evidence against a hypothesis of equal shape parameters β in the two groups. On the other hand, a difference in median remission times for the two groups is indicated by their respective confidence intervals. Survival probabilities also differ substantially in the two groups, with results shown for $S(10)$, the probability of remission lasting longer than 10 weeks.

4.2.4 Comparisons and Remarks on the Large-Sample Methods

Tests and confidence intervals can be based on either the limiting normal distributions of the m.l.e.'s, as described in Section 4.2.1, or on the likelihood ratio methods of Section 4.2.2. The approximations involved with the latter are likely to be better in small- or moderate-size samples than those involved with the former. On the other hand, the methods employing the m.l.e.'s are easier to use, especially in obtaining confidence intervals. To explore these points a little let us first compare the different methods in an example. For convenience all results are expressed in terms of the extreme value distribution.

Example 4.2.2 (Example 4.1.2 revisited) Consider the Type II censored sample with $n=40$ and $r=28$ discussed in Example 4.1.2. Since the sample is Type II censored, we can compare intervals from the two approximate methods with the exact confidence intervals available in this case. Note that in employing the methods of this section we have $k=r=28$, and all 12 individuals whose lifetimes are censored have a log censoring time $\eta_i=0.296$. The m.l.e.'s of the extreme value parameters u and b were found earlier to be $\hat{u}=0.1563$ and $\hat{b}=0.9104$. The observed information matrix $\mathbf{I}_0$

given by (4.2.10) is easily calculated, and the approximate covariance matrix $\mathbf{I}_0^{-1}$ for $(\hat{u}, \hat{b})$ easily found. Entries in $\mathbf{I}_0^{-1}$ are $\text{Asvar}(\hat{u})=0.032889$, $\text{Asvar}(\hat{b})=0.025654$, $\text{Ascov}(\hat{u}, \hat{b})=0.003101$.

To obtain confidence intervals for u and b by the methods of Section 4.2.1 we treat $(\hat{u}-u)/[\text{Asvar}(\hat{u})]^{1/2}$ and $(\hat{b}-b)/[\text{Asvar}(\hat{b})]^{1/2}$, respectively, as being $N(0,1)$. For example, to get an approximate .90 confidence interval for u we have

$$\Pr\left(-1.645 \leqslant \frac{\hat{u}-u}{[\text{Asvar}(\hat{u})]^{1/2}} \leqslant 1.645\right) \doteq .90$$

which gives $\{\hat{u}-1.645[\text{Asvar}(\hat{u})]^{1/2}, \hat{u}+1.645[\text{Asvar}(\hat{u})]^{1/2}\} = (-0.142, 0.455)$ as the desired confidence interval.

To get approximate confidence intervals for $x_p = u + w_p b$, where $w_p = \log[-\log(1-p)]$, we treat $(\hat{x}_p - x_p)/[\text{Asvar}(\hat{x}_p)]^{1/2}$ as $N(0,1)$, where

$$\text{Asvar}(\hat{x}_p) = \text{Asvar}(\hat{u}) + w_p^2 \text{Asvar}(\hat{b}) + 2w_p \text{Ascov}(\hat{u}, \hat{b}).$$

As an illustration of these methods, .90 confidence intervals for u, b, $x_{.10}$, and $x_{.01}$ are given in Table 4.2.1.

Confidence intervals obtained by the likelihood ratio methods of Section 4.2.2 and exact conditional confidence intervals found by the methods of Section 4.1.2b are also given in Table 4.2.1. Comparison of the results for this particular sample shows the likelihood ratio method to produce better approximations to the exact intervals than the other method. A slightly better agreement between the confidence intervals for b is obtained by treating $\log \hat{b}$ instead of $\hat{b}$, as approximately normal. This leads to a .90 confidence interval $0.68 \leqslant b \leqslant 1.22$, almost the same as that given by the likelihood ratio method. Simulation results mentioned below provide some evidence on the extent to which some of the trends in this example appear to be general.

Table 4.2.1 .90 Confidence Intervals Obtained by Three Methods

Confidence Intervals for	u	b	$x_{.10}$	$x_{.01}$
Via $\hat{u}, \hat{b}$ (observed information)	$(-0.14, 0.46)$	$(0.65, 1.17)$	$(-2.53, -1.26)$	$(-5.12, -2.69)$
Likelihood ratio method	$(-0.12, 0.48)$	$(0.70, 1.22)$	$(-2.62, -1.37)$	$(-5.44, -3.05)$
Exact method	$(-0.11, 0.51)$	$(0.72, 1.28)$	$(-2.71, -1.41)$	$(-5.67, -3.15)$

Small-Sample Coverage Properties

As always, it is desirable to have some idea of the adequacy of the approximations involved with the large-sample methods. Lawless (1979) gives results of some sampling experiments in which the true coverage probabilities were estimated for confidence intervals obtained using the three methods described.

The results indicated that for the methods using normal approximations for $(\hat{u}, \hat{b})$, two-sided confidence intervals for b, $x_{.10}$, and $x_{.01}$ tended to have both upper and lower limits that were substantially too high, even for samples as large as 60, with moderate Type I censoring. Confidence intervals for u had coverage probabilities fairly close to the nominal ones for light censoring, but not for heavier censoring. Coverage properties for intervals based on a χ^2 approximation for the likelihood ratio statistic Λ were markedly better. Even for samples as small as 20 the distributions of Λ appear to be fairly well approximated by $\chi^2_{(1)}$. With the number of observed lifetimes r less than 20 or so, lower and upper confidence limits for u, b, and x_p $(p \leqslant .5)$ tend to be slightly too high, but, overall, the limits appear to give coverage probabilities fairly close to the nominal ones.

These results suggest that unless sample sizes are quite large, likelihood ratio procedures are preferable to those based on the normal approximations to the distribution of $(\hat{u}, \hat{b})$.

4.3 COMPARISON OF WEIBULL OR EXTREME VALUE DISTRIBUTIONS

We turn now to the comparison of Weibull or extreme value distributions. If lifetimes in m populations are represented by m Weibull distributions, with survivor functions

$$S_i(t) = \exp\left[-\left(\frac{t}{\alpha_i}\right)^{\beta_i}\right] \qquad i = 1, \ldots, m$$

then the comparison of these distributions will involve comparisons of the parameters (α_i, β_i), $i = 1, \ldots, m$.

A first question in comparing Weibull distributions is whether the shape parameters $\beta_1, \ldots, \beta_m$ or, equivalently, the scale parameters $b_1, \ldots, b_m$ of the corresponding extreme value distributions can be assumed to be equal. Equality of Weibull shape parameters across different groups of individuals is an important and simplifying assumption in many applications. In regression problems involving a Weibull or extreme value distribution such an assumption is analogous to the assumption of constant variance in normal regression models (see Chapter 6). For example, Weibull models for lifetimes of manufactured items often assume that only the scale parameter α is altered by changes in the environment in which the items are used, and not the shape parameter β (e.g., Nelson, 1972a).

Let us look more closely at the comparison of two Weibull distributions. First, note that if t_{1p} and t_{2p} are the pth quantiles of two distributions with parameters (α_1, β_1) and (α_2, β_2), respectively, then

$$\frac{t_{1p}}{t_{2p}} = \frac{\alpha_1[-\log(1-p)]^{1/\beta_1}}{\alpha_2[-\log(1-p)]^{1/\beta_2}}. \tag{4.3.1}$$

If $\beta_1 = \beta_2$, then $t_{1p}/t_{2p} = \alpha_1/\alpha_2$ is constant; that is, the quantiles of the one distribution are proportional to those of the other, and comparing these quantiles is just a matter of comparing α_1 and α_2. The ratio of the means of the two distributions is also α_1/α_2.

Consider also the ratio of the two hazard functions,

$$\frac{h_1(t)}{h_2(t)} = \frac{(\beta_1/\alpha_1)(t/\alpha_1)^{\beta_1 - 1}}{(\beta_2/\alpha_2)(t/\alpha_2)^{\beta_2 - 1}}.$$

When $\beta_1 = \beta_2 = \beta$, we have $h_1(t)/h_2(t) = (\alpha_2/\alpha_1)^\beta$, that is, the ratio of the hazard functions is independent of t. In this case we speak of the two hazard functions being proportional; proportional hazards models are very important and are discussed at length in Chapters 6 and 7. Finally, when $\beta_1 = \beta_2 = \beta$, the survivor functions of the two distributions, $S_i(t) = \exp[-(t/\alpha_i)^\beta]$ $(i = 1, 2)$, are related by

$$S_1(t) = S_2(t)^\delta$$

where $\delta = (\alpha_2/\alpha_1)^\beta$, so that if $\alpha_2 \neq \alpha_1$, the survivor function of one distribution lies entirely below that of the other.

The comparison of Weibull distributions is obviously simpler and more meaningful when their shape parameters are equal. The first problem discussed will thus be the comparison of shape parameters.

4.3.1 Comparison of Weibull Shape (Extreme Value Scale) Parameters

It is once again convenient to work with the extreme value distribution. The problem is to compare the scale parameters $b_1, \ldots, b_m$ of m distributions, without any assumptions about the location parameters. The case $m=2$ will be considered first.

4.3.1a *Comparison of Two Shape Parameters With Type II Censored Samples*

For Type II censored or complete samples the approximations in Section 4.1.2d give simple methods of testing the hypothesis $H_0: b_1=b_2$ and obtaining confidence intervals for b_1/b_2. If $\tilde{b}_1$ and $\tilde{b}_2$ are equivariant estimators of b_1 and b_2, based on Type II censored samples from the two distributions, then

$$U=\frac{\tilde{b}_1/b_1}{\tilde{b}_2/b_2} \tag{4.3.2}$$

is a pivotal quantity. When $b_1=b_2$, U becomes $\tilde{b}_1/\tilde{b}_2$ and can be used to test H_0. More generally, U can be used to test hypotheses of the form $H_0: b_1=ab_2$ or to obtain confidence intervals for b_1/b_2.

The exact unconditional distribution of U is intractable for estimators such as the m.l.e.'s and b.l.i.e.'s and it is computationally unfeasible to use the conditional distribution of U, given the values of the ancillary statistics in the two samples. Percentage points for U can be determined by Monte Carlo methods, but since there are four factors r_1, n_1, r_2, n_2 involved, it is not feasible to produce tables covering any very wide range of sample sizes. However, a few tables exist for the case in which the m.l.e.'s $\hat{b}_1$ and $\hat{b}_2$ are used in (4.3.2). Thoman and Bain (1969) give percentage points of U for complete samples, with $n_1=n_2$ $(=r_1=r_2)$ ranging from 5 to 120. McCool (1974) gives tables of percentage points for 17 cases in which $n_1=n_2$ and $r_1=r_2$, namely, $n_1=5$ $(r_1=3,5)$, $n_1=10$ $(r_1=3,5,10)$, $n_1=15$ $(r_1=5,10,15)$, $n_1=20$ $(r_1=5,10,15,20)$, and $n_1=30$ $(r_1=5,10,15,20,30)$. Excerpts from these tables are given in McCool (1975b).

A simpler approach that is satisfactory in virtually all situations is to use χ^2 approximations for $\hat{b}/b$, given in Section 4.1.2d, that is to approximate the distribution of U by a F distribution. Only the m.l.e.'s are used here, though other estimators $\tilde{b}$ for which $\tilde{b}/b$ has approximately a χ^2 distribution could also be used (see Section 4.1.2d). For sample i $(i=1,2)$ (4.1.21) gives

$$\frac{g_i\hat{b}_i}{b_i} \sim \chi^2_{(h_i)}$$

where $g_i=g_i(r_i, n_i)$ and $h_i=h_i(r_i, n_i)$ are obtained from Table 4.1.2. Since

$\hat{b}_1$ and $\hat{b}_2$ are independent, this gives the approximation

$$\frac{g_1 h_2 \hat{b}_1 b_2}{g_2 h_1 \hat{b}_2 b_1} = \frac{g_1 h_2}{g_2 h_1} U \sim F_{(h_1, h_2)} \tag{4.3.3}$$

which can be used to test hypotheses or find confidence intervals for b_1/b_2.

The approximation (4.3.3) involves the F distribution with noninteger degrees of freedom. There are several fairly convenient methods for obtaining probabilities or percentage points in this situation. Mardia and Zemroch (1978) give $F_{(h_1, h_2)}$ percentage points for a wide range of h_1 and h_2 values (including noninteger values) and show how to interpolate in their tables for other (h_1, h_2) pairs. Pearson and Hartley (1966) and Johnson and Kotz (1970, Ch. 26) also discuss this. Mardia and Zemroch (1978) and Abramowitz and Stegun (1965, Ch. 26) give formulas for computing F probabilities. In applications of (4.3.3) h_1 and h_2 are usually large enough so that one of several good approximations can be used. In particular, for h_1 and h_2 greater than about 15, a very good approximation can be derived from the Wilson–Hilferty χ^2 approximation (4.1.22). With $F = F_{(h_1, h_2)}$, this is [see (B22) of appendix B]

$$\left[F^{1/3}\left(1 - \frac{2}{9b}\right) - \left(1 - \frac{2}{9a}\right)\right]\left(\frac{2}{9a} + \frac{2}{9b}F^{2/3}\right)^{-1/2} \simeq N(0,1). \tag{4.3.4}$$

This can be used to compute probabilities for $F_{(h_1, h_2)}$. To approximate the pth quantile F_p for $F_{(h_1, h_2)}$ we need to find the value of F that makes the left-hand side of (4.3.4) equal to N_p, the pth quantile for $N(0, 1)$. This can be done by solving the quadratic equation

$$Y^2\left[\frac{2}{9h_2}N_p^2 - \left(1 - \frac{2}{9h_2}\right)^2\right] + 2\left(1 - \frac{2}{9h_1}\right)\left(1 - \frac{2}{9h_2}\right)Y$$
$$+ \left[\frac{2}{9h_1}N_p^2 - \left(1 - \frac{2}{9h_2}\right)^2\right] = 0 \tag{4.3.5}$$

where $Y = F_p^{1/3}$. The larger root of (4.3.5) gives upper percentage points for $F_{(h_1, h_2)}$.

Ling (1978) discusses the accuracy of this approximation and several others.

Table 4.3.1 Exact and Approximate Quantiles for U

			Quantile		
			.75	.90	.95
$n_1=n_2=10$	Approximate	(4.3.3)	1.29	1.62	1.86
	Exact		1.29	1.66	1.90
$n_1=n_2=20$	Approximate	(4.3.3)	1.19	1.39	1.52
	Exact		1.19	1.40	1.53

Example 4.3.1 Let us compare percentage points for U obtained by using (4.3.3) with those given by Thoman and Bain (1969) for the cases $n_1=r_1=n_2=r_2=10$ and 20, respectively. For the first case one finds from Table 4.1.2 that $h_1=h_2=29.3$ and $g_1=g_2=31.3$; for the second case $h_1=h_2=62.4$ and $g_1=g_2=64.4$. Since $g_1=g_2$ and $h_1=h_2$, (4.3.3) gives $U \sim F_{(h_1,h_2)}$ in both cases. Table 4.3.1 shows .75, .90, and .95 quantiles of U calculated with (4.3.3) and those taken from Thoman and Bain's paper. The $\chi^2_{(h_1)}$ percentage points were calculated by using (4.3.5). The F approximation is very good, except for relatively small departures in the .90 and .95 quantiles for the $n_1=n_2=10$ case.

4.3.1b *Samples with Arbitrary Censoring: Likelihood Ratio Methods*

When the data are Type I censored, likelihood ratio methods can be used to test for equality of b_1 and b_2 or to obtain confidence intervals for b_1/b_2. This will be discussed along with the problem of comparing more than two Weibull shape parameters on the basis of either Type I or Type II censored data.

As in the treatment of one-sample likelihood ratio methods in Section 4.2.2, results will be presented in Weibull form. The basic problem is to test the hypothesis

$$H_0: \beta_1 = \cdots = \beta_m$$

with samples from the m distributions that may be either Type I or Type II censored. From (4.2.4), the log likelihood function from m independent samples is

$$\log L(\boldsymbol{\alpha}, \boldsymbol{\beta}) = \sum_{j=1}^{m} r_j \log \beta_j - \sum_{j=1}^{m} r_j \beta_j \log \alpha_j + \sum_{j=1}^{m} (\beta_j - 1) \sum_{i \in D_j} \log t_{ji} - \sum_{j=1}^{m} \sum_{i=1}^{n_j} \left(\frac{t_{ji}}{\alpha_i} \right)^{\beta_j} \tag{4.3.6}$$

where t_{ji} $(i=1,\ldots,n_j)$ are the lifetimes and censoring times in the sample from the jth distribution $(j=1,\ldots,m)$, r_j is the number of observed lifetimes in the jth sample, and D_j is the set of individuals in the jth sample whose lifetimes are observed. To test H_0 against the alternative that the β_j's are not all equal one can use the likelihood ratio statistic

$$\Lambda=-2\log L(\tilde{\alpha}_1,\ldots,\tilde{\alpha}_m,\tilde{\beta},\ldots,\tilde{\beta})+2\log L(\hat{\alpha}_1,\ldots,\hat{\alpha}_m,\hat{\beta}_1,\ldots,\hat{\beta}_m) \tag{4.3.7}$$

where $\hat{\alpha}_1,\ldots,\hat{\alpha}_m,\hat{\beta}_1,\ldots,\hat{\beta}_m$ are the unrestricted m.l.e.'s of the parameters and $\tilde{\alpha},\ldots,\tilde{\alpha}_m$ and $\tilde{\beta}_1=\cdots=\tilde{\beta}_m=\tilde{\beta}$ are the m.l.e.'s under H_0. In large samples the distribution of Λ under H_0 is approximately $\chi^2_{(m-1)}$.

The unrestricted m.l.e.'s are obtained by solving m pairs of equations of the forms (4.2.5) and (4.2.6). These are

$$\sum_{i=1}^{n_j} t_{ji}^{\hat{\beta}_j}\log t_{ji}\Big/\sum_{i=1}^{n_j} t_{ji}^{\hat{\beta}_j}-\frac{1}{\hat{\beta}_j}-\frac{1}{r_j}\sum_{i\in D_j}\log t_{ji}=0$$

$$\hat{\alpha}_j=\left(\frac{1}{r_j}\sum_{i=1}^{n_j} t_{ji}^{\hat{\beta}_j}\right)^{1/\hat{\beta}_j} \qquad j=1,\ldots,m.$$

For each pair the first equation is solved iteratively for $\hat{\beta}_j$, then $\hat{\alpha}_j$ is obtained from the second equation. To maximize (4.3.6) under H_0 we let $\beta_1=\cdots=\beta_m=\beta$ and set derivatives of $\log L$ with respect to $\alpha_1,\ldots,\alpha_m$ and β equal to zero. Rearranging the equations so formed, we find that $\tilde{\beta}$ satisfies the equation

$$\sum_{j=1}^{m} r_j\sum_{i=1}^{n_j} t_{ji}^{\tilde{\beta}}\log t_{ji}\Big/\sum_{i=1}^{n_j} t_{ji}^{\tilde{\beta}}-\sum_{j=1}^{m} r_j\Big/\tilde{\beta}-\sum_{j=1}^{m}\sum_{i\in D_j}\log t_{ji}=0. \tag{4.3.8}$$

Once $\tilde{\beta}$ is obtained from (4.3.8), $\tilde{\alpha}_1,\ldots,\tilde{\alpha}_m$ are given by

$$\tilde{\alpha}_j=\left(\frac{1}{r_j}\sum_{i=1}^{n_j} t_{ji}^{\tilde{\beta}}\right)^{1/\tilde{\beta}} \qquad j=1,\ldots,m. \tag{4.3.9}$$

The likelihood ratio test is straightforward to carry out. In the case in which all samples are Type II censored an even simpler test is available and an example is deferred until this has been discussed. Finally, confidence intervals for $a=\beta_1/\beta_2$ can be obtained by considering the likelihood ratio test for $H_0: \beta_1=a\beta_2$ versus $H_1: \beta_1 \neq a\beta_2$.

4.3.1c *A Simple Test When Censoring is All Type II*

For the case in which all censoring is Type II a simple test of equality of $\beta_1,\ldots,\beta_m$ can be obtained from the fact that the distributions of $\hat{b}_i/b_i = \beta_i/\hat{\beta}_i$ are well approximated by χ^2 distributions, as discussed in Section 4.1.2d. The problem of testing the equality of $\beta_1,\ldots,\beta_m$ (or $b_1,\ldots,b_m$) is therefore approximately the same as the problem of testing the equality of the scale parameters in m gamma distributions. As discussed in Section 3.3, Bartlett's test is the likelihood ratio test that arises in this situation. That is, if $g_i\hat{\theta}_i/\theta_i \sim \chi^2_{(h_i)}(i=1,\ldots,m)$, where g_i and h_i are known constants, then the likelihood ratio statistic for testing $H_0: \theta_1 = \cdots = \theta_m$ against the alternative that the θ_i's are not all equal is

$$h\log\theta^* - \sum_{i=1}^{m} h_i \log\left(\frac{g_i\hat{\theta}_i}{h_i}\right)$$

where $h=\Sigma h_i$ and $\theta^*=\Sigma g_i\hat{\theta}_i/h$. If $\hat{b}_i$ is the m.l.e. of b_i computed from a Type II censored sample consisting of the r_i smallest out of n_i observations, then $g_i\hat{b}_i/b_i$ is approximately distributed as $\chi^2_{(h_i)}$, where $g_i=g(r_i,n_i)$ and $h_i=h(r_i,n_i)$ are found from Table 4.1.2. To test $H_0: b_1=\cdots=b_m$ one might therefore consider the statistic

$$M=h\log b^* - \sum_{i=1}^{m} h_i \log\left(\frac{g_i\hat{b}_i}{h_i}\right) \tag{4.3.10}$$

where $h=\Sigma h_i$ and $b^*=\Sigma g_i\hat{b}_i/h$. The distribution of M under H_0 is approximately $\chi^2_{(m-1)}$ in large samples. A similar test statistic could also be based on one of the linear estimates $\tilde{b}$ of b, for which $\tilde{b}/b$ is well approximated by a χ^2 distribution (see Section 4.1.2d). Lawless and Mann (1976) examined the statistic M when it was based on certain linear estimators and found that even for fairly small samples the $\chi^2_{(m-1)}$ approximation to the distribution of M is suitable for most practical purposes. When one or more of the h_i's is fairly small, use of a correction factor analogous to that presented in Section 3.3 slightly improves the approximation. This involves treating CM as $\chi^2_{(m-1)}$, where

$$C^{-1}=1+\frac{1}{3(m-1)}\left(\sum_{i=1}^{m} h_i^{-1}-h^{-1}\right).$$

An example will illustrate this and the likelihood ratio test for the equality of Weibull shape parameters.

Example 4.3.2 Nelson (1970a) presents some data on the time to breakdown of a type of electrical insulating fluid subject to a constant voltage stress. (A portion of the data has been given in Example 1.1.5). The data, shown in Table 4.3.2, are breakdown times for seven groups of specimens, each group involving a different voltage level. A model suggested by engineering considerations is that, for a fixed voltage level, time to breakdown has a Weibull distribution. Furthermore, distributions corresponding to different voltage levels are thought to differ only with respect to their scale parameters α_i, the shape parameter β being the same for different levels. The data are uncensored, and times to breakdown are given in minutes.

A good initial step is to plot the seven samples on Weibull or extreme value probability paper. If the Weibull model is reasonable, the points from each sample should lie roughly on straight lines, and if Weibull distributions with a common shape parameter are reasonable, the plots for different samples should be roughly parallel. A plot of the data shows this pattern, though it is difficult to tell from the plot whether there is any evidence of a difference in shape parameters. This can be assessed by a formal test of the hypothesis $H_0: \beta_1 = \cdots = \beta_7$.

Table 4.3.3 presents the values of the unrestricted Weibull m.l.e.'s $\hat{\alpha}_i$ and $\hat{\beta}_i$ for each sample. Additional information necessary for carrying out the likelihood ratio test based on (4.3.7) or the approximate Bartlett's test based on (4.3.10) is also given. In particular, $\tilde{\beta}$ and $\tilde{\alpha}_1, \ldots, \tilde{\alpha}_7$ are the m.l.e.'s of β_i and $\alpha_1, \ldots, \alpha_7$ under $H_0: \beta_1 = \cdots = \beta_7$, and $b^* = \Sigma g_i \hat{b}_i / h_i$. The values of

Table 4.3.2 Insulating Fluid Failure Data

Voltage Level (kV)	n_i	Breakdown Times
26	3	5.79, 1579.52, 2323.7
28	5	68.85, 426.07, 110.29, 108.29, 1067.6
30	11	17.05, 22.66, 21.02, 175.88, 139.07, 144.12, 20.46, 43.40, 194.90, 47.30, 7.74
32	15	0.40, 82.85, 9.88, 89.29, 215.10, 2.75, 0.79, 15.93, 3.91, 0.27, 0.69, 100.58, 27.80, 13.95, 53.24
34	19	0.96, 4.15, 0.19, 0.78, 8.01, 31.75, 7.35, 6.50, 8.27, 33.91, 32.52, 3.16, 4.85, 2.78, 4.67, 1.31, 12.06, 36.71, 72.89
36	15	1.97, 0.59, 2.58, 1.69, 2.71, 25.50, 0.35, 0.99, 3.99, 3.67, 2.07, 0.96, 5.35, 2.90, 13.77
38	8	0.47, 0.73, 1.40, 0.74, 0.39, 1.13, 0.09, 2.38

Table 4.3.3 Estimates and Other Quantities Calculated From Voltage Breakdown Data[a]

Sample (kV)	n_i	$\hat{\beta}_i$	$\hat{b}_i = \hat{\beta}_i^{-1}$	$\hat{\alpha}_i$	$\tilde{\alpha}_i$	h_i
26	3	0.545	1.834	955.8	1168.4	6.5
28	5	0.978	1.022	352.5	319.1	12.9
30	11	1.058	0.945	77.58	69.11	32.6
32	15	0.561	1.781	28.94	33.72	45.9
34	19	0.771	1.297	12.22	11.46	59.1
36	15	0.889	1.125	4.291	3.692	45.9
38	8	1.362	0.734	1.001	0.839	22.7

[a] $\tilde{\beta}=0.787$, $h=\Sigma h_i=225.6$, $g_i=h_i+2$, $b^*=\Sigma g_i \hat{b}_i / h = 1.330$.

$h_i = h(n_i, n_i)$, taken from Table 4.1.2, are also shown in the table. The statistic b^* is a good estimate of the common scale parameter b under H_0 and is typically close to the m.l.e. $\tilde{b}=\tilde{\beta}^{-1}$ of b under H_0. In this case we note that $b^*=1.330$ and $\tilde{b}=0.787^{-1}=1.271$.

The likelihood ratio statistic (4.3.7) gives a value of $\Lambda=9.77$. Treating Λ as $\chi^2_{(6)}$ under H_0 then gives a significance level of about .14. The statistic (4.3.10), on the other hand, gives an M value of 8.35, which gives a significance level of about .22 when we treat M as $\chi^2_{(6)}$. The results of the two tests are in broad agreement, and it appears that there is no real evidence of a difference in shape parameters.

4.3.2 Comparison of Weibull Scale Parameters or Quantiles

Comparison of Weibull scale parameters is of interest mainly when the shape parameters of the distributions are equal: the ratio of the pth quantiles of two distributions then equals α_1/α_2, the ratio of their scale parameters. When the shape parameters are unequal, comparison of the scale parameters is less important, and in the discussion below shape parameters will be taken to be equal.

With just two distributions we may wish to test $H_0: \alpha_1=\alpha_2$ or obtain confidence intervals for α_1/α_2. We may also want confidence intervals for $(\alpha_1/\alpha_2)^\beta$, which is the ratio of the hazard functions for the two distributions, when $\beta_1=\beta_2=\beta$. In terms of the extreme value distribution the problem is to make inferences about $u_1-u_2=\log(\alpha_1/\alpha_2)$, or perhaps about $(u_1-u_2)/b=\beta\log(\alpha_1/\alpha_2)$. If the data from each distribution are Type II censored, a natural approach is to consider a pivotal quantity such as

$$W_1 = \frac{(\tilde{u}_1 - \tilde{u}_2) - (u_1 - u_2)}{\tilde{b}} = \tilde{\beta} \log\left(\frac{\tilde{\alpha}_1/\alpha_1}{\tilde{\alpha}_2/\alpha_2} \right)$$

where $\tilde{u}_1$, $\tilde{u}_2$, and $\tilde{b}$ are the m.l.e.'s of u_1, u_2, and b under the hypothesis $H_0: b_1 = b_2 = b$.

One cannot determine percentage points for W_1 analytically. Schafer and Sheffield (1976) have generated by Monte Carlo methods a table of percentage points for the case in which samples from the two distributions are uncensored and of the same size. Another approach is taken by Engelhardt and Bain (1979), who develop a rather involved F approximation for W_1, using ideas similar to those in Section 4.1.2d.

Thoman and Bain (1969) and McCool (1970, 1974, 1975b) consider the statistic

$$W_2 = \frac{\hat{\beta}_1 + \hat{\beta}_2}{2} \log\left(\frac{\hat{\alpha}_1/\alpha_1}{\hat{\alpha}_2/\alpha_2} \right) \tag{4.3.11}$$

instead of W_1, where $(\hat{\beta}_1, \hat{\alpha}_1)$ and $(\hat{\beta}_2, \hat{\alpha}_2)$ are the m.l.e.'s for β and α in Type II censored samples from the two distributions. The distribution of W_2 is also intractable, but Thoman and Bain (1969) and McCool (1974) give some Monte Carlo generated tables of percentage points. Thoman and Bain's tables cover complete samples, with $n_1 = n_2$, and McCool's cover 17 situations, with $(r_1, n_1) = (r_2, n_2)$, the (r_1, n_1) combinations being those mentioned in Section 4.3.1a.

If the data are not all Type II censored of if the sample sizes are not among those few covered in the aforementioned tables, then likelihood ratio methods can be employed. Consider the general situation in which there are m Weibull distributions with equal shape parameters $\beta_1 = \cdots = \beta_m = \beta$ and we wish to test $H_0: \alpha_1 = \cdots = \alpha_m$, the alternative being that the α_i's are not all equal. Suppose that from the jth distribution there is a sample that may be Type I or Type II censored, consisting of observed lifetimes or censoring times t_{ji} $(i = 1, \ldots, n_j)$, of which $r_j \geqslant 1$ are lifetimes. The log likelihood is exactly as given in (4.3.6), except that $\beta_1 = \cdots = \beta_m = \beta$. The likelihood ratio statistic for testing $H_0: \alpha_1 = \cdots = \alpha_m, \beta_1 = \cdots = \beta_m$ versus $H_1: \alpha_i$'s not all equal, $\beta_1 = \cdots = \beta_m$, is

$$\Lambda = -2\log L(\alpha^*, \ldots, \alpha^*, \beta^*, \ldots, \beta^*) + 2\log L(\tilde{\alpha}_1, \ldots, \tilde{\alpha}_m, \tilde{\beta}, \ldots, \tilde{\beta}) \tag{4.3.12}$$

where β^* and α^* are the m.l.e.'s of $\beta_1 = \cdots = \beta_m = \beta$ and $\alpha_1 = \cdots = \alpha_m = \alpha$ under H_0, and $\tilde{\beta}_1 = \cdots = \tilde{\beta}_m = \tilde{\beta}$ and $\tilde{\alpha}_1, \ldots, \tilde{\alpha}_m$ are the m.l.e.'s under H_1. The latter estimates were discussed in the preceding section and are found from equations (4.3.8) and (4.3.9). To find β^* and α^* one must maximize $\log L(\boldsymbol{\alpha}, \boldsymbol{\beta})$ with respect to $\beta_1 = \cdots = \beta_m = \beta$ and $\alpha_1 = \cdots = \alpha_m = \alpha$. It is easily seen that this leads to the equations

$$\sum_{j=1}^{m} \sum_{i=1}^{n_j} t_{ji}^{\beta^*} \log t_{ji} \Big/ \sum_{j=1}^{m} \sum_{i=1}^{n_j} t_{ji}^{\beta^*} - \frac{1}{\beta^*} - \frac{1}{r_.} \sum_{j=1}^{m} \sum_{i \in D_j} \log t_{ji} = 0 \qquad (4.3.13)$$

$$\alpha^* = \left(\frac{1}{r_.} \sum_{j=1}^{m} \sum_{i=1}^{n_j} t_{ji}^{\beta^*} \right)^{1/\beta^*} \qquad (4.3.14)$$

where $r_. = \Sigma r_j$. These are, of course, the maximum likelihood equations for a sample from a single Weibull distribution, consisting of observations on Σn_j individuals, with $r_.$ observed lifetimes. In large samples the distribution of Λ under H_0 is approximately $\chi^2_{(m-1)}$, and this can be used to calculate approximate significance levels.

Confidence Intervals for α_1/α_2

When two distributions are under consideration, confidence intervals for α_1/α_2 may be of interest. These can be obtained by considering the likelihood ratio test of $H_0: \alpha_1 = a\alpha_2, \beta_1 = \beta_2$ versus $H_1: \alpha_1 \neq a\alpha_2, \beta_1 = \beta_2$. A γ confidence interval for α_1/α_2 consists of those a values for which H_0 is not rejected at the $1-\gamma$ level of significance. With the approximation $\Lambda \sim \chi^2_{(1)}$, this entails finding the set of a values such that $\Lambda \leqslant \chi^2_{(1),\gamma}$. The following observation simplifies matters: note that under $H_0: \alpha_1 = a\alpha_2$ $(\beta_1 = \beta_2)$ the log likelihood (4.3.6), when rearranged slightly, is

$$\log L(a\alpha_2, \alpha_2, \beta, \beta) = (r_1 + r_2)\log\beta - r_1\beta\log\alpha_2 - r_2\beta\log\alpha_2 + (\beta - 1)$$

$$\times \left(\sum_{i \in D_1} \log\left(\frac{t_{1i}}{a}\right) + \sum_{i \in D_2} \log t_{2i} \right) - \sum_{i=1}^{n_1} \left(\frac{t_{1i}}{a\alpha_2}\right)^{\beta} - \sum_{i=1}^{n_2} \left(\frac{t_{2i}}{\alpha_2}\right)^{\beta} - r_1 \log a.$$

Aside from the additive constant $-r_1 \log a$, this is exactly the log likelihood that we would have if $\alpha_1 = \alpha_2, \beta_1 = \beta_2$, and the observations t_{1i} were all replaced by t_{1i}/a. Thus to test $H_0: \alpha_1 = a\alpha_2$ $(\beta_1 = \beta_2)$ we merely need to replace t_{1i} by t_{1i}/a $(i = 1, \ldots, n_1)$ and go through the procedure for testing that $\alpha_1 = \alpha_2$. This involves solving (4.3.13) and (4.3.14) with the redefined

t_{1i}'s to find α^* and β^*; then $\alpha_1^* = a\alpha^*, \alpha_2^* = \alpha^*, \beta_1^* = \beta_2^* = \beta^*$, and we can calculate $\Lambda = -2\log L(\alpha_1^*, \alpha_2^*, \beta^*, \beta^*) + 2\log L(\tilde{\alpha}_1, \tilde{\alpha}_2, \tilde{\beta}, \tilde{\beta})$.

Example 4.3.3 The data given below are the voltage levels at which failures occurred in two types of electrical cable insulation when specimens were subjected to an increasing voltage stress in a laboratory experiment. The test involved 20 specimens of each type, and the failure voltages (in kilovolts per millimeter) were

Type I Insulation	32.0, 35.4, 36.2, 39.8, 41.2, 43.3, 45.5, 46.0, 46.2, 46.4, 46.5, 46.8, 47.3, 47.3, 47.6, 49.2, 50.4, 50.9, 52.4, 56.3
Type II Insulation	39.4, 45.3, 49.2, 49.4, 51.3, 52.0, 53.2, 53.2, 54.9, 55.5, 57.1, 57.2, 57.5, 59.2, 61.0, 62.4, 63.8, 64.3, 67.3, 67.7

Engineering experience suggests that failure voltages for the two types of cable are adequately represented by Weibull distributions with a common shape parameter β; Stone and Lawless (1979) discuss this area. Probability plots of the two samples indicate that this assumption appears reasonable here, the points in the graph lying roughly along two parallel lines. The m.l.e.'s from the two samples are

$$\text{I}: \hat{\alpha}_1 = 47.781 \quad \hat{\beta}_1 = 9.383 \qquad \text{II}: \hat{\alpha}_2 = 59.125 \quad \hat{\beta}_2 = 9.141.$$

A test of equality of β_1 and β_2 reveals no evidence that they are different. Under the assumption that β_1 and β_2 are equal, m.l.e.'s for the common β value, α_1, and α_2 are found to be

$$\tilde{\beta} = 9.261 \qquad \tilde{\alpha}_1 = 47.753 \qquad \tilde{\alpha}_2 = 59.161$$

by solving equations (4.3.8) and (4.3.9).

To test $\alpha_1 = \alpha_2$ or to get confidence intervals for α_1/α_2, either likelihood ratio methods or the tables of Thoman and Bain (1969) or Schafer and Sheffield (1976) can be used. The likelihood ratio procedure gives a .90 confidence interval $0.762 \leqslant \alpha_1/\alpha_2 \leqslant 0.855$. This is found by determining the set of a values such that

$$\Lambda = -2\log L(a\alpha_2^*, \alpha_2^*, \beta^*, \beta^*) + 2\log L(\tilde{\alpha}_1, \tilde{\alpha}_2, \tilde{\beta}, \tilde{\beta}) \leqslant 2.706$$

where $2.706 = \chi^2_{(1),.90}$. Alternately, we can refer to Table 2 of Thoman et al. (1969), which gives percentage points for (4.3.11), and find that

$$\Pr(-0.593 \leqslant W_2 \leqslant 0.593) = .90$$

This gives the confidence interval $0.758 \leqslant \alpha_1/\alpha_2 \leqslant 0.861$. The two methods in this case give results that are in good agreement.

Final Remark

It is sometimes desired to compare quantiles of two Weibull distributions that do not have equal shape parameters. The ratio of the two pth quantiles in this case depends on p [see (4.3.1)]. Likelihood ratio procedures can be applied to give confidence intervals for this ratio, but often it will be sufficient to simply obtain and compare confidence intervals for the pth quantile in each distribution by using the methods of Sections 4.1 or 4.2.

4.4 ADDITIONAL PROBLEMS

Because of the importance of the Weibull model, it is worthwhile touching upon, at least briefly, two other problems. These are, respectively, methods for dealing with the inclusion of a threshold parameter and the planning of life test experiments under the Weibull model. The discussion also illustrates the difficulty in dealing satisfactorily with these problems in distributions other than the exponential distribution.

4.4.1 Inference With the Three-Parameter Weibull Distribution

The three-parameter Weibull distribution includes a threshold parameter μ and has p.d.f.

$$f(t;\mu,\alpha,\beta)=\frac{\beta}{\alpha}\left(\frac{t-\mu}{\alpha}\right)^{\beta-1}\exp\left[-\left(\frac{t-\mu}{\alpha}\right)^{\beta}\right] \qquad t \geqslant \mu.$$

The three-parameter model is appropriate when there is a time μ before which no deaths can occur. Often it is satisfactory to assume that μ is known and to then treat observations $t-\mu$ as samples from the two-parameter Weibull distribution and use methods discussed earlier in this chapter. By doing this for a few different μ values one can usually obtain a satisfactory picture of things. If μ is formally treated as an unknown parameter, matters are more difficult. This is not unusual: similar difficulties arise when a threshold parameter is introduced into other distributions such as the log-normal or gamma distribution. A brief discussion of some of these problems follows.

Maximum likelihood estimation for the three-parameter Weibull distribution has been discussed by many authors (e.g., Harter and Moore, 1967;

Lemon, 1975; Rockette et al., 1974). As usual, suppose that we have a possibly censored sample, and let T_i $(i=1,\ldots,n)$ and L_i $(i=1,\ldots,n)$ represent lifetimes and censoring times, respectively, for the n individuals involved. Let $t_i=\min(T_i, L_i)$ and $\delta_i=1$ (if $t_i=T_i$) or 0 (if $t_i=L_i$); the likelihood function is then of the form (1.4.9),

$$L(\mu,\alpha,\beta)=\prod_{i=1}^{n} f(t_i;\mu,\alpha,\beta)^{\delta_i} S(t_i;\mu,\alpha,\beta)^{1-\delta_i}$$

where $S(t;\mu,\alpha,\beta)=\exp\{-[(t-\mu)/\alpha]^{\beta}\}$ is the survivor function. The observed likelihood function is thus

$$L(\mu,\alpha,\beta)=\frac{\beta^r}{\alpha^r}\prod_{i\in D}\left(\frac{t_i-\mu}{\alpha}\right)^{\beta-1}\prod_{i=1}^{n}\exp\left[-\left(\frac{t_i-\mu}{\alpha}\right)^{\beta}\right]$$

where D is the set of individuals whose lifetimes are observed and $r=|D|$. The log likelihood is

$$\log L(\mu,\alpha,\beta)=r\log\beta-r\beta\log\alpha+(\beta-1)\sum_{i\in D}\log(t_i-\mu)-\sum_{i=1}^{n}\left(\frac{t_i-\mu}{\alpha}\right)^{\beta}. \tag{4.4.1}$$

Setting the derivatives of $\log L$ equal to zero produces a set of equations

$$\frac{\partial \log L}{\partial \alpha}=\frac{-r\beta}{\alpha}+\frac{\beta}{\alpha}\sum_{i=1}^{n}\left(\frac{t_i-\mu}{\alpha}\right)^{\beta}=0$$

$$\frac{\partial \log L}{\partial \beta}=\frac{r}{\beta}-r\log\alpha+\sum_{i\in D}\log(t_i-\mu)-\sum_{i=1}^{n}\left(\frac{t_i-\mu}{\alpha}\right)^{\beta}\log\left(\frac{t_i-\mu}{\alpha}\right)=0 \tag{4.4.2}$$

$$\frac{\partial \log L}{\partial \mu}=-(\beta-1)\sum_{i\in D}(t_i-\mu)^{-1}+\frac{\beta}{\alpha}\sum_{i=1}^{n}\left(\frac{t_i-\mu}{\alpha}\right)^{\beta-1}=0.$$

The main difficulty with maximum likelihood estimation is that $L(\mu,\alpha,\beta)$ is unbounded. In particular, for any $\beta<1$, $L(\mu,\beta,\alpha)\to\infty$ as μ approaches $\min_{i\in D}(t_i)$ from below. Therefore the solution to equations (4.4.2) does not produce a global maximum to the likelihood function. It is also possible for

(4.4.2) to not possess a solution. Results given by Rockette et al. (1974) and Pike (1966) show that (provided that $r \geqslant 1$)

1. The equations (4.4.2) can have two or no solutions. In the case in which there are two solutions, one is a local maximum of $L(\mu, \alpha, \beta)$ and the other is a saddle point. There cannot be just one solution to (4.4.2). It also appears from empirical evidence that there cannot be more than two solutions, though this has not been proven.
2. If attention is restricted to β values greater than or equal to unity, then when no solution to (4.4.2) exists, the maximum of $L(\mu, \alpha, \beta)$ occurs at

$$\hat{\mu} = \min_{i \in D} (t_i) \qquad \hat{\beta} = 1 \qquad \hat{\alpha} = \sum_{i=1}^{n} \frac{t_i - \hat{\mu}}{r}. \tag{4.4.3}$$

   In the case in which the equations (4.4.2) have a solution that is a local maximum of $L(\mu, \alpha, \beta)$, this maximum may not necessarily be the global maximum. It is possible that (4.4.3) gives the global maximum in this case.

In most situations where the Weibull model is used, β can be assumed to be $\geqslant 1$; only then is the failure rate increasing. The following procedure can be used in such cases:

1. Obtain the local maximum of $L(\mu, \alpha, \beta)$, if one exists, by solving (4.4.2), or otherwise, and compare it with (4.4.3). Choose whichever of these gives the largest value of $L(\mu, \alpha, \beta)$ as the estimate.
2. If there is no solution to (4.4.2), and hence no local maximum, choose (4.4.3) as the estimate.
3. In any case, the fit of the Weibull model with the estimated parameter values to the data should be examined. Some idea of the shape of the likelihood function is also helpful. In many instances the likelihood function yields little information on μ, and a wide range of μ values will yield similar estimates of $S(t)$.

A useful way of determining plausible values for μ and finding an m.l.e. is to obtain the maximized likelihood function for μ. This involves computing the m.l.e.'s $\hat{\alpha}(\mu)$ and $\hat{\beta}(\mu)$ of α and β for different fixed values of μ. The estimates are found in the usual way for the two-parameter Weibull distribution, replacing t_i by $t_i - \mu$ in equations (4.2.5) and (4.2.6). The maximized likelihood function for μ is then $L[\mu, \hat{\alpha}(\mu), \hat{\beta}(\mu)] = L_{\max}(\mu)$. A graph of this shows plausible values for μ, and in most instances allows $\hat{\mu}$ to be accurately determined. Note that only values of μ that are less than or equal to $\min_{i\in D}(t_i)$ need be considered. If the m.l.e. $(\hat{\mu}, \hat{\alpha}, \hat{\beta})$ is obtained, the maximized relative likelihood function $R_{\max}(\mu) = L[\mu, \hat{\alpha}(\mu), \hat{\beta}(\mu)]/L(\hat{\mu}, \hat{\alpha}, \hat{\beta})$ can also be calculated.

Concerning tests and interval estimation, several authors have discussed a few special problems. These include Mann and Fertig (1975b), who discuss tests and confidence bounds for μ, Klimko et al. (1975), and Engelhardt and Bain (1975), who consider tests of $\beta=1$ versus $\beta>1$. Problems arise in applying large-sample theory because of the presence of the threshold parameter (e.g., Harter and Moore, 1967; Lemon, 1975). The most sensible approach is usually to carry out inferences about β, quantiles, or other characteristics of the distribution, assuming μ to be known. This reduces the problems to those of the two-parameter Weibull distribution. Results can be obtained for different values of μ, determined by examining $L_{\max}(\mu)$. The following example illustrates this.

Example 4.4.1 (Example 2.4.1 revisited) Pike (1966) has given some data from a laboratory investigation of vaginal cancer in rats. In the experiment, vaginas of rats were painted with the carcinogen DMBA, and the number of days T until a carcinoma appeared was recorded. The data given T for a group of 19 rats (Group 1 in Pike's paper), of which 17 had developed carcinomas at the time the data were collected. There are consequently two censored observations, which are denoted by asterisks:

143, 164, 188, 188, 190, 192, 206, 209, 213, 216, 220, 227, 230, 234, 246, 265, 304, 216*, 244*

These data were examined in Example 2.4.1, where it was noted that when 100 was subtracted from each observation, the resultant values appeared consonant with a two-parameter Weibull model (see Figure 2.4.2). This means that a three-parameter Weibull model, with μ in the neighborhood of 100, is reasonable. Let us consider this further.

Table 4.4.1 shows some quantities calculated while determining the maximized relative likelihood function $R_{\max}(\mu)$. For each value of μ shown, m.l.e.'s $\hat{\alpha}(\mu)$ and $\hat{\beta}(\mu)$ were obtained by solving the first two equations of (4.4.2), and $\log L[\mu, \hat{\alpha}(\mu), \hat{\beta}(\mu)]$ was computed. From this we see that a local maximum of $L(\mu, \alpha, \beta)$ occurs at about $\mu=122$, $\beta=2.712$, $\alpha=108.4$, though the likelihood function is very flat in the region of this point. A graph of $R_{\max}(\mu)$ is given in Figure 4.4.1, where it is seen that no values of μ are particularly implausible, except for values of μ very close to 143. The likelihood function becomes infinite as μ approaches 143 if we allow β values less than unity. In the present situation it is reasonable to suppose

Table 4.4.1 M.l.e.'s and Relative Likelihood for Various Values of μ

μ	$\hat{\beta}(\mu)$	$\hat{\alpha}(\mu)$	$\log L[\mu, \hat{\alpha}(\mu), \hat{\beta}(\mu)]$	$R_{\max}(\mu)$
0	6.083	234.3	−88.233	.368
20	5.556	214.0	−88.118	.452
60	4.487	173.2	−87.833	.602
80	3.940	152.6	−87.656	.717
100	3.377	131.8	−87.465	.870
110	3.084	121.2	−87.380	.946
120	2.776	110.6	−87.326	.998
122	2.712	108.4	−87.324	1.000
125	2.614	105.2	−87.330	.994
130	2.440	99.65	−87.382	.944
135	2.244	94.02	−87.542	.804
140	1.985	88.00	−88.064	.477
142	1.795	85.16	−88.773	.235
143	1.0	81.06	−91.718	.012

that $\beta \geqslant 1$, and this will not concern us. The last row of Table 4.4.1 gives the relative likelihood for the estimate (4.4.3), obtained when $\beta = 1$.

It should be noted that although $\hat{\beta}$ and $\hat{\alpha}$ vary substantially with μ, estimates of characteristics such as quantiles of the distribution do not. For example, the m.l.e. of the pth quantile is $\hat{t}_p = \mu + \hat{\alpha}(\mu)[-\log(1-p)]^{1/\hat{\beta}(\mu)}$, with a particular value for μ. These are easily calculated from the results of Table 4.4.1. Results for $t_{.10}$, $t_{.50}$, and $t_{.90}$, for $\mu = 60$, 100, and 140, are, for

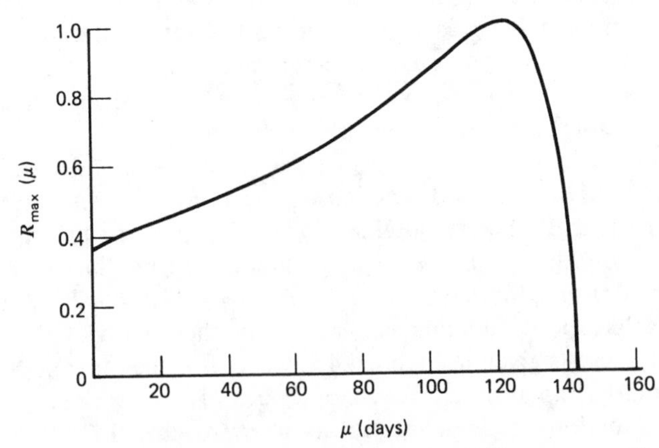

Figure 4.4.1 Maximized relative likelihood for Weibull threshold parameter μ (Example 4.4.1)

example, as follows:

μ	$\hat{t}_{.10}$	$\hat{t}_{.50}$	$\hat{t}_{.90}$
60	164.9	219.6	268.6
100	167.7	218.2	268.7
140	168.3	213.2	274.0

The m.l.e.'s of the quantiles are relatively unaffected by the assumed value of μ. One would suspect this, because plots of the data with the different μ values indicate an adequate fit to the Weibull model in each instance.

Finally, the value $\mu=0$ is often of special interest. With these data, $\mu=0$ is not implausible, the maximized relative likelihood being 0.368. It is not appropriate to treat $-2\log[R_{\max}(0)]$ as approximately $\chi^2_{(1)}$ under H_0, since μ is a threshold parameter, though under suitable conditions it can be treated as $\chi^2_{(2)}$ (see Section 3.5). If the data were Type II instead of Type I censored, a statistic given by Mann and Fertig (1975b) could be used to test that $\mu=0$ (see Problem 9.9).

4.4.2 Life Test Plans Under a Weibull Model

If there is choice in the selection of an experimental plan, then testing costs, time restrictions, and other physical constraints are important considerations. Some of these were discussed in Section 3.4 for the case of an underlying exponential distribution. Life test plans under the Weibull model have not been very thoroughly investigated, however, because of the complexity of the associated distributional problems. This is unfortunate and sometimes leads to the use of plans based on the exponential distribution when plans based on the Weibull distribution would be more appropriate. This is dangerous in view of the nonrobustness of exponential life test plans described in Section 3.6. A brief discussion of problems associated with Weibull life test plans follows. There are many possible types of life test plans, including the simple Type I or Type II censored plans discussed earlier in this chapter. Even for quite complicated plans the large-sample methods given for Type I censored data generally provide valid procedures. However, it is almost always impossible to determine exact small-sample properties or to make effective comparisons of plans, except by simulation. With a Type II censored plan involving no replacement of items, it is possible to examine properties of the test fairly easily, but for many plans even determination of the expected duration of the test can be difficult. Therefore little formal discussion of the merits of different types of plans has taken place when the underlying model was a Weibull model.

Formal Acceptance Procedures

To elaborate on these points in one special situation let us consider the formulation of life test acceptance plans as described in Section 3.4. When the underlying distribution is a Weibull distribution, such plans are difficult to derive, except for the case of Type II censored samples with no replacement of items. For this situation tests can be set up using the results given in Section 4.1, though power functions are usually not easy to obtain. Let us examine this briefly.

Hypothesis tests about β pose few problems, particularly in cases where the χ^2 approximations of Section 4.1.2d can be used to determine critical regions and power functions (see Problem 4.13). A more important problem from the point of view of acceptance sampling is the development of tests for the quantiles or the survivor function of the distribution. Consider, for example, the corresponding hypothesis for the extreme value distribution,

$$H_0: x_p = x_0 \quad \text{vs.} \quad H_1: x_p < x_0$$

where $x_p = u + b\log[-\log(1-p)]$ is the pth quantile of the distribution. Tests of H_0 versus H_1 are available from results about the pivotals $Z_p = (\tilde{u} - x_p)/\tilde{b}$ discussed in Section 4.1, but power functions for the tests require additional calculations. In particular, size γ tests of H_0 versus H_1 will have a rejection region of the form $(\tilde{u} - x_0)/\tilde{b} \leqslant z_{p,\gamma}$, where $z_{p,\gamma}$ is the γth quantile of Z_p; note that this gives $\Pr(\text{reject } H_0; x_p = x_0) = \Pr[(\tilde{u} - x_p)/\tilde{b} \leqslant z_{p,\gamma}] = \Pr(Z_p \leqslant z_{p,\gamma}) = \gamma$. The power function is then found as

$$\begin{aligned} P(x) &= \Pr(\text{reject } H_0; x_p = x) \\ &= \Pr\left(\frac{\tilde{u} - x_0}{\tilde{b}} \leqslant z_{p,\gamma}; x_p = x \right) \\ &= \Pr\left(Z_p + \frac{x - x_0}{\tilde{b}} \leqslant z_{p,\gamma} \right) \end{aligned}$$

which is not directly obtainable from percentage points of Z_p. Since the exact distributions of Z_p and $\tilde{b}$ are intractable, power functions have to be determined by Monte Carlo methods or through the use of approximations. Thoman and Bain (1969) give a few power functions obtained by Monte Carlo methods for tests about the location parameter u $(= x_{.632})$, with pivotals based on the m.l.e.'s, and McCool (1974) gives some similar results for $x_{.10}$ and $x_{.50}$. Mann and Fertig (1980) give an approximation to the distribution of Z_p when b.l.i.e.'s or b.l.u.e.'s are used, which makes power calculations possible. This approximation is used by Fertig and Mann (1980) to tabulate some Type II censored life test plans.

Further development of test plans under a Weibull model would be useful. Plans for comparing distributions are also of interest. Thoman and Bain (1969) present a few power functions for two-sample problems, but not a great deal else is available, except for the problem of comparing Weibull shape parameters. In this case power calculations for the approximate F tests discussed in Section 4.3.1 are easily carried out.

4.5 PROBLEMS AND SUPPLEMENTS

4.1 Linear estimation of location and scale parameters. Let $X_{(1)} \leqslant \cdots \leqslant X_{(n)}$ be the ordered observations in a random sample of n from a location–scale parameter distribution with p.d.f. of the form

$$f(x; u, b) = \frac{1}{b} g\left(\frac{x-u}{b}\right).$$

Let $Z_{(i)} = (X_{(i)} - u)/b, i = 1, \ldots, n$, be the standardized order statistics, and define

$$\alpha_i = E\left(Z_{(i)}\right) \qquad v_{ij} = \mathrm{Cov}\left(Z_{(i)}, Z_{(j)}\right) \qquad i, j = 1, \ldots, n.$$

a. Show that if $\mathbf{X} = (X_{(1)}, \ldots, X_{(n)})'$, then $E(\mathbf{X}) = \mathbf{A}\boldsymbol{\theta}$ and $\mathrm{Var}(\mathbf{X}) = b^2\mathbf{V}$, where

$$\mathbf{A} = \begin{bmatrix} 1 & \alpha_1 \\ 1 & \alpha_2 \\ \vdots & \vdots \\ 1 & \alpha_n \end{bmatrix} \qquad \boldsymbol{\theta} = (u, b)' \qquad \mathbf{V} = (v_{ij})_{n \times n}.$$

Thus show that the linear unbiased estimators of u and b that have minimum variance are given by

$$\tilde{\boldsymbol{\theta}} = (\tilde{u}, \tilde{b})' = (\mathbf{A}'\mathbf{V}^{-1}\mathbf{A})^{-1}\mathbf{A}'\mathbf{V}^{-1}\mathbf{X}$$

and that the covariance matrix for $(\tilde{u}, \tilde{b})'$ is $(\mathbf{A}'\mathbf{V}^{-1}\mathbf{A})^{-1}b^2$. Calculation of the best linear unbiased estimates of u and b for a given distribution therefore requires knowledge of the means, variances, and covariances of the standardized order statistics in samples from the distribution.

(Section 4.1.1; Lloyd, 1952; Kendall and Stuart, 1967, Ch. 19)

b. Let $\phi = l_1 u + l_2 b$ and $\tilde{\phi} = l_1 \tilde{u} + l_2 \tilde{b}$ and suppose that $\mathrm{Var}(\tilde{\phi}) = Ab^2$, $\mathrm{Var}(\tilde{b}) = Cb^2$, and $\mathrm{Cov}(\tilde{\phi}, \tilde{b}) = Bb^2$. Define new estimators

$$b^* = \frac{\tilde{b}}{1+C} \qquad \phi^* = \tilde{\phi} - \frac{B}{1+C}\tilde{b}.$$

Prove that the mean-square errors of b^* and ϕ^* are less than those of $\tilde{b}$ and $\tilde{\phi}$, respectively. In fact, it can be shown that b^* and ϕ^* are the best linear invariant estimators of b and ϕ.

(Section 4.1.1; Mann, 1969)

4.2 Equivariant estimators of location and scale parameters. Let u and b be location and scale parameters in a model with p.d.f. of the form

$$f(x; u, b) = \frac{1}{b} g\left(\frac{x-u}{b}\right) \qquad -\infty < x < \infty \tag{4.5.1}$$

as defined in Section 4.1.2a.

a. Consider linear estimators of u and b of the form (4.1.9), based on a Type II censored sample from (4.5.1). Determine necessary and sufficient conditions on the coefficients $a_i(n,r)$ and $c_i(n,r)$ so that the estimators are equivariant, that is, satisfy (4.1.11) and (4.1.12).
b. Show that the b.l.u.e.'s of u and b (see Problem 4.1) are equivariant (Hint: show that $\Sigma_{i=1}^{r} \alpha_i = 0$).
c. Let $\hat{u}$ and $\hat{b}$ be m.l.e.'s of u and b obtained from a Type II censored sample from (4.5.1). Show that $\hat{u}$ and $\hat{b}$ are equivariant. (Hint: consider the effect of data transformations on either the likelihood function or the maximum likelihood equations.)

(Sections 4.1.1, 4.1.2; Appendix G)

4.3

a. Determine the expected information matrix $\boldsymbol{I}(a, b)$ for u and b based on a complete sample of n observations from the extreme value distribution (4.0.2) by letting $c \to \infty$ and evaluating the expressions in (4.2.8). Thus show that the covariance matrix for the asymptotic normal distribution of $(\hat{u}, \hat{b})$ is

$$\mathbf{I}(u,b)^{-1} = \frac{b^2}{n}\begin{pmatrix} 1+\dfrac{6}{\pi^2}(1-\gamma)^2 & -\dfrac{6}{\pi^2}(1-\gamma) \\ -\dfrac{6}{\pi^2}(1-\gamma) & \dfrac{6}{\pi^2} \end{pmatrix}$$

where $\gamma = 0.5772\ldots$ is Euler's constant.

b. For $n=30$ compare exact percentage points of $\hat{b}/b$ given by Table 1 of Thoman et al. (1969) with approximate percentage points obtained from the asymptotic normal approximation $\hat{b}/b \sim N(1, 6/\pi^2 n)$ obtained from part (a). Compare the accuracy of this approximation with (1) that which treats $\sqrt{n}\log(\hat{b}/b) \sim N(0, 6/\pi^2)$, and (2) the χ^2 approximation (4.1.21). (Note: Thoman et al. give percentage points of $\hat{\beta}/\beta = b/\hat{b}$ in their table.)

c. McCool (1974) gives the following percentage points for $(\hat{x}_p - x_p)/\hat{b}$, for $n=r=30$ and $p=.10$:

1	5	10	90	95	99
−0.790	−0.567	−0.442	0.706	0.915	1.389

Compare these with approximate percentage points of $(\hat{x}_p - x_p)/\hat{b}$ obtained by using the asymptotic normal approximation $(\hat{u}, \hat{b}) \sim N_2[(u, b), I(u, b)^{-1}]$.

4.4 Show that the conditional methods developed in Section 4.1.2b apply to the case of progressively Type II censored data. Specifically, consider the sampling distribution (1.4.4) that applies to progressive censoring with two stages and show that the results developed in Appendix G are still valid.

(Section 4.1.2)

4.5 Bayesian methods with an improper prior distribution. Consider a Type II censored sample from the extreme value distribution (4.0.2). Determine the form of the posterior distribution for u and b if the prior distribution is

$$p(u, b) = \frac{1}{b} \qquad -\infty < u < \infty \quad b > 0.$$

Determine the marginal posterior distributions for u and b and show that posterior probability intervals for u and b obtained from these are numerically equivalent to conditional confidence intervals obtained by using (4.1.17) and (4.1.18). This shows that conditional confidence intervals can, if desired, be calculated as Bayesian posterior probability intervals.

(Section 4.1.2; Bogdanoff and Pierce, 1973)

4.6 Computer generation of a Type II censored Weibull sample. Let $T_{(1)} \leqslant \cdots \leqslant T_{(r)}$ be the r smallest observations in a sample of n from the Weibull distribution (4.0.1). By noting that $Y_{(i)} = (T_{(i)}/\alpha)^\beta$ are the ordered observations in a random sample from the standard exponential distribution, show how to use the result of Theorem 3.1.1 to generate $T_{(1)}, \ldots, T_{(n)}$

sequentially. Specifically, show that the $T_{(i)}$'s can be represented as

$$T_{(i)} = \alpha \left(\sum_{j=1}^{i} \frac{W_j}{n-j+1} \right)^{1/\beta}$$

where the W_j's are independent and have standard exponential distributions.

4.7 Prediction of a future observation. Suppose that $x_1 \leqslant \cdots \leqslant x_r$ are the r smallest observations in a sample of size n from the extreme value distribution (4.0.2) and let $\tilde{u}$ and $\tilde{b}$ be equivariant estimators of the parameters in the model, based on $x_1, \ldots, x_r$. If Y_1 is the smallest observation in a future sample of size m from the same distribution, then prediction intervals for Y_1 can be based on the quantity

$$U = \frac{Y_1 - \tilde{u}}{\tilde{b}}.$$

a. Show that U is a pivotal quantity.
b. Determine the joint distribution of $W_1 = (Y_1 - u)/b$, $Z_2 = \tilde{b}/b$, and $Z_3 = (\tilde{u} - u)/b$, conditional on the ancillary statistics $a_i = (x_i - \tilde{u})/\tilde{b}$ $(i = 1, \ldots, r)$. Noting that $U = (W_1 - Z_3)/Z_2$, show that the conditional survivor function for U, given $\mathbf{a}$, can be written as

$$\Pr(U \geqslant x | \mathbf{a}) = \int_0^\infty \left[t^{r-2} \exp\left(t \sum_{i=1}^{r} a_i \right) \Big/ \left(m e^{xt} + \sum_{i=1}^{r}{}^{*} e^{a_i t} \right)^r \right] dt.$$

This allows confidence intervals to be obtained for Y_1. Mann (1977) and Engelhardt and Bain (1979) discuss approximations to the unconditional distribution of U.

(Lawless, 1973; Section 4.1.2)

4.8 Consider the data in Example 4.2.1 concerning remission times for two groups of leukemia patients, one given a placebo and the other the drug 6-MP.

a. Supposing that the data arise from two Weibull distributions, test the hypothesis that the shape parameters in the two distributions are the same. If this hypothesis is reasonable, test the hypothesis that the shape parameter equals unity.

Table 4.5.1 Failure Times of Bearing Specimens

Type of Compound				
I	II	III	IV	V
3.03	3.19	3.46	5.88	6.43
5.53	4.26	5.22	6.74	9.97
5.60	4.47	5.69	6.90	10.39
9.30	4.53	6.54	6.98	13.55
9.92	4.67	9.16	7.21	14.45
12.51	4.69	9.40	8.14	14.72
12.95	5.78	10.19	8.59	16.81
15.21	6.79	10.71	9.80	18.39
16.04	9.37	12.58	12.28	20.84
16.84	12.75	13.41	25.46	21.51

b. Obtain confidence intervals for the ratio α_1/α_2 of the quantiles for the two distributions, assuming a common shape parameter.

c. Compare the confidence intervals for the ratio of two quantiles obtained in (b) with those obtained under the assumption that the two life distributions are exponential.

4.9 Table 4.5.1 shows results of an experiment designed to compare the performances of high-speed turbine engine bearings made out of five different compounds (McCool, 1979). The experiment tested 10 bearings of each type; the times to fatigue failure are given in units of millions of cycles.

a. Assuming that the failure times in each sample came from a Weibull distribution, obtain m.l.e.'s for α and β and find confidence intervals for the tenth percentile of each distribution ($x_{.10}$ is used as a rating life).

b. Carry out a comparison of the five failure time distributions and, in particular, the tenth percentiles of the distributions.

(Sections 4.1, 4.3)

4.10 Suppose that two Weibull distributions have the same shape parameter β but possibly different scale parameters α_1 and α_2. The ratio of the hazard functions then depends on $\delta=(\alpha_1/\alpha_2)^{\beta}$, and the survivor functions are related by $S_1(t)=S_2(t)^{\delta}$. Develop likelihood ratio tests for δ and show how to use these to obtain confidence intervals for δ.

(Section 4.3.2)

4.11 Consider the problem of testing

$$H_0: b=b_0 \quad \text{vs.} \quad H_1: b>b_0$$

where b is the scale parameter of an extreme value distribution. Consider size γ tests with critical regions of the form $\hat{b}/b_0 \geqslant z_{2,1-\gamma}$, where $\hat{b}$ is the m.l.e. of b from a Type II censored sample and $z_{2,1-\gamma}$ is the $(1-\gamma)$th quantile of $Z_2 = \hat{b}/b_0$ under H_0.

a. Show how to use (4.1.21) to calculate the approximate power function for such a test.

b. If $\gamma=0.05$ and $n=r=20$, use (4.1.21) to determine the critical region and calculate the power function. Repeat this for a sample with $n=20$ and $r=10$.

c. If $\gamma=0.05$, $b_0=1$, and the test is to have power .90 at $b=2$, how large should n be for a test with $r=n$?

(Section 4.4.2)

4.12 In Example 4.4.1 estimates of $t_{.50}$ were given for a lifetime distribution assumed to be three-parameter Weibull, with three different values $\mu=60$, 100, and 140 assumed for the threshold parameter. Obtain .95 confidence intervals for $t_{.50}$ in each of these situations via the likelihood ratio method of Section 4.2.3. Comment on the extent to which the intervals depend on μ.

(Sections 4.4.1, 4.2.3)

4.13 The data below are failure times for two types of polyethylene cable insulation, obtained from an accelerated life test. Of the 10 specimens of each type tested, 9 of each failed. Ordered failure times, in hours, are given below. The last time in each case is a censoring time.

Type I	5.1, 9.2, 9.3, 11.8, 17.7, 19.4, 22.1, 26.7, 37.3, 60*
Type II	11.0, 15.1, 18.3, 24.0, 29.1, 38.6, 44.2, 45.1, 50.9, 70*

Assuming that failures times for each type have a Weibull distribution, compare the two failure time distributions and assess the possible superiority of Type II insulation.

(Section 4.3)

CHAPTER 5

Inference Procedures for Some Other Models

Chapters 3 and 4 contained lengthy discussions of inference procedures for the exponential and Weibull distributions, the motivation for this being the importance of these models in lifetime distribution work. Several other parametric models also receive wide use, and we now consider inference under some of these. After the exponential and Weibull distributions, the two most frequently used models are the gamma and log-normal distributions; Sections 5.1 and 5.2 contain a fairly thorough discussion of inference problems for them. In addition, inference under the generalized gamma model is considered in Section 5.3. Discussion in each instance is limited to single-sample problems; problems involving two or more samples can be handled through regression models given in Chapter 6. Section 5.4 examines models with polynomial hazard functions and discrete mixture models, though in less detail than the gamma, log-normal, and generalized gamma distributions. These models have one or two features not possessed by the more common parametric distributions, for example, the flexibility to give a U-shaped hazard function. Finally, Section 5.5 briefly considers the estimation of parameters in a continuous parametric model on the basis of grouped data. This does not involve any new models, but is an important practical problem that warrants discussion.

5.1 THE GAMMA DISTRIBUTION

With uncensored data, some inference procedures for the gamma distribution are fairly straightforward and are discussed in a number of statistics textbooks. However, when data are censored, or when it is desired to obtain interval estimates of quantities such as quantiles or the survivor function of the distribution, matters are more complicated. This is one reason why the gamma distribution is less widely used as a lifetime distribution than is the Weibull distribution, or even the log-normal distribution. The gamma distribution is nevertheless a useful model, and a reasonably thorough presentation of inference procedures for it will be given. Properties and uses of the model were briefly discussed in Section 1.3.3.

5.1.1 Point Estimation

The two-parameter gamma distribution has p.d.f.

$$\frac{1}{\alpha\Gamma(k)}\left(\frac{t}{\alpha}\right)^{k-1}\exp\left(\frac{-t}{\alpha}\right) \qquad t>0 \tag{5.1.1}$$

where $\alpha>0$ and $k>0$ are unknown scale and shape parameters, respectively.

Uncensored Data

Estimation from a complete random sample $t_1,\ldots,t_n$ from (5.1.1) will be considered first, since it is straightforward. The likelihood function is

$$L(k,\alpha)=\frac{1}{\alpha^{nk}\Gamma(k)^n}\left(\prod_{i=1}^{n} t_i^{k-1}\right)\exp\left(-\sum_{i=1}^{n}\frac{t_i}{\alpha}\right).$$

Let

$$\bar{t}=\sum_{i=1}^{n}\frac{t_i}{n} \quad \text{and} \quad \tilde{t}=\left(\prod_{i=1}^{n} t_i\right)^{1/n}$$

represent the arithmetic and geometric means, and $L(k,\alpha)$ can be written as

$$L(k,\alpha)=\frac{\tilde{t}^{n(k-1)}}{\alpha^{nk}\Gamma(k)^n}\exp\left(\frac{-n\bar{t}}{\alpha}\right) \tag{5.1.2}$$

which shows that $\bar{t}$ and $\tilde{t}$ are jointly sufficient for k and α. The log likelihood function is

$$\log L(k,\alpha)=-nk\log\alpha-n\log\Gamma(k)+n(k-1)\log\tilde{t}-\frac{n\bar{t}}{\alpha}. \tag{5.1.3}$$

Setting $\partial\log L/\partial k$ and $\partial\log L/\partial\alpha$ equal to 0 and rearranging slightly, we get the likelihood equations

$$\hat{k}\hat{\alpha}=\bar{t} \tag{5.1.4}$$

$$\log \hat{k}-\psi(\hat{k})=\log\left(\frac{\bar{t}}{\tilde{t}}\right) \tag{5.1.5}$$

where $\psi(k)=d\log\Gamma(k)/dk=\Gamma'(k)/\Gamma(k)$ is the digamma function introduced in Section 1.3.3 (see Appendix B). It is a simple matter to solve (5.1.5) iteratively for $\hat{k}$. Values of $\psi(k)$ and the trigamma function $\psi'(k)=d\psi(k)/dk$, which are required if Newton's method is used to solve (5.1.5), are tabulated in Abramowitz and Stegun (1965, Ch. 6). Alternately, series expressions can be used to evaluate $\psi(k)$ or $\psi'(k)$. Useful approximations can be based on (see Appendix B)

$$\psi(k)=\log k-\frac{1}{2k}-\frac{1}{12k^2}+\frac{1}{120k^4}-\frac{1}{252k^6}+\cdots \tag{5.1.6}$$

$$\psi'(k)=\frac{1}{k}+\frac{1}{2k^2}+\frac{1}{6k^3}-\frac{1}{30k^5}+\frac{1}{42k^7}-\cdots. \tag{5.1.7}$$

These expressions, truncated at the last terms shown, are adequate unless k is fairly small. If k is small, the truncated series can be used in conjunction with the following recursion relations

$$\psi(z+1)=\psi(z)+\frac{1}{z} \qquad z>0 \tag{5.1.8}$$

$$\psi'(z+1)=\psi'(z)-\frac{1}{z^2} \qquad z>0 \tag{5.1.9}$$

which are derived directly from the well-known relationship $\Gamma(z+1)=z\Gamma(z)$ for the gamma function.

A very close approximation to $\hat{k}$ is given by the empirically determined formulas (see Johnson and Kotz, 1970, p. 189)

$$\hat{k}\doteq S^{-1}(0.5000876+0.1648852S-0.0544274S^2) \qquad 0<S\leqslant 0.5772$$

$$\hat{k}\doteq S^{-1}(17.79728+11.968477S+S^2)^{-1}$$

$$\times(8.898919+9.059950S+0.9775373S^2) \qquad 0.5772<S\leqslant 17 \tag{5.1.10}$$

where $S=\log(\bar{t}/\tilde{t})$. The relative error of these formulas is claimed to be less than 0.0001.

Many authors have studied maximum likelihood and other point estimation methods for the gamma distribution; Johnson and Kotz (1970, Ch. 17) give numerous references. From our point of view an extended discussion of point estimation in uncensored samples is unnecessary, however, and this topic will not be pursued further except to mention that moment estimates of k and α can be easily obtained. These are found by equating the mean $(k\alpha)$ and variance $(k\alpha^2)$ of (5.1.1) to the sample mean $\bar{t}$ and variance s^2, respectively, and solving to get $\tilde{\alpha}=s^2/t$ and $\tilde{k}=\bar{t}/\tilde{\alpha}$. These estimates are not appreciably less efficient than the m.l.e.'s in small- to moderate-size samples (see Mann et al., 1974, p. 262), though they are in large samples.

The usual large-sample normal approximations to the distribution of $(\hat{k}, \hat{\alpha})$ are easily obtained, though they are rather inaccurate unless n is very large. The second derivatives of $\log L$ are $\partial^2 \log L/\partial k^2 = -n\psi'(k)$, $\partial^2 \log L/\partial \alpha^2 = (nk\alpha - 2n\bar{t})/\alpha^3$, and $\partial^2 \log L/\partial k\, \partial \alpha = -n/\alpha$, from which the observed or expected information matrix can be determined. The expected information matrix is

$$\begin{pmatrix} n\psi'(k) & n/\alpha \\ n/\alpha & nk\alpha^2 \end{pmatrix}.$$

Example 5.1.1 The data below are survival times in weeks for 20 male rats that were exposed to a high level of radiation. These data were originally given by Furth et al. (1959) and have been discussed by Gross and Clark (1975, p. 104), Engelhardt and Bain (1977b) and others. The times are

$$152, 152, 115, 109, 137, 88, 94, 77, 160, 165,$$
$$125, 40, 128, 123, 136, 101, 62, 153, 83, 69$$

The arithmetic and geometric means of the 20 observations are $\bar{t}=113.45$ and $\tilde{t}=107.07$. To use (5.1.10) we first calculate that $S=\log(\bar{t}/\tilde{t})=0.0579$ and then find the approximation $\hat{k}=8.80$ from the first equation in (5.1.10). It is also found that $\hat{k}=8.80$ is the solution to (5.1.5), correct to two decimal places. The m.l.e.'s are thus $\hat{k}=8.80$ and $\hat{\alpha}=\bar{t}/\hat{k}=12.89$. The moment estimates, on the other hand, are $\tilde{k}=10.05$ and $\tilde{\alpha}=11.29$.

Censored Data

We now turn to the computationally more complicated case of censored data. Type I and Type II censoring will be considered simultaneously, since they give the same form of likelihood function. Suppose that from a random sample on n individuals, r lifetimes and $n-r$ censoring times are observed.

Both censoring times and lifetimes will be designated by t_i, and D and C will denote the sets of individuals for which t_i is a lifetime and a censoring time, respectively. The survivor function for the gamma distribution (5.1.1) is, from Section 1.3.3,

$$S(t)=Q\left(k,\frac{t}{\alpha}\right)$$

where

$$Q(k,x)=1-I(k,x)=\frac{1}{\Gamma(k)}\int_x^{\infty}u^{k-1}e^{-u}\,du. \qquad (5.1.11)$$

The likelihood function for the censored sample is then [see (1.4.9)]

$$L(k,\alpha)=\left[\prod_{i\in D}\frac{1}{\alpha\Gamma(k)}\left(\frac{t_i}{\alpha}\right)^{k-1}e^{-t_i/\alpha}\right]\left[\prod_{i\in C}Q\left(k,\frac{t_i}{\alpha}\right)\right]. \qquad (5.1.12)$$

Letting

$$\bar{t}=\sum_{i\in D}\frac{t_i}{r} \quad \text{and} \quad \tilde{t}=\left(\prod_{i\in D}t_i\right)^{1/r}$$

represent the arithmetic and geometric means of the observed lifetimes, we can write the log likelihood function as

$$\log L(k,\alpha)=-rk\log\alpha-r\log\Gamma(k)$$
$$+r(k-1)\log\tilde{t}-\frac{r\bar{t}}{\alpha}+\sum_{i\in C}\log\left[Q\left(k,\frac{t_i}{\alpha}\right)\right]. \qquad (5.1.13)$$

Since the derivatives of $\log L$ are bothersome to calculate, it is often simplest to maximize $\log L$ directly by a method that does not require formulas for derivatives (see Appendix F). Incomplete gamma integrals $Q(k,t_i/\alpha)$ must be calculated, but most computer installations have subroutines for doing this. See Appendix B for references pertaining to the computation of $Q(k,x)$ or $I(k,x)$.

Alternately, first derivatives of $\log L$ are

$$\frac{\partial\log L}{\partial k}=-r\log\alpha-r\psi(k)+r\log\tilde{t}+\sum_{i\in C}\frac{1}{Q(k,t_i/\alpha)}\frac{\partial Q(k,t_i/\alpha)}{\partial k} \qquad (5.1.14)$$

$$\frac{\partial\log L}{\partial\alpha}=-\frac{rk}{\alpha}+\frac{r\bar{t}}{\alpha}+\sum_{i\in C}\frac{1}{Q(k,t_i/\alpha)}\frac{\partial Q(k,t_i/\alpha)}{\partial\alpha} \qquad (5.1.15)$$

where $\psi(k)$ is the digamma function and

$$\frac{\partial Q(k, t_i/\alpha)}{\partial k} = \frac{1}{\Gamma(k)} \int_{t_i/\alpha}^{\infty} u^{k-1} (\log u) e^{-u}\, du - \frac{Q(k, t_i/\alpha)}{\psi(k)} \tag{5.1.16}$$

$$\frac{\partial Q(k, t_i/\alpha)}{\partial \alpha} = \frac{1}{\alpha \Gamma(k)} \left(\frac{t_i}{\alpha} \right)^k e^{-t_i/\alpha}. \tag{5.1.17}$$

Maximum likelihood equations are obtained by setting $\partial \log L/\partial k$ and $\partial \log L/\partial \alpha$ equal to 0. The quantities in (5.1.14) and (5.1.15) are all readily calculated except for (5.1.16), which involves an incomplete digamma function. In addition, if we want to use a procedure such as Newton's method to solve the maximum likelihood equations, second derivatives of $\log L(k, \alpha)$ are required. These are still more complicated and involve incomplete trigamma integrals. We shall not give the expressions here; they are straightforward but tedious to derive. Gross and Clark (1975, p. 123) and Harter and Moore (1967) give the formulas for the special case of Type II censoring.

If m.l.e.'s are to be found through (5.1.14) and (5.1.15), a good, simple approach is to solve the single equation $\partial \log L/\partial \alpha = 0$ to find $\tilde{\alpha}(k)$ for different values of k. This is easily done iteratively, since all the quantities in (5.1.15) are easily calculated. The maximized log likelihood function $\log L[k, \tilde{\alpha}(k)]$ is usually determined accurately enough after calculation with a few k values to allow $\hat{k}$ to be found graphically. A further iteration or two will give k to greater accuracy if this is required.

In the case of Type II censored data Wilk et al. (1962b) give tables that help to find $\hat{k}$ and $\hat{\alpha}$, but use of these tables does not entail much less work than the method of the previous paragraph. Wilk et al. (1962a) discuss the use of probability plots for estimating α, given the value of k. However, when both parameters are unknown, this method does not provide precise estimates of k. Although it requires a litte computation, the maximum likelihood method is to be preferred for estimating k and α.

Example 5.1.2 Wilk et al. (1962b) give data on the lifetimes of 34 transistors in an accelerated life test. The times, in weeks, are given below; three of the times are censoring times and are denoted by asterisks.

3, 4, 5, 6, 6, 7, 8, 8, 9, 9, 9, 10, 10, 11, 11, 11, 13, 13, 13, 13,
13, 17, 17, 19, 19, 25, 29, 33, 42, 42, 52, 52*, 52*, 52*

Although the data are heavily rounded off, let us treat them as a random sample from the two-parameter gamma distribution and obtain m.l.e.'s $\hat{k}$ and $\hat{\alpha}$.

Either of the two methods of finding $\hat{k}$ and $\hat{\alpha}$ recommended above work well here. The log likelihood (5.1.13) can be maximized directly using a search procedure such as the Nelder–Mead algorithm (see Appendix F) to give the m.l.e.'s $\hat{k}=1.625$ and $\hat{\alpha}=12.361$. Since the data in this sample are only lightly censored, a simple way of obtaining initial guesses at $\hat{k}$ and $\hat{\alpha}$ is for us to calculate moment estimates of k and α, pretending that the last three observations are uncensored. As an *ad hoc* adjustment the last three observations can be increased slightly before computing the estimates. Replacing the three starred observations with three 65s we get, for example, the moment estimates $\tilde{\alpha}=16.7$ and $\tilde{k}=1.2$. These are crude, but they are sufficiently close to $(\hat{k}, \hat{\alpha})$ to allow search procedures to converge quite quickly to the m.l.e.'s. Figure 5.1.2, given later as part of Example 5.1.5, shows a contour graph of the likelihood function for this example.

Determination of the m.l.e.'s by working with the maximized likelihood function is also straightforward. For a given value of k, $\log L(k, \alpha)$ can be maximized directly or by solving the equation $\partial \log L/\partial\alpha=0$ [see (5.1.15)]. This gives the value $\tilde{\alpha}(k)$, and the maximized log likelihood is then

$$\log L_{\max}(k)=\log L[k, \tilde{\alpha}(k)].$$

Figure 5.1.1 shows a graph of $\log L_{\max}(k)$, which was calculated by evaluating $\tilde{\alpha}(k)$ at a few different k values. The value of $\hat{k}$ can be estimated fairly accurately from the graph, though a couple of additional iterations will pin it down more closely.

5.1.2 Interval Estimates and Tests for Parameters: Uncensored Data

For complete samples inference procedures for k and α are fairly straightforward. The joint p.d.f. of a random sample of n observations is, according to (5.1.1),

$$\frac{1}{\alpha^{nk}\Gamma(k)^n}\left(\prod_{i=1}^{n} t_i^{k-1}\right)\exp\left(-\sum_{i=1}^{n}\frac{t_i}{\alpha}\right) \tag{5.1.18}$$

and the arithmetic and geometric means, denoted by $\bar{t}$ and $\tilde{t}$, respectively, are jointly sufficient for k and α. Tests and interval estimation can therefore be based on $\bar{t}$ and $\tilde{t}$ or, equivalently, on $\bar{t}$ and $W=\tilde{t}/\bar{t}$. To find the joint distribution of $\bar{t}$ and W, make the change of variables from $t_1,\ldots,t_n$ to $T=t_1+\cdots+t_n$, $y_2=t_2/T,\ldots, y_n=t_n/T$. The Jacobian of this transformation is $\partial(t_1,\ldots,t_n)/\partial(T, y_2,\ldots, y_n)=T^{n-1}$, and from (5.1.18) the joint p.d.f.

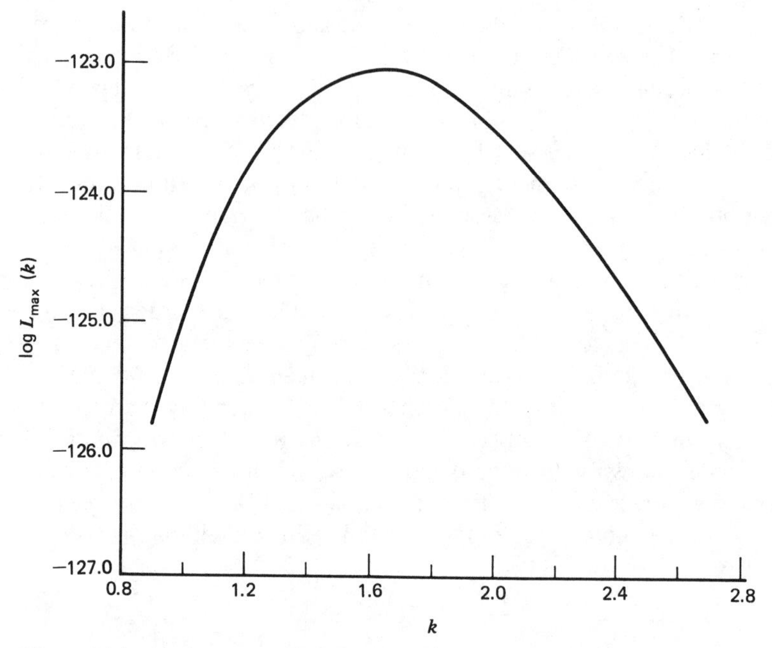

Figure 5.1.1 Maximized log likelihood for gamma parameter k (Example 5.1.2)

of $T, Y_2, \ldots, Y_n$ is found to be

$$\frac{T^{nk-1}e^{-T/\alpha}}{\alpha^{nk}\Gamma(nk)}\left[\frac{\Gamma(nk)}{\Gamma(k)^n}y_2^{k-1}\cdots y_n^{k-1}\left(1-\sum_{i=2}^{n}y_i\right)^{k-1}\right] \tag{5.1.19}$$

where $T>0$, $y_2>0, \ldots, y_n>0$, and $\Sigma_{i=2}^n y_i<1$. The first term in (5.1.19) is the p.d.f. of T; as is well known, T has a gamma distribution with index parameter nk and scale parameter α and is independent of $Y_2, \ldots, Y_n$. The variables $Y_2, \ldots, Y_n$ have a joint Dirichlet distribution. In addition, since $\bar{t}=T/n$ and

$$W=\frac{\tilde{t}}{\bar{t}}=n\left[Y_2\cdots Y_n\left(1-\sum_{i=2}^{n}Y_i\right)\right]^{1/n}$$

$\bar{t}$ and W are independently distributed. The exact distribution of W can be obtained from that of $(Y_2, \ldots, Y_n)$, but this is complicated. Glaser (1976a), Chao and Glaser (1978), and Dyer and Keating (1980) discuss this and give procedures for computing probabilities. For our purposes a very good χ^2

approximation to the distribution of W given by Bain and Engelhardt (1975) is satisfactory, so the exact distribution of W will not be discussed further.

Exact tests and interval estimates for the parameters k and α are available. Exact procedures are not available for the quantiles or survivor function, however; tests and estimation for these are treated in Section 5.1.3.

Inferences for k

For the gamma distribution shape parameter k first consider tests of the form

$$H_0: k=k_0 \quad \text{vs.} \quad H_1: k>k_0.$$

Scale-invariant tests of H_0 versus H_1 can be based on W, whose distribution is free of α. It is shown here that large values of W provide evidence against H_0, so that the significance level associated with an observed value w_{obs} of W is given by $\Pr(W \geqslant w_{\text{obs}};\ k=k_0)$. Alternatively, a size p hypothesis test of H_0 versus H_1 will reject H_0 whenever $W \geqslant w_p$, where w_p is such that $\Pr(W \geqslant w_p;\ k=k_0)=p$. A $1-p$ upper confidence limit for k is the largest value k_0 not rejected by this test at the p level of significance.

The test can be derived through the theory of invariant tests (e.g., Cox and Hinkley, 1974, Sec. 5.3). The most powerful invariant test of $H_0: k=k_0$ versus $H_A: k=k_1$ $(k_1>k_0)$ is based on the likelihood of the maximal invariant $(Y_2, \ldots, Y_n)$. The joint p.d.f. of $Y_2, \ldots, Y_n$ is the second term in (5.1.19), and the test rejects H_0 whenever the likelihood ratio

$$\frac{\Gamma(nk_0)\Gamma(k_1)^n}{\Gamma(nk_1)\Gamma(k_0)^n} W^{n(k_0-k_1)}$$

is sufficiently small or, in other words, when W is sufficiently large. The test given is thus a uniformly most powerful invariant test of H_0 versus H_1. Tests of $H_0: k=k_0$ versus $H_1: k<k_0$ are derived in the same way and lead to rejection regions based on small values of W.

Bain and Engelhardt (1975) give an approximation to the distribution of W that is adequate for most testing and estimation problems. This involves approximating the distribution of

$$S=-\log W$$

by a χ^2 distribution, by equating the first two moments of S to those of a multiple of a χ^2 random variable. Specifically, $g=g(k,n)$ and $h=h(k,n)$

are determined so that

$$g(2nkS)\sim\chi^2_{(h)} \tag{5.1.20}$$

approximately. It is easily shown (see Problem 5.1) that $E(S)=-\log n-\psi(k)+\psi(nk)$ and $\text{Var}(S)=(1/n)\psi'(k)-\psi'(nk)$, where $\psi(z)$ and $\psi'(z)$ are the digamma and trigamma functions. Equating the mean and variance of the two sides of (5.1.20), we get

$$g(k,n)=\frac{E(S)}{nk\,\text{Var}(S)}$$

$$=\frac{\psi(nk)-\psi(k)-\log n}{k\psi'(k)-nk\psi'(nk)} \tag{5.1.21}$$

$$h(k,n)=2nkE(S)g(k,n). \tag{5.1.22}$$

In calculating (5.1.21) and (5.1.22) it is helpful to rewrite $E(S)$ as

$$E(S)=\psi(nk)-\log(nk)+\log(nk)-\psi(k)-\log n$$

$$=[\psi(nk)-\log(nk)]-[\psi(k)-\log k]$$

with which (5.1.6) can be used when k is bigger than unity or so. Bain and Engelhardt give a table that provides values of $g(k,n)$ and $h(k,n)$ for a wide range of situations. In addition, Engelhardt and Bain (1978a) provide a table that can be used to obtain confidence limits for k; this is discussed in Example 5.1.3.

A simpler approximation to the distribution of W is available when k is not too small. This is obtained by noticing that W is closely related to Bartlett's statistic, first discussed in Section 3.3.1. In fact, if we replace each of $r_1,\dots,r_n$ with k, k with n, and $r_i\hat{\theta}_i$ with t_i in (3.3.3), Bartlett's statistic becomes $\Lambda=-2nk\log W$. The approximation discussed in Section 3.3.1, whereby $C\Lambda$ was treated as $\chi^2_{(k-1)}$, then gives

$$S\sim\frac{1+(n+1)/6nk}{2nk}\chi^2_{(n-1)}. \tag{5.1.23}$$

This approximation requires that k be sufficiently large in contrast to the Bain–Engelhardt approximation, which gets better as n increases. Linhart (1965) reports, however, that (5.1.23) works well in situations in which k is larger than about 2. This means that the approximation is often useful in obtaining upper confidence limits on k.

Example 5.1.3 (Example 5.1.1 revisited) In example 5.1.1 we had a complete sample with 20 observations that gave $\bar{t}=113.45$ and $\tilde{t}=107.07$ for the arithmetic and geometric means and thus $S=\log(\bar{t}/\tilde{t})=0.05788$. To illustrate the procedures outlined above, let us obtain a two-sided .95 confidence interval for k.

Approximation (5.1.23) is by far the easiest to use here. With $n=20$ this gives

$$\frac{40kS}{1+(21/120k)} \sim \chi^2_{(19)}.$$

Since $\Pr(8.91 \leqslant \chi^2_{(19)} \leqslant 32.85)$ and the observed value of $40S$ is $40(0.05788)=2.3152$, the approximate .95 confidence interval consists of those k values for which

$$8.91 \leqslant \frac{2.3152k}{1+(21/120k)} \leqslant 32.85.$$

This inequality is easily rearranged to give the confidence interval $4.02 \leqslant k \leqslant 14.36$. The k values in the interval are all well above 2, so that the approximation should be quite accurate.

Alternately, approximation (5.1.20) can be used in obtaining the desired confidence interval. In this case we need to find the values of k such that

$$\chi^2_{(h),.025} \leqslant 2.3152gk \leqslant \chi^2_{(h),.975} \tag{5.1.24}$$

where $h=h(k,20)$ and $g=g(k,20)$ are functions of k given by (5.1.21) and (5.1.22). Table 1 of Engelhardt and Bain (1975) gives values of g and h, and trial and error can be used to locate the appropriate values of k. This is not hard; we note that for k in the range 4 to 20, $h(k,20)$ changes very little and lies in the range 19.00 to 19.06. Thus the end points of (5.1.24) will be very nearly fixed at the values $\chi^2_{(19),.025}=8.91$ and $\chi^2_{(19),.975}=32.85$. By trial and error the values of k satisfying (5.1.24) are found to be those in the interval $4.03 \leqslant k \leqslant 14.40$, which is virtually identical to the interval given by the first method.

Engelhardt and Bain (1978a, Table 3) tabulate confidence limits as a function of S and n (actually, their table is given in terms of $t=1/2S$). Their table does not include values for .025 lower confidence or .975 upper confidence limits, however, so in this case one must still resort to trial and error, but in some situations the table allows one to get confidence limits directly.

Inferences for α

Inferences for α can be based on the conditional distribution of t, given $\tilde{t}$, or, equivalently, of $U=\bar{t}/\tilde{t}$, given $\tilde{t}$. The tests are uniformly most powerful unbiased (UMPU) tests (see Lehmann, 1959, p. 134 ff.) and are discussed in detail by Engelhardt and Bain (1977b); we only sketch the procedures here. The distribution of U, given $G=\tilde{t}/\alpha$, does not depend on k or α, and the UMPU test of, say,

$$H_0: \alpha=\alpha_0 \quad \text{vs.} \quad H_1: \alpha<\alpha_0$$

rejects H_0 for sufficiently small values of U. The conditional distribution of U, given G, is complicated but has been tabulated by Engelhardt and Bain (1977b). Specifically, they consider the standardized variate

$$Z(g_0)=\sqrt{ng_0}\left[U-E(U|g_0)\right]$$

where $g_0=\tilde{t}/\alpha_0$. They give percentage points $z_p(g_0)$ for $Z(g_0)$ and values of $E(U|g_0)$ for $n=5,10,20,30,40$, and ∞ and selected values of g_0. For values of g_0 and n not in the table linear interpolation can be used. To test H_0 versus H_1 at the $1-p$ level of significance we reject H_0 if $Z(g_0)$ is less than $z_p(g_0)$, the pth quantile of the distribution of $Z(g_0)$. A p upper confidence limit on α can be obtained by finding g^* that satisfies $Z(g^*)_{\text{obs}}=z_p(g^*)$; this gives the upper confidence limit as $\tilde{t}/g^*$. Tests of H_0 versus $H_1: \alpha>\alpha_0$ and lower confidence limits on α are determined in an analogous manner, with large values of U, in this case, providing evidence against H_0. Engelhardt and Bain (1978a), alternately, give a χ^2 approximation to the distribution of U, given g, that can be used to obtain percentage points $z_p(g_0)$.

Example 5.1.4 (Example 5.1.1 revisited) Consider once again the data of Example 5.1.1, where we had $\bar{t}=113.45$, $\tilde{t}=107.07$, and thus an observed U value of $113.45/107.07=1.06$. Let us determine a two-sided .95 confidence interval for α. To do this we find the set of values α_0 such that $H_0: \alpha=\alpha_0$ is not rejected at the .05 level of significance. Using a two-tailed test with probability 0.25 in each tail, we determine values g_0 such that $Z(g_0)\leqslant z_{.975}(g_0)$ and $Z(g_0)\geqslant z_{.025}(g_0)$. This must be done by trial and error, using Table 1 of Engelhardt and Bain (1977b). To illustrate, suppose we wish to know whether $\alpha_0=25$ is in the desired .95 confidence interval. With $\alpha_0=25$ we have $g_0=\tilde{t}/\alpha_0=4.28$. By interpolating from the table we then get $E(U|4.28)=1.125$, and then, since the observed U value is 1.060, the observed $Z(g_0)$ value is $Z(4.28)=\sqrt{20(4.28)}(1.060-1.125)=-1.252$. Also from Table 1, $z_{.025}(4.25)=-1.103$, and thus $\alpha_0=25$ is not inside the confidence interval. Proceeding in this way, we find by trial and error that the desired .95 confidence interval for α is $6.5 \leqslant \alpha \leqslant 23.7$.

An alternate way of obtaining tests or confidence intervals for α is to use likelihood ratio methods. This is discussed in the next section, as are tests and interval estimates for quantiles or the survivor function of the distribution.

5.1.3 Likelihood Ratio Procedures for Censored or Uncensored Data

When the data are censored, or when interval estimates or tests are wanted for the quantiles or survivor function of the gamma distribution, exact procedures are not available, but likelihood ratio methods can be used. Large-sample normal approximations to the distributions of the m.l.e.'s can also be used, though evidence indicates that unless sample sizes are very large the distribution of $(\hat{k}, \hat{\alpha})$ is not close to being a bivariate normal distribution. In addition, the observed information matrix required for these approximations is bothersome to compute when there is censoring, as noted in Section 5.1.1. Consequently, only likelihood ratio procedures are considered here.

A plot of the joint relative likelihood function for k and α provides an informative picture, as do plots of maximized relative likelihood functions for k and α. These, and corresponding likelihood ratio tests and confidence intervals for k and α, can be obtained with little difficulty. For example, to test the hypothesis $H_0: k = k_0$ versus $H_1: k \neq k_0$ one uses the likelihood ratio statistic

$$\Lambda = -2\log\left(\frac{L(k_0, \tilde{\alpha})}{L(\hat{k}, \hat{\alpha})}\right)$$

where $\tilde{\alpha} = \tilde{\alpha}(k_0)$ is the m.l.e. for α when $k = k_0$, obtained by solving $\partial \log L(k_0, \alpha)/\partial\alpha = 0$ [see (5.1.15)], and $\hat{k}$ and $\hat{\alpha}$ are the unrestricted m.l.e.'s of k and α. In large samples the distribution of Λ under H_0 is approximately $\chi^2_{(1)}$; it appears that this approximation is also reasonable for fairly small sample sizes. An approximate γ confidence interval consists of the set of values k_0 giving $\Lambda \leqslant \chi^2_{(1),\gamma}$. Wyckoff and Engelhardt (1979) propose alternate inference procedures for k when there is Type II censoring.

Tests and confidence intervals for α are obtained in a similar way. Getting tests and confidence intervals for quantiles or the survivor function is more complicated, however, mainly because the survivor function of the gamma distribution is not expressible in simple closed form. Suppose, for example, that a confidence interval for $S(t_0)$, the survivor function at time t_0, is wanted. Since $S(t_0) = Q(k, t_0/\alpha)$, where $Q(k, x)$ is given by (5.1.11), we consider hypotheses of the form

$$H_0: Q\left(k, \frac{t_0}{\alpha}\right) = Q_0 \tag{5.1.24}$$

where, in addition to t_0, Q_0 is also specified. If $\tilde{k}$ and $\tilde{\alpha}$ are the m.l.e.'s of k and α, subject to the restriction (5.1.24), then under H_0 the likelihood ratio statistic

$$\Lambda = -2\log\left(\frac{L(\tilde{k}, \tilde{\alpha})}{L(\hat{k}, \hat{\alpha})}\right)$$

is approximately $\chi^2_{(1)}$. Large values of Λ provide evidence against H_0, and an approximate γ confidence interval for $S(t_0)$ is obtained by finding the set of values Q_0 such that $\Lambda \leqslant \chi^2_{(1),\gamma}$.

Tests and confidence intervals for quantiles can be obtained in a similar way. The pth quantile of (5.1.1) is the value t_p such that $Q(k, t_p/\alpha) = 1 - p$. The hypothesis $H_0: t_p = t_0$ can thus be written in the form

$$H_0: Q\left(k, \frac{t_0}{\alpha}\right) = 1 - p$$

which is of exactly the same form as (5.1.24). Tests of H_0 are therefore carried out as for (5.1.24). To find a γ confidence interval for t_p we need to keep $Q_0 = 1 - p$ in (5.1.24) fixed and find the set of values t_0 such that $\Lambda \leqslant \chi^2_{(1),\gamma}$.

The main difficulty in using the likelihood ratio method with quantiles or the survivor function is the problem of maximizing $\log L(k, \alpha)$ subject to the restriction (5.1.24). One way to do this is as follows:

1. Define $M(k) = \log L[k, \alpha(k)]$, where, for given k, $\alpha(k)$ is defined implicitly by (5.1.24). Note that to find $\alpha(k)$ for a given k it is necessary to solve the equation

$$Q\left(k, \frac{t_0}{\alpha}\right) = Q_0$$

for α. This involves finding the $(1 - Q_0)$th quantile $t^*_{1-Q_0}$ of the one-parameter gamma distribution Ga(k). This satisfies

$$Q(k, t^*_{1-Q_0}) = Q_0 \tag{5.1.25}$$

and then $\alpha = t_0/t^*_{1-Q_0}$. Some comments on quantiles for Ga(k) are given in Appendix B.

2. Use an algorithm that does not require analytical first derivatives to maximize $M(k)$ for k (see Appendix F).

An alternative to the exact numerical calculation of tests and confidence intervals is the estimation of results by graphical means. This requires less formal programming than the numerical procedures yet portrays results of tests and gives satisfactory estimates of confidence intervals in most instances. The following procedure is illustrated in Example 5.1.5:

1. After determining $\hat{k}$ and $\hat{\alpha}$, plot contours of the relative likelihood function $R(k,\alpha) = L(k,\alpha)/L(\hat{k},\hat{\alpha})$; this is easily done on most computer systems. Besides providing a picture of plausible pairs of values (k,α), the plot can be used to obtain confidence intervals or to show the results of tests of significance. If it is desired to obtain an approximate γ confidence interval for a single parametric function of k and α, or to test hypotheses at significance level $1-\gamma$, the plot should include the contour

$$R(k,\alpha) = \exp\left(\frac{-\chi^2_{(1),\gamma}}{2}\right)$$

since all points (k,α) inside this contour will have $-2\log R(k,\alpha) \leqslant \chi^2_{(1),\gamma}$. It is useful to routinely plot contours such as $R(k,\alpha) = .258$, .147, and .036, which correspond to the quantiles $\chi^2_{(1),.90}$, $\chi^2_{(1),.95}$, and $\chi^2_{(1),.99}$.

2. Results of significance tests about k or α can be easily determined from the contour plots. For example, for the hypothesis $H_0: \alpha = \alpha_0$ versus $H_1: \alpha \neq \alpha_0$ the appropriate likelihood ratio statistic is

$$\Lambda = -2\log\left(\frac{L_{\max}(k,\alpha_0)}{L(\hat{k},\hat{\alpha})}\right)$$

$$= -2\log[R_{\max}(k,\alpha_0)]$$

where $R_{\max}(k,\alpha_0)$ is the maximum value of $R(k,\alpha)$ on the line $\alpha = \alpha_0$. If one draws in the line $\alpha = \alpha_0$ on the contour graph, this value is easily estimated. In particular, it is readily seen whether $R_{\max}(k,\alpha_0)$ is inside or outside a given contour, and hence whether H_0 is contradicted at a particular level of significance. An approximate γ confidence interval for α can be obtained graphically by locating the values α_0 such that $R_{\max}(k, \alpha_0) \leq \exp(-\chi^2_{(1),\gamma}/2)$.

To handle a hypothesis of the form $H_0: Q(k, t_0/\alpha) = Q_0$, where t_0 and Q_0 are specified, is a little more work. If we determine a few points (k, α) on the curve $Q(k, t_0/\alpha) = Q_0$, the curve can be sketched on the contour graph and the maximum value that $R(k, \alpha)$ takes on this curve determined approximately. Confidence intervals can be estimated by drawing in a few curves $Q(k, t_0/\alpha) = Q_0$, keeping either t_0 or Q_0 fixed and varying the other, depending on whether a confidence interval is wanted for $S(t_0)$ or t_{1-Q_0}, and then using graphical interpolation.

Example 5.1.5 (Example 5.1.2 revisited) Consider again the data of Example 5.1.2, for which $n=34$ and $r=31$ and the m.l.e.'s were $\hat{k}=1.625$ and $\hat{\alpha}=12.361$. Figure 5.1.2 shows a graph of the relative likelihood function, which includes the contours $R(k, \alpha) = .50$, .147, and .036; the last two contours correspond to the .95 and .99 quantiles of $\chi^2_{(1)}$, respectively. This graph by itself provides an informative picture of plausible pairs of values (k, α). If desired, approximate joint confidence regions for k and α can be obtained. By the likelihood ratio method an approximate γ confidence region for k and α consists of the set of points (k, α) satisfying $\Lambda = -2\log[R(k, \alpha)] \leq \chi^2_{(2),\gamma}$. For example, we have $\chi^2_{(2),.90} = 4.605$, which corresponds to the value $R(k, \alpha) = \exp(-4.605/2) = .100$. This contour has not been portrayed on the graph, but the approximate extent of the confidence region is evident.

Now suppose we wish to test $H_0: k=1$ versus $H_1: k \neq 1$; if $k=1$, the lifetime distribution is exponential. By drawing in the line $k=1$ on the contour graph, we find that it almost touches the $R=.147$ contour, implying that $R_{\max}(1, \alpha) \doteq .147$. Thus the significance level for H_0 is very close to .05, indicating that there is some evidence against H_0. To go further and obtain, say, a .95 confidence interval for k we need to determine the two lines $k=k_0$ and $k=k_1$ that are tangent to the $R=.147$ contour. The appropriate lines are seen to be approximately $k=1$ and $k=2.5$. This graphical determination is adequate for all practical purposes, though exact values could be determined numerically. Doing this here, we find in fact that the confidence interval is, to two decimal places, $1.00 \leq k \leq 2.50$.

Approximate confidence intervals for α can be similarly obtained. Determined numerically, the .95 confidence interval for α turns out to be $7.49 \leq \alpha \leq 23.05$, though graphical approximation is once again adequate.

Finally, consider the problem of obtaining an approximate .95 confidence interval for $S(15)$, the probability of surviving past $t=15$. To do this we

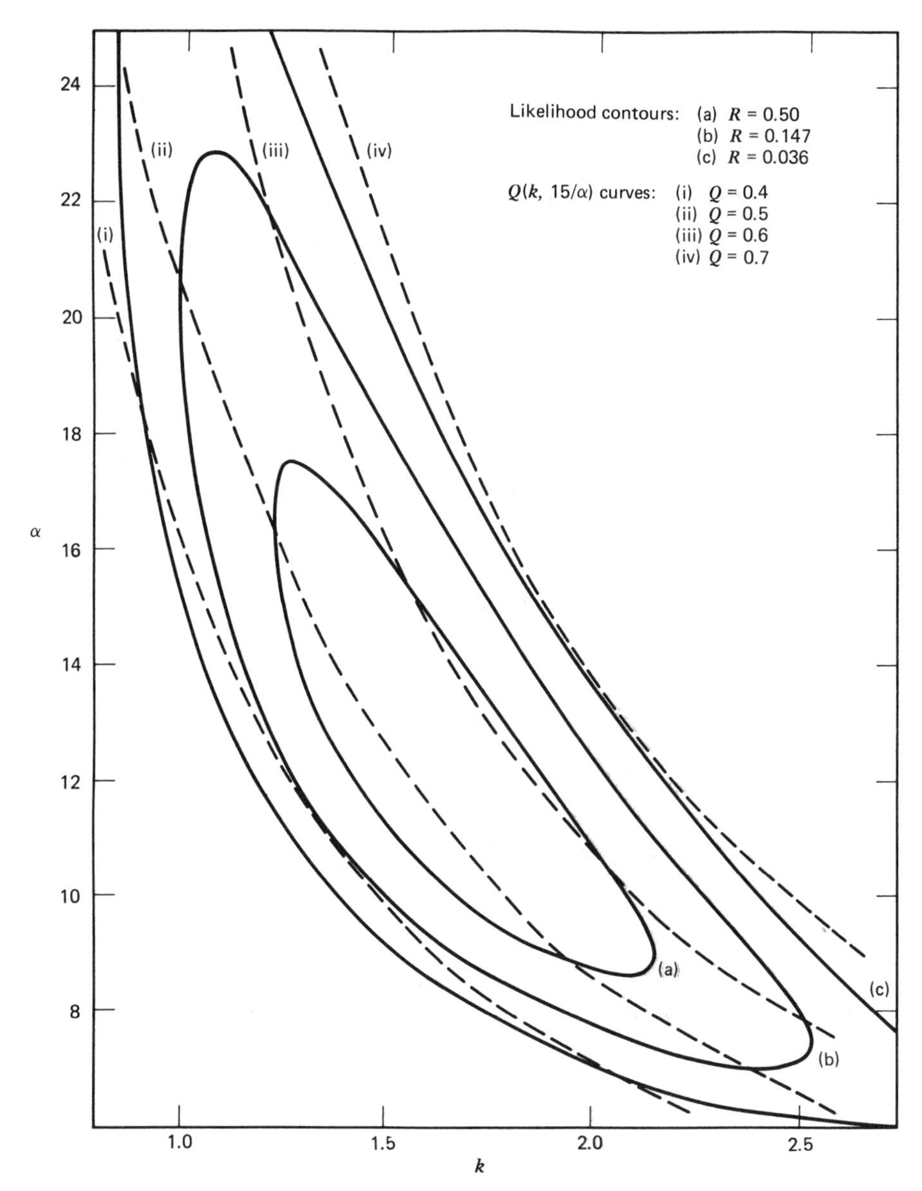

Figure 5.1.2 Likelihood contours and $Q(k, t/\alpha)$ curves for the gamma model with the data of Example 5.1.5.

consider hypotheses of the form

$$H_0: Q\left(k, \frac{15}{\alpha}\right) = Q_0$$

and find the set of values Q_0 for which H_0 is not contradicted at the .05 level of significance. The m.l.e. of $S(15)$ is $Q(\hat{k}, 15/\hat{\alpha}) = Q(1.625, 1.213) = .534$. Thus Q_0 values in the neighborhood of this value should be considered. The range of values that make the likelihood ratio statistic less than 3.841 can, if desired, be determined numerically, and it turns out to be $.40 \leqslant Q_0 \leqslant .66$, yielding the confidence interval $.40 \leqslant S(15) \leqslant .66$. Approximate graphical determination of the interval will usually be satisfactory, however. Figure 5.1.2 shows the four curves $Q(k, 15/\alpha) = .4$, .5, .6, and .7, respectively. To obtain the .95 confidence interval we need to estimate which Q_0 values yield curves that just touch the $R = .147$ contour. This can clearly be done with good accuracy with the four curves shown.

5.2 LOG-NORMAL AND NORMAL DISTRIBUTIONS

For the log-normal distribution log lifetimes are normally distributed. That is, if lifetime T has a log-normal distribution with p.d.f. (1.3.16), then $Y = \log T$ has a normal distribution with p.d.f.

$$f(y) = \frac{1}{(2\pi)^{1/2}\sigma} \exp\left(\frac{-(y-\mu)^2}{2\sigma^2}\right) \qquad -\infty < y < \infty. \tag{5.2.1}$$

Because of the well-known properties of the normal distribution and because it is a location–scale parameter model, we work with it throughout this section. Results are easily transformed to the log-normal distribution. Properties of the log-normal distribution were discussed in Section 1.3.4.

In this section the censored data situation is stressed, since methods for complete normal samples are well known and are a topic of almost every elementary statistics text. Only the two-parameter log-normal model is discussed here. For the case in which a threshold parameter is introduced into the model, problems arise that are similar to those discussed in Section 4.4 for the Weibull distribution (e.g., Johnson and Kotz, 1970, Ch. 14; Cohen and Whitten, 1980; Griffiths, 1980; see also Problem 5.8).

5.2.1 Point Estimation

When the data are uncensored, results are simple and well known. If $y_1,\ldots, y_n$ is a random sample from (5.2.1), the m.l.e.'s of μ and σ^2 are

$$\hat{\mu}=\bar{y} \qquad \hat{\sigma}^2=\frac{1}{n}\sum_{i=1}^{n}(y_i-\bar{y})^2.$$

An alternate estimate of σ^2 is the sample variance $s^2=\Sigma(y_i-\bar{y})^2/(n-1)$, which is unbiased for σ^2. The estimators $\bar{y}$ and s^2 can be shown to be minimum variance unbiased estimators of μ and σ^2, and $\bar{y}$ and $n\hat{\sigma}^2/(n+1)$ are minimum mean-square error estimators of μ and σ^2.

Maximum Likelihood Estimation With Censored Data

With censored data maximum likelihood estimation is a little more complicated. Let $\phi(z)$ and $Q(z)$ denote the p.d.f. and survivor function for the standard normal distribution,

$$\phi(z)=\frac{1}{(2\pi)^{1/2}}e^{-z^2/2} \quad \text{and} \quad Q(z)=\int_z^{\infty}\phi(x)\,dx \qquad -\infty<z<\infty. \tag{5.2.2}$$

If $Y\sim N(\mu,\sigma^2)$, then Y has p.d.f. $\sigma^{-1}\phi[(y-\mu)/\sigma]$ and survivor function $Q[(y-\mu)/\sigma]$. Consider a sample that may be Type I or Type II censored involving observations on the lifetimes of n individuals. We shall denote both log lifetimes and log censoring times as y_i $(i=1,\ldots, n)$ and let D be the set of individuals for which y_i is an observed log lifetime and C the set for which y_i is a log censoring time. The number of observed lifetimes is $r=|D|$.

The likelihood function is of the usual form (1.4.9), which gives

$$\prod_{i\in D}\frac{1}{\sigma}\phi\left(\frac{y_i-\mu}{\sigma}\right)\prod_{i\in C}Q\left(\frac{y_i-\mu}{\sigma}\right).$$

This gives

$$\log L(\mu,\sigma)=-r\log\sigma-\frac{1}{2\sigma^2}\sum_{i\in D}(y_i-\mu)^2+\sum_{i\in C}\log Q\left(\frac{y_i-\mu}{\sigma}\right).$$

The first derivatives of $\log L$ are

$$\frac{\partial \log L}{\partial \mu}=\frac{1}{\sigma^2}\sum_{i\in D}(y_i-\mu)+\frac{1}{\sigma}\sum_{i\in C}\phi\left(\frac{y_i-\mu}{\sigma}\right)\Big/Q\left(\frac{y_i-\mu}{\sigma}\right)$$

$$\frac{\partial \log L}{\partial \sigma}=-\frac{r}{\sigma}+\frac{1}{\sigma^3}\sum_{i\in D}(y_i-\mu)^2+\frac{1}{\sigma}\sum_{i\in C}\frac{y_i-\mu}{\sigma}\phi\left(\frac{y_i-\mu}{\sigma}\right)\Big/Q\left(\frac{y_i-\mu}{\sigma}\right).$$

Maximum likelihood equations are obtained by setting $\partial \log L/\partial\mu$ and $\partial \log L/\partial\sigma$ equal to zero. In handling these and other expressions it is helpful to introduce a little additional notation: let us write $z_i=(y_i-\mu)/\sigma$ $(i=1,\ldots,n)$ and define the functions

$$V(z)=-\frac{d}{dz}\log Q(z)=\frac{\phi(z)}{Q(z)}$$

$$\lambda(z)=\frac{d}{dz}V(z)=V(z)[V(z)-z]. \tag{5.2.3}$$

It will be noted that $V(z)$ is the hazard function for the normal distribution, and $\lambda(z)$ its derivative.

With (5.2.3) the maximum likelihood equations for μ and σ can be written as

$$\sum_{i\in D} z_i+\sum_{i\in C} V(z_i)=0 \tag{5.2.4}$$

$$-r+\sum_{i\in D} z_i^2+\sum_{i\in C} z_iV(z_i)=0. \tag{5.2.5}$$

Second derivatives of $\log L$ are

$$\frac{\partial^2 \log L}{\partial\mu^2}=-\frac{r}{\sigma^2}-\frac{1}{\sigma^2}\sum_{i\in C}\lambda(z_i)$$

$$\frac{\partial^2 \log L}{\partial\sigma^2}=\frac{r}{\sigma^2}-\frac{3}{\sigma^2}\sum_{i\in D} z_i^2-\frac{2}{\sigma^2}\sum_{i\in C} z_iV(z_i)-\frac{1}{\sigma^2}\sum_{i\in C} z_i^2\lambda(z_i) \tag{5.2.6}$$

$$\frac{\partial^2 \log L}{\partial\mu\,\partial\sigma}=-\frac{2}{\sigma^2}\sum_{i\in D} z_i-\frac{1}{\sigma^2}\sum_{i\in C} V(z_i)-\frac{1}{\sigma^2}\sum_{i\in C} z_i\lambda(z_i).$$

The Newton–Raphson method can be used to solve (5.2.4) and (5.2.5) for $\hat\mu$ and $\hat\sigma$ but occasionally fails to converge unless very good starting values are used. Negative σ values are sometimes encountered during the iterations, though this can usually be handled by a simple device such as replacing negative values by half the value of σ from the preceding iteration.

Another iterative procedure that can be used to get $\hat{\mu}$ and $\hat{\sigma}$ was first suggested by Sampford and Taylor (1959), and adopted by Wolynetz (1974, 1979). This depends on the observation that if $Y \sim N(\mu, \sigma^2)$, then

$$E(Y|Y \geqslant L) = \mu + \sigma V\left(\frac{L-\mu}{\sigma}\right). \tag{5.2.7}$$

To see this note that

$$E(Y|Y \geqslant L) = \int_L^\infty y \frac{1}{\sigma} \phi\left(\frac{y-\mu}{\sigma}\right) dy \Big/ Q\left(\frac{L-\mu}{\sigma}\right)$$

$$= \mu + \left[\sigma \Big/ Q\left(\frac{L-\mu}{\sigma}\right)\right] \int_{(L-\mu)/\sigma}^\infty z\phi(z)\, dz.$$

But

$$\int_a^\infty z\phi(z)\, dz = \int_a^\infty \frac{1}{(2\pi)^{1/2}} z e^{-z^2/2}\, dz$$

$$= \frac{1}{(2\pi)^{1/2}} e^{-a^2/2} = \phi(a)$$

and hence (5.2.7) holds.

Let Y_i^0 represent the actual (unobserved) log lifetime of individual i; then, by (5.2.7), $E(Y_i^0 | Y_i^0 \geqslant y_i) = \mu + \sigma V[(y_i - \mu)/\sigma]$. Now, define w_i's for all n individuals as follows:

$$w_i = y_i \qquad i \in D \tag{5.2.8}$$

$$w_i = \mu + \sigma V\left(\frac{y_i - \mu}{\sigma}\right)$$

$$= \mu + \sigma V(z_i) \qquad i \in C.$$

The maximum likelihood equations (5.2.4) and (5.2.5) can then be rewritten as

$$\sum_{i=1}^n \frac{w_i - \mu}{\sigma} = 0 \tag{5.2.9}$$

$$-r + \sum_{i=1}^n \left(\frac{w_i - \mu}{\sigma}\right)^2 - \sum_{i \in C} \lambda(z_i) = 0. \tag{5.2.10}$$

To get (5.2.10) we note that (5.2.5) gives

$$0=-r+\sum_{i=1}^{n}\left(\frac{w_i-\mu}{\sigma}\right)^2-\sum_{i\in C}\left(\frac{w_i-\mu}{\sigma}\right)^2+\sum_{i\in C}z_iV(z_i)$$

$$=-r+\sum_{i=1}^{n}\left(\frac{w_i-\mu}{\sigma}\right)^2-\sum_{i\in C}\left[V(z_i)^2-z_iV(z_i)\right].$$

If all of the w_i's and $\lambda(z_i)$'s in (5.2.9) and (5.2.10) were known, these equations could be solved analytically for μ and σ to get

$$\tilde{\mu}=\sum_{i=1}^{n}\frac{w_i}{n}$$

$$\tilde{\sigma}^2=\sum_{i=1}^{n}(w_i-\tilde{\mu})^2\Big/\left(r+\sum_{i\in C}\lambda(z_i)\right). \tag{5.2.11}$$

Thus define the following iteration scheme: obtain initial estimates of μ and σ and use these to calculate w_i $(i\in C)$ of (5.2.8) and $\lambda(z_i)=\lambda[(y_i-\mu)/\sigma]$, $i\in C$. Now, calculate new estimates $\tilde{\mu}$ and $\tilde{\sigma}$ by (5.2.11) and repeat the entire procedure. With reasonable initial estimates this iteration scheme will converge to the m.l.e.'s $\hat{\mu}$ and $\hat{\sigma}$. Wolynetz (1974) reports that this scheme converges somewhat more surely than Newton's method, though more slowly. Since the procedure is easily programmed, it provides a convenient method of obtaining the m.l.e.'s. This procedure is, incidentally, essentially an Expectation-Maximization algorithm (Dempster et al., 1977).

In finding the m.l.e.'s it is necessary to compute $Q(z)$ and $V(z)$. Most modern computer installations have good routines for evaluating $Q(z)$ or $1-Q(z)$. If direct calculation is desired, several useful formulas are given by Abramowitz and Stegun (1965, pp. 931–933). A particularly simple and useful one is the polynomial approximation

$$Q(z)=\phi(z)\left(a_1t+a_2t^2+a_3t^3\right)$$

where $t=(1+pz)^{-1}$, $p=.33267$, $a_1=.4361836$, $a_2=-.1201676$, and $z_3=.9372980$. The error in this approximation is less than 10^{-5} for all z. $V(z)$ can be computed directly as $\phi(z)/Q(z)$ if $|z|$ is not too large. If $|z|$ is large, it is preferable to use one of several special expressions for evaluating $V(z)$, such as

$$V(z)\doteq z\left(1-\frac{1}{z^2}+\frac{3}{z^4}-\frac{15}{z^6}\right)^{-1}$$

which is suitable for $|z|$ greater than 7 or so. Abramowitz and Stegun (1965, p. 932) and Johnson and Kotz (1970, Ch. 33) give other approximations.

For the case in which the data are Type II censored, several authors have given aids for finding the m.l.e.'s. Cohen (1961), for example, presents tables that can be used to obtain $\hat{\mu}$ and $\hat{\sigma}$, and Schmee and Nelson (1976) extend these.

Other Estimates of μ and σ

Other estimates of μ and σ are sometimes useful, either on their own or as initial approximations to the m.l.e.'s. One important pair is the best linear unbiased estimators of μ and σ, defined for Type II censored samples. Sarhan and Greenberg (1962) discuss these and provide tables for their calculation for samples with $r \leqslant n \leqslant 20$. Dixon (1960) discusses some other linear estimators. Several other methods of estimation have also been proposed in the literature (e.g., Tiku, 1967, 1970). We mention only one such method here; these estimators have been proposed by Persson and Rootzen (1977) and can be used with Type II or singly Type I censored data. In either of these cases all of the censored observations y_i $(i \in C)$ are equal, say, to y_L, so that the likelihood function becomes

$$L(\mu,\sigma)=\frac{1}{\sigma^r}\exp\left(-\frac{1}{2\sigma^2}\sum_{i\in D}(y_i-\mu)^2\right)\left[Q\left(\frac{y_L-\mu}{\sigma}\right)\right]^{n-r}.$$

Letting $x_i=y_i-y_L$ and $\theta=(y_L-\mu)/\sigma$, we can write

$$L(\mu,\sigma)=\frac{1}{\sigma^r}\exp\left(-\frac{1}{2\sigma^2}\sum_{i\in D}(x_i+\theta\sigma)^2\right)Q(\theta)^{n-r}. \tag{5.2.12}$$

A sensible estimate of $Q(\theta)$ is $(n-r)/n$, the proportion of censored lifetimes. Persson and Rootzen's proposal is to replace θ in (5.2.12) with $\theta^*=Q^{-1}[(n-r)/n]$, the (r/n)th quantile of the standard normal distribution. If this is done, (5.2.12) becomes a function of σ only,

$$L_1(\sigma)=\frac{1}{\sigma^r}\exp\left(-\frac{1}{2\sigma^2}\sum_{i\in D}(x_i+\theta^*\sigma)^2\right)\left(\frac{n-r}{n}\right)^{n-r}.$$

Maximizing L_1 with respect to σ $(\sigma>0)$, we get

$$\tilde{\sigma}=\frac{1}{2}\left\{\frac{\theta^*}{r}\sum_{i\in D}x_i+\left[\left(\frac{\theta^*}{r}\sum_{i\in D}x_i\right)^2+\frac{4}{r}\sum_{i\in D}x_i^2\right]^{1/2}\right\}. \tag{5.2.13}$$

We then estimate μ by

$$\tilde{\mu}=y_L-\theta^*\tilde{\sigma}. \tag{5.2.14}$$

Persson and Rootzen find that even in small samples $\tilde{\mu}$ and $\tilde{\sigma}$ have properties similar to those of the m.l.e.'s and are generally numerically close to the m.l.e.'s. As would be expected, they are relatively more efficient when censoring is fairly severe.

As a final method of obtaining estimates we mention plotting techniques. If the data are Type II censored, an ordinary probability plot, on normal or log-normal probability paper if it is available, will yield estimates of μ and σ. With Type I censored data, a linearized plot of the empirical survivor function or cumulative hazard function (see Section 2.4) will give estimates.

Example 5.2.1 We consider data from an accelerated life test experiment involving specimens of electrical insulation. In the experiment 10 specimens were put on test and the test was terminated at the time of the eighth failure. The eight observed log failure times were 6.00, 6.43, 6.77, 7.07, 7.40, 7.66, 8.10, and 8.40 (times are in hours). Since this is a Type II censored sample ($n=10$, $r=8$), we can illustrate several estimation procedures with it.

A plot of the data on normal probability paper (see Section 2.4) gives roughly a straight line, and estimates of μ and σ obtained from a line fitted by eye were found to be $\hat{\sigma}=1.07$ and $\hat{\mu}=7.60$.

B.l.u.e.'s of μ and σ can also be calculated. Using Table 10.C.1 of Sarhan and Greenberg (1962, p. 222) to obtain coefficients of the ordered log failure times $Y_1<\cdots<Y_8$, we find the b.l.u.e.'s to be $\mu^*=7.60$ and $\sigma^*=1.14$. Another possibility is to compute the estimates (5.2.13) and (5.2.14). In this case $y_L=8.40$ and $\theta^*=Q^{-1}(.20)=.84$, and the estimates are $\mu'=7.57$ and $\sigma'=0.99$.

Finally, with any of the other estimates used as starting values, either the Newton–Raphson or Sampford–Taylor iteration procedure gives the m.l.e.'s in a few iterations as $\hat{\mu}=7.58$ and $\hat{\sigma}=1.00$. The various estimates are thus in good agreement, in spite of the small sample size.

5.2.2 Interval Estimation and Tests

5.2.2a *Exact Methods for Complete Samples*

We continue to work with the normal distribution and will discuss tests and interval estimation for μ, σ, the survivor function $Q[(y-\mu)/\sigma]$, or the pth quantile $y_p=\mu+u_p\sigma$ of (5.2.1). Here u_p is the pth quantile of the standard normal distribution.

With complete samples, procedures for estimating or testing hypotheses about μ or σ are well known and are discussed in elementary statistics texts. This will not be reviewed, except to note that inferences on μ and σ can be based on the pivotal quantities $Z_1=(\bar{y}-\mu)\sqrt{n}/s$ and $Z_2=(n-1)s^2/\sigma^2$, where $\bar{y}=\Sigma y_i/n$ and $s^2=\Sigma(y_i-\bar{y})^2/(n-1)$ are the mean and variance of a random sample $y_1,\ldots,y_n$ from (5.2.1). It is well known that Z_1 has a Student-t distribution with $n-1$ degrees of freedom ($Z_1\sim t_{(n-1)}$) and that $Z_2\sim\chi^2_{(n-1)}$.

Estimation of quantiles or the survivor function of the normal distribution is not usually discussed in elementary statistics texts, however, so let us examine these problems, still for the case of complete samples. To estimate the pth quantile $y_p=\mu+u_p\sigma$ of (5.2.1), consider the pivotal

$$Z_p=\frac{(\bar{y}+u_p s)-y_p}{s}=\frac{\hat{y}_p-y_p}{s} \tag{5.2.15}$$

where $\hat{y}_p=\bar{y}+u_p s$ is an estimate of y_p. Since

$$\Pr(Z_p\leqslant a)=\Pr[y_p\geqslant\bar{y}+(u_p-a)s] \tag{5.2.16}$$

probability statements for Z_p produce confidence limits for y_p. To obtain probabilities for Z_p involves working with the noncentral t distribution; a noncentral t random variable with ν degrees of freedom and noncentrality parameter λ, denoted $t'_{(\nu)}(\lambda)$, arises by considering

$$t=\frac{Z+\lambda}{(W/\nu)^{1/2}}$$

where $Z\sim N(0,1)$ and $W\sim\chi^2_{(\nu)}$ are independent random variables and λ is a constant. The special case $\lambda=0$ yields the ordinary (or "central") t distribution with ν degrees of freedom. The noncentral t distribution is discussed at length by Johnson and Kotz (1970, Ch. 31), Owen (1968), and others.

Returning to the problem of evaluating (5.2.16) for a given a, we find that

$$\Pr(Z_p\leqslant a)=\Pr\left[\frac{\sigma}{s}\left(\frac{\bar{y}-\mu}{\sigma}-u_p\right)\leqslant a-u_p\right]$$

$$=\Pr\left(\frac{\sqrt{n}[(\bar{y}-\mu)/\sigma]-\sqrt{n}u_p}{s/\sigma}\leqslant(a-u_p)\sqrt{n}\right).$$

Since $\sqrt{n}(\bar{y}-\mu)/\sigma\sim N(0,1)$ and $(n-1)s^2/\sigma^2\sim\chi^2_{(n-1)}$ are independent, this implies that

$$\Pr(Z_p \leq a) = \Pr\left[t'_{(n-1)}(-\sqrt{n}u_p) \leq (a-u_p)\sqrt{n}\right]. \tag{5.2.17}$$

Thus, obtaining the γth quantile of Z_p is equivalent to finding the γth quantile of the noncentral t variate $t'_{(n-1)}(-\sqrt{n}u_p)$. There are several comprehensive sets of tables that give probabilities of quantiles for the noncentral t distribution, including those of Owen (1963), Resnikoff and Lieberman (1957), and Locks et al. (1963). Owen (1968) also gives a small table containing only .95 quantiles. These tables cover a wide range of situations: Owen (1963) gives values of $(\nu+1)^{-1/2}t'_{\nu,\alpha}[u_p(\nu+1)^{1/2}]$, where $t'_{\nu,\alpha}(\lambda)$ represents the αth quantile of $t'_{(\nu)}(\lambda)$. The tables cover $p=.75$ to .99, $\nu=1(1)200$, and $\alpha=.95$, .975, .99, and .995. Resnikoff and Lieberman (1957) tabulate $\nu^{-1/2}t'_{\nu,\alpha}[u_p(\nu+1)^{1/2}]$ for $p=.001$ to .25, $\nu=2(1)24(5)49$, and α ranging from .005 to .995. The tables of Locks et al. (1963) are similar to these, but more extensive. Kramer and Paik (1979) discuss approximations for finding quantiles or probabilities.

The tables mentioned can be used to obtain confidence limits for quantiles. According to (5.2.16) and (5.2.17) a lower γ confidence limit for y_p takes the form

$$y_p \geq \bar{y} - \frac{t'_{n-1,\gamma}(-\sqrt{n}u_p)}{\sqrt{n}}s.$$

The tables of Owen (1963) give the factor $t'_{n-1,\gamma}(-\sqrt{n}u_p)/\sqrt{n}$ directly, whereas those of Resnikoff and Lieberman (1957) and Locks et al. (1963) give $t'_{n-1,\gamma}(-\sqrt{n}u_p)/(n-1)^{1/2}$, from which the desired factor is easily calculated.

Example 5.2.2 The data given here arose in tests on the endurance of deep groove ball bearings. They were originally discussed by Lieblein and Zelen (1956), who assumed that the data came from a Weibull distribution. A probability plot of the data shows them to also be consonant with a log-normal model, and they will be used here to illustrate the methods described above. (Further examination of these data, including a comparison of inferences under the Weibull and log-normal models, is made in Example 5.3.1.) The data are the number of million revolutions before failure for each of the 23 ball bearings in the life test and they are 17.88, 28.92, 33.00, 41.52, 42.12, 45.60, 48.40, 51.84, 51.96, 54.12, 55.56, 67.80, 68.64, 68.64, 68.88, 84.12, 93.12, 98.64, 105.12, 105.84, 127.92, 128.04, and 173.40.

Suppose that we want .95 lower confidence limits for the .01 and .10 quantiles, $t_{.01}$ and $t_{.10}$, of the lifetime distribution. To work with the normal distribution the 23 failure times are converted to log failure times $y_1, \ldots, y_{23}$: 2.884, 3.365, ..., 5.156. The sample mean and variance for the y_i's are $\bar{y}=4.150$ and $s^2=.2841$ ($s=.534$). For a .95 lower confidence limit for the pth quantile y_p we see from (5.2.16) and (5.2.17) that the .95 quantile of $t'_{(22)}(-\sqrt{23}u_p)$ is needed. Since $u_{.10}=-1.28$, we must consider $t'_{(22)}(6.139)$. Using Table 5.1 of Owen (1968), we find $\Pr[t'_{(22)}(6.139) \leqslant 8.9655]=.95$. Thus, by (5.2.17), the .95 quantile a of $Z_{.10}$ satisfies

$$(a-u_{.10})\sqrt{23}=8.9655$$

which gives $u_{.10}-a=-1.869$. Hence the desired .95 lower confidence limit for $y_{.10}$ is $\bar{y}+(u_{.10}-a)s=\bar{y}-1.869s$. Alternately, we could find from Resnikoff and Lieberman (1957, p. 388) that $t'_{22,.95}(-\sqrt{23}u_{.10})/\sqrt{22}=1.911$, which also gives $\bar{y}-1.869s$. With the observed values of $\bar{y}$ and s, this gives the confidence interval $y_{.10} \geqslant 3.152$. Transforming this back to the log-normal lifetime distribution, we get $t_{.10} \geqslant \exp(3.152)$, that is, $t_{.10} \geqslant 23.38$.

Similarly, to obtain a .95 lower confidence limit for $y_{.01}$ we have $-\sqrt{23}u_{.01}=11.174$ and, from Owen's table, $\Pr[t'_{(22)}(11.174) \leqslant 15.3982]=.95$. This gives $\bar{y}-3.211s$ as the .95 lower confidence limit for $y_{.01}$, which, with the observed values of $\bar{y}$ and s, becomes 2.435. The corresponding confidence limit for $t_{.01}$ is 11.42.

A F Approximation

If tables like those mentioned are not accessible, a F approximation given by Mann (1977) can often be used to get confidence intervals. The approximation is

$$\frac{\bar{y}-y_p}{-u_p s/C_n} \sim F_{(v_1, v_2)} \tag{5.2.18}$$

where

$$C_n = E\left(\frac{s}{\sigma}\right) = \left(\frac{2}{n-1}\right)^{1/2} \frac{\Gamma(n/2)}{\Gamma[(n-1)/2]}$$

and $v_1=2nu_p^2$ and $v_2=2C_n^2/(1-C_n^2)$. This is similar in origin to (4.1.25) of Chapter 4 and gives very good approximations to the αth quantiles of $(\bar{y}-y_p)/(-u_p s/C_n)$ for $\alpha \geqslant .5$ when n is larger than about 9 and p is not too large. Mann (1977) reports it to be excellent for p in the range .0014 to .16. The approximation is very useful for life distribution work, since the ranges of α, p, and n for which it is good are those of main practical importance.

To illustrate the approximation let us use it to obtain a .95 lower confidence limit for $y_{.10}$ in Example 5.2.2. When $n=23$, $C_n=\sqrt{\frac{2}{22}}\Gamma(11.5)/\Gamma(11)=9.887$. Since $u_{.10}=-1.28$, the degrees of freedom in the F approximation are $v_1=75.4$ and $v_2=87.0$, and (5.2.18) becomes

$$\frac{\bar{y}-y_{.10}}{1.295s}\sim F_{(75.4,87.0)}.$$

As in Section 4.3, we require quantiles of a F distribution with nonintegral degrees of freedom. Using (4.3.6), we obtain $\Pr(F_{(75.4,87.0)}\leqslant 1.439)=.95$, which gives $y_{.10}\geqslant\bar{y}-1.295(1.439)s$, that is, $y_{.10}\geqslant\bar{y}-1.864s$ as the desired .95 lower confidence limit. This is very close to the exact limit $\bar{y}-1.869s$ calculated in the example.

Estimation of the Survivor Function

Confidence intervals for the survivor function $S(y_0)=Q[(y_0-\mu)/\sigma]$ can also be obtained. Since

$$\Pr[S(y_0)\geqslant p]=\Pr(y_{1-p}\geqslant y_0)$$

an α lower confidence limit for $S(y_0)$ requires finding the value of p such that y_0 is an α lower confidence limit for y_{1-p}. By (5.2.16) and (5.2.17), this is the value of p such that

$$\Pr\left(t'_{(n-1)}(-\sqrt{n}u_{1-p})\leqslant\frac{\sqrt{n}(\bar{y}-y_0)}{s}\right)=\alpha.$$

Owen (1968; Secs. 6,8) describes tables that can be used to obtain p directly; Owen and Hua (1977) present fairly extensive tables. One can also obtain p values by interpolation in tables of percentage points of $t'_{(n-1)}(\lambda)$.

For example, let us return once again to Example 5.2.2 and find a .95 lower confidence limit for $\Pr(T\geqslant 20)=S(\log 20)=S(2.996)$. Since $\sqrt{n}(\bar{y}-y_0)/s=10.383$, we need the value of $\lambda=-\sqrt{23}u_{1-p}$ such that $\Pr[t'_{(22)}(\lambda)\leqslant 10.383]=.95$. From Table 6.1 of Owen (1968) this value is found to be $\lambda=7.2723$, which gives $u_{1-p}=-1.516$. This is the .065 quantile for the standard normal distribution, and thus the desired .95 lower confidence limit for $S(\log 20)$ is $1-p=.935$.

5.2.2b *Exact Methods for Type II Censored Samples*

When the data are censored, exact methods are not generally available, but for Type II censored samples some exact procedures are available for a few special cases. Confidence intervals can be based on the pivotal quantities

$$Z_1 = \frac{\tilde{\mu}-\mu}{\tilde{\sigma}} \qquad Z_2 = \frac{\tilde{\sigma}}{\sigma} \qquad Z_p = \frac{(\tilde{\mu}+u_p\tilde{\sigma})-(\mu+u_p\sigma)}{\tilde{\sigma}}$$

where $\tilde{\mu}$ and $\tilde{\sigma}$ are equivariant estimators of the location parameter μ and scale parameter σ (see Appendix G). The exact distributions of Z_1, Z_2, and Z_p are, as usual, intractable for reasonable estimators $\tilde{\mu}$ and $\tilde{\sigma}$. However, one can obtain percentage points by Monte Carlo methods, as was done in the case of the extreme value distribution (see Section 4.1.2c). Nelson and Schmee (1979) have prepared tables of percentage points for pivotals Z_1, Z_2, and Z_p ($p=.01$, .05, and .10) based on the b.l.u.e.'s of μ and σ. At present these tables cover only samples with $r \leqslant n \leqslant 10$, however. Nelson and Schmee (1979) present charts that give confidence limits on the survivor function, and Schmee and Nelson (1977) give similar tables and charts for pivotals based on the m.l.e.'s.

Another approach to exact inferences with Type II censored data is through conditional confidence interval procedures, based on general results for location–scale parameter distributions (see Appendix G and Section 4.1.2a). Unfortunately, two-dimensional numerical integration is required to calculate conditional probabilities for Z_1, Z_2, or Z_p. Specifically, the joint p.d.f. of Z_1 and Z_2, given the ancillary statistics $a_i=(y_i-\tilde{\mu})/\tilde{\sigma}$ $(i=1,\ldots,r)$, is, according to (G5), of the form

$$k'(\mathbf{a},r,n)z_2^{r-1}\exp\left(-\frac{1}{2}\sum_{i=1}^{r}(a_iz_2+z_1z_2)^2\right)\left[Q(a_rz_2+z_1z_2)\right]^{n-r}. \qquad (5.2.19)$$

Neither z_1 nor z_2 can be integrated out of (5.2.19) analytically when $r<n$. Marginal p.d.f.'s of Z_1, Z_2, or Z_p can be calculated by numerical integration and probabilities for Z_1, Z_2, or Z_p obtained by numerical double integration. This is rather too time consuming to consider often, though in important situations it may be worthwhile.

5.2.2c *Approximate Methods for Censored Samples*

Approximate confidence intervals and tests can be based on large-sample properties of m.l.e.'s or likelihood ratio tests. In addition, various authors have attempted to develop other approximate procedures, though these attempts have met with only limited success. We shall mention two of these before going on to discuss large-sample methods.

A rather simple approximation that appears to be sufficiently accurate to recommend even with small sample sizes has been given by Wolynetz (1974). The approximation can be used for inferences on μ when censoring is either Type I or Type II, and it takes the form

$$\frac{\sqrt{n}(\hat{\mu}-\mu)}{\hat{\sigma}_1} \sim t_{(n-1)}. \tag{5.2.20}$$

In this, $\hat{\mu}$ and $\hat{\sigma}$ are the m.l.e.'s of μ and σ, $\hat{\sigma}_1^2 = n_1\hat{\sigma}^2/(n_1-1)$, and $n_1 = r + \Sigma_{i \in C}\lambda(\hat{z}_i)$, where $\lambda(z)$ is given by (5.2.3) and $\hat{z}_i = (y_i - \hat{\mu})/\hat{\sigma}$. Attempts to find similar, simple approximations that can be used for inferences for σ or y_p have, unfortunately, not been fruitful.

A second potentially useful class of approximations has been developed by Mann (1977). These are extensions of (5.2.18) to Type II censored samples and are similar to the F approximations given for the extreme value distribution in Section 4.1.2d. As in Section 4.1.2d, suppose that $\tilde{\mu}$ and $\tilde{\sigma}$ are unbiased estimators of μ and σ based on a Type II censored sample, and let

$$A(r,n) = \frac{\text{Var}(\tilde{\mu})}{\sigma^2} \qquad B(r,n) = \frac{\text{Cov}(\tilde{\mu},\tilde{\sigma})}{\sigma^2} \qquad C(r,n) = \frac{\text{Var}(\tilde{\sigma})}{\sigma^2}.$$

The approximation given by Mann (1977) is

$$(y^* - y_p)/\tilde{\sigma}\left(-\frac{B(r,n)}{C(r,n)} - u_p\right) \sim F_{(v_1, v_2)} \tag{5.2.21}$$

where $y^* = \tilde{\mu} - B(r,n)\tilde{\sigma}/C(r,n)$ and the degrees of freedom v_1 and v_2 are

$$v_1 = 2\left(-\frac{B(r,n)}{C(r,n)} - u_p\right)^2 \Big/ \left(A(r,n) - \frac{B(r,n)^2}{C(r,n)}\right)$$

$$v_2 = \frac{2}{C(r,n)}.$$

This approximation is potentially useful for obtaining confidence limits on quantiles, but at present tables of the constants $A(r,n)$, $B(r,n)$, and $C(r,n)$ do not exist for suitable unbiased estimators of μ and σ. These could, of course, be calculated, say, for the b.l.u.e.'s (Sarhan and Greenberg, 1962), though this would be a fairly big job. Mann (1977) reports work in progress that would provide suitable linear estimators and tables of $A(r,n)$, $B(r,n)$, and $C(r,n)$ so that (5.2.21) can be used with censored samples. In the complete sample case use of (5.2.21) with $\tilde{\mu}=\bar{x}$ and the unbiased estimator $\tilde{\sigma}=s/C_n$ yields (5.2.18).

Maximum Likelihood and Likelihood Ratio Methods

The simplest large-sample approach is to assume that the m.l.e.'s $(\hat{\mu},\hat{\sigma})$ are approximately bivariate normal with the mean (μ,σ) and covariance matrix $\mathbf{I}_0^{-1}$, where $\mathbf{I}_0$ is the observed information matrix

$$\mathbf{I}_0=\begin{pmatrix} -\partial^2\log L/\partial\mu^2 & -\partial^2\log L/\partial\mu\,\partial\sigma \\ -\partial^2\log L/\partial\mu\,\partial\sigma & -\partial^2\log L/\partial\sigma^2 \end{pmatrix}_{(\hat{\mu},\hat{\sigma})}.$$

The derivatives in $\mathbf{I}_0$ are given by (5.2.6). In addition to approximate confidence intervals for μ or σ, approximate confidence intervals for $y_p=\mu+u_p\sigma$ can be obtained by treating $\hat{y}_p=\hat{\mu}+u_p\hat{\sigma}$ as approximately normal, with mean y_p and variance $(1,u_p)\mathbf{I}_0^{-1}(1,u_p)'$. The main shortcoming of this method is that unless sample sizes are quite large, the normal approximations involved are not very good, except for the distribution of $\hat{\mu}$. Somewhat better results for σ are obtained by treating log $\hat{\sigma}$, instead of $\hat{\sigma}$, as approximately normal, but no completely satisfactory way of treating $\hat{y}_p$ is known.

Likelihood ratio methods are probably preferable to the methods just mentioned; the χ^2 approximations involved with them appear reasonable, provided that sample sizes are not too small. Let us consider the procedure for obtaining confidence intervals for $y_p=\mu+u_p\sigma$; confidence intervals for μ are given by the special case $u_p=0$.

The hypothesis $H_0\colon y_p=y_0$ is equivalent to

$$H_0\colon \frac{y_0-\mu}{\sigma}=u_p. \tag{5.2.22}$$

For a likelihood ratio test of H_0 versus $H_1\colon(y_0-\mu)/\sigma\neq u_p$ it is necessary to obtain the m.l.e.'s $\tilde{\mu}$ and $\tilde{\sigma}$ of μ and σ, subject to the restriction (5.2.22). The log likelihood function given in Section 5.2.1 can be rewritten as a function of σ only, by replacing μ by $y_0-u_p\sigma$, according to (5.2.22). This gives

$$\log L_1(\sigma)=-r\log\sigma-\frac{1}{2\sigma^2}\sum_{i\in D}(y_i-y_0+u_p\sigma)^2+\sum_{i\in C}\log Q\left(\frac{y_i-y_0}{\sigma}+u_p\right).$$

Setting $d\log L_1/d\sigma$ equal to zero, we get

$$-r+\sum_{i\in D}\frac{y_i-y_0}{\sigma}\left(\frac{y_i-y_0}{\sigma}+u_p\right)+\sum_{i\in C}\frac{y_i-y_0}{\sigma}V\left(\frac{y_i-y_0}{\sigma}+u_p\right)=0. \tag{5.2.23}$$

This can be solved iteratively for $\tilde{\sigma}$, and then $\tilde{\mu}=y_0-u_p\tilde{\sigma}$. The likelihood ratio statistic for testing H_0 versus H_1 is $\Lambda=-2\log[L(\tilde{\mu},\tilde{\sigma})/L(\hat{\mu},\hat{\sigma})]$, and under H_0, Λ is approximately $\chi^2_{(1)}$. Approximate γ confidence intervals for y_p consist of the set of values y_0 such that $\Lambda\leqslant\chi^2_{(1),\gamma}$. Conversely, an approximate γ confidence interval for the survivor function $S(y_0)=Q[(y_0-\mu)/\sigma]$ can be obtained by fixing y_0 in (5.2.22) and finding the set of p values giving $\Lambda\leqslant\chi^2_{(1),\gamma}$.

Tests and confidence intervals for σ are also readily obtained. To handle the hypothesis $H_0: \sigma=\sigma_0$ one maximizes the log likelihood $\log L(\mu,\sigma_0)$ with respect to μ. The maximizing value $\tilde{\mu}$ is found by solving equation (5.2.4), with $\sigma=\sigma_0$. The likelihood ratio statistic for testing H_0 versus $H_1: \sigma\neq\sigma_0$ is $\Lambda=-2\log[L(\tilde{\mu},\sigma_0)/L(\hat{\mu},\hat{\sigma})]$, and an approximate γ confidence interval for σ consists of those values σ_0 for which $\Lambda\leqslant\chi^2_{(1),\gamma}$.

The two examples that follow illustrate the methods given for the treatment of censored normal or log-normal data.

Example 5.2.3 (Example 5.2.1 revisited) Let us consider the sample of Example 5.2.1, for which $n=10$ and $r=8$. With the underlying life distribution assumed to be a log-normal distribution, m.l.e.'s and b.l.u.e.'s were obtained and were $\hat{\mu}=7.58$ and $\hat{\sigma}=1.00$ (m.l.e.'s) and $\tilde{\mu}=7.60$ and $\tilde{\sigma}=1.14$ (b.l.u.e.'s). Since the sample has $n\leqslant 10$, the tables produced by Nelson and Schmee (1979) can be used to obtain confidence intervals based on the b.l.u.e.'s. Alternately, tables given in Schmee and Nelson (1976) that employ the m.l.e.'s can be used. Let us obtain .90 confidence intervals for μ, σ, and the .10 quantile $y_{.10}$ of the normal distribution of the log failure time. From Tables 1, 2, and 3a of Nelson and Schmee (1979) we find $\Pr(0.564\leqslant\tilde{\sigma}/\sigma\leqslant 1.494)=.90$, $\Pr(-0.652\leqslant(\tilde{\mu}-\mu)/\tilde{\sigma}\leqslant 0.566)=.90$, and $\Pr(-2.401\leqslant(y_{.10}-\tilde{\mu})/\tilde{\sigma}\leqslant -0.688)=.90$. With the observed values $\tilde{\mu}=7.60$ and $\tilde{\sigma}=1.14$, these lead to confidence intervals $0.76\leqslant\sigma\leqslant 2.02$, $6.95\leqslant\mu\leqslant 8.34$, and $4.86\leqslant y_{.10}\leqslant 6.81$, respectively.

Confidence intervals can also be obtained by likelihood ratio methods or a normal approximation to the distribution of $(\hat{\mu},\hat{\sigma})$. This is unnecessary here, since the Nelson–Schmee tables are available, but for illustration let us compute a .90 confidence interval for $y_{.10}$, noting, however, that the large-sample approximations involved with these methods may not be very good with such a small sample size. Since $u_{.10}=-1.28$, using the likelihood ratio test procedure described in this section, we need to determine the set of values y_0 such that the hypothesis $H_0:(y_0-\mu)/\sigma=-1.28$ is not contradicted at the .10 level of significance. For a given y_0, the m.l.e.'s of μ and σ are found from (5.2.23) by iteration and then the associated value of the likelihood ratio statistic Λ is calculated. The set of y_0 values that give $\Lambda \leqslant \chi^2_{(1),.90}=2.706$ is found by trial and error to be $5.30 \leqslant y_0 \leqslant 6.88$, and thus $5.30 \leqslant y_{.10} \leqslant 6.88$ is the desired approximate .90 confidence interval for $y_{.10}$. The top limit for this interval is close to that for the interval obtained by the exact method, but the lower limit is substantially higher than the previous lower limit. All things considered, the agreement in the intervals is fair, considering the small sample size.

Let us also calculate an approximate confidence interval for $y_{.10}$ by using a normal approximation to the distribution of $(\hat{\mu}, \hat{\sigma})$. In the course of computing the m.l.e.'s $\hat{\mu}$ and $\hat{\sigma}$ the observed information matrix $\mathbf{I}_0$ with entries given by (5.2.6) is found, and its inverse is

$$\mathbf{I}_0^{-1}=\begin{pmatrix} 0.0285736 & 0.0013818 \\ 0.0013818 & 0.0675594 \end{pmatrix}$$

$$=\begin{pmatrix} \text{Asvar}(\hat{\mu}) & \text{Ascov}(\hat{\mu},\hat{\sigma}) \\ \text{Ascov}(\hat{\mu},\hat{\sigma}) & \text{Asvar}(\hat{\sigma}) \end{pmatrix}.$$

To obtain an approximate .90 confidence interval for $y_{.10}=\mu-1.28\sigma$ we treat $\hat{y}_{.10}=\hat{\mu}-1.28\hat{\sigma}$ as approximately normal, with mean $y_{.10}$ and variance equal to

$$\text{Asvar}(\hat{y}_{.10})=\text{Asvar}(\hat{\mu})-2.56\,\text{Ascov}(\hat{\mu},\hat{\sigma})+1.28^2\,\text{Asvar}(\hat{\sigma})$$

$$=0.135726.$$

An approximate .90 confidence interval for $y_{.10}$ is thus $(\hat{y}_{.10}-1.645\sqrt{0.135726},\ \hat{y}_{.10}+1.645\sqrt{0.135726})=(5.70, 6.91)$. Once again, agreement of the top limit with that obtained by the exact method is good, but agreement with the bottom limit is not. It can be noted that the agreement is substantially poorer than for the interval obtained by the likelihood ratio method.

Example 5.2.4 As a second example we consider a much larger sample, in which case only approximate methods are available. The data below have been discussed by Nelson and Schmee (1977) and show the number of thousand miles at which different locomotive controls failed in a life test involving 96 controls. The test was terminated after 135,000 miles, at which time 37 failures had occurred, so the data are Type I censored. The failure times for the 37 failed items are 22.5, 37.5, 46.0, 48.5, 51.5, 53.0, 54.5, 57.5, 66.5, 68.0, 69.5, 76.5, 77.0, 78.5, 80.0, 81.5, 82.0, 83.0, 84.0, 91.5, 93.5, 102.5, 107.0, 108.5, 112.5, 113.5, 116.0, 117.0, 118.5, 119.0, 120.0, 122.5, 123.0, 127.5, 131.0 132.5, and 134.0.

In addition, there are 59 censoring times, all equal to 135.0. We shall assume that log failure times are normally distributed, that is, that failure times are log-normal. The m.l.e.'s for μ and σ are found as described in Section 5.2.1, and they are $\hat{\mu}=4.117$ and $\hat{\sigma}=0.706$. In this situation the main point of interest is to find a confidence interval for the survivor function at $t=80$, since a warranty on the locomotive controls is to cover the first 80,000 miles. Either likelihood ratio methods or methods based on a normal approximation to $(\hat{\mu},\hat{\sigma})$ can be used to get this.

Let us use the likelihood ratio method to obtain an approximate .95 confidence interval for $S(\log 80)=S(4.382)=Q[(4.382-\mu)/\sigma]$. This involves finding the set of values z_0 such that $H_0:(4.382-\mu)/\sigma=z_0$ is not contradicted at the .05 level of significance, which requires $\Lambda \leqslant \chi^2_{(1),.95}=3.841$. The required set of z_0 values is found to be $-1.30 \leqslant z_0 \leqslant -0.79$. Since $Q(-1.30)=.903$ and $Q(-0.79)=.785$, the desired .95 confidence interval is $.785 \leqslant S(4.382) \leqslant .903$.

Alternately, an approximation to the distribution of $(\hat{\mu},\hat{\sigma})$ could be employed. The inverse of the observed information matrix is found to be

$$\mathbf{I}_0^{-1}=\begin{pmatrix} 0.0023196 & 0.0002393 \\ 0.0002393 & 0.0056850 \end{pmatrix}$$

$$=\begin{pmatrix} \text{Asvar}(\hat{\mu}) & \text{Ascov}(\hat{\mu},\hat{\sigma}) \\ \text{Ascov}(\hat{\mu},\hat{\sigma}) & \text{Asvar}(\hat{\sigma}) \end{pmatrix}.$$

As with the likelihood ratio procedure, the problem is to determine the set of values z_0 such that $H_0:(4.382-\mu)/\sigma=z_0$ is not contradicted at the .05 level of significance. Noting that H_0 implies that $\mu+\sigma z_0=4.382$, we can test H_0 for a given z_0 by considering

$$Z'=\frac{(\hat{\mu}+\hat{\sigma}z_0)-4.382}{[\text{Asvar}(\hat{\mu}+\hat{\sigma}z_0)]^{1/2}}$$

which under H_0 is approximately $N(0,1)$. Here

$$\begin{aligned}\text{Asvar}(\hat{\mu}+\hat{\sigma}z_0)&=\text{Asvar}(\hat{\mu})+2z_0\text{Ascov}(\hat{\mu},\hat{\sigma})+z_0^2\text{Asvar}(\hat{\sigma})\\&=0.0023196+2z_0(0.0002393)+0.005685z_0^2.\end{aligned}$$

We require the set of values z_0 that make $|Z'| \leq 1.96$, since H_0 is to be acceptable at the .05 level of significance. The required range of z_0 is found by solving a quadratic equation and is $-1.35 \leq z_0 \leq -0.83$. This gives associated values $Q(-1.35) = .912$ and $Q(-0.83) = .797$, so that the desired .95 confidence interval is $.797 \leq S(4.382) \leq .912$. This is close to the result obtained by the likelihood ratio method.

5.3 THE GENERALIZED GAMMA DISTRIBUTION

The generalized gamma distribution, introduced in Section 1.3.5, has p.d.f.

$$\frac{\beta}{\Gamma(k)} \frac{t^{\beta k-1}}{\alpha^{\beta k}} \exp\left[-\left(\frac{t}{\alpha}\right)^{\beta}\right] \qquad t>0 \tag{5.3.1}$$

where $\beta > 0$, $\alpha > 0$, and $k > 0$ are parameters. This is a fairly flexible family of distributions that includes as special cases the exponential ($\beta = k = 1$), Weibull ($k = 1$), and gamma ($\beta = 1$) distributions. The log-normal distribution also arises as a limiting form of (5.3.1), as shown in Section 1.3.5, and thus the generalized gamma model includes as special cases all of the most commonly used lifetime distributions. This makes it useful for discriminating among these other models. In addition, it can be used as a three-parameter distribution itself.

In early work with the generalized gamma distribution there was difficulty in developing inference procedures. For example, with the model written in the form (5.3.1), it is not uncommon, even with samples of 200 or 300, for algorithms for determining m.l.e.'s to fail to converge (e.g., Hager and Bain, 1970). With simulated samples from (5.3.1), Hager and Bain found that convergence problems were worse when k was large than when it was small. Several other authors (e.g., Parr and Webster, 1965; Harter, 1967; Stacy and Mihram, 1965) encountered similar problems with maximum likelihood estimation. In addition, it was found to be difficult to develop suitable tests or interval estimation procedures. Large-sample methods based on normal approximations to the distribution of the m.l.e.'s $\hat{\alpha}$, $\hat{\beta}$, and $\hat{k}$ are unsatisfactory, since the limiting normal distribution is approached very slowly. Hager and Bain found that normal approximations could be poor even for sample sizes as large as 400.

These problems originally limited the use of the generalized gamma distribution to a certain extent, but work by Prentice (1974) clarified matters a great deal and inference can now be fairly easily handled (e.g., Farewell and Prentice, 1977; Lawless, 1980). Much of the difficulty with the model arises because distributions (5.3.1) with very different sets of parameter values α, β, and k look alike. Prentice (1974) considered the distribution in a reparameterized form that reduces this effect and makes properties of the distribution much more transparent. This form of the model is examined below.

Reparameterized Model

Suppose that T has p.d.f. (5.3.1); we consider $Y=\log T$ instead of T and, in addition, we will reparameterize the model. If we set $u=\log\alpha$ and $b=\beta^{-1}$, it follows from (5.3.1) that

$$W_1=\frac{Y-u}{b}$$

has a log-gamma distribution [see (1.3.13)] with p.d.f.

$$\frac{1}{\Gamma(k)}\exp(kw_1-e^{w_1}) \qquad -\infty<w_1<\infty. \tag{5.3.2}$$

As $k\to\infty$ this distribution is shifted further and further to the right, and its mean and variance become infinite. However, as noted in Section 1.3.3, the distribution of $\sqrt{k}(w_1-\log k)$ converges to a standard normal distribution as $k\to\infty$. We therefore introduce a further reparameterization, obtained by considering

$$\begin{aligned} W&=\sqrt{k}(W_1-\log k)\\ &=\frac{Y-(u+b\log k)}{b/\sqrt{k}}=\frac{Y-\mu}{\sigma} \end{aligned} \tag{5.3.3}$$

where

$$\sigma=\frac{b}{\sqrt{k}} \qquad \mu=u+b\log k. \tag{5.3.4}$$

The p.d.f. of W is easily found from (5.3.2) to be, for finite $k>0$,

$$f(w;k)=\frac{k^{k-1/2}}{\Gamma(k)}\exp\left(\sqrt{k}w-ke^{w/\sqrt{k}}\right) \qquad -\infty<w<\infty. \tag{5.3.5}$$

Equivalently, the p.d.f. of $Y=\log T$ is

$$\frac{k^{k-1/2}}{\sigma\Gamma(k)}\exp\left(\sqrt{k}\frac{y-\mu}{\sigma}-ke^{(y-\mu)/\sigma\sqrt{k}}\right) \qquad -\infty<y<\infty. \tag{5.3.6}$$

As $k \to \infty$ (5.3.5) approaches the standard normal limiting form

$$f(w;\infty)=\frac{1}{(2\pi)^{1/2}}e^{-w^2/2} \qquad -\infty<w<\infty \tag{5.3.7}$$

as shown in Section 1.3.3.

With the model written in the form (5.3.6), the cases $k=1$ and $k=\infty$, respectively, give the extreme value and normal distributions for $Y=\log T$, or the Weibull and log-normal distributions for T. Figure 5.3.1 displays p.d.f.'s (5.3.5) for three finite values of k, as well as the standard normal p.d.f. corresponding to $k=\infty$. For k greater than about 20 the distribution

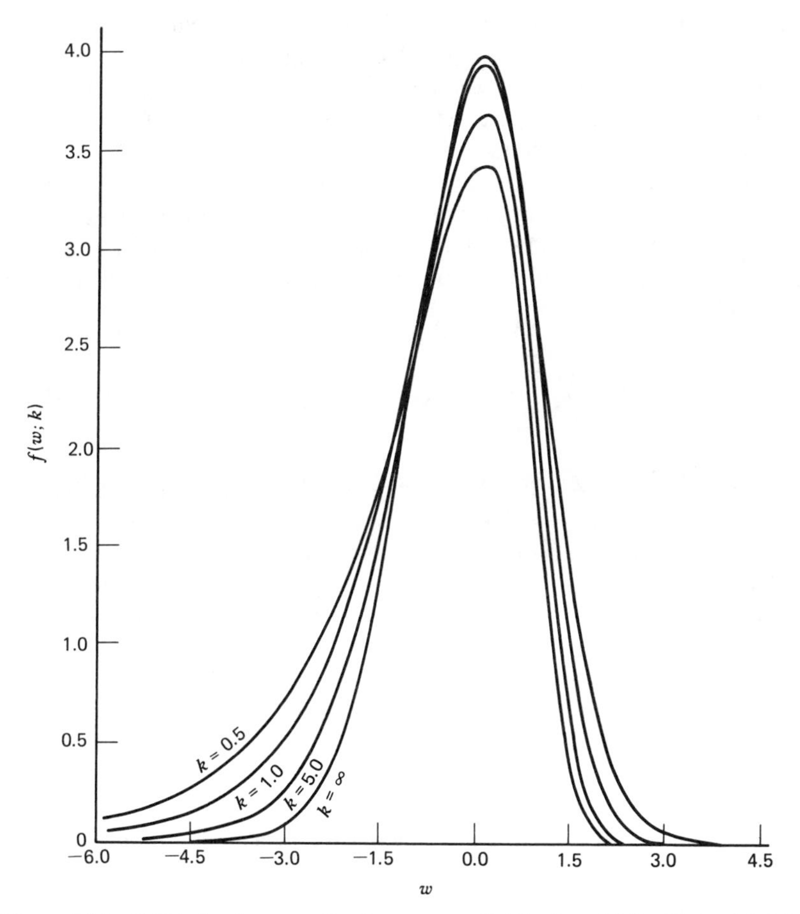

Figure 5.3.1 P.d.f.'s for the generalized log-gamma distribution (5.3.5) with $k=0.5$, 1.0, 5.0, and ∞.

changes very little, and even the extreme value distribution with $k=1$ does not differ drastically from the standard normal distribution. The similarity between the extreme value and normal distributions has been noted in the past (e.g., Irwin, 1942), and one frequently encounters situations in which both Weibull and log-normal models provide good fits to observed data.

The p.d.f. (5.3.6) of $Y=\log T$ changes relatively little as k increases from 1 to ∞, and unless a large amount of data is available, it will not usually be possible to estimate k very precisely. It is clear that the problems created by this are exaggerated if the original parameterizations, using either β and α or $b=\beta^{-1}$ and $u=\log\alpha$, are used, rather than μ and σ. Two distributions with similar μ and σ but very different k values can be very alike. In terms of the (u, b, k) parameterization, however, since $b=\sigma\sqrt{k}$ and $u=\mu-\sigma\sqrt{k}\log k$ [see (5.3.4)], the two distributions would have very different u and b values also. Thus, with the (u, b, k) or, equivalently, the (α, β, k) parameterization, difficulties in examining the likelihood function or obtaining m.l.e.'s are more serious than with the parameterization (μ, σ, k). Henceforth we shall almost always work with the model in the form (5.3.6).

A further useful reparameterization defines μ and σ as above but replaces k with $\lambda=k^{-1/2}$. The normal and extreme value models then correspond to $\lambda=0$ and 1, respectively. Prentice (1974) also extends the family (5.3.6) to include distributions with $\lambda<0$ by considering models in which $W=(Y-\mu)/\sigma$ has p.d.f.

$$\frac{|\lambda|}{\Gamma(\lambda^{-2})}(\lambda^{-2})^{\lambda^{-2}}\exp\left[\lambda^{-2}(\lambda w-e^{\lambda w})\right] \qquad -\infty<w<\infty \qquad (5.3.8)$$

where $-\infty<\lambda<\infty, \lambda\neq 0$. For $\lambda>0$ this is the family (5.3.5). Distributions with $\lambda<0$ correspond to situations in which $-W$ has p.d.f. (5.3.5), with $k=|\lambda|^{-2}$. Extending (5.3.5) in this way not only provides a wider class of models but also alleviates some technical difficulties for the important case $\lambda=0$, which is a boundary point if the parameter space has only nonnegative λ values.

With the model reformulation just given, maximum likelihood estimation and large-sample methods are manageable; these are discussed in Section 5.3.1. Exact procedures are given in Section 5.3.2 for the case of uncensored data.

5.3.1 Maximum Likelihood Estimation and Large-Sample Methods

Maximum Likelihood Estimation

We use the reparameterized form (5.3.6) of the model and proceed along lines similar to those of Farewell and Prentice (1977), who treat maximum likelihood estimation for regression models based on (5.3.6). The parameter k is used for discussion, since it retains its original meaning as in (5.3.1) and is more natural in some ways than $\lambda=k^{-1/2}$. On the other hand, in obtaining m.l.e.'s, plotting likelihoods, and so on it is convenient to work with λ and, where necessary, with the extended model (5.3.8).

The p.d.f of $W=(Y-\mu)/\sigma$ is (5.3.5) for the case in which k is finite, and (5.3.7) for the case where $k=\infty$. When $0<k<\infty$ the survivor function for W is

$$\Pr(W\geqslant w)=\int_w^\infty \frac{k^{k-1/2}}{\Gamma(k)}\exp\left(\sqrt{k}u-ke^{u/\sqrt{k}}\right)du$$

$$=\int_{k\exp(w/\sqrt{k})}^\infty \frac{x^{k-1}}{\Gamma(k)}e^{-x}\,dx=Q\left(k,ke^{w/\sqrt{k}}\right) \qquad (5.3.9)$$

where $Q(k,a)$ is the incomplete gamma integral defined in (5.1.11). For future reference we note that the pth quantile $w_{k,p}$ of W's distribution satisfies $Q(k,ke^{w_{k,p}/\sqrt{k}})=1-p$. Thus $ke^{w_{k,p}/\sqrt{k}}$ is the pth quantile of the one-parameter gamma distribution with index parameter k or, equivalently, half the pth quantile of $\chi^2_{(2k)}$. Solving for $w_{k,p}$, we therefore get

$$w_{k,p}=\sqrt{k}\log\left(\frac{1}{2k}\chi^2_{(2k),p}\right). \qquad (5.3.10)$$

The pth quantile for y is $y_p=\mu+w_{k,p}\sigma$, and the survivor function for y is $Q(k,ke^{(y-\mu)/\sigma\sqrt{k}})$.

For the special case $k=\infty$ the survivor function $Q(w)$ for W is that of the standard normal distribution, given by (5.2.2). The survivor function for Y is $Q[(y-\mu)/\sigma]$.

Consider a sample that may be Type I or Type II censored. As usual, let $y_1,\ldots,y_n$ represent observed log lifetimes or log censoring times of the n individuals; the set of individuals for which y_i is a log lifetime is denoted D and the set for which y_i is a log censoring time is denoted C. The number of observed lifetimes is $r=|D|$. Also, for ease of notation let $w_i=(y_i-\mu)/\sigma$. For models with $0<k<\infty$ the likelihood function is then

$$L(\mu,\sigma,k)=\prod_{i\in D}\frac{1}{\sigma}f(w_i;k)\prod_{i\in C}Q\left(k,ke^{w_i/\sqrt{k}}\right)$$

and thus

$$\log L(\mu,\sigma,k)=\sum_{i\in D}\left[\log f(w_i;k)-\log\sigma\right]+\sum_{i\in C}\log Q\left(k,ke^{w_i/\sqrt{k}}\right) \qquad (5.3.11)$$

where $f(w_i;\, k)$ is given by (5.3.5), and $Q(k, ke^{w_i/\sqrt{k}})$ by (5.3.9). For the case $k=\infty$ the likelihood and log likelihood functions are those given for the normal distribution in Section 5.2.1. In particular,

$$\log L(\mu,\sigma,\infty) = -r\log\sigma - \frac{1}{2\sigma^2}\sum_{i\in D}(y_i-\mu)^2 + \sum_{i\in C}\log Q\left(\frac{y_i-\mu}{\sigma}\right) \tag{5.3.12}$$

where Q is given by (5.2.2).

One approach to obtaining the m.l.e.'s $\hat{\mu}$, $\hat{\sigma}$, and $\hat{k}$ is to attempt to maximize $\log L(\mu,\sigma,k)$ by simultaneously solving the equations $\partial\log L/\partial\mu = 0, \partial\log L/\partial\sigma=0$, and $\partial\log L/\partial k=0$. This is troublesome because of the work required to calculate derivatives of $\log L$ with respect to k. Another approach is to attempt to maximize $\log L$ directly with a procedure that does not require the analytical determination of first or second derivatives. The simplest procedure, however, is to perform iterations in two stages, with k treated as fixed in the first stage. For a given value of k, one finds the values $\tilde{\mu}(k)$ and $\tilde{\sigma}(k)$ that maximize $\log L$ by solving $\partial\log L/\partial\mu=0$ and $\partial\log L/\partial\sigma=0$, with the specified value of k. If this is done for different values of k, the maximized likelihood function for k can be obtained and $\hat{k}$ located. This approach is taken here, so we shall give derivatives of $\log L$ only with respect to μ and σ. For convenience the contributions to $\log L$ and its derivatives from a single uncensored and a single censored observation will be written down separately.

Uncensored Observations

For an observed log lifetime y_i the contribution to the log likelihood function (5.3.11) and its derivatives are as follows:

$$\log L_i = (k-\tfrac{1}{2})\log k - \log\Gamma(k) - \log\sigma + \sqrt{k}w_i - ke^{w_i/\sqrt{k}}$$

$$\frac{\partial\log L_i}{\partial\mu} = -\frac{\sqrt{k}}{\sigma} + \frac{\sqrt{k}}{\sigma}e^{w_i/\sqrt{k}} \tag{5.3.13}$$

$$\frac{\partial\log L_i}{\partial\sigma} = -\frac{1}{\sigma} - \frac{\sqrt{k}}{\sigma}w_i + \sqrt{k}\frac{w_i}{\sigma}e^{w_i/\sqrt{k}}$$

$$\sigma^2\frac{\partial^2\log L_i}{\partial\mu^2} = -e^{w_i/\sqrt{k}}$$

$$\sigma^2\frac{\partial^2\log L_i}{\partial\mu\,\partial\sigma} = \sqrt{k} - \sqrt{k}e^{w_i/\sqrt{k}} - w_i e^{w_i/\sqrt{k}} \tag{5.3.14}$$

$$\sigma^2\frac{\partial^2\log L_i}{\partial\sigma^2} = 1 + 2\sqrt{k}w_i - \sqrt{k}w_i e^{w_i/\sqrt{k}} - w_i^2 e^{w_i/\sqrt{k}}$$

where

$$w_i = (y_i - \mu)/\sigma.$$

Censored Observations

For a log censoring time y_i the contributions to $\log L$ and its derivatives are given below. For convenience in writing the derivatives we define $x = k\exp(w/\sqrt{k}) = k\exp[(y-\mu)/\sigma\sqrt{k}]$ and note the following:

$$\frac{\partial x}{\partial \mu} = \frac{-x}{\sqrt{k}\sigma} \qquad \frac{\partial x}{\partial \sigma} = \frac{-wx}{\sqrt{k}\sigma}$$

$$\frac{\partial^2 x}{\partial \mu^2} = \frac{x}{k\sigma^2} \qquad \frac{\partial^2 x}{\partial \sigma^2} = \frac{xw^2}{k\sigma^2} + \frac{2xw}{\sqrt{k}\sigma^2} \qquad \frac{\partial^2 k}{\partial \mu\, \partial \sigma} = \frac{xw}{k\sigma^2} + \frac{x}{\sqrt{k}\sigma^2}$$

$$\frac{\partial Q(k,x)}{\partial x} = -\frac{x^{k-1}e^{-x}}{\Gamma(k)} \qquad \frac{\partial^2 Q}{\partial x^2} = \frac{e^{-x}x^{k-2}(x-k+1)}{\Gamma(k)}$$

$$\frac{\partial \log Q(k,x)}{\partial x} = \frac{1}{Q(k,x)}\frac{\partial Q}{\partial x}$$

$$\frac{\partial^2 \log Q}{\partial x^2} = \frac{1}{Q(k,x)}\frac{\partial^2 Q}{\partial x^2} - \frac{1}{Q^2(k,x)}\left(\frac{\partial Q}{\partial x}\right)^2.$$

The desired quantities are then

$$\log L_i = \log Q(k, ke^{w_i/\sqrt{k}})$$

$$\frac{\partial \log L_i}{\partial \mu} = \frac{\partial \log Q}{\partial x_i}\frac{\partial x_i}{\partial \mu}$$

$$\frac{\partial \log L_i}{\partial \sigma} = \frac{\partial \log Q}{\partial x_i}\frac{\partial x_i}{\partial \sigma} \tag{5.3.15}$$

$$\frac{\partial^2 \log L_i}{\partial \mu^2} = \frac{\partial^2 \log Q}{\partial x_i^2}\left(\frac{\partial x_i}{\partial \mu}\right)^2 + \frac{\partial \log Q}{\partial x_i}\frac{\partial^2 x_i}{\partial \mu^2}$$

$$\frac{\partial^2 \log L_i}{\partial \mu\, \partial \sigma} = \frac{\partial^2 \log Q}{\partial x_i^2}\frac{\partial x_i}{\partial \mu}\frac{\partial x_i}{\partial \sigma} + \frac{\partial \log Q}{\partial x_i}\frac{\partial^2 x_i}{\partial \mu\, \partial \sigma} \tag{5.3.16}$$

$$\frac{\partial^2 \log L_i}{\partial \sigma^2} = \frac{\partial^2 \log Q}{\partial x_i^2}\left(\frac{\partial x_i}{\partial \sigma}\right)^2 + \frac{\partial \log Q}{\partial x_i}\frac{\partial^2 x_i}{\partial \sigma^2}.$$

Derivatives of $\log L$ for the case in which $k=\infty$ have been given in Section 5.2.1.

The log likelihood function is of the form (for $0<k<\infty$)

$$\log L(\mu,\sigma,k)=\sum_{i\in D}\log L_i+\sum_{i\in C}\log L_i$$

where $\log L_i$ is given by the expressions above for $i\in D$ and for $i\in C$. The estimates $\tilde{\mu}(k)$ and $\tilde{\sigma}(k)$ can be found by solving the equations $\partial\log L/\partial\mu=0$ and $\partial\log L/\partial\sigma=0$, say, by Newton–Raphson iteration. The expressions involved look complicated but are easily programmed. For the case $k=\infty$ the determination of $\tilde{\mu}(\infty)$ and $\tilde{\sigma}(\infty)$ has been discussed in Section 5.1.2.

Simplification for Uncensored Data

When the data are uncensored, things simplify considerably. In this case, for $0<k<\infty$,

$$\log L(\mu,\sigma,k)$$

$$=n\left(k-\tfrac{1}{2}\right)\log k-n\log\Gamma(k)-n\log\sigma+\sqrt{k}\sum_{i=1}^{n}\frac{y_i-\mu}{\sigma}-k\sum_{i=1}^{n}e^{(y_i-\mu)/\sigma\sqrt{k}}$$

and from (5.3.13) the maximum likelihood equations for $\tilde{\mu}(k)$ and $\tilde{\sigma}(k)$ are

$$\sum_{i=1}^{n}e^{(y_i-\mu)/\sigma\sqrt{k}}=n \tag{5.3.17}$$

$$\sum_{i=1}^{n}\frac{y_i-\mu}{\sigma}e^{(y_i-\mu)/\sigma\sqrt{k}}-\sum_{i=1}^{n}\frac{y_i-\mu}{\sigma}=\frac{n}{\sqrt{k}}. \tag{5.3.18}$$

It is actually only necessary to solve a single equation by iterative means; from (5.3.17) it follows that

$$e^{\tilde{\mu}}=\left(\frac{1}{n}\sum_{i=1}^{n}e^{y_i/\tilde{\sigma}\sqrt{k}}\right)^{\tilde{\sigma}\sqrt{k}}. \tag{5.3.19}$$

Substituting this into (5.3.18) and simplifying slightly, we get

$$\sum_{i=1}^{n}y_ie^{y_i/\tilde{\sigma}\sqrt{k}}\Big/\sum_{i=1}^{n}e^{y_i/\tilde{\sigma}\sqrt{k}}-\bar{y}-\frac{\tilde{\sigma}}{\sqrt{k}}=0. \tag{5.3.20}$$

Thus (5.3.20) can be solved iteratively for $\tilde{\sigma}(k)$ and then $\tilde{\mu}(k)$ determined by (5.3.19).

Maximized Relative Likelihood for k

With $\tilde{\mu}(k)$ and $\tilde{\sigma}(k)$ calculated for a series of k values, the maximized likelihood function

$$L_{\max}(k)=L[\tilde{\mu}(k),\tilde{\sigma}(k),k] \tag{5.3.21}$$

can be determined sufficiently accurately to obtain the m.l.e. $\hat{k}$, which is the value maximizing $L_{\max}(k)$. This gives the m.l.e.'s $\hat{\mu}=\tilde{\mu}(\hat{k})$, $\hat{\sigma}=\tilde{\sigma}(\hat{k})$, and $\hat{k}$ for the three parameters, and the maximized relative likelihood function

$$R_{\max}(k)=\frac{L_{\max}(k)}{L(\hat{\mu},\hat{\sigma},\hat{k})}$$

can be determined. A graph of $R_{\max}(k)$ portrays plausible k values and is useful with likelihood ratio tests, described below.

It is usually sufficient to consider about 8 or 10 different k values in finding $L_{\max}(k)$. It is generally advisable to work with $\lambda=k^{-1/2}$ when computing and portraying $L_{\max}$ and when locating $\hat{k}=\hat{\lambda}^{-1/2}$. Values of λ close to zero (corresponding to large values of k) are common in lifetime applications, and the important extreme value and normal models correspond to $\lambda=1$ and $\lambda=0$. It is possible to have $\hat{\lambda}=0(\hat{k}=\infty)$; this happens when $\hat{\lambda}$ in the extended model (5.3.8) is less than or equal to zero. In such cases, and in situations where $\lambda=0$ is merely a plausible value, one should consider the extended model rather than (5.3.6). Example 5.3.1 illustrates this approach to maximum likelihood estimation.

Tests and Interval Estimates

One use of the generalized gamma model is in assessing special-case two-parameter lifetime distributions such as the Weibull and log-normal distributions. Likelihood ratio procedures can be used to test such models. For tests of the form $H_0: k=k_0$ versus $H_1: k\neq k_0$ we can use the likelihood ratio statistic

$$\Lambda=-2\log R_{\max}(k_0).$$

For finite k_0 the distribution of Λ under H_0 is asymptotically $\chi^2_{(1)}$. A slight technical difficulty arises in testing the log-normal model in this way, since $k_0=\infty(\lambda=0)$ is on the boundary of the parameter space. This is easily overcome by working with the extended family (5.3.8), in which case $\lambda=0$ is an interior point of the parameter space and the necessary conditions for the application of large-sample methods are met (Prentice, 1974).

Other parametric hypotheses can be tested with the likelihood ratio method. For example, the two-parameter gamma distribution is represented by $H_0: \beta=1$, where we revert to the original parameterization used in (5.3.1). To test H_0 we need to obtain the maximum, $L_{max}(H_0)$, of the likelihood function $L(\mu, \sigma, k)$ under H_0. This is discussed in Section 5.1.1. The likelihood ratio statistic for testing H_0 versus $H_1: \beta \neq 1$ is then $\Lambda = -2\log[L_{max}(H_0)/L(\hat{\mu}, \hat{\sigma}, \hat{k})]$, and in large samples Λ can be treated as approximately $\chi^2_{(1)}$ under H_0.

The generalized gamma family can also be used to examine the effect of assuming a particular model on inferences about the underlying life distribution. For example, if the .10 quantile is of particular interest, this can be estimated with different values of k, and the degree to which k affects the inference determined. With censored data, approximate confidence intervals for quantiles or other characteristics of the distribution can be obtained using likelihood ratio methods. This is discussed in detail for the special cases $k=1$ and $k=\infty$ in Sections 4.2.2 and 5.2.2c, respectively; similar procedures can be used for other values of k. For complete samples, exact methods are available and they are discussed in the next section.

In some situations we may want to obtain confidence intervals with all three parameters μ, σ, and k treated as unknowns. The likelihood ratio method can be used to obtain interval estimates, but a great deal of computation may be necessary. Alternately, one can use large-sample normal approximations to the distribution of $(\hat{\mu}, \hat{\sigma}, \hat{k})$, but this is also computationally forbidding and, at any rate, from investigations with complete samples appears to be inaccurate unless sample sizes are quite large (e.g., Prentice, 1974). These should not be major drawbacks for the use of the model, however. In most situations it is acceptable to make inferences conditional on k, as long as care is taken to consider the effect that k has on the inferences. This is illustrated in Example 5.3.2.

Example 5.3.1 (Example 5.2.2 revisited) Let us reconsider the data of Example 5.2.2, giving the number of million revolutions to failure for each of a group of 23 ball bearings in a fatigue test. These data, which were originally treated by Lieblein and Zelen (1956) as arising from a Weibull distribution, were considered in Example 5.2.2 under a log-normal model. Probability plots of the data on both Weibull and log-normal paper indicate that both models appear reasonably consonant with the data.

The 23 log lifetimes are 2.884, 3.365, 3.497, 3.726, 3.741, 3.820, 3.881, 3.948, 3.950, 3.991, 4.017, 4.217, 4.229, 4.229, 4.232, 4.432, 4.534, 4.591,

4.655, 4.662, 4.851, 4.852, and 5.156. Since this is a complete sample, m.l.e.'s $\tilde{\mu}(k)$ and $\tilde{\sigma}(k)$ can be found for different k values ($0<k<\infty$) by solving (5.3.20) and then using (5.3.19). Once $\tilde{\mu}(k)$ and $\tilde{\sigma}(k)$ are obtained, $\log L_{max}(k)$ is calculated as in (5.3.21). In calculating $\log L_{max}(k)$ the series expression [see (B4) of Appendix B]

$$(k-\tfrac{1}{2})\log k-\log\Gamma(k)=k-\tfrac{1}{2}\log(2\pi)-\frac{1}{12k}+\frac{1}{360k^3}-\cdots$$

is useful if k is large. For $k=\infty$ the maximum likelihood estimates for $\tilde{\mu}$ and $\tilde{\sigma}$ are those for the normal distribution. Since there is no censoring, these are $\tilde{\mu}=\bar{y}$ and $\tilde{\sigma}=[\Sigma(y_i-\bar{y})^2/n]^{1/2}$. Table 5.3.1 shows results obtained by considering a number of different k values.

The likelihood is maximized very close to $k=10.6(\lambda=.31)$ but decreases little as $k\to\infty$. The final column of Table 5.3.1 gives values of $R_{max}(k)=L_{max}(k)/L(\hat{\mu},\hat{\sigma},\hat{k})$, where $\hat{\mu}=4.230$, $\hat{\sigma}=.510$, and $\hat{k}=10.6$. Table 5.3.1 or a graph of $R_{max}(k)$, shown in Figure 5.3.2, gives a good picture of plausible values of k. To test $H_0: k=k_0$ versus $H_1: k\neq k_0$ the likelihood ratio statistic $\Lambda=-2\log R_{max}(k_0)$ can be used, with Λ treated as being approximately $\chi^2_{(1)}$ under H_0. For k_0 to be contradicted at the .05 level of significance would require $\Lambda\geq\chi^2_{(1),.95}=3.841$, or $R_{max}(k_0)\leq.147$. From Figure 5.3.2 we thus see that in this example an approximate .95 confidence interval for k includes all k values greater than about .4. Both the normal and extreme value models are well supported by the data, with the normal model the better supported of the two. Note that there are obviously members of the extended family (5.3.8) with $\lambda<0$ that are very well supported by the data, though maximized likelihoods have not been given for any of these models.

Table 5.3.1 Determination of $\log L_{max}(k)$ and $R_{max}(k)$

k	$\lambda=k^{-1/2}$	$\tilde{\mu}(k)$	$\tilde{\sigma}(k)$	$\log L_{max}(k)$	$R_{max}(k)$
0.3	1.826	4.604	.417	−20.138	.072
0.5	1.414	4.507	.449	−19.142	.216
1	1.000	4.405	.476	−18.231	.485
4	0.500	4.279	.502	−17.568	.942
9	0.333	4.237	.509	−17.509	.999
10.6	0.307	4.230	.510	−17.5075	1.000
12	0.289	4.225	.511	−17.508	1.000
400	0.050	4.164	.520	−17.619	.895
∞	0.0	4.150	.522	−17.680	.839

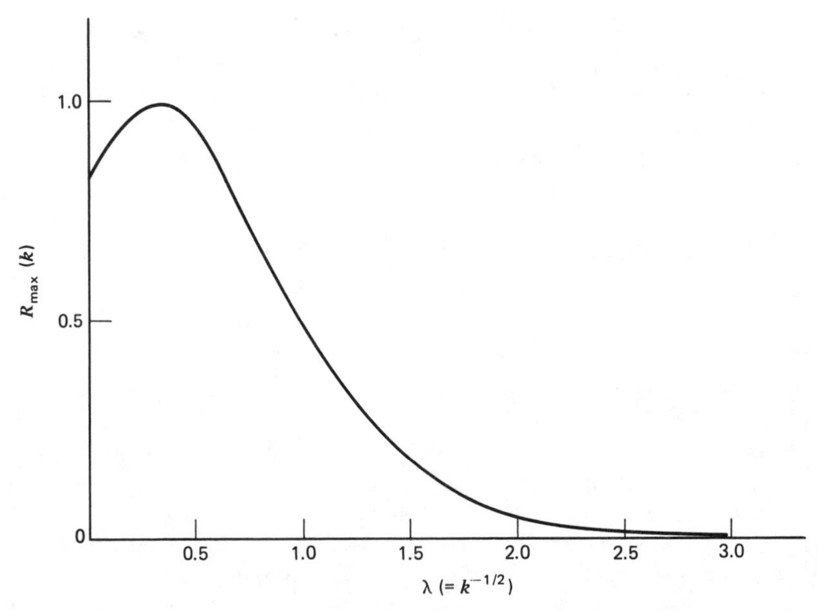

Figure 5.3.2 Maximized relative likelihood for log-gamma parameter k (Example 5.3.1)

This example is examined further in the next section, where confidence intervals for quantiles of the distribution are obtained for different values of k.

5.3.2 Exact Methods for Complete Samples

If samples are uncensored and k is assumed to be known, exact inference procedures are available. The procedures described in this section are conditional methods based on the general results given for estimation of location and scale parameters in Appendix G, and they are more exhaustively discussed by Lawless (1980). The procedures give confidence intervals for parameters, quantiles, or the survivor function of (5.3.6) or (5.3.8).

Let $y_1, \ldots, y_n$ be a random sample from (5.3.6), where k is assumed to be known and μ and σ are unknown. Note that μ and σ are a location and scale parameter, respectively. Let $\tilde{\mu}$ and $\tilde{\sigma}$ be a pair of equivariant estimators. (We shall drop the dependence on k in our notation for $\tilde{\mu}$ and $\tilde{\sigma}$.) Convenient estimators are the m.l.e.'s $\hat{\mu}(k)$ and $\hat{\sigma}(k)$, obtained by solving equations (5.3.19) and (5.3.20) for the given k value, or the sample mean and scaled standard deviation, $\tilde{\mu}=\bar{y}$ and $\tilde{\sigma}=[\Sigma(y_i-\bar{y})^2/n]^{1/2}$, which are also the m.l.e.'s of μ and σ for the special case $k=\infty$. By Theorem G2 of

Appendix G,

$$Z_1=\frac{\tilde{\mu}-\mu}{\tilde{\sigma}} \qquad Z_2=\frac{\tilde{\sigma}}{\sigma} \qquad Z_p=\frac{\tilde{\mu}-y_p}{\tilde{\sigma}} \tag{5.3.22}$$

are pivotal quantities. In (5.3.22) $y_p=\mu+w_{k,p}\sigma$ is the pth quantile of the distribution (5.3.6), and $w_{k,p}$ is the pth quantile in the standardized distribution of $W=(Y-\mu)/\sigma$, given by (5.3.10). Note that the pivotal quantity $Z_p=(\tilde{\mu}-y_p)/\tilde{\sigma}$ can be written in terms of Z_1 and Z_2 as

$$Z_p=Z_1-w_{k,p}Z_2^{-1}. \tag{5.3.23}$$

The quantities $a_i=(y_i-\tilde{\mu})/\tilde{\sigma}, i=1,\ldots,n$ are ancillary statistics, and confidence intervals and tests for μ, σ, or y_p can be based on the conditional distributions of Z_1, Z_2, and Z_p, given $\mathbf{a}$. The required conditional distributions are easily obtained: by Theorem G3 the joint p.d.f. of Z_1 and Z_2, given $\mathbf{a}$, is of the form

$$g(z_1,z_2|\mathbf{a}:k)=C'(\mathbf{a},n,k)z_2^{n-1}\exp\left(\sqrt{k}\sum_{i=1}^{n}(a_i+z_1)z_2-k\sum_{i=1}^{n}e^{(a_i+z_1)z_2/\sqrt{k}}\right) \tag{5.3.24}$$

where we bear in mind that the sample is uncensored and that the p.d.f. of $(Y_i-\mu)/\sigma$ is given by (5.3.5). Proceeding along lines exactly like those in Theorem 4.1.3 (or see Lawless, 1980), we obtain the following results:

1. The marginal p.d.f. of Z_2, given $\mathbf{a}$, is

$$g_2(z|\mathbf{a};k)=C(\mathbf{a},n,k)z^{n-2}\exp\left(\sqrt{k}(z-1)\sum_{i=1}^{n}a_i\right)\Big/\left(\frac{1}{n}\sum_{i=1}^{n}e^{a_iz/\sqrt{k}}\right)^{n} \qquad z>0. \tag{5.3.25}$$

2. The marginal distribution function of Z_p, given $\mathbf{a}$, is

$$\Pr(Z_p\leqslant t|\mathbf{a};k)=\int_0^{\infty}g_2(z|\mathbf{a};k)I\left(nk,ke^{(tz+w)/\sqrt{k}}\sum_{i=1}^{n}e^{a_iz/\sqrt{k}}\right)dz \tag{5.3.26}$$

where $I(r,s)$ is the incomplete gamma function (B12) and we write w for $w_{k,p}$.

Both (5.3.25) and the integral in (5.3.26) must be integrated numerically. The normalizing constant $C(\mathbf{a}, n, k)$ in (5.3.25) can be determined from the fact that

$$\int_0^\infty g_2(z|\mathbf{a}; k)\, dz = 1.$$

The integrals involved are of the same type as those encountered in Section 4.1.2b, and the discussion of computational procedures given there is useful here as well. The results above apply for finite k. For the normal model case $k=\infty$ the distributions of Z_1, Z_2, and Z_p are well known and are discussed in Section 5.2.2a.

A convenient approach with the generalized gamma model is to examine $R_{\max}(k)$ and determine plausible values of k. Then confidence intervals or tests can be obtained for quantities of interest, for different fixed values of k. This shows the effect of k on any inferences made and provides a good picture of the information about different characteristics of the life distribution. The following example illustrates these procedures.

Example 5.3.2 (Example 5.3.1 continued) In Example 5.3.1 we examined a complete sample of 23 observations and, on the assumption that the lifetimes arose from a generalized gamma distribution, found that a wide range of k values was supported by the data. In fact, an approximate .95 confidence interval for k included all values greater than about 0.4. The weibull ($k=1$) and log-normal ($k=\infty$) models for lifetimes were both well supported, with $R_{\max}(1)=.485$ and $R_{\max}(\infty)=.839$, respectively. To see the effect of k on inferences let us determine confidence intervals for the three quantiles $y_{.01}$, $y_{.10}$, and $y_{.50}$ of the log lifetime distribution for several different values of k. Table 5.3.2 shows some results for $k=0.5$, 1, 10, and ∞; these values more or less cover the range of plausible k values. The table gives two-sided .90 confidence intervals for each quantile. For the normal distribution ($k=\infty$) confidence intervals are obtained as described in Section 5.2.2a. In fact, lower confidence limits for $y_{.01}$ and $y_{.10}$ were calculated in Example 5.2.2. For the other three values of k (5.3.26) is used to obtain values l_1 and l_2 such that $\Pr(Z_p \leqslant l_1|\mathbf{a})=.05$ and $\Pr(Z_p \leqslant l_2|\mathbf{a})=.95$. Since $Z_p=(\hat{\mu}-y_p)/\hat{\sigma}$, this gives $\hat{\mu}-l_2\hat{\sigma} \leqslant y_p \leqslant \hat{\mu}-l_1\hat{\sigma}$ as the desired .90 confidence interval. Table 5.3.2 displays the m.l.e. $\hat{y}_p=\hat{\mu}+w_{k,p}\hat{\sigma}$ of y_p, the lower confidence limit (LCL) $\hat{\mu}-l_2\hat{\sigma}$, and the upper confidence limit (UCL) $\hat{\mu}-l_1\hat{\sigma}$ for each quantile and value of k.

The calculations used to obtain the confidence intervals will be described for the case where $k=10$ and $p=.50$. First, for the value of k assumed, the m.l.e.'s $\hat{\mu}$ and $\hat{\sigma}$ are obtained. This has previously been done and results given in Example 5.3.1; for $k=10$ the m.l.e.'s are $\hat{\mu}=4.232$ and $\hat{\sigma}=.510$. The

Table 5.3.2 M.l.e.'s and .90 Confidence Intervals for Quantiles for Different k Values

		$k=0.5$	$k=1$	$k=10$	$k=\infty$
$y_{.50}$	UCL	4.45	4.42	4.36	4.38
	$\hat{y}_{.50}$	4.26	4.23	4.18	4.15
	LCL	4.01	4.01	3.98	3.92
$y_{.10}$	UCL	3.56	3.63	3.69	3.68
	$\hat{y}_{.10}$	3.19	3.33	3.46	3.48
	LCL	2.56	2.83	3.10	3.15
$y_{.01}$	UCL	2.36	2.71	3.12	3.20
	$\hat{y}_{.01}$	1.73	2.22	2.81	2.94
	LCL	0.54	1.29	2.21	2.44

values $w_{k,p}$ that give the quantiles $y_p=\mu+w_{k,p}$ and that are used in (5.3.26) are calculated using (5.3.10). We find $w_{10,.50}=\sqrt{10}\log(\chi^2_{(20),.50}/20)=-.107$, whence $y_{.50}=\mu-.107\sigma$ for the distribution with $k=10$. The associated m.l.e. is $\hat{y}_{.50}=\hat{\mu}-.107\hat{\sigma}=4.177$. The next problem is to obtain the necessary percentage points l_1 and l_2 from the distribution (5.3.26) of $Z_{.50}$, given $\mathbf{a}$. The ancillaries $a_1,\ldots,a_{23}$ are calculated as $a_i=(y_i-\hat{\mu})/\hat{\sigma}=(y_i-4.232)/.510$. Next, the constant $C(\mathbf{a},n,10)$ in (5.3.25) is found by numerical integration to be 2.6522. Probabilities for $Z_{.50}$ can now be obtained from (5.3.26) by further numerical integration. With $n=23$, $k=10$, $w=-.107$ and the given values of $a_1,\ldots,a_{23}$, we find that $\Pr(z_{.50}\leqslant-.26|\mathbf{a})=.05$ and $\Pr(Z_{.50}\leqslant.49|\mathbf{a})=.95$. This gives the confidence interval $3.98\leqslant y_{.50}\leqslant 4.36$ shown in Table 5.3.2.

As would be expected, there is relatively broad agreement in the estimates of the median $y_{.50}$ for different values of k. However, as we move to the tail of the distribution different models give somewhat different results. The tails of a distribution are of much interest in lifetime distribution work, so this is important. For example, for some types of manufactured items the .10 quantile is used in rating items. If the lifetimes in this example are analyzed as coming from a log-normal distribution, the .95 lower confidence limit on $y_{.10}$ is 3.15 (corresponding to a limit of $\exp 3.15=23.34$ for the distribution of lifetimes), whereas it is 2.83 (corresponding to $\exp 2.83=16.95$ for lifetimes) if the data are assumed to arise from a Weibull distribution. This difference may or may not be of practical importance, depending on the situation, but the difference is fairly substantial relative to the actual lengths of the .90 confidence intervals. The differences in the estimates of $y_{.01}$ are larger still, to the extent that the m.l.e. for $y_{.01}$ under the normal model is not even inside the .90 interval for $y_{.01}$ under the extreme value model. A practical consequence of this is that the real precision with which we can estimate $y_{.01}$ is much less than that implied by an analysis based on a particular distribution, such as the normal or extreme value distribution, provided, of course, that we do not know from external information that one particular model is the correct one.

We conclude this section with two additional remarks. The first is that confidence intervals for the survivor function $S(t)$ can be obtained from Z_p by exploiting the relationship between $S(t)$ and the quantiles of the distribution. This is done exactly as described for the case of the extreme value distribution, in Section 4.1.2e. Finally, confidence intervals and tests can be based on unconditional distributions of the pivotals Z_1, Z_2, and Z_p, as was done in Section 4.1.2c for the extreme value distribution. The difficulty with this is the usual one of determining percentage points for the pivotals. Lawless (1980) shows how to obtain good approximations to percentage points, and Hager and Bain (1970) also discuss this problem briefly.

5.4 POLYNOMIAL HAZARD FUNCTION AND OTHER MODELS

The models discussed in Chapters 3 to 5, namely, the exponential, Weibull, gamma, log-normal, and generalized gamma distributions, are the most frequently used lifetime models. Reasons for the popularity of these models include their ability to fit different types of lifetime data well, their mathematical and statistical tractability, and the fact that there are failure or aging models that lead to them. Other parametric models are also used from time to time; a few of these were mentioned in Section 1.3.6. It is not feasible to treat other models in any detail, but this section provides a brief discussion of a few problems. First, the family of models for which the hazard function is a polynomial is examined. These models are capable of giving shapes of hazard function that the more common models cannot, in particular the U-shaped hazard function. In addition, these models are sometimes used to provide essentially distribution-free methods of estimating survivor or hazard functions. Section 5.4.2 discusses situations in which none of the common models are appropriate and briefly examines discrete mixture models. There are no special test or interval estimation procedures available for the models in this section, but one can use the usual large-sample likelihood methods. Only point estimation is discussed here.

5.4.1 Models With Polynomial Hazard Functions

Consider models for which the hazard function can be written as a polynomial,

$$h(t)=\beta_0+\beta_1 t+\cdots+\beta_{m-1}t^{m-1} \qquad t\geqslant 0. \tag{5.4.1}$$

Equivalently, the cumulative hazard function $H(t)=\int_0^t h(u)\,du$ is of the form

$$H(t)=\alpha_1 t+\alpha_2 t^2+\cdots+\alpha_m t^m \tag{5.4.2}$$

where

$$\alpha_j=\beta_{j-1}/j \qquad j=1,\dots,m.$$

The survivor function for the distribution is $S(t)=\exp[-H(t)]$ and the p.d.f. is $f(t)=h(t)S(t)$. Since $H(t)$ must be an increasing function, with $H(0)=0$ and $H(\infty)=\infty$, the parameters $\alpha_1,\dots,\alpha_m$ must satisfy certain constraints. In particular, it is necessary for $\alpha_1,\dots,\alpha_m$ to be such that $\alpha_1\geqslant 0$, $\alpha_m\geqslant 0$, and $H'(t)\geqslant 0$ for $t\geqslant 0$.

The model (5.4.1) with $m=1$ is the exponential distribution. That with $m=2$ is often called the linear hazard rate distribution (Kodlin, 1967; Bain, 1974). Distributions with $m>2$ are sometimes useful, though it is rarely worth considering models with m greater than about 4. Distributions with $m>2$ can have nonmonotone hazard functions; an example of this is given in Section 5.5. Distributions with $m\geqslant 2$ are discussed by Bain (1974), Krane (1963), Canfield and Borgman (1975), and others.

One use of this class of models is in providing adequate fits to data not readily handled by one of the more common models. The models are also sometimes used to give smooth estimates of the hazard and survivor functions in situations where strong assumptions about a parametric model are undesirable. Krane (1963) gives an example that involves the fitting of survival curves to the service life of pieces of equipment in a business.

Estimation of Parameters

Maximum likelihood estimation is somewhat cumbersome for distributions with polynomial hazard functions, and least squares estimation is often used instead. Consider, as usual, a sample that may be Type I or Type II censored and let t_i represent either the observed lifetime or censoring time for the ith individual ($i=1,\dots,n$). The set of individuals whose lifetimes are observed is denoted D and that consisting of those whose lifetimes are censored is C. The likelihood function is

$$L(\alpha_1,\dots,\alpha_m)=\prod_{i\in D} f(t_i)\prod_{i\in C} S(t_i)$$

$$=\left(\prod_{i\in D}\left(\alpha_1+2\alpha_2 t_i+\cdots+m\alpha_m t_i^{m-1}\right)\right)\exp\left(-\sum_{i=1}^{n}\sum_{j=1}^{m}\alpha_j t_i^j\right). \tag{5.4.3}$$

Maximizing $L(\alpha_1,\ldots,\alpha_m)$ can be difficult when m is larger than unity. As an illustration we shall consider the simplest case after the exponential model, that with $m=2$. Maximum likelihood estimation for this model has been examined by Kodlin (1967), Bain (1974), and Gross and Clark (1975, pp. 151–159).

When $m=2$, the log likelihood function (5.4.3) becomes

$$\log L(\alpha_1,\alpha_2)=\sum_{i\in D}\log(\alpha_1+2\alpha_2 t_i)-\alpha_1\sum_{i=1}^{n}t_i-\alpha_2\sum_{i=1}^{n}t_i^2. \tag{5.4.4}$$

The maximum likelihood equations are

$$\begin{aligned}\frac{\partial\log L}{\partial\alpha_1}&=\sum_{i\in D}(\alpha_1+2\alpha_2 t_i)^{-1}-\sum_{i=1}^{n}t_i=0\\ \frac{\partial\log L}{\partial\alpha_2}&=\sum_{i\in D}2t_i(\alpha_1+2\alpha_2 t_i)^{-1}-\sum_{i=1}^{n}t_i^2=0.\end{aligned} \tag{5.4.5}$$

The second derivatives of $\log L$ are

$$\frac{\partial^2\log L}{\partial\alpha_1^2}=-\sum_{i\in D}(\alpha_1+2\alpha_2 t_i)^{-2}$$

$$\frac{\partial^2\log L}{\partial\alpha_1\partial\alpha_2}=-\sum_{i\in D}2t_i(\alpha_1+2\alpha_2 t_i)^{-2}$$

$$\frac{\partial^2\log L}{\partial\alpha_2^2}=-\sum_{i\in D}4t_i^2(\alpha_1+2\alpha_2 t_i)^{-2}.$$

With this model both α_1 and α_2 must be nonnegative. However, it is possible for $\hat{\alpha}_1$ or $\hat{\alpha}_2$ to be 0, in which case the m.l.e. may not be a solution to (5.4.5). In addition, it is possible for there to be more than one solution to (5.4.5), so care is needed in obtaining the m.l.e.'s. Bain (1974) discusses these difficulties and gives a convenient procedure for finding $\hat{\alpha}_1$ and $\hat{\alpha}_2$.

As a simpler method of estimation, Bain and others have proposed least squares procedures of the type discussed in Section 2.5. Recall that these apply when there is a transformation of the survivor function that is linear in the unknown parameters. Here we have

$$H(t)=-\log S(t)=\alpha_1 t+\cdots+\alpha_m t^m.$$

Suppose that $Y_i = \tilde{H}(t_i)$ is an estimate of $H(t_i)$, for $i \in D$. Then estimates of $\alpha_1, \ldots, \alpha_m$ can be obtained by minimizing

$$\sum_{i \in D} \left[Y_i - (\alpha_1 t_i + \cdots + \alpha_m t_i^m) \right]^2 \tag{5.4.6}$$

as suggested in Section 2.5. The estimates $\tilde{\alpha}_1, \ldots, \tilde{\alpha}_m$ are the solution to the system of linear equations

$$\mathbf{A}\boldsymbol{\alpha} = \mathbf{B} \tag{5.4.7}$$

where $\boldsymbol{\alpha} = (\alpha_1, \ldots, \alpha_m)'$, $\mathbf{A}_{m \times m}$ has (k, l) entry $\Sigma_{i \in D} t_i^{k+l}$, and $\mathbf{B}_{m \times 1}$ has kth entry $\Sigma_{i \in D} t_i^k Y_i$. Bain (1974) considers only Type II censored data with observed lifetimes $t_1 \leqslant t_2 \leqslant \cdots \leqslant t_r$, so that $D = \{1, 2, \ldots, r\}$, and employs the values $Y_i = -\log[1 - i/(n+1)]$. These Y_i values, which can be thought of as arising from estimating $S(t_i)$ by $1 - i/(n+1)$, correspond to commonly used probability plotting positions. An alternate choice is $Y_i = -\log[1 - (i - 0.5)/n]$. With Type I censored data the Y_i's can be based on the empirical hazard function or the product-limit estimate of the survivor function, as described in Section 2.5. An example involving this family of models is deferred to Section 5.5, where the quadratic model with $h(t) = \beta_0 + \beta_1 t + \beta_2 t^2$ is considered.

In addition to the models just discussed, other models in which parameters enter $h(t)$ linearly are also occasionally useful. Shaked (1977), for example, considers models with $h(t) = \beta_1 h_1(t) + \beta_2 h_2(t)$, where $h_1(t)$ and $h_2(t)$ are known functions of t and β_1 and β_2 are parameters. This includes models such as $h(t) = \beta_1 + \beta_2 t^\gamma$, where γ is known. Models for which $\log h(t)$ is a polynomial can also be considered; Gehan and Siddiqui (1973) present least squares methods and compare them with maximum likelihood estimation for several models of this type (see also Problem 5.13).

5.4.2 Mixtures and Other Models

The models of the last three chapters are especially useful in situations where the lifetime distribution appears to be unimodal or to have a monotone hazard function. On the other hand, these models are generally unsuitable when the hazard function is U shaped or when the lifetime distribution is multimodal. Such situations are more difficult to handle than those considered in most of the examples thus far in the book. This is in part because it is harder to motivate models from physical considerations in these situations and in part because of problems in applying statistical methods. Parametric models capable of giving U-shaped hazard functions or multimodal p.d.f.'s typically involve three or more parameters, and unless there are large amounts of data, even point estimation often raises substantial problems.

Some families of models capable of giving U-shaped hazard functions were mentioned in Section 1.3.6. Somewhat informal statistical methods are frequently helpful in coming to grips with these models, and in this and the following section we briefly examine two examples of this. In this section we consider a discrete mixture model, and in Section 5.5.2 we consider a model with a quadratic hazard function. For another example involving a "difficult" model, see Problem 5.13.

Consider now discrete mixture models, described in Section 1.3.7. There is a large literature on mixtures of normal distributions; see, for example, Quandt and Ramsey (1978) and the ensuing discussion, Folkes (1979), or Aitkin and Wilson (1980) for recent results and many references to this area. Mixtures of exponential (e.g., Mendenhall and Hader, 1958; Tallis and Light, 1968), Weibull (e.g., Kao, 1959; Falls, 1970) and gamma distributions (e.g., Dickinson, 1974) have also been considered by numerous authors. Mixture models are generally rather difficult to handle statistically. In particular, formal methods of estimation often run into problems, primarily because there are usually many unknown parameters in the model and the likelihood function is flat in certain regions. Relatively *ad hoc* methods often help in examining a set of data; the following example portrays such an approach.

Example 5.4.1 The data below show the numbers of cycles to failure for a group of 60 electrical appliances in a life test. The failure times have been ordered for convenience.

14	34	59	61	69	80	123	142	165	210
381	464	479	556	574	839	917	969	991	1064
1088	1091	1174	1270	1275	1355	1397	1477	1578	1649
1702	1893	1932	2001	2161	2292	2326	2337	2628	2785
2811	2886	2993	3122	3248	3715	3790	3857	3912	4100
4106	4116	4315	4510	4584	5267	5299	5583	6065	9701

Table 5.4.1 is a frequency distribution obtained from the data. The table also shows some expected frequencies, calculated later. The data suggest that the failure time distribution might be bimodal. At any rate, the common distributions like the Weibull, log-normal, and gamma distributions do not appear to be appropriate. One might try a polynomial hazard function model. Another possibility, discussed here, is a mixture model with

Table 5.4.1 Frequency Distribution for Appliance Failure Data

Cycles to Failure	0–100	100–200	200–500	500–1000	1000–1500	1500–2000	2000–3000	3000–5000	5000–10,000
Observed frequency	6	3	4	6	9	5	10	12	5
Expected frequency	7.1	4.3	2.6	4.1	4.8	5.5	10.4	14.0	7.1

two components. A mixture model may in fact be physically meaningful, for example, if some items have defects that cause them to fail early, whereas items without defects are susceptible to a more gradual wear out.

Let us consider as a possible model a mixture of two Weibull distributions, with p.d.f. and survivor function

$$f(t)=p\frac{\beta_1}{\alpha_1}\left(\frac{t}{\alpha_1}\right)^{\beta_1-1}\exp\left[-\left(\frac{t}{\alpha_1}\right)^{\beta_1}\right]$$

$$+(1-p)\frac{\beta_2}{\alpha_2}\left(\frac{t}{\alpha_2}\right)^{\beta_2-1}\exp\left[-\left(\frac{t}{\alpha_2}\right)^{\beta_2}\right] \tag{5.4.8}$$

$$S(t)=p\exp\left[-\left(\frac{t}{\alpha_1}\right)^{\beta_1}\right]+(1-p)\exp\left[-\left(\frac{t}{\alpha_2}\right)^{\beta_2}\right].$$

Given, say, a complete sample $t_1,\ldots,t_n$ from this distribution, the likelihood function $L(p,\beta_1,\alpha_1,\beta_2,\alpha_2)=\prod f(t_i)$ involves five parameters. It is difficult to maximize L, particularly if the two components in the mixture are not well separated. Other methods of estimation are consequently often used with mixtures, though these frequently run into difficulties as well. Falls (1970) discusses moment estimation and notes problems such as nonunique solutions to the estimating equations.

Really precise estimation of parameters in a mixture model is not possible with only moderate amounts of data, but a main concern is to determine whether some model in the given class produces a reasonable fit to the data. Relatively informal methods are often helpful in examining this possibility, particularly if the data appear separable into two or more fairly distinct parts. Consider the data given above, for example, and suppose for the moment that the value of p is known. In a sample of n one would expect approximately np observations from the first component in (5.4.8) and $n(1-p)$ from the second. If the two distributions are fairly well separated and p is not too close to 0 or 1, the smallest observations will be almost exclusively from one component, and the largest observations from the other. Thus one might treat the r_1 smallest observations, say, as the first r_1 in a sample of size np from the first distribution, and the r_2 largest as the largest observations in a sample of size $n(1-p)$ from the second distribution, and estimate parameters of the two distributions separately from these two parts of the data. Probability plots of the two sets of observations are especially useful, providing both parameter estimates and a check on the assumed form of the two component distributions. This approach assumes p to be known, but it usually happens that the procedure is not too sensitive to the exact value of p used and one can informally estimate p sufficiently well to give useful results.

The data in the example suggest that if a two-component mixture is appropriate, one component has a fairly small proportion of the observations and most of its probability is fairly close to 0. For exploratory purposes let us assume a trial value of $p=.20$ for the leftmost component in the mixture and treat the sample as consisting of 12 and 48 observations from the leftmost and the rightmost component distributions, respectively. A Weibull probability plot of the first six or eight observations, treating these as the smallest observations in a sample of size 12, produces a reasonably straight plot and gives estimates of α_1 and β_1 in the region of $\tilde{\alpha}_1=110$ and $\tilde{\beta}_1=1.5$. Treating the 15 or 20 largest observations as the largest observations in a sample of size 48 similarly gives a reasonably straight Weibull probability plot and estimates of about $\tilde{\alpha}_2=3400$ and $\tilde{\beta}_2=1.65$. One can investigate the fit of the resulting model to the data: the estimated survivor function is

$$\tilde{S}(t)=.20\exp\left[-\left(\frac{t}{110}\right)^{1.5}\right]+.80\exp\left[-\left(\frac{t}{3400}\right)^{1.65}\right].$$

The expected frequencies in the Table 5.4.1 are calculated from this model and show fair agreement with the observed frequencies. This indicates that there are mixtures of two Weibull distributions that fit the data. If desired, one could use the estimates just obtained as initial estimates in an attempt to obtain m.l.e.'s or in some other formal estimation procedure.

In concluding this rather brief section we mention two points. The first is a reiteration of our earlier remark that mixtures and other univariate models with several parameters are often difficult to handle with regard to formal estimation and tests and that informal procedures can be helpful. The second point is that distribution-free methods can often be used in situations like those discussed here, according to the aims of the analysis. Nonparametric and distribution-free procedures are discussed in Chapters 2, 7, and 8.

5.5 GROUPED DATA FROM CONTINUOUS MODELS

When data from a continuous life distribution are grouped, estimation can be based on the exact likelihood function for the grouped data. Sometimes an alternative is to use least squares procedures similar to those described in Section 5.4.1 for the case of ungrouped data. Each approach will be discussed briefly.

5.5.1 Maximum Likelihood

Suppose, as in Chapter 2, that lifetimes are grouped into $k+1$ intervals $I_j=[a_{j-1}, a_j)$, $j=1,\ldots,k+1$, where $0=a_0<a_1<\cdots<a_k<a_{k+1}=\infty$. Let d_j be the number of lifetimes observed to fall into I_j, from a sample involving n individuals. The underlying distribution has survivor function $S(t)$ and p.d.f. $f(t)$; these involve one or more unknown parameters that are to be estimated.

If no censoring can occur, except possibly in the last interval, then $(d_1,\ldots,d_k)$ has a multinomial probability function

$$\frac{n!}{d_1!\cdots d_k!d_{k+1}!}\pi_1^{d_1}\cdots\pi_k^{d_k}\pi_{k+1}^{d_{k+1}}$$

where $\pi_j=S(a_{j-1})-S(a_j)$ is the probability of an individual's lifetime falling into I_j. The likelihood function for the unknown parameters $\boldsymbol{\theta}$ in the parametric model can thus be taken as

$$L(\boldsymbol{\theta})=\prod_{j=1}^{k+1}\left[S(a_{j-1})-S(a_j)\right]^{d_j}. \tag{5.5.1}$$

Maximizing $L(\boldsymbol{\theta})$ is sometimes awkward, but usually estimates can be obtained by one of the standard methods.

Example 5.5.1 Consider estimation of the parameters of a Weibull distribution. In this case the survivor function is of the form $S(t)=\exp[-(t/\alpha)^\beta]$, so that (5.5.1) gives

$$L(\alpha,\beta)=\prod_{j=1}^{k+1}\left\{\exp\left[-\left(\frac{a_{j-1}}{\alpha}\right)^\beta\right]-\exp\left[-\left(\frac{a_j}{\alpha}\right)^\beta\right]\right\}.$$

Let $\psi_j=(a_j/\alpha)^\beta$, $j=1,\cdots,k+1$. Then maximum likelihood equations are

$$\frac{\partial\log L}{\partial\alpha}=\frac{\beta}{\alpha}\sum_{j=1}^{k+1}\frac{\psi_{j-1}\exp[-\psi_{j-1}]-\psi_j\exp[-\psi_j]}{\exp[-\psi_{j-1}]-\exp[-\psi_j]}=0$$

$$\frac{\partial\log L}{\partial\beta}=-\sum_{j=1}^{k+1}\frac{\psi_{j-1}\log(a_{j-1}/\alpha)\exp[-\psi_{j-1}]-\psi_j\log(a_j/\alpha)\exp[-\psi_j]}{\exp[-\psi_{j-1}]-\exp[-\psi_j]}=0.$$

These equations can be simplified slightly, but must obviously be solved iteratively. Initial estimates of $\hat{\alpha}$ and $\hat{\beta}$ can be obtained via the least squares procedures described in the next section or by some other informal procedure. $L(\hat{\alpha}, \hat{\beta})$ could also be maximized directly without solving these equations.

If censoring times occur in intervals other than the last (i.e., there are withdrawals in some intervals), the exact likelihood function cannot be written down without further assumptions. One possibility is to assume that all withdrawals occur at the ends of intervals. In this case, if w_j represents the number of withdrawals in I_j, the likelihood function would be

$$L(\boldsymbol{\theta}) = \prod_{j=1}^{k+1} \left[S(a_{j-1}) - S(a_j)\right]^{d_j} S(a_j)^{w_j}.$$

Alternately, all censoring times might be taken to be equal to the midpoint of the interval in which they occur. Such *ad hoc* modifications are often satisfactory. A more formal approach would require explicit assumptions about the censoring and lifetime processes.

If class intervals are narrow, another possibility is to treat the data as continuous and assume that all observations in the jth interval occur, say, at the interval midpoint. This will be acceptable if intervals are quite narrow, but otherwise the exact likelihood should be used. Problem 3.3 of Chapter 3 examines the effect of grouping in the case of data from the exponential distribution.

5.5.2 Least Squares Methods

For models in which some transform of the survivor function is linear in the parameters, least squares estimation can be used in a way similar to that discussed for ungrouped data earlier in the book (see Sections 2.5, 5.4.1). Suppose that a function $g[S(t)]$ of $S(t)$ can be written as a linear function of the unknown parameters $\alpha_1, \ldots, \alpha_m$, say

$$g[S(t)] = \alpha_1 l_1(t) + \cdots + \alpha_m l_m(t)$$

where $l_1(t), \ldots, l_m(t)$ are known linearly independent functions of t. Let $\hat{S}(a_i), i = 1, \ldots, k$, be estimates of the $S(a_i)$'s and let

$$Y_i = g[\hat{S}(a_i)] \qquad i = 1, \ldots, k.$$

Provided that $m \leqslant k$, least squares estimates of $\alpha_1, \ldots, \alpha_m$ can be obtained by minimizing

$$\sum_{i=1}^{k} \left[Y_i - \alpha_1 l_1(a_i) - \cdots - \alpha_m l_m(a_i)\right]^2 = (\mathbf{Y} - \mathbf{T}\boldsymbol{\alpha})'(\mathbf{Y} - \mathbf{T}\boldsymbol{\alpha}) \quad (5.5.2)$$

where

$$\mathbf{Y} = (Y_1, \ldots, Y_k)' \qquad \boldsymbol{\alpha} = (\alpha_1, \ldots, \alpha_m)' \qquad \mathbf{T} = \begin{bmatrix} l_1(a_1) \cdots l_m(a_1) \\ \vdots \\ l_1(a_k) \cdots l_m(a_k) \end{bmatrix}_{k \times m}.$$

The estimates that minimize (5.5.2) are given by

$$\tilde{\boldsymbol{\alpha}} = (\mathbf{T}'\mathbf{T})^{-1}\mathbf{T}'\mathbf{Y}. \quad (5.5.3)$$

This ordinary least squares estimate does not take into account the variance–covariance structure of $Y_1, \ldots, Y_k$. If desired, generalized least squares estimates that do this can be obtained. Of course, $E(\mathbf{Y}) \neq \mathbf{T}\boldsymbol{\alpha}$ here, so that neither type of least squares possesses the optimality properties associated with estimation in linear models, but taking account of the covariance structure of $Y_1, \ldots, Y_k$ may be sensible. Section 2.2.2 gives variances and covariances of the $\hat{S}(a_i)$'s that are based on the standard life table estimates (2.2.3), and from these, approximate variances and covariances can be obtained for $Y_1, \ldots, Y_k$. Let us recall the life table notation of Chapter 2, where $P_i = S(a_i)$, $i = 1, \ldots, k$, and let $\hat{P}_i = \hat{S}(a_i)$ be the standard life table estimate of P_i based on (2.2.3). Two cases are distinguished:

1. If there is no censoring except in the last interval, then $\hat{P}_i = N_{i+1}/n$, where N_{i+1} is the number of individuals alive at time a_i. Exact variances and covariances of the $\hat{P}_i$'s are given by (2.2.9) and (2.2.10), and they are

$$\operatorname{Var}(\hat{P}_i) = \frac{P_i(1-P_i)}{n}$$

$$\operatorname{Cov}(\hat{P}_i, \hat{P}_j) = \frac{P_j(1-P_i)}{n} \qquad i<j \quad (5.5.4)$$

where $N_1 = n$ is the number of individuals alive at time 0. The covariance matrix $\mathbf{V}$ of $(\hat{P}_1, \ldots, \hat{P}_k)$ can be estimated by estimating the P_i's in (5.5.4) by the $\hat{P}_i$'s.

2. If there is censoring in intervals other than the last one, then the variances and covariances of the $\hat{P}_i$'s are not available. Estimates of the variances and covariances are given by Greenwood's formula [see (2.2.4) and (2.2.19)],

$$\widehat{\text{Cov}}\left(\hat{P}_i, \hat{P}_j\right) = \hat{P}_i \hat{P}_j \sum_{l=1}^{i} \frac{\hat{q}_l}{\hat{p}_l N_l'} \qquad i \leqslant j \tag{5.5.5}$$

where $N_l' = N_l - \frac{1}{2} W_l$ is the number alive at a_{l-1} minus half the number of withdrawals in the lth interval and $\hat{q}_l = 1 - \hat{p}_l = d_l / N_l'$.

Let $\hat{\mathbf{V}}$ be the estimate of the covariance matrix of $\hat{P}_1, \ldots, \hat{P}_k$, and let $\mathbf{G}$ be the $k \times k$ matrix with entries

$$G_{ij} = \frac{\partial Y_i}{\partial \hat{P}_j} \qquad i = 1, \ldots, k \quad j = 1, \ldots, k.$$

Then (see Appendix C) an estimate of the covariance matrix of $Y_1, \ldots, Y_k$ is

$$\hat{\mathbf{W}} = \mathbf{G} \hat{\mathbf{V}} \mathbf{G}'.$$

$\mathbf{G}$ here is a diagonal matrix, since Y_i is a function of $\hat{P}_i$ only, making calculation of $\hat{\mathbf{W}}$ especially easy. With $\hat{\mathbf{W}}$ as an approximate covariance matrix for $Y_1, \ldots, Y_k$, generalized least squares can be used to estimate $\alpha_1, \ldots, \alpha_m$. This requires the minimization of $(\mathbf{Y} - \mathbf{T}\boldsymbol{\alpha})' \hat{\mathbf{W}}^{-1} (\mathbf{Y} - \mathbf{T}\boldsymbol{\alpha})$ and gives the estimate

$$\tilde{\boldsymbol{\alpha}} = (\mathbf{T}' \hat{\mathbf{W}}^{-1} \mathbf{T})^{-1} \mathbf{T}' \hat{\mathbf{W}}^{-1} \mathbf{Y}. \tag{5.5.6}$$

Least squares methods like these can be used with many common models, including the exponential, Weibull, and log-normal models and those with polynomial hazard functions. The estimates provided are suitable for some purposes, although where precise estimation is important, one should compute maximum likelihood or other more precise estimates.

Example 5.5.2 The data given in Table 5.5.1 are from results concerning the time to second failure for 104 bus motors (Davis, 1952), the basic observations being the number of thousand miles to failure. The data suggest a model with a nonmonotone hazard function. One possible family of models is that for which the hazard function is a quadratic polynomial in t, in which case the cumulative hazard function is of the form $H(t) = \alpha_1 t +$

Table 5.5.1 Frequency Distribution for Bus Motor Failure Data

Number of Thousand Miles	0–20	20–40	40–60	60–80	80–100	100–120	≥120
Observed frequency	19	13	13	15	15	18	11
Expected frequency	21.74	10.92	10.50	15.18	18.60	15.78	11.34

$\alpha_2 t^2 + \alpha_3 t^3$ and the survivor function is $S(t)=\exp(-\alpha_1 t - \alpha_2 t^2 - \alpha_3 t^3)$. A quick way to fit such a model is to use the least squares methods described above, with $Y_i = H(a_i) = -\log[\hat{S}(a_i)]$. Here, $a_1=20$, $a_2=40$, $a_3=60$, $a_4=80$, $a_5=100$, and $a_6=120$, and since there is no censoring, the $\hat{S}(a_i)$'s are $\hat{S}(a_1)=.817$, $\hat{S}(a_2)=.692$, $\hat{S}(a_3)=.567$, $\hat{S}(a_4)=.423$, $\hat{S}(a_5)=.279$, and $\hat{S}(a_6)=.106$. The vector $\mathbf{Y}=(Y_1,\ldots,Y_6)'$ is thus

$$\mathbf{Y}=(0.202, 0.368, 0.567, 0.860, 1.277, 2.246).$$

For convenience we rescale the a_i's by dividing by 10 and write the cumulative hazard as $H(t)=\alpha_1'(t/10)+\alpha_2'(t/10)^2+\alpha_3'(t/10)^3$. We have

$$\mathbf{T}=\begin{bmatrix} a_1/10 & (a_1/10)^2 & (a_1/10)^3 \\ a_2/10 & (a_2/10)^2 & (a_2/10)^3 \\ \vdots & \vdots & \vdots \\ a_6/10 & (a_6/10)^2 & (a_6/10)^3 \end{bmatrix} = \begin{bmatrix} 2 & 4 & 8 \\ 4 & 16 & 64 \\ 6 & 36 & 216 \\ 8 & 64 & 512 \\ 10 & 100 & 1000 \\ 12 & 144 & 1728 \end{bmatrix}$$

$$\mathbf{T'Y}=\begin{bmatrix} 51.880 \\ 533.272 \\ 5746.048 \end{bmatrix} \qquad \mathbf{T'T}=\begin{bmatrix} 364 & 3,528 & 36,400 \\ 3,528 & 36,400 & 390,432 \\ 36,400 & 390,432 & 4,298,944 \end{bmatrix}.$$

Solving the equations $(\mathbf{T'T})\hat{\boldsymbol{\alpha}}'=\mathbf{T'Y}$, we get $\hat{\alpha}_1'=0.159148$, $\hat{\alpha}_2'=-0.025445$, and $\hat{\alpha}_3'=0.002300$, so that the estimated cumulative hazard function is

$$\hat{H}(t)=0.159148\frac{t}{10}-0.025445\left(\frac{t}{10}\right)^2+0.002300\left(\frac{t}{10}\right)^3. \qquad (5.5.7)$$

To assess the agreement between the estimated model and the data we can calculate expected frequencies for the classes in the frequency distribution. The expected frequency for $I_j=[a_{j-1}, a_j)$ is

$$104\left[\hat{S}(a_{j-1})-\hat{S}(a_j)\right] \qquad j=1,\ldots,7$$

where $\hat{S}(a_j)=\exp[-\hat{H}(a_j)]$. Calculating these using (5.5.7), we get the expected frequencies shown in Table 5.5.1. The agreement between expected and observed frequencies is adequate. A "chi square" statistic $\Sigma(\text{obs}-\text{exp})^2/\text{exp}$ gives a value of 2.36, though it should be noted that this value cannot be assessed against the percentage points of $\chi^2_{(3)}$, since estimates were obtained by least squares. If, however, estimates were obtained by maximum likelihood, then the statistic would be asymptotically $\chi^2_{(3)}$ under the assumed family of models. Finally, generalized least squares estimates could have been obtained here, but there is little to be gained by doing this.

In summary, least squares methods often provide simple and fairly effective ways of obtaining estimates with grouped data, though they are not a substitute for efficient methods of estimation, such as maximum likelihood, when precision is important. The procedures described were based on transforms of $\hat{S}(t)$ values, but other approaches are also possible for certain types of models. Gehan and Siddiqui (1973), for example, consider models for which $h(t)$ or some transform of it can be written as a linear function of unknown parameters, and they base least squares procedures on estimates of $h(t)$ at the midpoints of intervals in the life table. They demonstrate that for the exponential $[h(t)=\alpha]$, linear hazard rate $[h(t)=\alpha+\beta t]$, Gompertz $[h(t)=\exp(\alpha+\beta t)]$, and Weibull $[h(t)=(\beta/\alpha)(t/\alpha)^{\beta-1}]$ models the least squares procedure is in fact quite efficient. Their paper can be consulted for details; also see Problem 5.13.

5.6 PROBLEMS AND SUPPLEMENTS

5.1 Consider the gamma distribution (5.1.1) and let $W=\log(\bar{x}/\tilde{x})$, where $\bar{x}$ and $\tilde{x}$ are the arithmetic and geometric means, respectively, from a complete sample of size n. Show that the cumulant generating function of W is

$$K(\theta)=\log\left[E(e^{\theta W})\right]$$

$$=-\theta\log n+n\log\Gamma\left(k-\frac{\theta}{n}\right)-\log\Gamma(nk-\theta)-n\log\Gamma(k)+\log\Gamma(nk).$$

Thus show that

$$E(S)=-\log n-\psi(k)+\psi(nk)$$

$$\text{Var}(S)=\frac{1}{n}\psi'(k)-\psi'(nk)$$

and verify (5.1.21) and (5.1.22).

(Section 5.1.2; Bartlett, 1937)

5.2 The following observations are failure times (in minutes) for a sample of 15 electronic components in an accelerated life test:

1.4, 5.1, 6.3, 10.8, 12.1, 18.5, 19.7, 22.2, 23.0, 30.6, 37.3, 46.3, 53.9, 59.8, 66.2

a. Assuming that the data came from a gamma distribution, obtain the m.l.e.'s $\hat{k}$ and $\hat{\alpha}$ of the shape and scale parameters.

b. Let $Q_p(k, \alpha)$ represent the pth quantile of the two-parameter gamma distribution, given k and α. That is, $Q_p(k, \alpha)$ satisfies

$$I\left(k, \frac{Q_p}{\alpha}\right) = p$$

where $I(k, x)$ is the incomplete gamma function. Examine the adequacy of the gamma model by plotting the points

$$\left[Q_{(i-.5)/n}(\hat{k}, \hat{\alpha}), t_{(i)}\right] \qquad i = 1, \ldots, n$$

where $t_{(i)}$ is the ith smallest observation in the sample of n. (This type of Q–Q plot is useful with distributions that are not simply location–scale parameter models.)

(Section 5.1; Wilk et al., 1962a)

5.3 Consider the data in Problem 3.5. Fit via maximum likelihood (1) a two-parameter gamma distribution and (2) a linear hazard rate distribution to the data, and test within each family of models whether the exponential distribution is appropriate.

(Sections 5.1, 5.4)

5.4 Suppose that independent observations $t_1, \ldots, t_n$ that come from a gamma distribution (5.1.1) are subject to uniform rounding error. Examine the extent to which this can distort the likelihood function and m.l.e.'s; pay special attention to the t_i's near zero.

(Section 5.1)

5.5 The data below show the number of cycles to failure for 100-centimeter specimens of yarn, tested at a particular strain level:

86, 146, 251, 653, 98, 175, 176, 76, 264, 15, 157, 220, 42, 321, 180, 198, 38, 20, 61, 121, 282, 224, 149, 180, 325

Assuming that these data come from a log-normal distribution, determine .95 lower confidence limits for the .01 and .10 quantiles of the failure time distribution.

(Section 5.2.2)

5.6 *Data with both left and right censoring.*

a. Modify the discussion of Section 5.2.1 to allow the data to be left censored as well as right censored.

b. The data in Table 5.6.1 are log (to base 10) survival times y of rats poisoned with carbon tetrachloride in a laboratory experiment; time was measured in minutes. Pairing was used in the experiment: one rat from a pair of litter mates was injected with vitamin B_{12}, and the other received no additional drug. Observation was suspended after 16 hours ($y = 2.98$).

   Because of the censoring, differences in log survival times can be both left and right censored. Assuming that differences in log survival times are normally distributed, obtain m.l.e.'s of the parameters and test the hypothesis that the mean of the distribution is zero.

(Section 5.2.1; Sampford and Taylor, 1959; Wolynetz, 1979)

Table 5.6.1 Log_{10} **Survival Times for Rats**

$y_1(B_{12})$	y_2 (Control)	$z = y_1 - y_2$
2.73	>2.98	$<-.25$
2.80	>2.98	$<-.18$
2.01	2.84	$-.83$
2.19	2.76	$-.57$
2.34	2.83	$-.49$
2.61	2.73	$-.12$
2.51	2.62	$-.11$
2.65	2.70	$-.05$
2.72	2.76	$-.04$
2.79	2.82	$-.03$
2.90	2.79	.11
2.78	2.64	.14
2.78	2.48	.30
2.97	2.64	.33
2.74	2.31	.43
2.96	2.51	.45
>2.98	2.68	$>.30$

5.7 Prove that if $(\hat{\mu}, \hat{\sigma})$ is a convergent solution to the iteration scheme based on the estimates (5.2.11), then $\hat{\mu}$ and $\hat{\sigma}$ are the m.l.e.'s of μ and σ.

(Section 5.2.1; Sampford and Taylor, 1959; Dempster et al., 1977)

5.8 The three-parameter log-normal distribution. For the three-parameter log-normal distribution, $\log(t-\gamma)$ is normally distributed with mean μ and variance σ^2, where $t \geqslant \gamma$ and $\gamma \geqslant 0$ is a threshold parameter.

a. If γ is known, determine the m.l.e.'s $\hat{\mu}(\gamma)$ and $\hat{\sigma}(\gamma)$ of μ and σ from a complete sample of size n. Thus obtain the maximized likelihood function $L_{\max}(\gamma)$. Show that

$$\lim_{\gamma \to t_{(1)}^-} L_{\max}(\gamma) = \infty.$$

Consider the ramifications of this for maximum likelihood estimation.

b. Treating the two censored observations as uncensored, consider the rat tumor data of Example 4.4.1 as having arisen from a three-parameter log-normal distribution. Compute and examine the maximized likelihood function $L_{\max}(\gamma)$. Obtain the value $\hat{\gamma}$ that gives a local maximum of $L_{\max}(\gamma)$. Treating this as the m.l.e., estimate all three parameters. Determine a range of plausible values for γ and examine the effect of γ on estimates of distribution quantiles.

(Hill, 1963a; Griffiths, 1980)

5.9 For the yarn failure data in Problem 5.5 use the generalized log-gamma distribution to examine whether the log-normal, Weibull, or other distributions of the log-gamma family (5.3.6) are plausible. Assess the effect of the choice of model on estimation of the probability of surviving past 50 cycles.

(Section 5.3)

5.10 Mixtures with known components. Suppose that X has p.d.f.

$$f(x; p) = pf_1(x) + (1-p)f_2(x)$$

where $0 < p < 1$ and f_1 and f_2 are completely specified p.d.f.'s.

a. Show that the expected information from a complete sample of size n from the distribution is

$$I(p) = \frac{n}{p(1-p)}\left(1 - \int_{-\infty}^{\infty} \frac{f_1(x) f_2(x)}{f(x; p)}\, dx\right).$$

Note that this reduces to the binomial information $n[p(1-p)]^{-1}$ when f_1 and f_2 do not overlap.

b. If

$$f_1(x)=\frac{1}{(2\pi)^{1/2}}e^{-(x-\mu)^2/2} \quad \text{and} \quad f_2(x)=\frac{1}{(2\pi)^{1/2}}e^{-x^2/2}$$

evaluate $I(.5)$ numerically for several values of μ. Comment on the precision with which one can estimate p as a function of μ.

(Section 5.4.2; Hill, 1963b)

5.11 Fit a polynomial hazard function model to the data in Example 5.4.4, using least squares, and compare the fit with that of the mixture model examined in the example. Try to obtain m.l.e.'s under each model by using a search-type maximization procedure (see Appendix F).

(Section 5.4.1)

5.12 ***Grouped exponential data.*** Davis (1952) gives the following data on the lifetimes of a random sample of 100 V600 vacuum tubes.

Lifetime (hours)	0 –100	100 –200	200 –300	300 –400	400 –600	600 –800	≥800
Frequency observed	29	22	12	10	10	9	8

a. Assuming that the data came from an exponential distribution with mean θ, obtain the likelihood function for θ. Show that the m.l.e. of $\beta=\exp(-100/\theta)$ can be found as the solution of a quadratic equation.

b. Obtain a lower .95 confidence limit for $S(200)=\exp(-200/\theta)$ by likelihood ratio methods. Compare with the binomial confidence limit obtained by using only the fact that 49 lifetimes in the sample exceed 200 hours.

c. Assess the fit of the exponential distribution to these data.

(Section 5.5)

5.13 The life table data in Table 5.6.2 are from a study involving 112 patients with plasma cell myeloma treated at the National Cancer Institute (Carbone et al., 1967).

a. Fit Weibull and linear hazard function models to these data, using the methods of Section 5.5.2. Assess the fits of the two models. Try to repeat this using maximum likelihood estimation (Section 5.5.1).

Table 5.6.2 Survival Times for Patients With Plasma Cell Myeloma

Interval (Months)	Number at Risk at Start (n_i)	Number of Withdrawals (w_i)
[0,5.5)	112	1
[5.5,10.5)	93	1
[10.5,15.5)	76	3
[15.5,20.5)	55	0
[20.5,25.5)	45	0
[25.5,30.5)	34	1
[30.5,40.5)	25	2
[40.5,50.5)	10	3
[50.5,60.5)	3	2
[60.5,∞)	0	0

b. Gehan and Siddiqui (1973) use least squares to fit models to grouped data, but they base this on estimates of the hazard function at the midpoints of intervals rather than on the survivor function estimates used in Section 5.5. Note that $h(t)=\beta_0+\beta_1 t$ for the linear hazard function distribution, and that for the Weibull distribution, $\log h(t)$ can be expressed as $\log h(t)=\beta_0+\beta_1\log t$. Using the estimate $\hat{h}_1(t)$ of Problem 2.1, obtain least squares estimates of parameters in the two models by considering the "models" $\hat{h}_1(t)=\beta_0+\beta_1 t+\text{error}$ and $\log\hat{h}_1(t)=\beta_0+\beta_1\log t+\text{error}$, respectively. Compare these with the estimates obtained in (a).

c. Fit a Gompertz distribution, which has $h(t)=\exp(\beta_0+\beta_1 t)$, to these data, using the method in (b).

(Section 5.5; Gehan and Siddiqui, 1973)

5.14 The observations below are survival times (in weeks) of male mice exposed to a 240-roentgen dose of gamma radiation (Furth et al., 1959; Kimball, 1960). Brackets after a value indicate the number of observations with that value.

40	48	50	54	56	59
62	63	67(2)	69	70	71
73(2)	76	77	80	81(2)	82
83	84	86(2)	87	88(5)	89
90(2)	91	93	94	95	96
97(2)	98	99(2)	100(4)	101(3)	102(2)
103(5)	104(3)	105(2)	106(3)	107	108

109(2)	110(3)	111(3)	112	113(2)	114(2)
115	116(2)	117	118(3)	119(2)	120(3)
121(2)	123(2)	124(3)	125(2)	126(5)	127(4)
128(4)	129(6)	130(4)	131(2)	132	133(3)
134(4)	135(3)	136(4)	137(3)	138	139(2)
140(2)	141(5)	142	144(5)	145(2)	146(4)
147(4)	148(4)	149	150	151(4)	152(2)
153	155	156	157	158(2)	160
161	162(2)	163(2)	164	165(2)	166
168	169	171(2)	172(2)	174	177(2)

Examine these data, with a view to later comparisons with other sets of data of a similar type. Use both parametric and nonparametric methods.

CHAPTER 6

Parametric Regression Models

6.1 INTRODUCTION

Until this point we have dealt almost exclusively with univariate samples from a single lifetime distribution. In practice many situations involve heterogeneous populations, and it is important to consider the relationship of lifetime to other factors. One way to do this is through regression models, in which the dependence of lifetime on concomitant variables is explicitly recognized. Regression methods for lifetime distribution work are treated in detail in this chapter and in Chapter 7. We start with a few examples of situations where regression methods would be appropriate.

Example 6.1.1 (Example 1.1.5 revisited) Example 1.1.5 described a situation in which the lifetime of a particular type of electrical insulation is affected by the voltage level the insulation is subject to while in use. The tendency is the higher the voltage, the shorter the lifetime of the insulation. In the experiment in question numerous specimens of insulation were tested at seven different voltage levels, with the aim being to investigate how the lifetime distribution for specimens of insulation depends on voltage.

Example 6.1.2 Zelen (1959) describes a similar experiment in which the lifetimes of glass capacitors were examined over a range of operating temperatures and voltage levels. Tests were run at eight different voltage–temperature combinations; at each combination several capacitors were put on test, and their lifetimes observed. Here lifetime is the "response" variable and temperature and voltage are concomitant variables.

Example 6.1.3 Krall et al. (1975) discuss a situation in which survival times for 65 multiple myeloma patients were recorded and related to a number of factors. A total of 16 concomitant variables was considered, including physiological measures such as the amount of hemoglobin in the blood and the white blood count at diagnosis, qualitative factors such as the presence or absence of infection at diagnosis, and the personal characteristics of each individual, such as sex and age. A main problem in the study was to assess which concomitant variables were strongly related to survival time.

Example 6.1.4 (Example 1.1.8 revisited) Example 1.1.8 presented some data given by Prentice (1973) concerning the survival experience of a group of 40 advanced lung cancer patients. Treatment was an important factor, some of the patients having been given one treatment and some another. Other concomitant variables such as the age and condition of the patient were also recorded. A main problem was to compare the effect of the two treatments on survival time while taking into account the other variables.

One way to examine the relationship of concomitant variables to lifetime is through a regression model in which lifetime has a distribution that depends upon the concomitant, or regressor, variables. This involves specifying a model for the distribution of T, given $\mathbf{x}$, where T represents lifetime and $\mathbf{x}$ is a vector of regressor variables for an individual. Two approaches to regression are discussed in this book. One employs parametric families of lifetime distributions and extends models such as the exponential, Weibull, and log-normal models to include regressor variables. This is the main topic of the present chapter. The second approach is distribution-free and assumes less about the underlying distributions than do the parametric methods. This is discussed in Chapters 7 and 8. Both approaches are extremely valuable in the analysis of lifetime data.

The most important parametric regression models are extensions of the distributions discussed in Chapters 3, 4, and 5. To introduce a few basic ideas let us consider models based on the Weibull distribution. The two-parameter Weibull distribution (4.0.1) has a scale parameter α and a shape parameter δ and can be extended to a regression model by allowing α and δ to depend on $\mathbf{x}$, where $\mathbf{x}$ is a $1\times p$ vector of regressor variables, or covariates. The most commonly used Weibull regression models are those for which just α, and not δ, depends on $\mathbf{x}$. These have survivor function for T, given $\mathbf{x}$, of the form

$$S(t|\mathbf{x})=\exp\left[-\left(\frac{t}{\alpha(\mathbf{x})}\right)^{\delta}\right] \qquad t\geqslant 0 \qquad (6.1.1)$$

where $\alpha(\mathbf{x})$ is a function of $\mathbf{x}$ that typically involves unknown parameters. To examine these models a little more closely note that the hazard function of T, given $\mathbf{x}$, is

$$h(t|\mathbf{x})=\frac{-S'(t|\mathbf{x})}{S(t|\mathbf{x})}$$

$$=\frac{\delta}{\alpha(\mathbf{x})}\left(\frac{t}{\alpha(\mathbf{x})}\right)^{\delta-1}. \tag{6.1.2}$$

Thus the ratio of the hazard functions for two individuals with covariate vectors $\mathbf{x}=\mathbf{x}_1$ and $\mathbf{x}=\mathbf{x}_2$ is

$$\frac{h(t|\mathbf{x}_1)}{h(t|\mathbf{x}_2)}=\left(\frac{\alpha(\mathbf{x}_2)}{\alpha(\mathbf{x}_1)}\right)^{\delta}$$

which does not depend on t. In other words, different individuals have proportional hazard functions under (6.1.1). The class of models in which different individuals have proportional hazard functions is a very important one and is discussed further below.

It is also instructive to view (6.1.1) through the distribution of $Y=\log T$, given $\mathbf{x}$. The p.d.f. of Y, given $\mathbf{x}$, is of extreme value form,

$$f(y|\mathbf{x})=\frac{1}{\sigma}\exp\left[\frac{y-\mu(\mathbf{x})}{\sigma}-\exp\left(\frac{y-\mu(\mathbf{x})}{\sigma}\right)\right] \qquad -\infty<y<\infty \tag{6.1.3}$$

where $\sigma=1/\delta$ and $\mu(\mathbf{x})=\log\alpha(\mathbf{x})$. Written another way, (6.1.3) is

$$Y=\mu(\mathbf{x})+\sigma Z \tag{6.1.4}$$

where Z has a standard extreme value distribution with p.d.f. $\exp(z-e^z)$, $-\infty<z<\infty$. This is a location–scale regression model with error Z. The constancy of δ in (6.1.1) corresponds to the constancy of σ in (6.1.4), so $\log T$ has constant variance. A variety of functional forms for $\alpha(\mathbf{x})$ or $\mu(\mathbf{x})$ are often employed with (6.1.1) or (6.1.3). The most useful form is perhaps the log-linear one, for which

$$\mu(\mathbf{x})=\mathbf{x}\boldsymbol{\beta} \tag{6.1.5}$$

where $\mathbf{x}=(x_1,\dots,x_p)$ is the $1\times p$ vector of regressor variables and $\boldsymbol{\beta}=(\beta_1,\dots,\beta_p)'$ is a $p\times 1$ vector of regression coefficients. This model is discussed in Section 6.4.

Two broad classes of regression models have been mentioned: proportional hazards models for T and location–scale models for $\log T$. Most regression analyses of lifetime data work with one or the other of these families, and we now examine them in a little more detail.

6.1.1 Proportional Hazards Models

A proportional hazards family is a class of models with the property that different individuals have hazard functions which are proportional to one another. That is, the ratio $h(t|\mathbf{x}_1)/h(t|\mathbf{x}_2)$ of the hazard functions for two individuals with regression vectors $\mathbf{x}_1$ and $\mathbf{x}_2$ does not vary with t. This implies that the hazard function of T, given $\mathbf{x}$, can be written in the form

$$h(t|\mathbf{x})=h_0(t)g(\mathbf{x}). \tag{6.1.6}$$

Both h_0 and g may involve unknown parameters; $h_0(t)$ can be thought of as a baseline hazard function, being the hazard function for an individual for whom $g(\mathbf{x})=1$.

The Weibull model represented by (6.1.2) is a proportional hazards model, and other life distributions considered in previous chapters can also be extended to proportional hazards models in various ways. A particularly useful family of models is obtained from a univariate lifetime model with hazard function $h_0(t)$ by defining

$$h(t|\mathbf{x})=h_0(t)e^{\mathbf{x}\boldsymbol{\beta}} \tag{6.1.7}$$

where $\mathbf{x}\boldsymbol{\beta}=x_1\beta_1+\cdots+x_p\beta_p$ and the β_i's are unknown regression coefficients. This model is natural and sufficiently flexible for many purposes. Since $e^{\mathbf{x}\boldsymbol{\beta}}$ is always positive, $h(t|\mathbf{x})$ is automatically nonnegative for all $\mathbf{x}$ and $\boldsymbol{\beta}$. The exponential and Weibull regression models discussed in Sections 6.3 and 6.4 are of the form (6.1.7), and in Chapter 7 distribution-free methods based on (6.1.7) are presented.

Let us note the effect of $\mathbf{x}$ on the survivor function in the family of models (6.1.6). Since

$$S(t|\mathbf{x})=\exp\left(-\int_0^t h(u|\mathbf{x})\,du\right)$$

it follows that the survivor function of T, given $\mathbf{x}$, is

$$S(t|\mathbf{x})=S_0(t)^{g(\mathbf{x})} \tag{6.1.8}$$

where

$$S_0(t)=\exp\left(-\int_0^t h_0(u)\,du\right)$$

is the baseline survivor function for an individual with $g(\mathbf{x})=1$. There is thus an ordering of survivor functions in a proportional hazards family; for two individuals with regression vectors $\mathbf{x}_1$ and $\mathbf{x}_2$ either $S(t|\mathbf{x}_1) \geqslant S(t|\mathbf{x}_2)$ for all t or vice-versa.

Proportional hazards models assume that concomitant variables have a multiplicative effect on the hazard function. This assumption appears to be reasonable in many situations. Some examples and references to this in the biomedical area are contained, for example, in Breslow (1975), Armitage and Gehan (1974), and Prentice and Kalbfleisch (1979). Similarly, proportional hazards models in engineering contexts are considered by Nelson (1972a), Lawless (1976), Mann (1978), and many others. Methods for assessing whether a proportional hazards model is plausible are discussed in Section 6.2 and Chapter 7.

6.1.2 Location–Scale Models for log T

A second important class of models is that for which the log lifetime $Y=\log T$, given $\mathbf{x}$, has a distribution with a location parameter $\mu(\mathbf{x})$ and a constant scale parameter σ. These models can be written in the form

$$Y=\mu(\mathbf{x})+\sigma e \tag{6.1.9}$$

where $\sigma>0$ and e has a distribution that is independent of $\mathbf{x}$. The family of models for which e has a standard normal distribution is familiar, but models in which e has some other distribution are also important.

The family of models (6.1.9) is essentially quite distinct from the proportional hazards family. In fact, the Weibull model discussed earlier in this section is the only model in both classes, though exponential models are also included as special cases of the Weibull model. It is interesting to compare the effect of $\mathbf{x}$ in the two families: in (6.1.9) $\mathbf{x}$ acts linearly on Y, and hence multiplicatively on T. The survivor function for Y, given $\mathbf{x}$, is of the form $G\{[y-\mu(\mathbf{x})]/\sigma\}$, where $G(e)$ is the survivor function for e. The survivor function for $T=\exp Y$ is thus of the form

$$S(t|\mathbf{x})=G\left(\frac{\log t-\mu(\mathbf{x})}{\sigma}\right)$$

$$=S_1\left[\left(\frac{t}{\alpha(\mathbf{x})}\right)^{\delta}\right]$$

where $\alpha(\mathbf{x})=\exp[\mu(\mathbf{x})]$, $\delta=1/\sigma$, and $S_1(w)=G(\log w)$. Alternately, this can be written as

$$S(t|\mathbf{x}) = S_0\left(\frac{t}{\alpha(\mathbf{x})}\right) \tag{6.1.10}$$

where $S_0(w) = S_1(w^\delta)$. The hazard function of T, given $\mathbf{x}$, is then easily found to be $h(t|\mathbf{x}) = \alpha(\mathbf{x})^{-1} h_0[t/\alpha(\mathbf{x})]$, where $h_0(t)$ is the hazard function for an individual with $\alpha(\mathbf{x}) = 1$. In short, the regression variable $\mathbf{x}$ acts multiplicatively on T in the location–scale model, whereas in the proportional hazards model it acts multiplicatively on the hazard function for T.

This chapter deals with the statistical analysis of data under several regression models, all of which are one of these two types. In addition to the Weibull and extreme value models (6.1.1) and (6.1.3), which include exponential distribution models, we examine the following models:

1. The log-normal or normal model, in which $Y = \log T$, given $\mathbf{x}$, is normally distributed with mean $\mu(\mathbf{x})$ and variance σ^2.
2. The generalized gamma or log-gamma model, in which $[T/\alpha(\mathbf{x})]^\delta$, given $\mathbf{x}$, has a one-parameter gamma distribution (1.3.12). Equivalently,

$$Y = \mu(\mathbf{x}) + \sigma e_k$$

where $Y = \log T$ and e_k has a standard log-gamma distribution with p.d.f. (5.3.5).

Both of these models are of the form (6.1.9), but they are not proportional hazards models.

The remainder of this chapter deals with the analysis of data under the various parametric models. Section 6.2 examines a few general methods of assessing regressor variables and of checking model assumptions; it is convenient to discuss these before considering specific models in detail. Sections 6.3 through 6.6 deal with statistical analysis under exponential, Weibull, log-normal, and generalized gamma models. All of these models can be put into the location–scale form (6.1.9). The methodology is mostly based on maximum likelihood estimation and associated large-sample procedures, though least squares and other methods are also introduced. Section 6.7 specifically examines least squares estimation.

6.2 RESIDUAL ANALYSIS AND OTHER MODEL CHECKS

Graphical and other relatively informal methods serve several purposes in the analysis of data. They can be used to summarize information, suggest possible models, highlight special features of the data, and provide checks on models considered for adoption. A few procedures that are useful in the regression analysis of lifetime data are presented here. Examples involving these procedures are given in Sections 6.3 through 6.6.

6.2.1 Looking for Models

A first problem encountered when dealing with concomitant variables is that of assessing the rough nature of the relationship between lifetime and these variables. In simple situations involving a single regressor variable x, plots of time or log time against x provide a clear picture, but if there are many regressor variables, some iteration between informal methods and formal tests and estimation with specific models is usually needed. Knowledge of the particular area under study often suggests relationships between lifetime and other variables. Computation of the mean or median lifetimes or log lifetimes at different levels of a regressor variable is usually helpful, although if there are several regressor variables, it may not be possible to get a clear picture from this. Examination of mean or median lifetimes for different combinations of levels for two regressor variables is similarly useful.

In many situations the first models considered are of the proportional hazards type (6.1.6) or of the location–scale type (6.1.9). Most people are familiar with approaches to data analysis under location–scale models, particularly under the normal linear model (e.g., Draper and Smith, 1966). When there are several observations at $\mathbf{x}_i$, probability or other plots of the data at $\mathbf{x}_i$ can be used to examine particular models for e. More generally, the data can be partitioned or stratified according to the values of the regressor variables. Models fitted to the different parts can be examined for constancy of σ and other parameters and for specific error distributions. In preliminary investigations least squares estimation can be useful if there is not too much censoring; this is discussed in Section 6.7.

The proportional hazards family is less familiar as a regression model than is the location–scale family, so we shall note some features that can be exploited in analyzing data and checking models. From (6.1.8) the relationship for the survivor function of T, given $\mathbf{x}$, can be expressed as

$$\log[-\log S(t|\mathbf{x})] = \log g(\mathbf{x}) + \log[-\log S_0(t)]. \tag{6.2.1}$$

In other words, the log($-$log) survivor functions of T, given $\mathbf{x}$, are parallel for different regression vectors $\mathbf{x}=\mathbf{x}_1$ and $\mathbf{x}=\mathbf{x}_2$. This suggests that if there are numerous observations for each distinct $\mathbf{x}_i$, a check on the proportional hazards assumption can be made by computing product-limit estimates of the survivor function $\hat{S}_i(t)=\hat{S}(t|\mathbf{x}_i)$ from the data at each $\mathbf{x}_i$. Plots of

$\log[-\log \hat{S}_i(t)]$ versus t should be roughly parallel if a proportional hazards model is apppropriate. A useful alternative is to plot $\log[-\log \hat{S}_i(t)]$ versus $\log t$, since if a Weibull proportional hazards model is appropriate, these plots should be roughly linear as well as roughly parallel.

One will also want to examine the effect of particular regressor variables in relation to the proportional hazards assumption. The relationship (6.2.1), with $\log g(\mathbf{x})=\mathbf{x}\boldsymbol{\beta}$, for example, gives

$$\log[-\log S(t|\mathbf{x})]=\mathbf{x}\boldsymbol{\beta}+\log[-\log S_0(t)]. \tag{6.2.2}$$

Suppose that it is desired to consider the effect of a particular variable, say x_1, and suppose that x_1 takes on just two values x_{11} and x_{12}. If $x_1=x_{1j}$, then

$$\begin{aligned}\log[-\log S(t|\mathbf{x})]&=x_1\beta_1+\mathbf{x}^-\boldsymbol{\beta}^-+\log[-\log S_0(t)]\\ &=\mathbf{x}^-\boldsymbol{\beta}^-+\log[-\log S_{0j}(t)]\end{aligned} \tag{6.2.3}$$

where $\mathbf{x}^-=(x_2,\ldots,x_p)$, $\boldsymbol{\beta}^-=(\beta_2,\ldots,\beta_p)$, and $\log[-\log S_{0j}(t)]=x_{1j}\beta_1+\log[-\log S_0(t)]$, $j=1,2$. If separate models of the form (6.1.8) are fitted to the two parts of the data having the values x_{11} and x_{12} for x_1, plots of $\log[-\log \hat{S}_{0j}(t)]$ versus t for the two parts should be roughly parallel. The same idea can be used when x_1 takes on several values; unless the number of values taken on is small, it will be necessary to group together observations with similar x_1 values. Besides looking for roughly parallel plots of $\log[-\log \hat{S}_{0j}(t)]$ versus t, one can examine the estimates of $\beta_2,\ldots,\beta_p$ in the different groups for constancy as a check on whether x_1 enters the model additively, as in (6.2.2). It may also be useful to carry out similar checks by grouping the data according to the joint values of two or more regressor variables.

6.2.2 Examination of Residuals

The examination of residuals from a fitted model is an important way of checking assumptions in the model and of revealing special features of the data, such as extreme observations. First, let us consider the concept of a residual when the model is not necessarily of location–scale form. Suppose that Y_i is a response variable and $\mathbf{x}_i$ is an associated vector of regressor variables. The distribution of Y_i, given $\mathbf{x}_i$, is specified except for a vector $\boldsymbol{\theta}$ of unknown parameters, and we assume that the model can be represented in terms of quantities

$$e_i=g_i(Y_i,\boldsymbol{\theta},\mathbf{x}_i) \tag{6.2.4}$$

that are i.i.d. and whose distribution is known. If $\hat{\boldsymbol{\theta}}$ is the m.l.e. of $\boldsymbol{\theta}$, determined from data $(y_i, \mathbf{x}_i)$, $i=1,\ldots,n$, residuals $\hat{e}_i$ are then defined by

$$\hat{e}_i = g_i(y_i, \hat{\boldsymbol{\theta}}, \mathbf{x}_i). \tag{6.2.5}$$

These residuals are often considered as behaving approximately like a random sample of size n from the distribution of e_i. The residuals are, of course, not independent and are not in general identically distributed, though for plots there is often little harm in treating them as such. Cox and Snell (1968) derive first-order corrections to the means, variances, and covariances of residuals that can be used to calculate adjusted residuals with properties closer to those of the e_i's.

Example 6.2.1 Consider the Weibull regression model in which the distribution of T, given $\mathbf{x}$, has survivor function

$$S(t|\mathbf{x}) = \exp\left[-\left(te^{-\mathbf{x}\boldsymbol{\beta}}\right)^{\delta}\right]. \tag{6.2.6}$$

Suppose that a random sample $t_1,\ldots,t_n$ is taken, corresponding to given regression vectors $\mathbf{x}_1,\ldots,\mathbf{x}_n$, respectively, and let $\hat{\boldsymbol{\beta}}$ and $\hat{\delta}$ be the m.l.e.'s of $\boldsymbol{\beta}$ and δ obtained from these data (see Section 6.3.1). Since the quantities $e_i = (t_i e^{-\mathbf{x}_i\boldsymbol{\beta}})^{\delta}$ are i.i.d. with standard exponential distributions, residuals could be defined by

$$\hat{e}_i = \left(t_i e^{-\mathbf{x}_i\hat{\boldsymbol{\beta}}}\right)^{\hat{\delta}} \qquad i=1,\ldots,n. \tag{6.2.7}$$

These behave approximately like a random sample from the standard exponential distribution.

Residuals could be defined in other ways as well. For example, if (6.2.6) is represented as a linear model for $Y = \log T$, then

$$Y_i = \mathbf{x}_i\boldsymbol{\beta} + \sigma e_i'$$

where $\sigma = \delta^{-1}$ and e_i' has a standard extreme value distribution. A natural way, then, to define residuals is by

$$\hat{e}_i' = \frac{(y_i - \mathbf{x}_i\hat{\boldsymbol{\beta}})}{\hat{\sigma}} \qquad i=1,\ldots,n. \tag{6.2.8}$$

These are, of course, related to the $\hat{e}_i$'s by $\hat{e}_i' = \log \hat{e}_i$. Depending on what is to be done with the residuals, one may prefer to work with either (6.2.7) or (6.2.8).

Residuals have many uses: probability plots of $\hat{e}_i$'s provide checks on distributional assumptions about e, plots of residuals against individual regressor variables or other factors can bring out inadequacies in the model, and so on. Residual analysis for normal regression models is discussed by many authors (e.g., Draper and Smith, 1966; Seber, 1977), and Cox and Snell give examples employing generalized residuals (6.2.5). When the data are censored, modifications are necessary: if one observes a censoring time t_i^* rather than a lifetime, the corresponding residual is censored as well. One approach in this situation is to treat the observed residuals, both censored and uncensored, as a censored sample from the distribution of e_i. The product-limit estimate or empirical hazard function can be calculated from the residuals for an estimate of the underlying survivor function of the e_i's. Plots of this estimate can be used to assess the underlying distribution.

Another procedure that is convenient when there are only a few censored observations, or when it is desired to plot residuals against other factors, is to adjust the censored residuals and to then treat these as if they were uncensored. For example, suppose that the distribution of T_i, given $\mathbf{x}_i$, were exponential, with mean $\theta_i = \exp(\mathbf{x}_i\boldsymbol{\beta})$. It is easily shown that $E(T_i|T_i \geqslant t_i^*) = t_i^* + \theta_i$, and thus $E(\theta_i^{-1}T_i|T_i \geqslant t_i^*) = (t_i^*/\theta_i) + 1$. If residuals are defined by

$$\hat{e}_i = \frac{t_i}{\hat{\theta}_i} = t_i e^{-\mathbf{x}_i\hat{\boldsymbol{\beta}}} \tag{6.2.9}$$

for uncensored observations t_i, then a reasonable adjustment for a censored observation t_i^* would be to define

$$\hat{e}_i = t_i^* e^{-\mathbf{x}_i\hat{\boldsymbol{\beta}}} + 1. \tag{6.2.10}$$

A general procedure for deriving an adjustment of this sort is as follows: suppose that the cumulative hazard function of T_i, given $\mathbf{x}_i$, is $H(t_i|\mathbf{x}_i)$. Since $S(t_i|\mathbf{x}_i) = \exp[-H(t_i|\mathbf{x}_i)]$, $i = 1, \dots, n$, are i.i.d. random variables uniformly distributed on $(0, 1)$, the $H(t_i|\mathbf{x}_i)$'s are i.i.d. standard exponential random variables. If residuals for uncensored observations $t_1, \dots, t_n$ are defined by

$$\hat{e}_i = \hat{H}(t_i|\mathbf{x}_i) \tag{6.2.11}$$

where $\hat{H}(t_i|\mathbf{x}_i)$ involves the m.l.e.'s of unknown parameters, then, as a first approximation, $\hat{e}_1, \dots, \hat{e}_n$ can be treated as a random sample from the standard exponential distribution. In the case of a censored observation t_i^*

(6.2.10) suggests defining adjusted residuals as

$$\hat{e}_i = \hat{H}(t_i^* | \mathbf{x}_i) + 1. \tag{6.2.12}$$

This definition of residuals is particularly useful with proportional hazards models. For the exponential and Weibull models referred to earlier, (6.2.11) gives the residuals (6.2.9) and (6.2.7) defined earlier. If residuals are based on (6.2.11) and (6.2.12), plots of $-\log[\hat{S}_1(e)]$ versus e should give roughly a straight line with slope 1 when the model is adequate, where $\hat{S}_1(e)$ is the product-limit estimate of the survivor function for the e_i's.

6.2.3 Discussion

Examples involving the examination of residuals and other facets of regression analysis are given in Sections 6.3 and 6.4. Some interesting case studies and further discussion of the regression models of this chapter and of Chapter 7 are given by several authors, including Aitkin and Clayton (1980), Armitage and Gehan (1974), Breslow (1975), Kay (1977), and Ware and Byar (1979).

Finally, the treatment of parametric regression models in the remainder of this chapter considers only the most basic situations. In practice one frequently needs to modify or extend this treatment. For example, in situations involving models of the form $Y = \mu(\mathbf{x}) + \sigma e$ it is common to find that σ is not entirely independent of $\mathbf{x}$. This can be handled in various ways, according to the occasion, but it is not feasible to discuss these problems in any detail here.

6.3 EXPONENTIAL REGRESSION MODELS

When individuals have constant hazard functions that may depend on concomitant variables, an exponential regression model is appropriate. These are a special case of the Weibull model (6.1.1) with $\beta = 1$; the p.d.f. of T, given $\mathbf{x}$, is

$$f(t|\mathbf{x}) = \theta_{\mathbf{x}}^{-1} \exp\left(\frac{-t}{\theta_{\mathbf{x}}}\right) \qquad t > 0. \tag{6.3.1}$$

Here $\mathbf{x}$ is a vector of regressor variables and $\theta_{\mathbf{x}} = E(T|\mathbf{x})$. Exponential regression models are employed in applications ranging from accelerated life testing (e.g., Lawless, 1976; Zelen, 1959) to the analysis of survival data on patients suffering from chronic diseases (e.g., Prentice, 1973; Feigl and Zelen, 1965; Krall et al., 1975).

Various functional forms for $\theta_{\mathbf{x}}$ are possible, but the most useful is

$$\theta_{\mathbf{x}} = \exp(\mathbf{x}\boldsymbol{\beta}) \tag{6.3.2}$$

where $\mathbf{x} = (x_1, \ldots, x_p)$ and $\boldsymbol{\beta} = (\beta_1, \ldots, \beta_p)'$ is a vector of regression parameters. This form of the model has been discussed by Glasser (1967), Cox and Snell (1968), Prentice (1973), Lawless (1976), and others. Other frequently used forms are $\theta_{\mathbf{x}} = \mathbf{x}\boldsymbol{\beta}$ (e.g., Feigl and Zelen, 1965; Zippin and Armitage, 1966) and $\theta_{\mathbf{x}} = (\mathbf{x}\boldsymbol{\beta})^{-1}$ (Greenberg et al., 1974). An advantage of (6.3.2) is that the requirement $\theta_{\mathbf{x}} > 0$ is automatically satisfied for all $\mathbf{x}$ and $\boldsymbol{\beta}$, whereas with the other two parameterizations this requirement places restrictions on $\boldsymbol{\beta}$. This sometimes leads to numerical or statistical problems (e.g., Mantel and Myers, 1971). Frequently there is no physical reason to choose one of these parameterizations over the others, and then it is convenient to use (6.3.2), or perhaps some nonlinear analog with $\theta_{\mathbf{x}} = \exp[g(\mathbf{x}, \boldsymbol{\beta})]$. In this section only (6.3.2) is considered explicitly, though similar procedures can be given for models in which $\theta_{\mathbf{x}}$ is some other function of $\mathbf{x}$.

The model (6.3.1) is a proportional hazards model. In addition, it can be viewed as a location–scale model for $Y = \log T$. From (6.3.1) and (6.3.2) the p.d.f. of Y, given $\mathbf{x}$, is

$$\exp\left[(y - \mathbf{x}\boldsymbol{\beta}) - \exp(y - \mathbf{x}\boldsymbol{\beta})\right] \qquad -\infty < y < \infty. \tag{6.3.3}$$

Alternately, we can write

$$Y = \mathbf{x}\boldsymbol{\beta} + z \tag{6.3.4}$$

where z has a standard extreme value distribution with p.d.f. $\exp(z - e^z)$, $-\infty < z < \infty$. Most results below are given in terms of (6.3.3), but there is no difficulty in reverting back to the exponential form.

6.3.1 Maximum Likelihood Methods

In most situations one has to rely on maximum likelihood methods of analysis. Suppose that associated with each individual is a lifetime or censoring time t_i and a regression vector $\mathbf{x}_i = (x_{i1}, \ldots, x_{ip})$. The notation $i \in D$ and $i \in C$ will be used to refer to individuals i for which t_i is a lifetime and a censoring time, respectively. Since we work with log times, $y_i = \log t_i$ correspondingly represents a log lifetime or log censoring time, as the case may be. Log lifetime Y has p.d.f. and survivor functions

$$f(y|\mathbf{x}) = \exp\left[(y - \mathbf{x}\boldsymbol{\beta}) - \exp(y - \mathbf{x}\boldsymbol{\beta})\right]$$

and

$$S(y|\mathbf{x})=\exp[-\exp(y-\mathbf{x}\boldsymbol{\beta})]$$

respectively. The likelihood function for a censored sample based on n individuals is

$$L(\boldsymbol{\beta})=\prod_{i\in D}\exp[(y_i-\mathbf{x}_i\boldsymbol{\beta})-\exp(y_i-\mathbf{x}_i\boldsymbol{\beta})]\prod_{i\in C}\exp[-\exp(y_i-\mathbf{x}_i\boldsymbol{\beta})]$$

and thus

$$\log L(\boldsymbol{\beta})=\sum_{i\in D}(y_i-\mathbf{x}_i\boldsymbol{\beta})-\sum_{i=1}^{n}\exp(y_i-\mathbf{x}_i\boldsymbol{\beta}). \tag{6.3.5}$$

The first and second derivatives of $\log L$ are

$$\frac{\partial \log L}{\partial \beta_r}=-\sum_{i\in D}x_{ir}+\sum_{i=1}^{n}x_{ir}\exp(y_i-\mathbf{x}_i\boldsymbol{\beta}) \qquad r=1,\dots,p \tag{6.3.6}$$

$$\frac{\partial^2 \log L}{\partial \beta_r\,\partial \beta_s}=-\sum_{i=1}^{n}x_{ir}x_{is}\exp(y_i-\mathbf{x}_i\boldsymbol{\beta}) \qquad r,s=1,\dots,p. \tag{6.3.7}$$

The maximum likelihood equations $\partial \log L/\partial\beta_r=0$ $(r=1,\dots,p)$ are readily solved by the Newton–Raphson method or some other iterative method to get the m.l.e. $\hat{\boldsymbol{\beta}}$. When there is a constant term in the model, so that $\mathbf{x}_i\boldsymbol{\beta}=\alpha+\beta_2x_{i2}+\cdots+\beta_px_{ip}$, it is possible to solve for α in terms of $\beta_2,\dots,\beta_p$ from the equation $\partial\log L/\partial\alpha=0$, and if we substitute this into the remaining equations, it is necessary to solve just $p-1$ equations iteratively for $\beta_2,\dots,\beta_p$. If there is little or no censoring, initial estimates can be obtained by using least squares with (6.3.4), ignoring censoring.

The $p\times p$ observed information matrix is $\mathbf{I}_0=(-\partial^2\log L/\partial\beta_r\,\partial\beta_s)_{\hat{\boldsymbol{\beta}}}$. In situations in which a fixed censoring time L_i is known for each individual, the expected information matrix can be calculated. A straightforward evaluation along the lines used in Section 3.2 gives

$$E\left(-\frac{\partial^2\log L}{\partial\beta_r\,\partial\beta_s}\right)=\sum_{i=1}^{n}x_{ir}x_{is}\left[1-\exp\left(-L_ie^{-\mathbf{x}_i\boldsymbol{\beta}}\right)\right] \qquad r,s=1,\dots,p.$$

The expected information is particularly simple when there is no censoring. Letting $L_1,\dots,L_n$ tend to infinity, we see that the expected information matrix in the uncensored case has entries

$$I_{rs} = \sum_{i=1}^{n} x_{ir}x_{is} \qquad r,s=1,\ldots,p. \tag{6.3.8}$$

Approximate tests and interval estimates for $\boldsymbol{\beta}$ can be based on likelihood ratio methods or on normal approximations for $\hat{\boldsymbol{\beta}}$. When there is censoring, it is most convenient to use $\hat{\boldsymbol{\beta}}\sim N(\boldsymbol{\beta},\mathbf{I}_0^{-1})$. It appears (Singhal, 1978) that this normal approximation to the distribution of $\hat{\boldsymbol{\beta}}$ is quite good if the sample size is not too small. Likelihood ratio methods require a little more computation, but they are also straightforward.

Let us specifically consider one important problem. Many hypotheses about $\boldsymbol{\beta}$ can be put in the form $H_0:\boldsymbol{\beta}_1=\boldsymbol{\beta}_1^0$, where $\boldsymbol{\beta}'$ is partitioned as $(\boldsymbol{\beta}_1,\boldsymbol{\beta}_2)'$ and where $\boldsymbol{\beta}_1$ is $k\times 1$ $(k<p)$ and $\boldsymbol{\beta}_1^0$ is a specified vector. To test H_0 against $H_1:\boldsymbol{\beta}_1\neq\boldsymbol{\beta}_1^0$ one can use the likelihood ratio statistic

$$\Lambda=-2\log\left(\frac{L(\boldsymbol{\beta}_1^0,\tilde{\boldsymbol{\beta}}_2)}{L(\hat{\boldsymbol{\beta}}_1,\hat{\boldsymbol{\beta}}_2)}\right)$$

where $\tilde{\boldsymbol{\beta}}_2$ is the m.l.e. of $\boldsymbol{\beta}_2$ under H_0 and $\hat{\boldsymbol{\beta}}=(\hat{\boldsymbol{\beta}}_1,\hat{\boldsymbol{\beta}}_2)$ is the unconstrained m.l.e. The estimate $\tilde{\boldsymbol{\beta}}_2$ is found by solving the system of equations $\partial\log L/\partial\beta_r=0$ $(r=k+1,\ldots,p)$, with $\boldsymbol{\beta}_1$ equal to $\boldsymbol{\beta}_1^0$. Large values of Λ give evidence against H_0, and significance levels can be calculated via the asymptotic approximation $\Lambda\sim\chi^2_{(k)}$. An alternate statistic for testing H_0, based on the asymptotic normal approximation $\hat{\boldsymbol{\beta}}\sim N(\boldsymbol{\beta},\mathbf{I}_0^{-1})$, is

$$\Lambda_1=(\hat{\boldsymbol{\beta}}_1-\boldsymbol{\beta}_1^0)'\mathbf{C}_{11}^{-1}(\hat{\boldsymbol{\beta}}_1-\boldsymbol{\beta}_1^0) \tag{6.3.9}$$

where $\mathbf{C}_{11}$ is $k\times k$ and $\mathbf{C}=\mathbf{I}_0^{-1}$ is partitioned as

$$\mathbf{C}=\begin{pmatrix}\mathbf{C}_{11} & \mathbf{C}_{12}\\ \mathbf{C}_{12}' & \mathbf{C}_{22}\end{pmatrix}.$$

That is, $\mathbf{C}_{11}$ is the asymptotic covariance matrix for $\hat{\boldsymbol{\beta}}_1$. Under H_0, Λ_1 is also approximately $\chi^2_{(k)}$ in large samples. The χ^2 approximation for Λ is typically somewhat better than that for Λ_1 in small- and moderate-size samples. As mentioned above, however, work by Singhal (1978) suggests that the normal approximation for $\hat{\boldsymbol{\beta}}$ and the consequent χ^2 approximation for Λ_1 are usually satisfactory for this model, even for n as small as 25 or so.

Inference under the exponential regression model thus poses few problems. An example will illustrate the use of the model.

Example 6.3.1 (Example 1.1.8 revisited) Example 1.1.8 described lung cancer survival data for patients assigned to one of two chemotherapy treatments. The data, given in Table 6.3.1, include observations on 40 patients: 21 were given one treatment (standard), and 19 the other (test). Several factors thought to be relevant to an individual's prognosis were also recorded for each patient. These include performance status (x_1) at diagnosis (a measure of the general medical condition on a scale of 0 to 100), the age of the patient (x_2), and the number of months from diagnosis of cancer (x_3) to entry into the study. In addition, tumors were classified into four types: squamous, small, adeno, and large. Censored observations are starred.

Preliminary analysis suggests that treatment, tumor type, and performance status are the more important factors. Table 6.3.2*a* shows, for example, estimates of mean survival time for the eight treatment–tumor type combinations. Estimates are calculated as though the observations for each combination came from an exponential distribution as the total survival time divided by the number of uncensored survival times r. The value of r is shown in brackets in each case. Table 6.3.2*b* shows similar estimates $\hat{\theta}$ calculated for individuals with different performance status values x_1.

As a first step in model fitting, we consider an exponential regression model of the form (6.3.1) and (6.3.2), starting with a model involving all of the factors. This can be done by associating a regression vector $\mathbf{x}=(x_0, x_1, \ldots, x_7)$ with each individual, where

$$
\begin{aligned}
x_0 &= 1\\
x_1 &= \text{Performance status}\\
x_2 &= \text{Age}\\
x_3 &= \text{Months from diagnosis to entry into the study}\\
x_4 &= 1 \text{ if tumor type is squamous, 0 otherwise}\\
x_5 &= 1 \text{ if tumor type is small, 0 otherwise}\\
x_6 &= 1 \text{ if tumor type is adeno, 0 otherwise}\\
x_7 &= 0 \text{ if treatment is test, 1 if it is standard.}
\end{aligned}
$$

It is wise to center the regressor variables; we have centered just x_1, x_2, and x_3 here and work with the model for which

$$\log \theta_{\mathbf{x}} = \beta_0 + \beta_1(x_1 - \bar{x}_1) + \beta_2(x_2 - \bar{x}_2) + \beta_3(x_3 - \bar{x}_3) + \sum_{i=4}^{7} \beta_i x_i.$$

Derivatives of the log likelihood function are given by (6.3.6) and (6.3.7), in terms of $y = \log t$. Maximum likelihood estimates for $\beta_0, \ldots, \beta_7$ are readily

Table 6.3.1 Lung Cancer Survival Data[a]

t	x_1	x_2	x_3
Standard, Squamous			
411	70	64	5
126	60	63	9
118	70	65	11
92	40	69	10
8	40	63	58
25*	70	48	9
11	70	48	11
Standard, Small			
54	80	63	4
153	60	63	14
16	30	53	4
56	80	43	12
21	40	55	2
287	60	66	25
10	40	67	23
Standard, Adeno			
8	20	61	19
12	50	63	4
Standard, Large			
177	50	66	16
12	40	68	12
200	80	41	12
250	70	53	8
100	60	37	13
Test, Squamous			
999	90	54	12
231*	50	52	8
991	70	50	7
1	20	65	21
201	80	52	28
44	60	70	13
15	50	40	13
Test, Small			
103*	70	36	22
2	40	44	36
20	30	54	9
51	30	59	87
Test, Adeno			
18	40	69	5
90	60	50	22
84	80	62	4
Test, Large			
164	70	68	15
19	30	39	4
43	60	49	11
340	80	64	10
231	70	67	18

[a] Days of survival t, performance status x_1, age in years x_2, and number of months from diagnosis to entry into study x_3.

Table 6.3.2 Sample Means for Patients Classified by Treatment by Tumor Type and by Performance Status

a. Treatment by Tumor Type: Sample Means $\hat{\theta}$

	Squamous	Small	Adeno	Large
Standard	130.2(6)	85.3(7)	10.0(2)	147.8(5)
Test	413.7(6)	58.7(3)	64.0(3)	159.4(5)

b. Performance Status: Sample Means $\hat{\theta}$

x_1	20	30	40	50	60	70	80	90
$\hat{\theta}(r)$	4.5 (2)	26.5 (4)	21.9 (7)	145.0 (3)	120.4 (7)	329.1 (7)	155.8 (6)	999 (1)

obtained by Newton–Raphson iteration. M.l.e.'s and their estimated standard deviations (in brackets), as given by the inverse of the observed information matrix, are as follows:

$\hat{\beta}_0 = 4.722(0.405)$ $\quad\hat{\beta}_1 = 0.0544(0.011)$ $\quad\hat{\beta}_2 = 0.0085(0.020)$
$\hat{\beta}_3 = 0.0034(0.012)$ $\quad\hat{\beta}_4 = 0.3448(0.445)$ $\quad\hat{\beta}_5 = -0.1210(0.486)$
$\hat{\beta}_6 = -0.8634(0.586)$ $\quad\hat{\beta}_7 = -0.2778(0.388)$

This suggests that x_1 (performance status) is very important but that x_2 and x_3 are not. In addition, once performance status has been accounted for, treatment and tumor type do not appear to have sizeable effects, though there is a suggestion that the adeno tumor type is important. Tests about individual regression coefficients can be based on normal approximations to the distribution of $\hat{\boldsymbol{\beta}}$. Alternately, likelihood ratio tests can be carried out, though this requires remaximizing the likelihood function under other models. Table 6.3.3 shows likelihood ratio statistic values Λ for tests of various submodels against the full model with all seven variables present. In each case $\Lambda = -2\log[L(\tilde{\boldsymbol{\beta}})/L(\hat{\boldsymbol{\beta}})]$, where $\hat{\boldsymbol{\beta}}$ is the unconstrained m.l.e. of $\boldsymbol{\beta}$ and $\tilde{\boldsymbol{\beta}}$ is the m.l.e. of $\boldsymbol{\beta}$ under the model being tested. The column "d.f." indicates the degrees of freedom d in the approximation $\Lambda \sim \chi^2_{(d)}$ that holds under the hypothesized model. The table also shows statistic values Λ_1 as in (6.3.9), obtained from the normal approximation $\hat{\boldsymbol{\beta}} \sim N(\boldsymbol{\beta}, \mathbf{I}_0^{-1})$. The values of Λ and Λ_1 are in good agreement and it clearly does not matter which statistic is used here. Note that the values of Λ_1 for β_1, β_2, β_3, and β_7 are calculated as $\hat{\beta}_i^2/I_0^{ii}$ $(i=1,2,3,7)$, where the (i,i) entry I_0^{ii} in $\mathbf{I}_0^{-1}$ is the estimated variance of $\hat{\beta}_i$. The statistic Λ_1 for row 5 is $(\hat{\beta}_4\hat{\beta}_5\hat{\beta}_6)\mathbf{C}_{11}^{-1}(\hat{\beta}_4\hat{\beta}_5\hat{\beta}_6)'$, where $\mathbf{C}_{11}$ is the estimated covariance matrix for $(\hat{\beta}_4\hat{\beta}_5\hat{\beta}_6)$, obtained from $\mathbf{I}_0^{-1}$. The values in the table differ slightly from those calculated with the values of $\hat{\beta}_i$ and their estimated standard deviations given above because the latter quantities were rounded off.

Treatment–subject interactions are very common in situations like this one. In this instance treatment–tumor type and treatment–performance

Table 6.3.3 Tests for Factors in the Exponential Regression Model

Model	$\log L$	Λ	d.f.	Λ_1
Full model ($\beta_0, \beta_1, \ldots, \beta_7$ present)	-56.852			
$\beta_1 = 0$	-67.702	21.7	1	25.2
$\beta_2 = 0$	-56.944	0.18	1	0.19
$\beta_3 = 0$	-56.897	0.09	1	0.09
$\beta_4 = \beta_5 = \beta_6 = 0$	-58.863	4.02	3	4.60
$\beta_7 = 0$	-57.110	0.52	1	0.51

status interactions appear to be possibilities. This can be assessed by introducing additional regressor variables $x_8=x_1x_7$, $x_9=x_4x_7$, $x_{10}=x_5x_7$, and $x_{11}=x_6x_7$ into the model. The model with all of $x_0,\dots,x_{11}$ present gives a maximum log likelihood of -55.755. The model with only $x_0,\dots,x_7$ present can be tested against this by using the likelihood ratio statistic, which has an observed value of $-2(-56.852+55.755)=2.19$. This is non-significant on four degrees of freedom, and consequently no interaction term appears to be necessary.

Residual plots can be used to help assess the models fitted. Defining residuals as in (6.2.9) and (6.2.10), we have

$$\hat{e}_i=t_i\exp\left(-\hat{\beta}_0-\hat{\beta}_1(x_{i1}-\bar{x}_1)-\hat{\beta}_2(x_{i2}-\bar{x}_2)-\hat{\beta}_3(x_{i3}-\bar{x}_3)-\sum_{j=4}^{7}\hat{\beta}_jx_{ij}\right)$$

as the residual for an uncensored survival time t_i and $\hat{e}_i+1$ as the residual for a censoring time. No really striking features appear to manifest themselves in the residuals. Figure 6.3.1 shows a plot of the ordered residuals

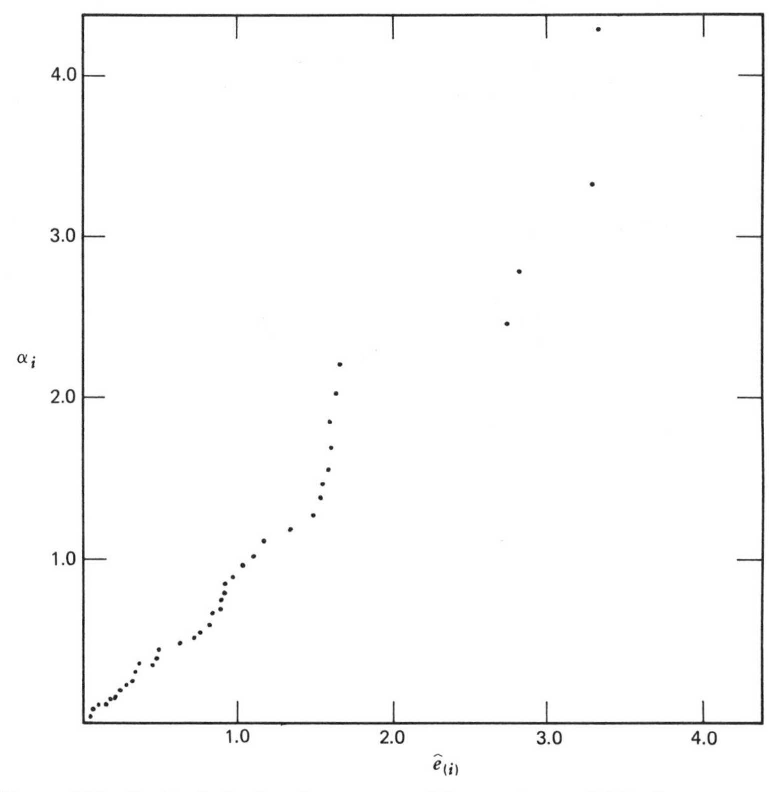

Figure 6.3.1 Residual plot based on exponential regression model for lung cancer survival data (Example 6.3.1)

against the standard exponential order statistics $\alpha_i = \Sigma_{j=1}^{i}(40-j+1)^{-1}$ (see Section 2.4.2) and there does not appear to be any strong suggestion in this that the exponential model is inappropriate.

A second example involving an exponential regression model is given in Example 6.3.2.

6.3.2 Exact Conditional Procedures*

Model (6.3.3) is a linear regression model with scale parameter $\sigma=1$, and consequently it is possible with uncensored data to derive exact conditional confidence interval procedures from the results given in Appendix G. These procedures require numerical integration and are not feasible when there are several regressor variables. They are useful when there is a single regressor variable, however, and shed light on the adequacy of large-sample normal approximations for $\hat{\boldsymbol{\beta}}$. With no censoring the joint p.d.f. of the log lifetimes $Y_1,\ldots,Y_n$, given $\mathbf{x}_1,\ldots,\mathbf{x}_n$, is

$$\prod_{i=1}^{n} \exp[(y_i-\mathbf{x}_i\boldsymbol{\beta})-\exp(y_i-\mathbf{x}_i\boldsymbol{\beta})]=\exp\left(\sum_{i=1}^{n}(y_i-\mathbf{x}_i\boldsymbol{\beta}) - \sum_{i=1}^{n}\exp(y_i-\mathbf{x}_i\boldsymbol{\beta})\right).$$

Let $\tilde{\boldsymbol{\beta}}$ be any estimator of $\boldsymbol{\beta}$ that satisfies the invariance conditions given by (G7) in Appendix G. (It can be shown that the m.l.e.'s, least squares, and certain other linear estimators of $\boldsymbol{\beta}$ satisfy them.) The residuals

$$a_i = y_i - \mathbf{x}_i\tilde{\boldsymbol{\beta}} \qquad i=1,\ldots,n$$

are ancillary statistics, any $n-p$ of which form a functionally independent set, and $\mathbf{Z}=\tilde{\boldsymbol{\beta}}-\boldsymbol{\beta}$ is a pivotal quantity. Conditional procedures are based on the conditional distribution of $\tilde{\boldsymbol{\beta}}$ or $\mathbf{Z}$, given the observed value of the residual vector $\mathbf{a}=(a_1,\ldots,a_n)$. The exponential model is of the form (G6), with $b=1$. In this case the p.d.f. of $\mathbf{Z}$, given $\mathbf{a}$, is given by (G9) of Appendix G, with $z_2=1$. This gives

$$f(\mathbf{z}|\mathbf{a})=k(\mathbf{a},\mathbf{X},\mathbf{n})\exp\left(\sum_{i=1}^{n}(a_i+\mathbf{x}_i\mathbf{z})-\sum_{i=1}^{n}\exp(a_i+\mathbf{x}_i\mathbf{z})\right) \qquad (6.3.10)$$

where $k(\mathbf{a},\mathbf{X},n)$ is a normalizing constant.

Probability statements about $\mathbf{Z}$ give confidence intervals and tests for $\boldsymbol{\beta}$. Since (6.3.10) must be integrated numerically, conditional procedures are usually not feasible unless there is only a single regressor variable, however. In this case, let us rewrite $\mathbf{x}_i$ as $(1, x_i)$ and let $\boldsymbol{\beta}=(\beta_0, \beta_1)$ so that $\log\theta_x = \beta_0+\beta_1 x$. Then (6.3.10) becomes

$$f(z_0, z_1|\mathbf{a})=k(\mathbf{a},\mathbf{x},n)\exp\left(nz_0+\sum_{i=1}^{n}(a_i+z_1x_i)-e^{z_0}\sum_{i=1}^{n}e^{a_i+z_1x_i}\right) \quad (6.3.11)$$

where $Z_0=\tilde{\beta}_0-\beta_0$, $Z_1=\tilde{\beta}_1-\beta_1$, and $\mathbf{x}=(x_1,\ldots,x_n)$. The marginal distribution of Z_1, given $\mathbf{a}$, is used to make inferences about β_1. To obtain this we note that z_0 can be integrated out of (6.3.11) analytically: we find

$$\int_{-\infty}^{\infty} f(z_0, z_1|\mathbf{a})\,dz_0=\left[k(\mathbf{a},\mathbf{x},n)\exp\left(\sum_{i=1}^{n}(a_i+z_1x_i)\right)\Big/\left(\sum_{i=1}^{n}\exp(a_i+z_1x_i)\right)^{n}\right]\int_0^{\infty}u^{n-1}e^{-u}\,du$$

using the transformation

$$u=e^{z_0}\sum\exp(a_i+z_1x_i).$$

This gives the p.d.f. of Z_1, given $\mathbf{a}$, as

$$h(z_1|\mathbf{a})=k_1(\mathbf{a},\mathbf{x},n)\exp(n\bar{x}z_1)\Big/\left(\sum_{i=1}^{n}\exp(a_i+z_1x_i)\right)^{n} \qquad -\infty<z_1<\infty \quad (6.3.12)$$

where for convenience we have defined a new normalizing constant $k_1(\mathbf{a},\mathbf{x},n)=(n-1)!k(\mathbf{a},\mathbf{x},n)\exp(\Sigma a_i)$, and $\bar{x}=\Sigma x_i/n$. The constant $k_1(\mathbf{a},\mathbf{x},n)$ can be determined numerically from the fact that

$$\int_{-\infty}^{\infty} h(z_1|\mathbf{a})\,dz_1=1,$$

which gives

$$k_1^{-1}(\mathbf{a},\mathbf{x},n)=\int_{-\infty}^{\infty}\left[\exp(n\bar{x}z_1)\Big/\left(\sum_{i=1}^{n}\exp(a_i+z_1x_i)\right)^{n}\right]dz_1. \quad (6.3.13)$$

It is often desired to estimate θ_x for some specified x; θ_x is the mean lifetime and determines the pth quantile $-\theta_x \log(1-p)$ of T, given x. Consider $\log \theta_x = \beta_0 + \beta_1 x$ and let

$$Z_2 = (\tilde{\beta}_0 + \tilde{\beta}_1 x) - (\beta_0 + \beta_1 x)$$

$$= \log \tilde{\theta}_x - \log \theta_x.$$

This quantity is pivotal, since $Z_2 = Z_0 + xZ_1$. The marginal distribution function of Z_2, given **a**, is easily obtained: by direct integration

$$\Pr(Z_2 \leqslant t|\mathbf{a}) = \int_{-\infty}^{\infty} \int_{-\infty}^{t-xz_1} f(z_0, z_1|\mathbf{a})\, dz_0\, dz_1$$

$$= \int_{-\infty}^{\infty} \int_0^{t^*} \left[k(\mathbf{a}, \mathbf{x}, n) \exp\left(\sum_{i=1}^{n} (a_i + z_1 x_i) \right) u^{n-1} e^{-u} \Big/ \left(\sum_{i=1}^{n} \exp(a_i + z_1 x_i) \right)^n \right] du\, dz_1$$

where the transformation $u = \exp(z_0) \Sigma \exp(a_i + z_1 x_i)$, $z_1 = z_1$, has been used, and where $t^* = [\exp(t - xz_1)] \Sigma \exp(a_i + z_1 x_i)$. This gives

$$\Pr(Z_2 \leqslant t|\mathbf{a}) = \int_{-\infty}^{\infty} h(z_1|\mathbf{a}) I\left(n, e^{t-xz_1} \sum_{i=1}^{n} e^{a_i + x_i z_1} \right) dz_1 \qquad (6.3.14)$$

where $I(n, s)$ is the incomplete gamma function (1.3.11).

The one-dimensional numerical integrations required to calculate probabilities for Z_1 and Z_2 are straightforward, so when there is a single regressor variable and no censoring, these methods are feasible. On the other hand, large-sample procedures are easier to use, and Singhal (1978) has shown that even for moderate-size samples the normal approximation $\hat{\boldsymbol{\beta}} \sim N(\boldsymbol{\beta}, \mathbf{I}_0^{-1})$ provides a good approximation to the exact conditional distribution of $\hat{\boldsymbol{\beta}}$. Thus it seems unnecessary to go to the trouble of employing the exact procedures except for small sample sizes.

The following example illustrates the conditional methods. Further discussion and examples are given in Lawless (1976).

Example 6.3.2 In Table 6.3.4 Feigl and Zelen (1965) give the data for 17 patients with acute myelogenous leukemia. The times t_i are survival times, in weeks. A single covariate x is associated with each patient: this is the base

Table 6.3.4 Survival Time Data for 17 Leukemia Patients

t_i	WBC	x_i	$\hat{e}_i$	t_i	WBC	x_i	$\hat{e}_i$
65	2,300	3.36	0.562	143	7,000	3.85	2.130
156	750	2.88	0.793	56	9,400	3.97	0.953
100	4,300	3.63	1.167	26	32,000	4.51	0.805
134	2,600	3.41	1.225	22	35,000	4.54	0.704
16	6,000	3.78	0.221	1	100,000	5.00	0.053
108	10,000	4.02	1.943	1	100,000	5.00	0.053
121	10,000	4.00	2.129	5	52,000	4.72	0.195
4	17,000	4.23	0.091	65	100,000	5.00	3.466
39	5,400	3.73	0.509				

10 logarithm of white blood count (WBC) measured at the time of diagnosis. Also shown in the table are residuals $\hat{e}_i$, calculated below.

Let us consider an exponential model with θ_x of the form (6.3.2). Working with centered x's, we have

$$\theta_{x_i} = \exp[\beta_0 + \beta_1(x_i - \bar{x})]$$

$$= \exp(\beta_0 + \beta_1 x_i')$$

where $\bar{x}=4.0959$ is the average x value for the group of 17 patients. From (6.3.6) the maximum likelihood estimates $\hat{\beta}_0$ and $\hat{\beta}_1$ are readily found as solutions to the equations

$$\sum_{i=1}^{n} t_i x_i' \exp(-\beta_1 x_i') = 0$$

$$\beta_0 = \log\left(\frac{1}{n} \sum_{i=1}^{n} t_i \exp(-\beta_1 x_i')\right).$$

We shall continue here to write expressions in terms of the exponential lifetimes t_i rather than in terms of log lifetimes as in (6.3.6). Solving the first equation iteratively, we find $\hat{\beta}_1 = -1.109$, and then $\hat{\beta}_0$ is found from the second equation as $\hat{\beta}_0 = 3.934$. The observed and expected information matrices are also easily calculated. From (6.3.7) the observed information matrix is

$$\mathbf{I}_0 = \begin{pmatrix} \Sigma \hat{e}_i & \Sigma x_i' \hat{e}_i \\ \Sigma x_i' \hat{e}_i & \Sigma (x_i')^2 \hat{e}_i \end{pmatrix} = \begin{pmatrix} 17 & -0.002042 \\ -0.002042 & 5.846231 \end{pmatrix}$$

where

$$\hat{e}_i = t_i \exp\left(-\hat{\beta}_0 - \hat{\beta}_1 x_i'\right).$$

The expected information matrix is, from (6.3.9),

$$\mathbf{I} = \begin{pmatrix} n & 0 \\ 0 & \Sigma(x_i')^2 \end{pmatrix} = \begin{pmatrix} 17 & 0 \\ 0 & 6.260812 \end{pmatrix}.$$

The residuals $\hat{e}_i$ correspond to the standard exponential variates $e_i = t_i \exp(-\beta_0 - \beta_1 x_i')$; their values are given with the data.

Let us obtain a .95 confidence interval for β_1, using the conditional method. The first step is to calculate $k_1(\mathbf{a}, \mathbf{x}, n)$ via (6.3.13), noting that $a_i = \log \hat{e}_i$. This gives $k_1(\mathbf{a}, \mathbf{x}, n) = .9597$ and gives us (6.3.12), from which probabilities for $Z_1 = \hat{\beta}_1 - \beta_1$ can be calculated by numerical integration. We find that $\Pr(Z_1 \leqslant -.834 | \mathbf{a}) = .025$ and $\Pr(Z_1 \leqslant .806 | \mathbf{a}) = .975$, and from this the .95 confidence interval $-1.92 \leqslant \beta_1 \leqslant -0.28$ is obtained. There is clearly evidence against the hypothesis that $\beta_1 = 0$, though the data do not estimate β_1 too precisely. It is interesting to compare this with what is obtained by the usual large-sample methods. To use the approximation $\hat{\boldsymbol{\beta}} \sim \mathrm{N}(\boldsymbol{\beta}, \mathbf{I}_0^{-1})$ we require

$$\mathbf{I}_0^{-1} = \begin{pmatrix} 0.058824 & 0.000021 \\ 0.000021 & 0.171050 \end{pmatrix}$$

from which one obtains the approximation $Z_1 = \hat{\beta}_1 - \beta_1 \sim N(0, 0.171050)$. This yields the .95 confidence interval $-1.92 \leqslant \beta_1 \leqslant -0.30$, which is almost identical to that obtained by the exact method.

Let us, in addition, estimate the mean survival time θ_x at $x = 4.0$. From (6.3.14) we find that $\Pr(Z_2 \leqslant -0.585 | \mathbf{a}) = .025$ and $\Pr(Z_2 \leqslant 0.406 | \mathbf{a}) = .975$, where $Z_2 = \log(\hat{\theta}_4 / \theta_4)$. Since $\log \hat{\theta}_4 = \hat{\beta}_0 + \hat{\beta}_1(4 - \bar{x}) = 4.0404$, this gives the .95 confidence interval $3.634 \leqslant \log \theta_4 \leqslant 4.625$, which gives $37.9 \leqslant \theta_4 \leqslant 102.0$. Alternately, we could use the large-sample approximation $\hat{\boldsymbol{\beta}} \sim N(\boldsymbol{\beta}, \mathbf{I}_0^{-1})$; this gives $\widehat{\mathrm{Var}}(\log \hat{\theta}_4) = \widehat{\mathrm{Var}}[\hat{\beta}_0 + \hat{\beta}_1(4 - \bar{x})] = 0.060393$, which leads to the .95 confidence interval $35.0 \leqslant \theta_4 \leqslant 92.0$.

In this case the large-sample method based on $\hat{\boldsymbol{\beta}} \sim N(\boldsymbol{\beta}, \mathbf{I}_0^{-1})$ gives a very close approximation to the exact conditional confidence interval for β_1, and a slightly poorer approximation to the exact confidence interval for θ_4. Singhal (1978) demonstrates that this is usually the case for sample sizes greater than about 25 or so. The approximation $\hat{\boldsymbol{\beta}} \sim N(\boldsymbol{\beta}, \mathbf{I}^{-1})$ provides slightly poorer approximations. It is therefore generally suitable to use large-sample methods with this model, unless sample sizes are fairly small.

For checks on various model assumptions we can examine the residuals $\hat{e}_i$. Figure 6.3.2 shows a plot of the points $(\hat{e}_{(i)}, \alpha_i)$, where $\alpha_i = \Sigma_{j=1}^{i}(17 - j +$

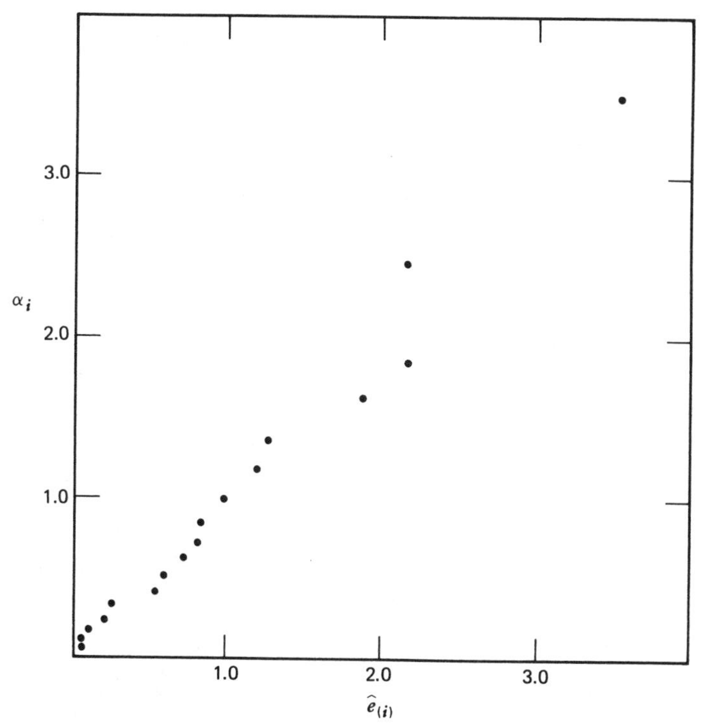

Figure 6.3.2 Residual plot based on exponential model for leukemia survival data (Example 6.3.2)

$1)^{-1}$ is the expected ith standard exponential order statistic in a sample of 17. The points lie roughly along a straight line with unit slope and do not provide any evidence against the assumption of exponentiality. Plots of the $(\log \hat{e}_i)$'s against x_i and aginst $\log t_i$ can also be examined; these similarly do not exhibit any surprising features.

6.3.3 Data With Several Observations at Each x

In some situations, such as industrial life test experiments, several observations may be taken at each separate regression vector. This case is, of course, covered by the results in Section 6.3.1, but simple least squares procedures are also sometimes useful.

Suppose that observations are taken at m distinct vectors $\mathbf{x}_i=(x_{i1},\dots,x_{ip})$ for the regressor variables $(i=1,\dots,m)$, with observations on n_i individuals at $\mathbf{x}_i$. We will allow the sample at $\mathbf{x}_i$ to be Type II censored, in which case the data consist of the r_i smallest lifetimes $t_{i1}\leqslant t_{i2}\leqslant\cdots\leqslant t_{ir_i}(r_i\leqslant n_i)$ at each

$\mathbf{x}_i$. Letting

$$T_i=\sum_{j=1}^{r_i} t_{ij}+(n_i-r_i)t_{ir_i} \qquad i=1,\dots,m$$

we note that for the model (6.3.1) with $\theta_{\mathbf{x}_i}=\exp(\mathbf{x}_i\boldsymbol{\beta})$, the quantities $T_1,\dots,T_m$ are sufficient for $\boldsymbol{\beta}$. Since $(T_i/\theta_{\mathbf{x}_i})$ has a one-parameter gamma distribution (1.3.12) with index parameter r_i, the joint p.d.f. of $T_1,\dots,T_m$ is

$$\prod_{i=1}^{m}\frac{T_i^{r_i-1}}{\theta_{\mathbf{x}_i}^{r_i}(r_i-1)!}\exp\left(\frac{-T_i}{\theta_{\mathbf{x}_i}}\right).$$

Equivalently, the joint p.d.f. of $Y_1,\dots,Y_m$, where $Y_i=\log T_i$, is

$$\left(\prod_{i=1}^{m}\frac{\exp[r_i(Y_i-\mathbf{x}_i\boldsymbol{\beta})]}{(r_i-1)!}\right)\exp\left(-\sum_{i=1}^{m}\exp(Y_i-\mathbf{x}_i\boldsymbol{\beta})\right). \qquad (6.3.15)$$

Deleting the constant term $\prod 1/(r_i-1)!$, we get the log likelihood

$$\log L(\boldsymbol{\beta})=\sum_{i=1}^{m} r_i(y_i-\mathbf{x}_i\boldsymbol{\beta})-\sum_{i=1}^{m}\exp(y_i-\mathbf{x}_i\boldsymbol{\beta}). \qquad (6.3.16)$$

Except for an additive constant, this is the same expression as is obtained from (6.3.5). Maximum likelihood estimates, tests, and interval estimates are found exactly as described in Section 6.3.1. Exact conditional methods apply, but these are too computationally laborious to be of much use except when there is a single regressor variable. In the case of a single regressor variable, when $\theta_{x_i}=\beta_0+\beta_1 x_i$, expressions for the distributions of $Z_1=\hat{\beta}_1-\beta$ and $Z_2=(\hat{\beta}_0+\hat{\beta}_1 x)-(\beta_0+\beta_1 x)$ are given by (6.3.12) and (6.3.14), with $\bar{x}$ now equal to $\Sigma r_i x_i/\Sigma r_i$. Examples of the conditional procedure used with Type II censored data are given by Lawless (1976).

Least squares methods also provide simple, fairly efficient analyses in the present situation. Since $T_i/\theta_{\mathbf{x}_i}$ has a one-parameter gamma distribution, the model in (6.3.15) can be written as

$$Y_i-\mathbf{x}_i\boldsymbol{\beta}=e_i' \qquad i=1,\dots,m$$

where e_i' has a log-gamma distribution (5.3.2). The mean and variance of e_i' are, from Section 1.3.3, $E(e_i')=\psi(r_i)$ and $\mathrm{Var}(e_i')=\psi'(r_i)$, where $\psi(\cdot)$ and $\psi'(\cdot)$ are the digamma and trigamma functions. The model can be rewritten as

$$Y_i'=Y_i-\psi(r_i)=\mathbf{x}_i\boldsymbol{\beta}+e_i \qquad (6.3.17)$$

where $e_i = e_i' - \psi(r_i)$ has mean 0 and variance $\psi'(r_i)$. Weighted least squares can be used with (6.3.17); approximate tests and interval estimates can be obtained by treating the e_i's as normally distributed. Although the least squares estimates are less efficient than the m.l.e.'s, they have good efficiency if the r_i's are bigger than about 4 (see Section 6.7). In addition, the log-gamma distribution of e_i approaches normality fairly rapidly as r_i increases, so that interval estimates and tests based on normal theory are satisfactory unless the r_i's are small. There is little harm in replacing $\psi(r_i)$ and $\psi'(r_i)$ by their first-order approximations $\log r_i - 1/2r_i$ and $1/r_i$, respectively, unless the r_i's are quite small.

Example 6.3.3 Factorial life test experiments are often used to study the effect of factors on the lifetimes of some type of item. For example, suppose that a complete factorial experiment is run in which two factors A and B appear at a and b distinct levels, respectively. Suppose that n_{ij} items are tested at each combination (i, j) of levels of A and B and that testing is terminated at the r_{ij}th failure, so the data consist of lifetimes $t_{ij1} \leqslant \cdots \leqslant t_{ijr_{ij}}$ for each (i, j). If lifetimes at (i, j) are assumed to come from an exponential distribution with mean θ_{ij}, the log likelihood is

$$\log L = -\sum_{i=1}^{a}\sum_{j=1}^{b} r_{ij}\log\theta_{ij} - \sum_{i=1}^{a}\sum_{j=1}^{b}\frac{T_{ij}}{\theta_{ij}} \tag{6.3.18}$$

where

$$T_{ij} = \sum_{k=1}^{r_{ij}} t_{ijk} + (n_{ij} - r_{ij})t_{ijr_{ij}}.$$

The unrestricted m.l.e.'s of the θ_{ij}'s are $\hat{\theta}_{ij} = T_{ij}/r_{ij}$. In this situation a variety of models might be of interest. Zelen (1959, 1960) and Lawless and Singhal (1980) consider analysis of variance-type models and compare maximum likelihood and least squares methods. The least squares method is carried out with the model

$$Y_{ij} = \log\theta_{ij} + e_{ij}$$

where $Y_{ij} = \log\hat{\theta}_{ij} + \log r_{ij} - \psi(r_{ij})$ and where e_{ij} has a log-gamma distribution with mean 0 and variance $\psi'(r_{ij})$. One family of models that would often be of interest is the two-way fixed effects family with, say,

$$\log\theta_{ij} = \mu + \alpha_i + \beta_j + \gamma_{ij}$$

where the α_i's, β_j's, and γ_{ij}'s satisfy some linear restrictions. The least squares method uses weights $1/\psi'(r_{ij})$; details concerning such an analysis are given in standard references on experimental design or regression, such as Kempthorne (1952, pp. 79–91) or Kendall and Stuart (1968, Ch. 35). Here least squares requires about the same amount of work as maximum likelihood. Lawless and Singhal give details and analyze a set of data using both methods.

In concluding this section we remind the reader that many inference procedures for exponential models are nonrobust. If there is doubt about the appropriateness of the model, other models or methods should be considered. The Weibull model discussed in the next section has, for example, much better robustness properties than the exponential distribution. In this connection it is preferable to carry out least squares analyses assuming that e_i in (6.3.17) has variance $\sigma^2\psi'(r_i)$, where σ will be estimated from the data, if there is any doubt about the appropriateness of the exponential model.

6.4 WEIBULL AND EXTREME VALUE REGRESSION MODELS

The family of exponential regression models requires that each individual have a constant hazard function. Weibull models are more flexible than this and are useful in many regression situations. The Weibull distribution can be extended to include regressor variables in different ways. However, the most commonly used models are those given by (6.1.1), for which the p.d.f. of lifetime, given the vector $\mathbf{x}$ of regressor variables, is of the form

$$\frac{\delta}{\alpha(\mathbf{x})}\left(\frac{t}{\alpha(\mathbf{x})}\right)^{\delta-1}\exp\left[-\left(\frac{t}{\alpha(\mathbf{x})}\right)^{\delta}\right] \qquad t\geq 0. \qquad (6.4.1)$$

In (6.4.1) only the scale parameter α depends on $\mathbf{x}$. Some properties of this model were noted in Section 6.1. In particular, the fact that δ does not depend on $\mathbf{x}$ implies proportional hazards for lifetimes and constant variance for log lifetimes of individuals. This assumption is reasonable in many situations; Pike (1966), Peto and Lee (1973), and Nelson (1972a), for example, discuss this in specific contexts.

This section deals only with (6.4.1), though there are obviously situations where one will want to relax the assumption that δ is the same for all individuals. We shall often work with log lifetime: the p.d.f. of $Y=\log T$, given $\mathbf{x}$, is

$$f(y|\mathbf{x})=\frac{1}{\sigma}\exp\left[\frac{y-\mu(\mathbf{x})}{\sigma}-\exp\left(\frac{y-\mu(\mathbf{x})}{\sigma}\right)\right] \qquad -\infty<y<\infty \quad (6.4.2)$$

where $\mu(\mathbf{x})=\log \alpha(\mathbf{x})$. The most frequently used model is the linear one, with

$$\mu(\mathbf{x})=\mathbf{x}\boldsymbol{\beta}$$

where $\mathbf{x}=(x_1,\ldots,x_p)$ and $\boldsymbol{\beta}=(\beta_1,\ldots,\beta_p)'$. Discussion will be confined to this model, but other models can be handled in a similar manner.

The model (6.4.2), with $\mu(\mathbf{x})=\mathbf{x}\boldsymbol{\beta}$, can be written as

$$Y=\mathbf{x}\boldsymbol{\beta}+\sigma z \tag{6.4.3}$$

where z has a standard extreme value distribution with p.d.f. $\exp(z-e^z)$, $-\infty<z<\infty$. This is a linear regression model for which exact methods of analysis are in principle available (see Appendix G) but are not usually feasible because of the computations required. Least squares methods can be used, though these can be somewhat inefficient; least squares for (6.4.3) and other linear models is discussed in Section 6.7. More efficient maximum likelihood and likelihood ratio methods are examined here. Sometimes other simple methods can be used, and these are discussed where appropriate.

6.4.1 Maximum Likelihood Methods

Expressions will be given in terms of log lifetime Y and its p.d.f. (6.4.2), with $\mu(\mathbf{x})=\mathbf{x}\boldsymbol{\beta}$. Suppose, as usual, that independent observations $(y_i,\mathbf{x}_i)$, $i=1,\ldots,n$ are available, where y_i is either a log lifetime or a log censoring time; D and C denote the sets of individuals for which y_i is a log lifetime and a log censoring time, respectively. The likelihood function is then

$$L(\boldsymbol{\beta},\sigma)=\prod_{i\in D}\frac{1}{\sigma}\exp\left[\frac{y_i-\mathbf{x}_i\boldsymbol{\beta}}{\sigma}-\exp\left(\frac{y_i-\mathbf{x}_i\boldsymbol{\beta}}{\sigma}\right)\right]\prod_{i\in C}\exp\left[-\exp\left(\frac{y_i-\mathbf{x}_i\boldsymbol{\beta}}{\sigma}\right)\right]$$

and thus

$$\log L(\boldsymbol{\beta},\sigma)=-r\log\sigma+\sum_{i\in D}\frac{y_i-\mathbf{x}_i\boldsymbol{\beta}}{\sigma}-\sum_{i=1}^{n}\exp\left(\frac{y_i-\mathbf{x}_i\boldsymbol{\beta}}{\sigma}\right) \tag{6.4.4}$$

where r is the observed number of lifetimes. If we let $z_i=(y_i-\mathbf{x}_i\boldsymbol{\beta})/\sigma$, first and second derivatives of $\log L$ are

$$\frac{\partial \log L}{\partial \beta_l}=-\frac{1}{\sigma}\sum_{i\in D}x_{il}+\frac{1}{\sigma}\sum_{i=1}^{n}x_{il}e^{z_i} \qquad l=1,\ldots,p \tag{6.4.5}$$

$$\frac{\partial \log L}{\partial \sigma}=-\frac{r}{\sigma}-\frac{1}{\sigma}\sum_{i\in D}z_i+\frac{1}{\sigma}\sum_{i=1}^{n}z_ie^{z_i} \tag{6.4.6}$$

$$\frac{\partial^2 \log L}{\partial \beta_l \partial \beta_s} = -\frac{1}{\sigma^2} \sum_{i=1}^n x_{il} x_{is} e^{z_i} \qquad l, s = 1, \ldots, p$$

$$\frac{\partial^2 \log L}{\partial \sigma^2} = \frac{r}{\sigma^2} + \frac{2}{\sigma^2} \sum_{i \in D} z_i - \frac{2}{\sigma^2} \sum_{i=1}^n z_i e^{z_i} - \frac{1}{\sigma^2} \sum_{i=1}^n z_i^2 e^{z_i} \tag{6.4.7}$$

$$\frac{\partial^2 \log L}{\partial \beta_l \partial \sigma} = \frac{1}{\sigma^2} \sum_{i \in D} x_{il} - \frac{1}{\sigma^2} \sum_{i=1}^n x_{il} e^{z_i} - \frac{1}{\sigma^2} \sum_{i=1}^n x_{il} z_i e^{z_i} \qquad l = 1, \ldots, p.$$

The maximum likelihood equations $\partial \log L/\partial \beta_l = 0$ $(l = 1, \ldots, p)$ and $\partial \log L/\partial \sigma = 0$ can usually be solved with no trouble by the Newton–Raphson method or some other iterative procedure. The observed information matrix $\mathbf{I}_0$ is $(p+1) \times (p+1)$ and of the partitioned form

$$\mathbf{I}_0 = -\begin{pmatrix} \dfrac{\partial^2 \log L}{\partial \beta_l \partial \beta_s} & \dfrac{\partial^2 \log L}{\partial \beta_l \partial \sigma} \\ \dfrac{\partial^2 \log L}{\partial \beta_s \partial \sigma} & \dfrac{\partial^2 \log L}{\partial \sigma^2} \end{pmatrix}_{(\hat{\boldsymbol{\beta}}, \hat{\sigma})} \tag{6.4.8}$$

with the appropriate partial derivatives given by (6.4.7). $\mathbf{I}_0$ can be simplified slightly by using the fact that the m.l.e.'s satisfy $\partial \log L/\partial \beta_l = 0$ and $\partial \log L/\partial \sigma = 0$. For example, for the important case of a single regressor variable, where $\mathbf{x}_i \boldsymbol{\beta} = \beta_0 + \beta_1 x_i$, the observed information matrix becomes

$$\mathbf{I}_0 = \frac{1}{\hat{\sigma}^2} \begin{bmatrix} r & \sum_{i \in D} x_i & r + \sum_{i \in D} \hat{z}_i \\ \sum_{i \in D} x_i & \sum_{i=1}^n x_i^2 e^{\hat{z}_i} & \sum_{i=1}^n x_i \hat{z}_i e^{\hat{z}_i} \\ r + \sum_{i \in D} \hat{z}_i & \sum_{i=1}^n x_i \hat{z}_i e^{\hat{z}_i} & r + \sum_{i=1}^n \hat{z}_i^2 e^{\hat{z}_i} \end{bmatrix}$$

where $\hat{z}_i = (y_i - x_i \hat{\boldsymbol{\beta}})/\hat{\sigma}$. In writing this down we have used the maximum likelihood equations to note that

$$r = \sum_{i=1}^n e^{\hat{z}_i}$$

$$\sum_{i \in D} x_i = \sum_{i=1}^n x_i e^{\hat{z}_i}$$

and

$$r+\sum_{i\in D}\hat{z}_i=\sum_{i=1}^{n}\hat{z}_i e^{\hat{z}_i}.$$

These are used to simplify the second derivatives in (6.4.7), evaluated at $(\hat{\boldsymbol{\beta}},\hat{\sigma})$.

Uncensored Data

When fixed censoring times are known for all individuals, the expected information matrix $\mathbf{I}$ can also be obtained. This is tedious to compute, since several definite integrals have to be evaluated numerically. There is no theoretical advantage to using $\mathbf{I}$ as opposed to $\mathbf{I}_0$, and the form of $\mathbf{I}$ for the general censored data case will not be given here. When there is no censoring in the data, the expected information matrix is very simple, however, and we will give it for this case. To evaluate the expectations of the second derivatives in (6.4.7), the following are required, where z_i has a standard extreme value distribution with p.d.f. $\exp(z-e^z)$, $-\infty<z<\infty$:

$$\begin{aligned}
&\text{(i)} && E(z_i)=-\gamma\\
&\text{(ii)} && E(e^{z_i})=1\\
&\text{(iii)} && E(z_i e^{z_i})=1-\gamma\\
&\text{(iv)} && E(z_i^2 e^{z_i})=\frac{\pi^2}{6}+\gamma^2-2\gamma.
\end{aligned} \tag{6.4.9}$$

These expectations are related to the gamma function and its derivatives (see Section 1.3.2 and Appendix B); $\gamma=0.5772\ldots$ is Euler's constant. From these results the expected information matrix in the case of no censoring is found to be

$$\mathbf{I}=\frac{1}{\sigma^2}\begin{pmatrix}\mathbf{X}'\mathbf{X} & \mathbf{m}\\ \mathbf{m}' & n\left[\pi^2/6+(1-\gamma)^2\right]\end{pmatrix} \tag{6.4.10}$$

where $\mathbf{m}$ is a $p\times 1$ vector with lth entry $(1-\gamma)\Sigma_{i=1}^{n}x_{il}$ and $\mathbf{X}'\mathbf{X}$ is the usual $p\times p$ cross-products matrix with (l,s) entry $\Sigma_{i=1}^{n}x_{il}x_{is}$.

The matrix (6.4.10) simplifies when regressor variables are centered about their means. Supposing that a constant term is included in the model, we write

$$\mathbf{x}_i\boldsymbol{\beta}=\beta_1+\beta_2x_{i2}+\cdots+\beta_p x_{ip}$$

and assume that the x_{il}'s $(l=2,\ldots,p)$ are centered so that $\Sigma x_{il}=0$ for each

l. Then (6.4.10) becomes

$$\mathbf{I}=\frac{1}{\sigma^2}\begin{pmatrix} n & \mathbf{0} & n(1-\gamma) \\ \mathbf{0} & \mathbf{X}_1'\mathbf{X}_1 & \mathbf{0} \\ n(1-\gamma) & \mathbf{0} & n\left[\pi^2/6+(1-\gamma)^2\right] \end{pmatrix}$$

where $\mathbf{X}_1'\mathbf{X}_1$ is the $(p-1)\times(p-1)$ cross-products matrix involving $x_2,\ldots,x_p$. This implies that $(\hat{\beta}_2,\ldots,\hat{\beta}_p)$ and $(\hat{\beta}_1,\hat{\sigma})$ are asymptotically independent; further, the asymptotic covariance matrix for $(\hat{\beta}_2,\ldots,\hat{\beta}_p)$ is $\sigma^2(\mathbf{X}_1'\mathbf{X}_1)^{-1}$.

Tests and Interval Estimates

Tests and interval estimates for parameters can be obtained either by using likelihood ratio methods or the approximate normality of the m.l.e.'s in large samples. In the latter case it is most convenient to use the approximation $(\hat{\boldsymbol{\beta}},\hat{\sigma})\sim N((\boldsymbol{\beta},\sigma),\mathbf{I}_0^{-1})$, unless there is no censoring, in which case the expected information matrix (6.4.10) can be substituted for $\mathbf{I}_0$, with σ estimated by $\hat{\sigma}$.

Two important inference problems will be mentioned explicitly. The first concerns the regression coefficients $\boldsymbol{\beta}$. Hypotheses about $\boldsymbol{\beta}$ can frequently be put in the form $H_0:\boldsymbol{\beta}_1=\boldsymbol{\beta}_1^0$, with $\boldsymbol{\beta}$ partitioned as $\boldsymbol{\beta}'=(\boldsymbol{\beta}_1,\boldsymbol{\beta}_2)'$, where $\boldsymbol{\beta}_1$ is $k\times 1$ $(k<p)$ and $\boldsymbol{\beta}_1^0$ is a specified vector. To test H_0 against the alternative that $\boldsymbol{\beta}_1\neq\boldsymbol{\beta}_1^0$ one can use the likelihood ratio statistic

$$\Lambda=-2\log\left(\frac{L(\boldsymbol{\beta}_1^0,\tilde{\boldsymbol{\beta}}_2,\tilde{\sigma})}{L(\hat{\boldsymbol{\beta}}_1,\hat{\boldsymbol{\beta}}_2,\hat{\sigma})}\right) \tag{6.4.11}$$

where $\tilde{\boldsymbol{\beta}}_2$ and $\tilde{\sigma}$ are the m.l.e.'s of $\boldsymbol{\beta}_2$ and σ under H_0, and $\hat{\sigma}$ and $(\hat{\boldsymbol{\beta}}_1,\hat{\boldsymbol{\beta}}_2)=\hat{\boldsymbol{\beta}}$ are the m.l.e.'s of σ and $\boldsymbol{\beta}$ under the full model. The estimates $\tilde{\boldsymbol{\beta}}_2$ and $\tilde{\sigma}$ are obtained by solving equations (6.4.6) and (6.4.5) with $l=k+1,\ldots,p$, and $\beta_1,\ldots,\beta_k$ fixed at their hypothesized values. Large values of Λ provide evidence against H_0, and approximate significance levels can be calculated by using the fact that in large samples Λ is approximately distributed as $\chi^2_{(k)}$ under H_0.

An alternate procedure for testing $H_0:\boldsymbol{\beta}_1=\boldsymbol{\beta}_1^0$ versus $H_1:\boldsymbol{\beta}_1\neq\boldsymbol{\beta}_1^0$ is to use

$$\Lambda_1=(\hat{\boldsymbol{\beta}}_1-\boldsymbol{\beta}_1^0)'\mathbf{C}_{11}^{-1}(\hat{\boldsymbol{\beta}}_1-\boldsymbol{\beta}_1^0) \tag{6.4.12}$$

as the test statistic. Here $\mathbf{C}=\mathbf{I}_0^{-1}$ is partitioned as

$$\mathbf{C}=\begin{pmatrix} \mathbf{C}_{11} & \mathbf{C}_{12} \\ \mathbf{C}_{12}' & \mathbf{C}_{22} \end{pmatrix}$$

where $\mathbf{C}_{11}$ is the $k\times k$ asymptotic covariance matrix for $\boldsymbol{\beta}_1$. Under H_0, Λ_1 is approximately $\chi^2_{(k)}$ in large samples. The statistics (6.4.11) and (6.4.12) are asymptotically equivalent, but for small- to moderate-size samples it is usually preferable to use (6.4.11), even though it requires a second maximization of the likelihood function in order to obtain $\tilde{\sigma}$ and $\tilde{\boldsymbol{\beta}}_2$. These procedures are illustrated in Example 6.4.1.

A second problem that is important in life testing and other applications is that of estimating quantiles or the survivor function of the distribution. Let us consider estimation of the pth quantile of Y, given $\mathbf{x}$, which is

$$\begin{aligned} y_p(\mathbf{x}) &= \mathbf{x}\boldsymbol{\beta}+\sigma\log[-\log(1-p)] \\ &= \mathbf{x}\boldsymbol{\beta}+\sigma w_p. \end{aligned} \tag{6.4.13}$$

Tests or confidence intervals for $y_p(\mathbf{x})$ can be obtained by either the likelihood ratio or maximum likelihood methods. Procedures based on the asymptotic normality of $(\hat{\boldsymbol{\beta}},\hat{\sigma})$ require much less computation, though the normal approximations involved may not be very good unless the sample size is large. This involves treating $\hat{y}_p(\mathbf{x})=\mathbf{x}\hat{\boldsymbol{\beta}}+\hat{\sigma}w_p$ as normally distributed, with mean $y_p(\mathbf{x})$ and variance

$$(\mathbf{x},w_p)\mathbf{I}_0^{-1}(\mathbf{x},w_p)'.$$

Alternately, confidence intervals for $y_p(\mathbf{x})$ can be obtained via the likelihood ratio method by considering

$$H_0: y_p(\mathbf{x})=Q_0 \quad \text{vs.} \quad H_1: y_p(\mathbf{x})\neq Q_0.$$

The statistic

$$\Lambda=-2\log\left(\frac{L(\tilde{\boldsymbol{\beta}},\tilde{\sigma})}{L(\hat{\boldsymbol{\beta}},\hat{\sigma})}\right)$$

is used to test this, where $(\tilde{\boldsymbol{\beta}},\tilde{\sigma})$ is the m.l.e. of $(\boldsymbol{\beta},\sigma)$ under the restriction that $y_p(\mathbf{x})=\mathbf{x}\boldsymbol{\beta}+\sigma w_p=Q_0$. One can find $(\tilde{\boldsymbol{\beta}},\tilde{\sigma})$ by using the fact that $\mathbf{x}\boldsymbol{\beta}+w_p\sigma=Q_0$ to rewrite the log likelihood (6.4.4) as a function of p parameters only. This is illustrated in Example 6.4.2.

We shall now illustrate the procedures discussed.

Example 6.4.1 (Example 6.3.1 revisited) In Example 6.3.1 we examined some lung cancer survival data under an exponential regression model in which the log mean was a function of the regressor variables. The Weibull model (6.4.1) with shape parameter δ and

$$\log \alpha(\mathbf{x}) = \beta_0 + \beta_1(x_1 - \bar{x}_1) + \beta_2(x_2 - \bar{x}_2) + \beta_3(x_3 - \bar{x}_3) + \sum_{i=4}^{7} \beta_i x_i$$

is a natural extension of this, the exponential model being given by the special case $\delta = 1$.

Maximum likelihood estimates for $\beta_0, \beta_1, \ldots, \beta_7$ and $\sigma = \delta^{-1}$ are readily found as described above. The estimates and their estimated standard deviations (in brackets) as obtained from the observed information matrix are

$\hat{\sigma} = 0.872$ (0.115)

$\hat{\beta}_0 = 4.744$ (0.360) $\quad \hat{\beta}_1 = 0.0542$ (0.010) $\quad \hat{\beta}_2 = 0.0092$ (0.018)

$\hat{\beta}_3 = 0.0041$ (0.010) $\quad \hat{\beta}_4 = 0.3812$ (0.395) $\quad \hat{\beta}_5 = -0.1256$ (0.425)

$\hat{\beta}_6 = -0.8759$ (0.513) $\quad \hat{\beta}_7 = -0.2655$ (0.346)

The hypothesis that $\sigma = 1$ can be tested with the statistic $z = (\hat{\sigma} - 1)/0.115$; this gives a z^2 value of 1.25, which is insignificant on the $\chi^2_{(1)}$ distribution. Alternately, $\sigma = 1$ can be tested with a likelihood ratio test: the maximum of the log likelihood for the fitted Weibull model is -56.346, whereas it was -56.852 for the exponential model ($\sigma = 1$) in Example 6.3.1. The observed value of the likelihood ratio statistic for testing $\sigma = 1$ versus $\sigma \neq 1$ is therefore $\Lambda = -2(-56.346 + 56.852) = 1.01$. This agrees fairly well with the z^2 value 1.25, and there is clearly no evidence against the exponential model.

The present estimates of the β_i's, and their standard errors, are not much different from the estimates obtained under the exponential model in Example 6.3.1, and the evidence about various factors is essentially the same as in the previous analysis.

Example 6.4.2 (quantile estimation with a single regressor variable) To illustrate the estimation of quantiles let us consider the case of a single regressor variable. If $\mathbf{x}_i\boldsymbol{\beta} = \beta_0 + \beta_1 x_i$, then the observed information matrix, given earlier for this case, is

$$\mathbf{I}_0 = \frac{1}{\hat{\sigma}^2}\begin{bmatrix} r & \sum_{i\in D} x_i & r+\sum_{i\in D}\hat{z}_i \\ \sum_{i\in D} x_i & \sum_{i=1}^{n} x_i^2 e^{\hat{z}_i} & \sum_{i=1}^{n} x_i \hat{z}_i e^{\hat{z}_i} \\ r+\sum_{i\in D}\hat{z}_i & \sum_{i=1}^{n} x_i\hat{z}_i e^{\hat{z}_i} & r+\sum_{i=1}^{n}\hat{z}_i^2 e^{\hat{z}_i} \end{bmatrix}. \tag{6.4.14}$$

Letting the inverse of $\mathbf{I}_0$ be

$$\mathbf{I}_0^{-1} = \hat{\sigma}^2 \begin{pmatrix} A_{11} & A_{12} & A_{13} \\ A_{21} & A_{22} & A_{23} \\ A_{31} & A_{32} & A_{33} \end{pmatrix}$$

we use the asymptotic normal approximation $(\hat{\beta}_0, \hat{\beta}_1, \hat{\sigma}) \sim N[(\beta_0, \beta_1, \sigma), \mathbf{I}_0^{-1}]$. The estimate of the pth quantile of Y, given x, is

$$\hat{y}_p(x) = \hat{\beta}_0 + \hat{\beta}_1 x + w_p \hat{\sigma}$$

where $w_p = \log[-\log(1-p)]$. Confidence intervals for $y_p(x)$ can be obtained by treating $\hat{y}_p(x)$ as approximately normally distributed with mean $y_p(x)$ and variance

$$V = \hat{\sigma}^2\left[A_{11} + A_{22}x^2 + A_{33}w_p^2 + 2A_{12}x + 2A_{13}w_p + 2A_{23}xw_p\right].$$

The likelihood ratio procedure is more involved. It is necessary to obtain the m.l.e. $(\tilde{\beta}_0, \tilde{\beta}_1, \tilde{\sigma})$ under the restriction $H_0: \beta_0 + \beta_1 x + w_p\sigma = Q_0$. One way to do this is to rewrite β_0 in terms of β_1 and σ as $\beta_0 = Q_0 - \beta_1 x - w_p\sigma$. Then $\log L$ can be written as a function of β_1 and σ only; from (6.4.4) this becomes

$$\log L_1 = -r\log\sigma + \sum_{i\in D}\left(\frac{y_i - Q_0 - \beta_1(x_i - x)}{\sigma} + w_p\right)$$

$$- \sum_{i=1}^{n}\exp\left(\frac{y_i - Q_0 - \beta_1(x_i - x)}{\sigma} + w_p\right).$$

Setting $\partial \log L_1/\partial\beta_1=0$ and $\partial \log L_1/\partial\sigma=0$, we get the pair of equations

$$-\sum_{i\in D}(x_i-x)+\sum_{i=1}^{n}(x_i-x)e^{z_i+w_p}=0$$

$$-r-\sum_{i\in D}z_i+\sum_{i=1}^{n}z_ie^{z_i+w_p}=0 \qquad (6.4.15)$$

where $z_i=[y_i-Q_0-\beta_1(x_i-x)]/\sigma$. These equations can be solved to give $\tilde{\beta}_1$ and $\tilde{\sigma}$, and then $\tilde{\beta}_0=Q_0-\tilde{\beta}_1x-w_p\tilde{\sigma}$. To test H_0: $y_p(x)=Q_0$ we use the likelihood ratio statistic

$$\Lambda=-2\log\left(\frac{L(\tilde{\beta}_0,\tilde{\beta}_1,\tilde{\sigma})}{L(\hat{\beta}_0,\hat{\beta}_1,\hat{\sigma})}\right),$$

treating Λ as approximately $\chi^2_{(1)}$ when H_0 is true. An approximate γ confidence interval for $y_p(x)$ consists of values Q_0 for which $\Lambda\leqslant\chi^2_{(1),\gamma}$.

An application of this to some life test data is discussed in the next section.

Discussion

The approach in this section has emphasized maximum likelihood methods. Two other approaches might be mentioned, both of which unfortunately apply only to uncensored data. Prentice and Shillington (1975) present least squares methods for uncensored Weibull data. Least squares is discussed in Section 6.7, but only briefly; the main drawbacks of the methods for Weibull data are that they are rather inefficient and do not apply to censored data. Williams (1978) presents a least squares type of procedure that has higher asymptotic relative efficiency but that likewise cannot be used with censored data.

We also remark that Peto and Lee (1973) discuss a Weibull regression model with a threshold parameter present. These models are obtained from (6.4.1) by replacing t with $t-\mu$ ($t\geqslant\mu$), where $\mu\geqslant 0$ is the threshold parameter. Here the best approach is that described in Section 4.4.1, namely, to treat μ as known but to carry out the analysis for a few different values of μ.

6.4.2 Data With Several Observations at Each x

Sometimes, for example, in life test experiments to investigate the effect of a small number of factors on lifetime, there may be several observations at each **x**. Relatively simple methods are often useful in such situations and are briefly discussed here.

Suppose that observations are taken on n_i individuals at $\mathbf{x}_i$ $(i=1,\ldots,m)$; we will allow the sample at $\mathbf{x}_i$ to be possibly Type II censored, so that just the first r_i ordered lifetimes $t_{i1}\leqslant t_{i2}\leqslant\cdots\leqslant t_{ir_i}$ out of the total of n_i are observed. Let $y_{ij}=\log t_{ij}$; the likelihood function for the model in extreme value form is

$$\prod_{i=1}^{m}\frac{n_i!}{(n_i-r_i)!}\left\{\prod_{j=1}^{r_i}\frac{1}{\sigma}\exp\left[\frac{y_{ij}-\mathbf{x}_i\boldsymbol{\beta}}{\sigma}-\exp\left(\frac{y_{ij}-\mathbf{x}_i\boldsymbol{\beta}}{\sigma}\right)\right]\right\}$$

$$\times\left\{\exp\left[-\exp\left(\frac{y_{ir_i}-\mathbf{x}_i\boldsymbol{\beta}}{\sigma}\right)\right]\right\}^{n_i-r_i}. \quad (6.4.16)$$

Dropping the proportionality constant $\prod n_i!/(n_i-r_i)!$, we can take the log likelihood function as

$$\log L(\boldsymbol{\beta},\sigma)=-r\log\sigma+\sum_{i=1}^{m}\sum_{j=1}^{r_i}\left(\frac{y_{ij}-\mathbf{x}_i\boldsymbol{\beta}}{\sigma}\right)-\sum_{i=1}^{m}\sum_{j=1}^{r_i}{}^{*}\exp\left(\frac{y_{ij}-\mathbf{x}_i\boldsymbol{\beta}}{\sigma}\right)$$

$$(6.4.17)$$

where $r=\sum r_i$ and we use the convenient notation

$$\sum_{j=1}^{r_i}{}^{*}a_{ij}=\sum_{j=1}^{r_i}a_{ij}+(n_i-r_i)a_{ir_i}.$$

This log likelihood is of the form (6.4.4), with the sample at $\mathbf{x}_i$ contributing n_i-r_i censoring times t_{ir_i}; tests and estimation can be handled by the maximum likelihood methods of Section 6.4.1.

If there are enough observations at each $\mathbf{x}_i$, other simple procedures are feasible. Information about the possible constancy of σ and the adequacy of the formulation $\mu(\mathbf{x}_i)=\mathbf{x}_i\boldsymbol{\beta}$ is obtained by computing separate estimates $\tilde{\mu}_i$ and $\tilde{\sigma}_i$ at each $\mathbf{x}_i$. Likelihood ratio tests can be used to provide formal assessments of hypotheses; tests for the equality of the σ_i's were discussed earlier, in Section 4.3.1. If the model with $\mu_i=\mathbf{x}_i\boldsymbol{\beta}$ and $\sigma_i=\sigma$ is appropriate, least squares procedures can be used. These are less efficient than maximum likelihood methods, but they are easy to work with. Suppose, in particular, that $\tilde{\mu}_i$ is an unbiased estimate of μ_i, with mean $\mu_i=\mathbf{x}_i\boldsymbol{\beta}$ and variance $c_i\sigma^2$, where c_i depends on n_i and r_i. One such estimate is the best linear unbiased estimate of Section 4.1; other estimates are given in Section 4.1.2d. A weighted least squares estimate of $\boldsymbol{\beta}$ is obtained by minimizing $\sum_{i=1}^{m}w_i(\tilde{\mu}_i-\mathbf{x}_i\boldsymbol{\beta})^2$, where $w_i=c_i^{-1}=\sigma^2/\mathrm{Var}(\tilde{\mu}_i)$. The estimate is

$$\tilde{\boldsymbol{\beta}}=(\mathbf{X}'\mathbf{W}\mathbf{X})^{-1}\mathbf{X}'\mathbf{W}\tilde{\boldsymbol{\mu}} \quad (6.4.18)$$

where

$$\mathbf{X}_{m\times p}=\begin{pmatrix}\mathbf{x}_1\\ \vdots\\ \mathbf{x}_m\end{pmatrix}\qquad \mathbf{W}_{m\times m}=\operatorname{diag}(w_1,\ldots,w_m)\qquad \tilde{\boldsymbol{\mu}}=\begin{pmatrix}\tilde{\mu}_1\\ \vdots\\ \tilde{\mu}_m\end{pmatrix}.$$

This estimate is unbiased for $\boldsymbol{\beta}$ and has covariance matrix

$$\boldsymbol{\Sigma}_{\tilde{\boldsymbol{\beta}}}=\sigma^2(\mathbf{X}'\mathbf{W}\mathbf{X})^{-1}. \tag{6.4.19}$$

We can get approximate tests and confidence intervals by treating $\tilde{\boldsymbol{\beta}}$ as normally distributed. To estimate σ we can use the overall m.l.e. $\hat{\sigma}$, if it has been calculated, or calculate an estimate from separate estimates $\tilde{\sigma}_1,\ldots,\tilde{\sigma}_m$ of σ at each $\mathbf{x}_i$. With the latter approach, suppose that $\tilde{\sigma}_i$ is an unbiased estimator of σ with variance $B_i^{-1}\sigma^2$; such estimates are discussed in Section 4.1.2d. The minimum variance estimate of σ that is a linear function of $\tilde{\sigma}_1,\ldots,\tilde{\sigma}_m$ is then

$$\tilde{\sigma}=\sum_{i=1}^{m} B_i\tilde{\sigma}_i \Big/ \sum_{i=1}^{m} B_i. \tag{6.4.20}$$

The m.l.e.'s $\hat{\sigma}_i$ can also be used in this way, though they are not unbiased nor are their variances known exactly. However, the χ^2 approximation (4.1.21) states that $g_i\hat{\sigma}_i/\sigma_i\sim\chi^2_{(h_i)}$, where g_i and h_i are functions of r_i and n_i given in Table 4.1.2. Thus $E(\hat{\sigma}_i/\sigma_i)\doteq h_i/g_i$, $\operatorname{Var}(\hat{\sigma}_i/\sigma_i)\doteq 2h_i/g_i^2$, and so $\tilde{\sigma}_i=g_i\hat{\sigma}_i/h_i$ is approximately unbiased for σ_i, with approximate variance $2/h_i$. The estimate (6.4.20) then becomes

$$\tilde{\sigma}=\sum_{i=1}^{m}\frac{h_i\tilde{\sigma}_i}{h}=\sum_{i=1}^{m}\frac{g_i\hat{\sigma}_i}{h} \tag{6.4.21}$$

where $h=\Sigma h_i$.

Nelson (1970a, 1972a) and Nelson and Hahn (1972) survey methods of this type and present examples, dealing mainly with the case of a single regressor variable. In addition to the simple methods above, best linear unbiased estimation of $\boldsymbol{\beta}$ and σ is also possible, though more complicated (Nelson and Hahn, 1973). This is discussed briefly in Section 6.7. An alternate approach to confidence interval estimation in the model with a single regression variable, given by Mann (1978), is also based on linear estimators.

The following example illustrates in some detail the least squares and maximum likelihood approaches.

Example 6.4.3 (Example 4.3.2 revisited) In Example 4.3.2 the results of an accelerated life test experiment on a type of electrical insulation were presented, the data being times to breakdown for specimens of insulating fluid subjected to fixed voltages. In the experiment specimens were tested at seven voltage levels, 26, 28, 30, 32, 34, 36, and 38 kilovolts (kV), respectively.

Background for this problem suggested what engineers refer to as a Weibull power law model: this stipulates that if items are subjected to a constant voltage v_i, then the lifetime distribution for the items is a Weibull distribution with shape parameter δ and scale parameter $\alpha_i = cv_i^p$. Letting $x_i = \log v_i$, we can write this as $\alpha_i = \exp(\beta_0 + \beta_1 x_i)$, where $\beta_0 = \log c$ and $\beta_1 = p$. In terms of log lifetime this model is of the form (6.4.3), with

$$\mu_i = \log \alpha_i = \beta_0 + \beta_1 x_i. \qquad (6.4.22)$$

A useful preliminary procedure is to plot the samples at the different voltage levels on Weibull probability paper; Figure 6.4.1 shows part of such a plot. If the stated model is appropriate, the plots for the different voltage levels should be roughly linear and roughly parallel, since the Weibull shape parameter δ is assumed to be the same at each voltage level. The plots mildly suggest one or two possible model departures, but we must remember that some of the plots have very few points. A formal test of equality of shape parameters was carried out in Example 4.3.2, where no evidence against the hypothesis of equality was found.

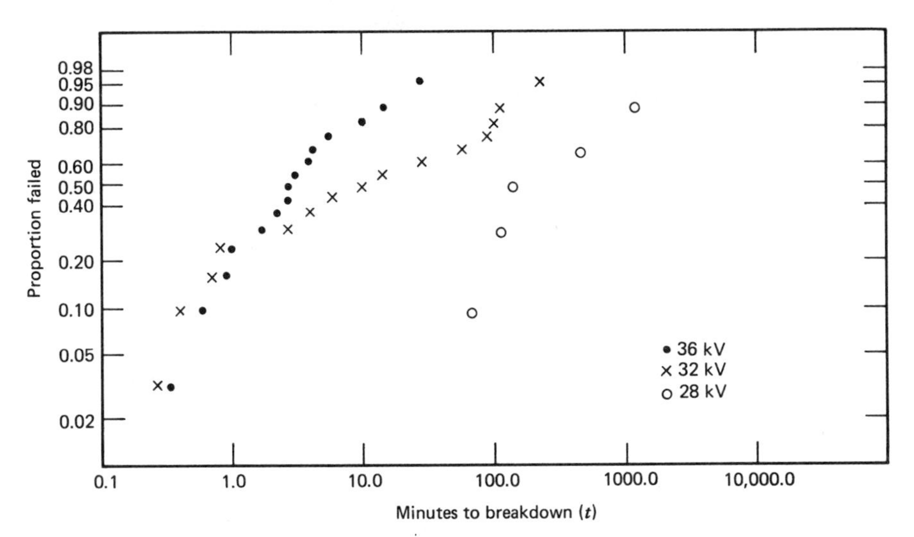

Figure 6.4.1 Probability plots of electrical insulation failure data at three voltage levels (Example 6.4.3)

Least Squares Methods

Let us consider the problem of assessing (6.4.22), and estimating β_0 and β_1. A good approach is to first obtain estimates of $\mu_i = \log \alpha_i$ and $\sigma_i = 1/\delta_i$ at each voltage level, under the assumption that lifetimes at v_i follow a Weibull distribution with survivor function $\exp[-(t/\alpha_i)^{\delta_i}]$. Either the b.l.u.e.'s $\tilde{\mu}_i$ and $\tilde{\sigma}_i$ or the m.l.e.'s $\hat{\mu}_i$ and $\hat{\sigma}_i$ can be readily obtained, as described in Section 4.1. Table 6.4.1 shows both sets of estimates as well as variances for $\tilde{\mu}_i$ and $\tilde{\sigma}_i$, obtained from Mann (1967b). A third possibility is to use simple linear estimates of μ and σ discussed in Section 4.1.2d. A plot of estimates $\tilde{\mu}_i$ or $\hat{\mu}_i$ against $x_i = \log v_i$ suggests that the linear relationship (6.4.22) is not unreasonable.

One way to estimate β_0 and β_1 is via least squares. Using (6.4.18) with the b.l.u.e.'s $\tilde{\mu}_i$ and their variances, one has $W = \text{diag}(w_1, \ldots, w_7)$, where $w_i = \sigma^2/\text{Var}(\mu_i)$, $\tilde{\boldsymbol{\mu}} = (\tilde{\mu}_1, \ldots, \tilde{\mu}_7)$, and

$$\mathbf{X} = \begin{pmatrix} 1 & x_1 \\ 1 & x_2 \\ \vdots & \vdots \\ 1 & x_7 \end{pmatrix}.$$

This gives estimates $\tilde{\beta}_0 = 64.792$ and $\tilde{\beta}_1 = -17.706$. The covariance matrix for $(\tilde{\beta}_0, \tilde{\beta}_1)$ is found, from (6.4.19), to have entries

$$\text{Var}(\tilde{\beta}_0) = 19.9314\sigma^2 \qquad \text{Var}(\tilde{\beta}_1) = 1.62847\sigma^2$$

$$\text{Cov}(\tilde{\beta}_0, \tilde{\beta}_1) = -5.69504\sigma^2.$$

Table 6.4.1 Quantities Obtained From Electrical Insulation Data

Voltage v_i	$x_i = \log v_i$	n_i	$\hat{\mu}_i$	$\hat{\sigma}_i$	$\tilde{\mu}_i$	$\tilde{\sigma}_i$	$\text{Var}(\tilde{\mu}_i)/\sigma^2$	$\text{Var}(\tilde{\sigma}_i)/\sigma^2$
26	3.2581	3	6.863	1.834	7.125	2.345	0.4029	0.3471
28	3.3322	5	5.865	1.022	5.957	1.224	0.2314	0.1667
30	3.4012	11	4.351	0.945	4.373	0.987	0.1025	0.06417
32	3.4675	15	3.256	1.781	3.310	1.898	0.07481	0.04534
34	3.5264	19	2.503	1.297	2.531	1.353	0.05890	0.03502
36	3.5835	15	1.457	1.125	1.473	1.154	0.07481	0.04534
38	3.6376	8	0.0009	0.734	0.0542	0.836	0.1420	0.09292

The pooled estimate of σ obtained from (6.4.20) is $\tilde{\sigma}=1.348$. Approximate confidence intervals can be obtained by assuming that $(\tilde{\beta}_0,\tilde{\beta}_1)$ is normally distributed with mean (β_0,β_1) and the variances given above, with $\tilde{\sigma}$ estimating σ. For example, to get confidence intervals for β_1 one assumes that $\tilde{\beta}_1 \sim N(\beta_1, 2.9591)$, and this gives an approximate two-sided .95 confidence interval for β_1 of $-21.08 \leqslant \beta_1 \leqslant -14.33$.

Confidence intervals can similarly be calculated for $\mu_x=\beta_0+\beta_1 x$. One reason for this experiment was to estimate lifetimes at the standard operating voltage of 20 kilovolts, which corresponds to $x=\log 20=2.9957$. An estimate of μ_{20} is $\tilde{\mu}_{20}=\tilde{\beta}_0+\tilde{\beta}_1(2.9957)=11.752$. To get confidence intervals for μ_{20} we assume $\tilde{\mu}_{20}$ to be normally distributed with mean μ_{20} and variance

$$V=\operatorname{Var}(\tilde{\beta}_0)+2(2.9957)\operatorname{Cov}(\tilde{\beta}_0,\tilde{\beta}_1)+(2.9957^2)\operatorname{Var}(\tilde{\beta}_1)$$

with $\tilde{\sigma}$ estimating σ. This gives $V=0.7711$. The approximation $\tilde{\mu}_{20} \sim N(\mu_{20}, 0.7711)$ gives, for example, a two-sided .95 confidence interval for μ_{20} as $10.03 \leqslant \mu_{20} \leqslant 13.47$. Note that μ_{20} is the .632 quantile of the distribution at 20 kilovolts. The confidence interval for the .632 quantile of the lifetime distribution at 20 kilovolts is (exp 10.03, exp 13.47), which is approximately (22,700, 707,000). This confidence interval is very wide, as we would expect in extrapolating rather far outside of the range of x values at which observations were taken. The usual cautions regarding extrapolation apply here.

Confidence intervals for other quantiles of the lifetime distribution may be desired for a specified x value. The pth quantile for log lifetime is, from (6.4.13),

$$y_p(x)=\beta_0+\beta_1 x+w_p\sigma$$

where $w_p=\log[-\log(1-p)]$. This quantity is estimated by $\tilde{y}_p(x)=\tilde{\beta}_0+\tilde{\beta}_1 x + w_p\tilde{\sigma}$. Confidence intervals for $y_p(x)$ can be obtained by treating $y_p(x)$ as normally distributed, though this approximation may not be very good when sample sizes are small. Note that to get the variance of $\tilde{y}_p(x)$, the covariances of $\tilde{\sigma}$ and $\tilde{\beta}_0$ and $\tilde{\sigma}$ and $\tilde{\beta}_1$ must be calculated. This can be done with the known covariances of $\tilde{\sigma}_i$ and $\tilde{\mu}_i$ (see Section 4.1.2d) but is tedious. An alternate method of obtaining confidence intervals for arbitrary quantiles is illustrated below.

Use of m.l.e.'s instead of the b.l.u.e.'s gives similar results. The m.l.e.'s $\hat{\mu}_i$ are not strictly unbiased, but as an approximation we can suppose that they are. In addition, when there is no censoring, an asymptotic normal approximation to the distribution of $\sqrt{n_i}(\hat{\mu}_i-\mu_i)$ is (see Problem 4.3) $\sqrt{n_i}(\hat{\mu}_i-$

$\mu_i)\sim N\{0,\sigma^2[1+6(1-\gamma)^2/\pi^2]\}$. From this the variance of $\hat{\mu}_i$ is approximated by $1.109\sigma^2/n_i$. The least squares procedure is then carried out with weights $w_i=1/n_i$ and gives estimates $\hat{\beta}_0=63.743$ and $\hat{\beta}_1=-17.419$. The pooled estimate (6.4.21) of σ was calculated previously in Example 4.3.2 and is $\hat{\sigma}=1.331$. Confidence intervals for β_1 and μ_{20} are obtained as before; symmetric two-sided .95 intervals are found to be $-20.86\leqslant\beta_1\leqslant-13.98$ and $9.91\leqslant\mu_{20}\leqslant13.22$, which are in reasonably good agreement with the estimates based on the b.l.u.e.'s.

Maximum Likelihood Methods

Another approach to estimation in (6.4.22) is via maximum likelihood, as described in Section 6.4.1; the special case of a single regressor variable was discussed in Example 6.4.2. In the present situation there is no censoring and there are n_i observations at x_i $(i=1,\ldots,7)$. The observed information matrix given by (6.4.14) then becomes

$$\mathbf{I}_0=\frac{1}{\hat{\sigma}^2}\begin{pmatrix} n & \sum_i n_i x_i & n+\sum_i\sum_j \hat{z}_{ij} \\ \sum_i n_i x_i & \sum_i\sum_j x_i^2\exp\hat{z}_{ij} & \sum_i\sum_j x_i\hat{z}_{ij}\exp\hat{z}_{ij} \\ n+\sum_i\sum_j \hat{z}_{ij} & \sum_i\sum_j x_i\hat{z}_{ij}\exp\hat{z}_{ij} & n+\sum_i\sum_j \hat{z}_{ij}^2\exp\hat{z}_{ij} \end{pmatrix}$$

where i ranges over $1,\ldots,7$, and j over $1,\ldots,n_i$ for each i, and where $n=\Sigma n_i=76$ and $\hat{z}_{ij}=(y_{ij}-\hat{\beta}_0-\hat{\beta}_1x_i)/\hat{\sigma}$, with y_{ij} being the jth log lifetime at voltage v_i. The m.l.e.'s $\hat{\beta}_0$, $\hat{\beta}_1$, and $\hat{\sigma}$ are found by iteration to be $\hat{\beta}_0=64.842$, $\hat{\beta}_1=-17.728$, and $\hat{\sigma}=1.288$. The large-sample normal approximation $(\hat{\beta}_0,\hat{\beta}_1,\hat{\sigma})\sim N[(\beta_0,\beta_1,\sigma),\mathbf{I}_0^{-1}]$ requires $\mathbf{I}_0^{-1}$, which is found to be

$$\mathbf{I}_0^{-1}=\begin{pmatrix} 31.5922 & -9.0295 & -0.008976 \\ -9.0295 & 2.5828 & 0.000949 \\ -0.008976 & 0.000949 & 0.012851 \end{pmatrix}. \qquad (6.4.23)$$

Alternately, since there is no censoring, a normal approximation employing the expected information matrix [see (6.4.10)] can be used. This has the slight advantage that if regressor variables x_i are centered so that $\Sigma n_i x_i=0$, then $\hat{\beta}_1$ and $(\hat{\beta}_0,\hat{\sigma})$ are asymptotically uncorrelated.

Approximate confidence intervals can be obtained by assuming that $(\hat{\beta}_0,\hat{\beta}_1,\hat{\sigma})$ is normally distributed. For example, (6.4.23) gives the approximation $\hat{\beta}_1\sim N(\beta_1,2.5828)$, which produces the approximate .95 confidence interval $-20.88 \leqslant \beta_1 \leqslant -14.58$. This is in fairly good agreement with the least squares result. To get approximate confidence intervals for the pth quantile of the distribution at a specified x value, we treat $y_p(x)=\hat{\beta}_0+\hat{\beta}_1 x + w_p\hat{\sigma}$ as normally distributed. For example, confidence intervals for the .632 quantile at $v=20$ ($x=2.9957$) are found by treating $\hat{\mu}_{20}=\hat{\beta}_0+\hat{\beta}_1(2.9957)$ as being normally distributed with mean μ_{20} and variance 0.6715, which is obtained from (6.4.23). Since $\hat{\mu}_{20}=11.734$, this gives the two-sided .95 confidence interval $10.13 \leqslant \mu_{20} \leqslant 13.34$; this is in good agreement with the estimate obtained by the least squares methods.

Discussion

The interval estimation procedures described employ large-sample approximations, and some idea of the adequacy of these would be desirable. In the example a check on the approximate methods could be made by calculating exact conditional confidence intervals (see Problem 6.13). This requires two-dimensional numerical integration and is too laborious for frequent use, but for illustration the exact two-sided symmetric .95 confidence interval for β_1 was obtained. The resulting interval was $-21.09 \leqslant \beta_1 \leqslant -14.63$, which is in good agreement with intervals obtained by the other methods. An investigation of greater interest would be to check the adequacy of the large-sample normal approximations to the distribution of $\hat{y}_p(x)$ for various combinations of p and x.

6.5 NORMAL AND LOG-NORMAL REGRESSION MODELS

We shall consider regression models in which lifetime is log-normal or, equivalently, log lifetime Y is normally distributed. The most important model is the usual linear one with constant variance, for which the distribution of Y, given the vector $\mathbf{x}$ of regressor variables, is given by

$$Y \sim N(\mathbf{x}\boldsymbol{\beta}, \sigma^2) \tag{6.5.1}$$

where $\mathbf{x}=(x_1,\ldots,x_p)$ and $\boldsymbol{\beta}=(\beta_1,\ldots,\beta_p)'$. This is the only model discussed in this section, though some of the treatment carries over in a fairly obvious way to other normal models.

Many books discuss statistical methods based on uncensored data from (6.5.1) in great detail (e.g., Draper and Smith, 1966; Seber, 1977). This well-known body of material is not reviewed here; we concern ourselves mainly with the problems of inference with censored data. There are many examples of the use of log-normal regression models with lifetime data in the literature, such as Boag (1949), Feinleib (1960), Glasser (1965), Nelson and Hahn (1972, 1973), and Whittemore and Altschuler (1976).

6.5.1 Maximum Likelihood Estimation (Censored Data)

In (6.5.1) the p.d.f. and survivor function of Y_i, given $\mathbf{x}_i$, are

$$f(y_i|\mathbf{x}_i)=\frac{1}{\sigma}\phi\left(\frac{y_i-\mathbf{x}_i\boldsymbol{\beta}}{\sigma}\right)$$

$$S(y_i|\mathbf{x}_i)=Q\left(\frac{y_i-\mathbf{x}_i\boldsymbol{\beta}}{\sigma}\right)$$

where $\phi(z)$ and $Q(z)$ are the standard normal p.d.f. and survivor functions given by (5.2.2). We consider the general case in which some censoring may be present, so that in a random sample of n individuals, r log lifetimes and $n-r$ log censoring times are observed. As usual, y_i will be used to denote either the log lifetime or log censoring time for the ith individual, with D and C denoting the sets of individuals for whom lifetimes and censoring times, respectively, are observed.

The likelihood function from such a sample is

$$L(\boldsymbol{\beta},\sigma)=\prod_{i\in D}\frac{1}{\sigma}\phi\left(\frac{y_i-\mathbf{x}_i\boldsymbol{\beta}}{\sigma}\right)\prod_{i\in C}Q\left(\frac{y_i-\mathbf{x}_i\boldsymbol{\beta}}{\sigma}\right)$$

and the log likelihood is

$$\log L(\boldsymbol{\beta},\sigma)=-r\log\sigma-\frac{1}{2\sigma^2}\sum_{i\in D}(y_i-\mathbf{x}_i\boldsymbol{\beta})^2+\sum_{i\in C}\log Q\left(\frac{y_i-\mathbf{x}_i\boldsymbol{\beta}}{\sigma}\right). \tag{6.5.2}$$

The first derivatives of $\log L$ are

$$\frac{\partial\log L}{\partial\beta_l}=\frac{1}{\sigma^2}\sum_{i\in D}x_{il}(y_i-\mathbf{x}_i\boldsymbol{\beta})+\frac{1}{\sigma}\sum_{i\in C}x_{il}\frac{\phi[(y_i-\mathbf{x}_i\boldsymbol{\beta})/\sigma]}{Q[(y_i-\mathbf{x}_i\boldsymbol{\beta})/\sigma]}\qquad l=1,\ldots,p$$

$$\frac{\partial\log L}{\partial\sigma}=-\frac{r}{\sigma}+\frac{1}{\sigma^3}\sum_{i\in D}(y_i-\mathbf{x}_i\boldsymbol{\beta})^2$$

$$+\frac{1}{\sigma}\sum_{i\in C}\frac{[(y_i-\mathbf{x}_i\boldsymbol{\beta})/\sigma]\phi[(y_i-\mathbf{x}_i\boldsymbol{\beta})/\sigma]}{Q[(y_i-\mathbf{x}_i\boldsymbol{\beta})/\sigma]}. \tag{6.5.3}$$

Expressions for maximum likelihood equations are simplified if we employ the functions $V(z)=\phi(z)/Q(z)$ and $\lambda(z)=V(z)[V(z)-z]$ defined in (5.2.3).

Doing this, and letting $z_i=(y_i-\mathbf{x}_i\boldsymbol{\beta})/\sigma$, we get the maximum likelihood equations

$$\frac{\partial \log L}{\partial \beta_l}=\frac{1}{\sigma}\sum_{i\in D} x_{il}z_i+\frac{1}{\sigma}\sum_{i\in C} x_{il}V(z_i)=0 \qquad l=1,\ldots,p \tag{6.5.4}$$

$$\frac{\partial \log L}{\partial \sigma}=-\frac{r}{\sigma}+\frac{1}{\sigma}\sum_{i\in D} z_i^2+\frac{1}{\sigma}\sum_{i\in C} z_iV(z_i)=0. \tag{6.5.5}$$

Second derivatives of $\log L$ are

$$\frac{\partial^2 \log L}{\partial \beta_l\,\partial \beta_s}=-\frac{1}{\sigma^2}\sum_{i\in D} x_{il}x_{is}-\frac{1}{\sigma^2}\sum_{i\in C} x_{il}x_{is}\lambda(z_i) \qquad l,s=1,\ldots,p$$

$$\frac{\partial \log L}{\partial \beta_l\,\partial \sigma}=-\frac{2}{\sigma^2}\sum_{i\in D} x_{il}z_i-\frac{1}{\sigma^2}\sum_{i\in C} x_{il}V(z_i)-\frac{1}{\sigma^2}\sum_{i\in C} x_{il}z_i\lambda(z_i)$$

$$l=1,\ldots,p \tag{6.5.6}$$

$$\frac{\partial^2 \log L}{\partial \sigma^2}=\frac{r}{\sigma^2}-\frac{3}{\sigma^2}\sum_{i\in D} z_i^2-\frac{2}{\sigma^2}\sum_{i\in C} z_iV(z_i)-\frac{1}{\sigma^2}\sum_{i\in C} z_i^2\lambda(z_i).$$

One way to get the m.l.e.'s is to use Newton–Raphson iteration to solve (6.5.4) and (6.5.5) for $\hat{\boldsymbol{\beta}}$ and $\hat{\sigma}$. However, if censoring is fairly heavy, this method sometimes fails to converge unless initial parameter estimates are very close to the m.l.e.'s. Negative values of σ can arise during the iteration procedure too, though this can be handled by replacing a negative value with one-half the σ value in the previous iteration. An alternative to the Newton–Raphson procedure is to use a method that requires only first derivatives of $\log L$, or one that does not require any derivatives. Wolynetz (1974) reports that Powell's method (see Appendix F) works very well in this situation.

Another good approach is to use a generalization of the method of Sampford and Taylor (1959), discussed previously in Section 5.2.1. This works well here, as it did in the univariate case. Let us define quantities w_i by

$$w_i=y_i \qquad i\in D$$

$$w_i=\mathbf{x}_i\boldsymbol{\beta}+\sigma V(z_i) \qquad i\in C. \tag{6.5.7}$$

The maximum likelihood equations (6.5.4) and (6.5.5) can then be written as

$$\frac{1}{\sigma}\sum_{i=1}^{n}(w_i-\mathbf{x}_i\boldsymbol{\beta})x_{il}=0 \qquad l=1,\ldots,p \qquad (6.5.8)$$

$$-r+\sum_{i=1}^{n}\left(\frac{w_i-\mathbf{x}_i\boldsymbol{\beta}}{\sigma}\right)^2-\sum_{i\in C}\lambda(z_i)=0. \qquad (6.5.9)$$

If the w_i's were known numbers, the solution to (6.5.8) would be

$$\tilde{\boldsymbol{\beta}}=(\mathbf{X}'\mathbf{X})^{-1}\mathbf{X}'\mathbf{W} \qquad (6.5.10)$$

where $\mathbf{W}=(w_1,\ldots,w_n)'$; also, (6.5.9) would than give

$$\tilde{\sigma}^2=\sum_{i=1}^{n}(w_i-\mathbf{x}_i\tilde{\boldsymbol{\beta}})^2\Big/\Big(r+\sum_{i\in C}\lambda(z_i)\Big) \qquad (6.5.11)$$

if the $\lambda(z_i)$'s $(i\in C)$ were also known. The following iterative procedure is thus suggested:

1. Obtain initial estimates $\tilde{\boldsymbol{\beta}}$ and $\tilde{\sigma}$ of $\boldsymbol{\beta}$ and σ.
2. Using these, calculate for $i\in C$

$$\tilde{z}_i=\frac{y_i-\mathbf{x}_i\tilde{\boldsymbol{\beta}}}{\tilde{\sigma}}$$

$$w_i=\mathbf{x}_i\tilde{\boldsymbol{\beta}}+\tilde{\sigma}V(\tilde{z}_i)$$

$$\lambda(z_i)=V(\tilde{z}_i)^2-\tilde{z}_iV(\tilde{z}_i).$$

3. Use (6.5.10) and (6.5.11) to calculate new estimates $\tilde{\boldsymbol{\beta}}$ and $\tilde{\sigma}$ of $\boldsymbol{\beta}$ and σ. The entire process is now repeated.

Provided that suitable initial estimates $\tilde{\boldsymbol{\beta}}$ and $\tilde{\sigma}$ are used, this procedure converges to the m.l.e.'s $\hat{\boldsymbol{\beta}}$ and $\hat{\sigma}$. It is easily programmed and appears to give fewer convergence problems than Newton–Raphson iteration (Wolynetz, 1974, 1979). Remarks made in Section 5.2.1 about the computation of $Q(z)$, $V(z)$, and $\lambda(z)$ also apply here.

An example involving censored log-normal data is given in the next section.

6.5.2 Interval Estimation and Tests of Hypotheses

When the data are censored, tests and interval estimates can be based on the usual large-sample methods. In addition, a few simpler procedures are available for certain problems.

Inferences About $\boldsymbol{\beta}$

Inferences about $\boldsymbol{\beta}$ can be based on likelihood ratio statistics or on the asymptotic normal distribution of the m.l.e.'s. Consider to start tests about $\boldsymbol{\beta}$ of the form $H_0: \boldsymbol{\beta}_1 = \boldsymbol{\beta}_1^0$ versus $H_1: \boldsymbol{\beta}_1 \neq \boldsymbol{\beta}_1^0$, where $\boldsymbol{\beta} = (\boldsymbol{\beta}_1, \boldsymbol{\beta}_2)'$ and $\boldsymbol{\beta}_1^0$ is a specified $k \times 1$ vector. The likelihood ratio statistic for testing H_0 versus H_1 is

$$\Lambda = -2\log\left(\frac{L(\tilde{\boldsymbol{\beta}}, \tilde{\sigma})}{L(\hat{\boldsymbol{\beta}}, \hat{\sigma})}\right)$$

where $(\tilde{\boldsymbol{\beta}}, \tilde{\sigma}) = (\boldsymbol{\beta}_1^0, \tilde{\boldsymbol{\beta}}_2, \tilde{\sigma})$ is the m.l.e of $(\boldsymbol{\beta}, \sigma)$ under H_0. In large samples Λ is approximately $\chi^2_{(k)}$ under H_0. This approximation is reasonably accurate if n is about 40 or larger, but appears to somewhat underestimate significance levels for smaller n (Wolynetz, 1974) when samples are subject to moderate (less than 30%) censoring. When there is no censoring, it can be shown that Λ is related to a F variate by

$$\Lambda = n\log\left(1 + \frac{kF_{(k,n-p)}}{n-p}\right)$$

and calculations show that treating Λ as $\chi^2_{(k)}$ underestimates the significance level.

Tests and estimates for $\boldsymbol{\beta}$ can also be based on the asymptotic normal approximation $(\hat{\beta}, \hat{\sigma}) \sim N[(\boldsymbol{\beta}, \sigma), \mathbf{I}_0^{-1}]$, where $\mathbf{I}_0$ is the $(p+1)\times(p+1)$ observed information matrix

$$\mathbf{I}_0 = \begin{pmatrix} -\dfrac{\partial^2 \log L}{\partial \beta_l\, \partial \beta_s} & -\dfrac{\partial^2 \log L}{\partial \beta_l\, \partial \sigma} \\ -\dfrac{\partial^2 \log L}{\partial \beta_s\, \partial \sigma} & -\dfrac{\partial^2 \log L}{\partial \sigma^2} \end{pmatrix}_{(\hat{\boldsymbol{\beta}}, \hat{\sigma})}$$

with entries given by the expressions in (6.5.6). Wolynetz (1974) has examined this approximation in a few cases and finds it satisfactory with respect to the distribution of $\hat{\boldsymbol{\beta}}$ for samples with n greater than 30 or so.

Another approach that appears satisfactory even for fairly small sample sizes and moderate censoring is to use the approximation

$$\hat{\boldsymbol{\beta}} \sim N\left[\boldsymbol{\beta}, \sigma^2(\mathbf{X}'\mathbf{X})^{-1}\right] \tag{6.5.12}$$

in conjunction with the estimate

$$\hat{\sigma}_u^2 = \frac{n_1 \hat{\sigma}^2}{n_1^*} \tag{6.5.13}$$

where $n_1 = r + \Sigma_{i\in C}\lambda(\hat{z}_i)$ and $n_1^* = n_1 - p$. A reasonably good approximation to the distribution of $\hat{\sigma}_u^2$ is given by $n_1^* \hat{\sigma}_u^2 \sim \chi^2_{(n_1^*)}$, and $\hat{\sigma}_u^2$ is nearly independent of $\hat{\boldsymbol{\beta}}$. Thus, for example, to test $H_0: \boldsymbol{\beta}_1 = \boldsymbol{\beta}_1^0$ versus $H_1: \boldsymbol{\beta}_1 \neq \boldsymbol{\beta}_1^0$ one can use the statistic

$$F = \frac{(\hat{\boldsymbol{\beta}}_1 - \boldsymbol{\beta}_1^0)'\mathbf{C}_{11}^{-1}(\hat{\boldsymbol{\beta}}_1 - \boldsymbol{\beta}_1^0)/k}{\hat{\sigma}_u^2}$$

where $\mathbf{C}_{11}$ is the $k \times k$ submatrix of $(\mathbf{X}'\mathbf{X})^{-1}$ that corresponds to $\boldsymbol{\beta}_1$. Under H_0, F is approximately distributed as $F_{(k, n_1^*)}$.

Wolynetz (1974), Sampford and Taylor (1959), Taylor (1973), and others discuss similar methods. It can be noted that the procedures mentioned are precisely those obtained if one considers the "model"

$$\hat{w}_i = \mathbf{x}_i\boldsymbol{\beta} + e_i \qquad i = 1, \ldots, n \tag{6.5.14}$$

where $e_i \sim N(0, \sigma^2)$, except that the effective sample size is taken to be $n_1 = r + \Sigma_{i\in C}\lambda(\hat{z}_i)$ instead of n. To see this it is only necessary to note that $\hat{\boldsymbol{\beta}}$ is the least squares estimate of $\boldsymbol{\beta}$, which comes out of (6.5.14) and that $\hat{\sigma}_u^2$ is the residual mean square estimate of σ^2, with n_1 as the effective sample size. That this is so follows from the fact that the m.l.e.'s $\hat{\boldsymbol{\beta}}$, $\hat{\sigma}$, and $\hat{w}_i$ $(i = 1, \ldots, n)$ satisfy (6.5.10) and (6.5.11).

Example 6.5.1 The data in Table 6.5.1 are from Glasser (1965). The response variable y is the logarithm of survival time (in days) for patients with primary lung tumors. Two covariables are represented in the data: x_1 is the age of the patient and x_2 is a performance status rating (divided by 10) similar to that introduced in Example 1.1.8. Starred y values are censored observations.

Let us consider the normal regression model

$$Y_i = \beta_0 + \beta_1(x_{i1} - \bar{x}_1) + \beta_2 x_{i2} + e_i$$

Table 6.5.1 Log Survival Times for Lung Cancer Patients

y (Log Survival Time)	x_1 (Age)	x_2 (Performance Status)
1.94	42	4
2.23	67	6
1.94	62	4
1.98	52	6
2.23	57	5
1.59	58	6
2.13	55	6
1.80	63	7
2.32	44	5
1.92	62	7
2.15*	51	7
2.05*	64	10
2.48*	54	8
2.42*	64	3
2.56*	54	9
2.56*	57	9

where $e_i \sim N(0, \sigma^2)$ are independent. The m.l.e.'s are obtained with no difficulty with either the Newton–Raphson or Sampford–Taylor iteration scheme. Satisfactory initial estimates can be found by least squares, with the censored observations treated as being uncensored. The m.l.e.'s are

$$\hat{\beta}_0 = 1.614 \qquad \hat{\beta}_1 = -0.006 \qquad \hat{\beta}_2 = 0.102 \qquad \hat{\sigma} = 0.356$$

Tests about the regression coefficients can be based on (6.5.12) and (6.5.13) or on $\hat{\boldsymbol{\beta}}$ and the observed information matrix. For the former we require

$$(\mathbf{X}'\mathbf{X})^{-1} = \begin{pmatrix} 0.7833 & 0.004800 & -0.1131 \\ 0.004800 & 0.001348 & -0.000753 \\ -0.1131 & -0.000753 & 0.01774 \end{pmatrix}$$

and $n_1 = r + \sum_{i \in C} \lambda(\hat{z}_i) = 13.512$. This gives the estimate $\hat{\sigma}_u = (n_1/n_1^*)^{1/2}\hat{\sigma} = 0.404$. Estimated standard deviations of $\hat{\beta}_0$, $\hat{\beta}_1$, and $\hat{\beta}_2$ are then 0.358, 0.015, and 0.054, respectively. If the approximation (6.5.12) is suitable, there is clearly no evidence against the hypothesis that $\beta_1 = 0$, whereas performance status (x_2) is significant at about the .03 level on a one-tail test. The observed information matrix gives very similar results, the estimated standard deviations of $\hat{\beta}_1$ and $\hat{\beta}_2$, for example, being 0.014 and 0.060, respectively.

Residual plots can be based on residuals defined for uncensored observations y_i as $\hat{e}_i=(y_i-\mathbf{x}_i\hat{\boldsymbol{\beta}})/\hat{\sigma}$, whereas for a censored observation y_i^*, $\hat{e}_i^*=(y_i^*-\mathbf{x}_i\hat{\boldsymbol{\beta}})/\hat{\sigma}$ is treated as a censored residual. Alternately, adjusted residuals can be calculated: the relationship (5.2.7) suggests that a reasonable adjusted residual for y_i^* would be $V(\hat{z}_i)=V[(y_i^*-\mathbf{x}_i\hat{\boldsymbol{\beta}})/\hat{\sigma}]$. Then residuals for censored and uncensored observations are defined as

$$\hat{e}_i=\begin{cases}\dfrac{(y_i-\mathbf{x}_i\hat{\boldsymbol{\beta}})}{\hat{\sigma}} & i\in D\\[2ex] V\left(\dfrac{y_i-\mathbf{x}_i\hat{\boldsymbol{\beta}}}{\hat{\sigma}}\right) & i\in C.\end{cases}$$

These are particularly convenient for plotting against covariables, the response variable, and so on, though if there are many censored observations, it is preferable, at least for plots of the survivor function, to leave censored residuals unadjusted and to treat all the residuals together as a censored normal sample.

Residual plots for this example do not reveal anything of special interest when we bear in mind the small sample size.

Confidence Intervals for Quantiles

In some situations, particularly in life testing, confidence intervals for distribution quantiles are wanted for a specific $\mathbf{x}$. The pth quantile of Y in the model (6.5.1) is

$$y_p(\mathbf{x})=\mathbf{x}\boldsymbol{\beta}+u_p\sigma$$

where u_p is the pth quantile of the standard normal distribution.

Let us consider uncensored data to start. In this case the method described in Section 5.2.2a for obtaining confidence intervals for quantiles in the nonregression case applies, with slight modification. Letting $\hat{y}_p(\mathbf{x})=\mathbf{x}\hat{\boldsymbol{\beta}}+u_p s$, where $s^2=\Sigma(y_i-\mathbf{x}_i\hat{\boldsymbol{\beta}})^2/(n-p)$ is the usual unbiased estimate of σ^2, we use the pivotal quantity

$$Z_p=\frac{\hat{y}_p(\mathbf{x})-y_p(\mathbf{x})}{s}$$

to get confidence intervals for $y_p(\mathbf{x})$. Since $(n-p)s^2/\sigma^2\sim\chi^2_{(n-p)}$, $\mathbf{x}\hat{\boldsymbol{\beta}}\sim$

$N[\mathbf{x}\boldsymbol{\beta}, \sigma^2 A^2(\mathbf{x})]$, where $A^2(\mathbf{x}) = \mathbf{x}'(\mathbf{X}'\mathbf{X})^{-1}\mathbf{x}$ and $\hat{\boldsymbol{\beta}}$ and s^2 are independent, it follows that

$$\Pr(Z_p \leqslant a) = \Pr\left(\frac{(\mathbf{x}\hat{\boldsymbol{\beta}} - \mathbf{x}\boldsymbol{\beta})/A(\mathbf{x})\sigma - u_p/A(\mathbf{x})}{s/\sigma} \leqslant \frac{a - u_p}{A(\mathbf{x})} \right)$$

$$= \Pr\left[t'_{(n-p)}\left(\frac{-u_p}{A(\mathbf{x})} \right) \leqslant \frac{a - u_p}{A(\mathbf{x})} \right] \tag{6.5.15}$$

where $t'_{(\nu)}(\lambda)$ represents the noncentral χ^2 distribution with ν degrees of freedom and noncentrality parameter λ, defined in Section 5.2.2a. The expression (6.5.15) is essentially the same as (5.2.17) of Section 5.2.2a, except that $A(\mathbf{x})^{-1}$ replaces $\sqrt{n}$ and $n-p$ replaces $n-1$. Therefore percentage points of Z_p can be found from the percentage points of noncentral t, as described in Section 5.2.2a. This allows confidence limits for $y_p(\mathbf{x})$ to be obtained: the probability statement $\Pr(Z_p \leqslant a) = \gamma$ gives $\hat{y}_p(\mathbf{x}) - as$ as a lower γ confidence limit for $y_p(\mathbf{x})$.

For data that are not too heavily censored (less than about 30%) one can use (6.5.12) and (6.5.13) with this method to give approximate confidence intervals. The only adjustments in this case are that $\hat{\sigma}_u$ replaces s in Z_p and that in (6.5.15) n_1^* replaces $n-p$.

For more heavily censored data likelihood ratio methods can be used: we consider the hypothesis $H_0: \mathbf{x}\boldsymbol{\beta} + u_p\sigma = Q_0$ versus $H_1: \mathbf{x}\boldsymbol{\beta} + u_p\sigma \neq Q_0$, where Q_0 is a specified value for the pth quantile. The likelihood ratio statistic for testing H_0 versus H_1 is $\Lambda = -2\log[L(\tilde{\boldsymbol{\beta}}, \tilde{\sigma})/L(\hat{\boldsymbol{\beta}}, \hat{\sigma})]$, where $\tilde{\boldsymbol{\beta}}$ and $\tilde{\sigma}$ are the m.l.e.'s of $\boldsymbol{\beta}$ and σ under H_0. Under H_0, Λ is asymptotically $\chi^2_{(1)}$; this approximation is reasonable provided that sample sizes are moderately large. For quite large samples, inferences can alternately be based on the large-sample normal approximation $(\hat{\boldsymbol{\beta}}, \hat{\sigma}) \sim N[(\boldsymbol{\beta}, \sigma), \mathbf{I}_0^{-1}]$, which gives the approximation $\hat{y}_p(\mathbf{x}) \sim N[y_p(\mathbf{x}), (\mathbf{x}, u_p)\mathbf{I}_0^{-1}(\mathbf{x}, u_p)']$. The sample size has to be fairly large for this approximation to be reasonable, however.

6.5.3 Other Methods

If there are several observations taken at each $\mathbf{x}_i$, simple methods analogous to those described in Section 6.4.2 for the extreme value regression model can be used. The same approach applies here as there, except, of course, that one now employs (preferably unbiased) estimates of the normal mean $\mu_i = \mathbf{x}_i\boldsymbol{\beta}$ and standard deviation σ from the observations at $\mathbf{x}_i$. Estimates from censored normal samples have been discussed in Section 5.2.1.

In situations for which there is Type II censoring at each $\mathbf{x}_i$, best linear unbiased estimation of both $\boldsymbol{\beta}$ and σ is possible. This approach is described in Section 6.7, where least squares and linear estimation methods are considered for arbitrary linear models. Nelson and Hahn (1973) discuss an example involving censored normal regression data, and Nelson and Hahn (1972) discuss the same example, using the methods referred to in the preceding paragraph.

6.6 GAMMA AND LOG-GAMMA REGRESSION MODELS

The generalized log-gamma distribution discussed in Section 5.3 includes the extreme value and normal distributions as special cases. Its use in discriminating between extreme value and normal distributions or, equivalently, between Weibull and log-normal distributions was noted there. It can also be used to examine the effect on inferences of moderate departures from a Weibull or log-normal model and is itself a flexible three-parameter model. Regression models based on the log-gamma distribution serve similar useful purposes.

The model discussed here is the linear model generalization of (5.3.6). The distribution of log lifetime Y, given the $1\times p$ vector $\mathbf{x}$ of regressor variables, can be represented as

$$Y=\mathbf{x}\boldsymbol{\beta}+\sigma z \tag{6.6.1}$$

where $\boldsymbol{\beta}=(\beta_1,\ldots,\beta_p)'$, $\sigma>0$ is a scale parameter, and z has a log-gamma distribution with p.d.f. $(-\infty<z<\infty)$

$$f(z;k)=\begin{cases}\dfrac{k^{k-1/2}}{\Gamma(k)}\exp\left(\sqrt{k}z-ke^{z/\sqrt{k}}\right) & 0<k<\infty\\[2ex] \dfrac{1}{(2\pi)^{1/2}}\exp\left(\dfrac{-z^2}{2}\right) & k=\infty\end{cases}. \tag{6.6.2}$$

The extreme value regression models of Section 6.4 are given as the special case $k=1$ of (6.6.1) and (6.6.2), and the limiting case $k=\infty$ gives the normal models of Section 6.5. The location parameter in (6.6.1) does not, of course, need to be a linear function of $x_1,\ldots,x_p$, though this is the most frequently used model and the only one considered here.

We can also consider the extended log-gamma family represented by (5.3.8). In this case the parameter k is replaced by $\lambda=k^{-1/2}$, and the model is extended to include distributions with both positive and negative values of λ. The p.d.f. of z is then of the form

$$f(z;\lambda)=\begin{cases}\dfrac{|\lambda|}{\Gamma(\lambda^{-2})}(\lambda^{-2})^{\lambda^{-2}}\exp\left[\lambda^{-2}(\lambda z-e^{\lambda z})\right] & \lambda\neq 0\\[2ex] \dfrac{1}{(2\pi)^{1/2}}\exp\left(\dfrac{-z^2}{2}\right) & \lambda=0.\end{cases} \tag{6.6.3}$$

The model (6.6.3) also arises if one allows either z or $-z$ to have p.d.f. (6.6.2).

Maximum likelihood estimation for (6.6.1) with (6.6.3) is described by Farewell and Prentice (1977). The approach is similar to that taken in Section 5.3 for the ordinary univariate model. This involves solving for the m.l.e.'s $\tilde{\boldsymbol{\beta}}(k)$ and $\tilde{\sigma}(k)$ for different k values or, if (6.6.3) is used, solving for $\tilde{\boldsymbol{\beta}}(\lambda)$ and $\tilde{\sigma}(\lambda)$ for different λ values. If we solve for a number of k or λ values, the maximized log likelihood function, say,

$$\log L_{\max}(k)=\log L\left[\tilde{\boldsymbol{\beta}}(k),\tilde{\sigma}(k),k\right]$$

can be determined, and the m.l.e. $\hat{k}$ obtained as the value that maximizes $\log L_{\max}(k)$. The m.l.e.'s of $\boldsymbol{\beta}$ and σ are $\hat{\boldsymbol{\beta}}=\tilde{\boldsymbol{\beta}}(\hat{k})$ and $\hat{\sigma}=\tilde{\sigma}(\hat{k})$.

Consider, as usual, a possibly censored sample involving n individuals and let y_i represent either the observed log lifetime or log censoring time for the ith individual. The set of individuals for which y_i is a log lifetime is denoted as D, and the set for which y_i is a log censoring time is denoted as C. For notational convenience we write $z_i=(y_i-\mathbf{x}_i\boldsymbol{\beta})/\sigma$. As in Section 5.3, we shall work with the parameter k and explicitly consider just the model (6.6.2), though the methods also apply directly to (6.6.3). To deal with negative λ values in (6.6.3) it is merely necessary to replace z_i with $-z_i$ and use the value $k=|\lambda|^{-2}$ in the expressions given below.

The log likelihood function is, as in (5.3.12),

$$\log L(\boldsymbol{\beta},\sigma,k)=\sum_{i\in D}\left[\log f(z_i;k)-\log\sigma\right]+\sum_{i\in C}\log Q(k,ke^{z_i/\sqrt{k}}) \tag{6.6.4}$$

for $0<k<\infty$. Here $f(z_i;k)$ is given by (6.6.2) and $Q(k,a)$ is the incomplete gamma integral

$$Q(k,a)=\int_a^\infty \frac{x^{k-1}}{\Gamma(k)}e^{-x}\,dx=1-I(k,a).$$

For $k=\infty$ the log likelihood function is that for the normal regression model, given by (6.5.2). Estimation for this model was discussed in detail in Section 6.5.1, and only the case of finite k will be considered here.

We shall proceed as in Section 5.3 and write down the separate contributions to $\log L$ and its derivatives from a censored and an uncensored observation. Since k is being treated as fixed, it is only necessary to give derivatives with respect to $\beta_1, \ldots, \beta_p$ and σ.

Uncensored Observations

If y_i is an observed log lifetime, the contribution to the log likelihood (6.6.4) and its derivatives are as follows [note that $z_i = (y_i - \mathbf{x}_i \boldsymbol{\beta})/\sigma$]:

$$\log L_i = (k - \tfrac{1}{2}) \log k - \log \Gamma(k) - \log \sigma + \sqrt{k}\, z_i - k e^{z_i/\sqrt{k}}$$

$$\frac{\partial \log L_i}{\partial \beta_s} = -x_{is}\frac{\sqrt{k}}{\sigma} + x_{is}\frac{\sqrt{k}}{\sigma} e^{z_i/\sqrt{k}} \qquad s=1,\ldots,p$$

$$\frac{\partial \log L_i}{\partial \sigma} = -\frac{1}{\sigma} - \frac{\sqrt{k}}{\sigma} z_i + \frac{\sqrt{k}}{\sigma} z_i e^{z_i/\sqrt{k}} \tag{6.6.5}$$

$$\sigma^2 \frac{\partial^2 \log L_i}{\partial \beta_s \partial \beta_t} = -x_{is} x_{it} e^{z_i/\sqrt{k}} \qquad s,t=1,\ldots,p$$

$$\sigma^2 \frac{\partial^2 \log L_i}{\partial \beta_s \partial \sigma} = x_{is}\sqrt{k} - x_{is}\sqrt{k}\, e^{z_i/\sqrt{k}} - x_{is} z_i e^{z_i/\sqrt{k}} \qquad s=1,\ldots,p \tag{6.6.6}$$

$$\sigma^2 \frac{\partial^2 \log L_i}{\partial \sigma^2} = 1 + 2\sqrt{k}\, z_i - 2\sqrt{k}\, z_i e^{z_i/\sqrt{k}} - z_i^2 e^{z_i/\sqrt{k}}.$$

Censored Observations

For a log censoring time the contributions to $\log L$ and its derivatives are given below. For convenience in writing out the derivatives, we define $m_i = k \exp[(y_i - \mathbf{x}_i \boldsymbol{\beta})/\sigma\sqrt{k}]$ and note the following results:

$$\frac{\partial m_i}{\partial \beta_s} = \frac{-x_{is} m_i}{\sqrt{k}\,\sigma} \qquad \frac{\partial m_i}{\partial \sigma} = \frac{-z_i m_i}{\sqrt{k}\,\sigma}$$

$$\frac{\partial^2 m_i}{\partial \beta_s \partial \beta_t} = \frac{x_{is} x_{it} m_i}{k\sigma^2} \qquad \frac{\partial^2 m_i}{\partial \sigma^2} = \frac{2 z_i m_i}{\sqrt{k}\,\sigma^2} + \frac{z_i^2 m_i}{k\sigma^2}$$

$$\frac{\partial^2 m_i}{\partial\beta_s\,\partial\sigma}=\frac{x_{is}m_i}{\sqrt{k}\,\sigma^2}+\frac{x_{is}m_iz_i}{k\sigma^2}$$

$$\frac{\partial Q(k,x)}{\partial x}=\frac{-x^{k-1}e^{-x}}{\Gamma(k)}\qquad \frac{\partial^2 Q}{\partial x^2}=\frac{e^{-x}x^{k-2}(x-k+1)}{\Gamma(k)}$$

$$\frac{\partial\log Q(t,x)}{\partial x}=\frac{1}{Q(k,x)}\frac{\partial Q}{\partial x}$$

$$\frac{\partial^2\log Q(k,x)}{\partial x^2}=\frac{1}{Q(k,x)}\frac{\partial^2 Q}{\partial x^2}-\frac{1}{Q^2(k,x)}\left(\frac{\partial Q}{\partial x}\right)^2.$$

The desired quantities are then

$$\log L_i=\log Q(k,ke^{z_i/\sqrt{k}})$$

$$\frac{\partial\log L_i}{\partial\beta_s}=\frac{\partial\log Q}{\partial m_i}\frac{\partial m_i}{\partial\beta_s}\qquad s=1,\dots,p$$

$$\frac{\partial\log L_i}{\partial\sigma}=\frac{\partial\log Q}{\partial m_i}\frac{\partial m_i}{\partial\sigma}\tag{6.6.7}$$

$$\frac{\partial^2\log L_i}{\partial\beta_s\,\partial\beta_t}=\frac{\partial^2\log Q}{\partial m_i^2}\frac{\partial m_i}{\partial\beta_s}\frac{\partial m_i}{\partial\beta_t}+\frac{\partial\log Q}{\partial m_i}\frac{\partial^2 m_i}{\partial\beta_s\,\partial\beta_t}\qquad s,t=1,\dots,p$$

$$\frac{\partial^2\log L_i}{\partial\beta_s\,\partial\sigma}=\frac{\partial^2\log Q}{\partial m_i^2}\frac{\partial m_i}{\partial\beta_s}\frac{\partial m_i}{\partial\sigma}+\frac{\partial\log Q}{\partial m_i}\frac{\partial^2 m_i}{\partial\beta_s\,\partial\sigma}\qquad s=1,\dots,p\tag{6.6.8}$$

$$\frac{\partial^2\log L_i}{\partial\sigma^2}=\frac{\partial^2\log Q}{\partial m_i^2}\left(\frac{\partial m_i}{\partial\sigma}\right)^2+\frac{\partial\log Q}{\partial m_i}\frac{\partial^2 m_i}{\partial\sigma^2}.$$

The log likelihood function (6.6.4) is calculated as

$$\log L(\boldsymbol{\beta},\sigma,k)=\sum_{i\in D}\log L_i+\sum_{i\in C}\log L_i$$

where $\log L_i$ is given by the respective expressions for $i\in D$ and for $i\in C$. Maximum likelihood estimates $\tilde{\boldsymbol{\beta}}(k)$ and $\tilde{\sigma}(k)$ for a given value of k can usually be conveniently found by solving the likelihood equations $\partial\log L/\partial\beta_s=0$ $(s=1,\dots,p)$ and $\partial\log L/\partial\sigma=0$ via Newton–Raphson iteration. Though the expressions involved look complicated, they are easily programmed. Search procedures (see Appendix F) also often work well.

The suggestions made in Section 5.3.1 for examining $\log L_{\max}(k)$ are pertinent here as well. Usually it is sufficient to obtain m.l.e.'s of $\boldsymbol{\beta}$ and σ for about 8 or 10 k values, after which it is possible to locate $\hat{k}$ to acceptible accuracy. In view of the flatness of the likelihood function as k increases, it is helpful to work with $\lambda=k^{-1/2}$, and since $\log L_{\max}(\lambda)$ may be maximized near 0 or at a negative value of λ, it may also be necessary to consider the extended model (6.6.3).

A plot of the maximized relative likelihood function $R_{\max}(k)=L_{\max}(k)/L(\hat{\boldsymbol{\beta}},\hat{\sigma},\hat{k})$ portrays plausible values of k. In addition, the likelihood ratio test of $H_0: k=k_0$ versus $H_1: k\neq k_0$ employs the statistic $\Lambda=-2\log R_{\max}(k_0)$, which can be treated as approximately $\chi^2_{(1)}$ under H_0. A technical difficulty arises in testing $k=\infty$ (i.e., $\lambda=0$) unless the extended model (6.6.3) is used, since $\lambda=0$ is a boundary point of the parameter space. With (6.6.3) this difficulty disappears and the usual large-sample theory applies (Prentice, 1974).

The simplest approach to inference about $\boldsymbol{\beta}$ or σ is to give results for different plausible fixed values of k by using likelihood ratio methods or procedures based on the approximate normality of $\tilde{\boldsymbol{\beta}}(k)$ and $\tilde{\sigma}(k)$. An alternative is to use large-sample methods that treat all of $\boldsymbol{\beta}$, σ, and k as unknowns, but the complexity of derivatives of $\log L$ with respect to k makes this unattractive. In many instances a satisfactory picture is gained by giving inferences conditional on k for several k values, and if results independent of k are required, by averaging conditional inferences in some more or less *ad hoc* way.

Example 6.6.1 (Example 6.4.3 revisited) In Example 6.4.3 we considered data from an accelerated life test on electrical insulating fluid. The data consist of times to breakdown for specimens tested at seven voltage levels. The model employed in Example 6.4.3 was such that at voltage level v_i, times to breakdown had a Weibull distribution with shape parameter δ and scale parameter $\alpha_i=cv_i^p$. Equivalently, log times y_{ij} were assumed to have an extreme value distribution with scale parameter $\sigma=\delta^{-1}$ and location parameter $\alpha+\beta x_i$, where $x_i=\log v_i$. We generalize this here and suppose that the log times y_{ij} $(j=1,\ldots,n_i)$ at voltage v_i follow a log-gamma regression model of the form (6.6.1), with $\mathbf{x}\boldsymbol{\beta}=\alpha+\beta x_i$. The extreme value model used in Example 6.4.3 is given as the special case $k=1$. An analysis under the log-gamma model provides a check on the appropriateness of the Weibull or extreme value model and indicates the extent to which inferences depend upon the model.

The observations given in Example 6.4.3 are all uncensored and $\mathbf{x}_i\boldsymbol{\beta}=\alpha+\beta x_i$; thus, for given k, the log likelihood and its first derivatives are, from

(6.6.5),

$$= n\left[(k - \tfrac{1}{2})\log k - \log\Gamma(k)\right] - n\log\sigma + \sqrt{k}\sum_{i=1}^{7}\sum_{j=1}^{n_i} z_{ij} - k\sum_{i=1}^{7}\sum_{j=1}^{n_i} e^{z_{ij}/\sqrt{k}}$$

$$\frac{\partial \log L}{\partial\alpha} = -\frac{n\sqrt{k}}{\sigma} + \frac{\sqrt{k}}{\sigma}\sum_{i=1}^{7}\sum_{j=1}^{n_i} e^{z_{ij}/\sqrt{k}}$$

$$\frac{\partial \log L}{\partial\beta} = -\frac{\sqrt{k}}{\sigma}\sum_{i=1}^{7} n_i x_i + \frac{\sqrt{k}}{\sigma}\sum_{i=1}^{7}\sum_{j=1}^{n_i} x_i e^{z_{ij}/\sqrt{k}}$$

$$\frac{\partial \log L}{\partial\sigma} = -\frac{n}{\sigma} - \frac{\sqrt{k}}{\sigma}\sum_{i=1}^{7}\sum_{j=1}^{n_i} z_{ij} + \frac{\sqrt{k}}{\sigma}\sum_{i=1}^{7}\sum_{j=1}^{n_i} z_{ij} e^{z_{ij}/\sqrt{k}}$$

where $z_{ij}=(y_{ij}-\alpha-\beta x_i)/\sigma$. The m.l.e.'s $\tilde{\alpha}(k)$, $\tilde{\beta}(k)$, and $\tilde{\sigma}(k)$ are readily obtained by solving the equations $\partial\log L/\partial\alpha=0$, $\partial\log L/\partial\beta=0$, and $\partial\log L/\partial\sigma=0$ by Newton–Raphson iteration. Calculating these m.l.e.'s and the associated maximized log likelihood values $\log L_{\max}(k)=\log L[\tilde{\alpha}(k),\tilde{\beta}(k),\tilde{\sigma}(k),k]$ for a range of k values, one finds that $\log L_{\max}(k)$ is maximized close to $k=1.83$ ($\lambda=0.74$) and that $\tilde{\alpha}(1.83)=2.689$, $\tilde{\beta}(1.83)=-17.628$, and $\tilde{\sigma}(1.83)=1.349$. Figure 6.6.1 gives the maximized relative

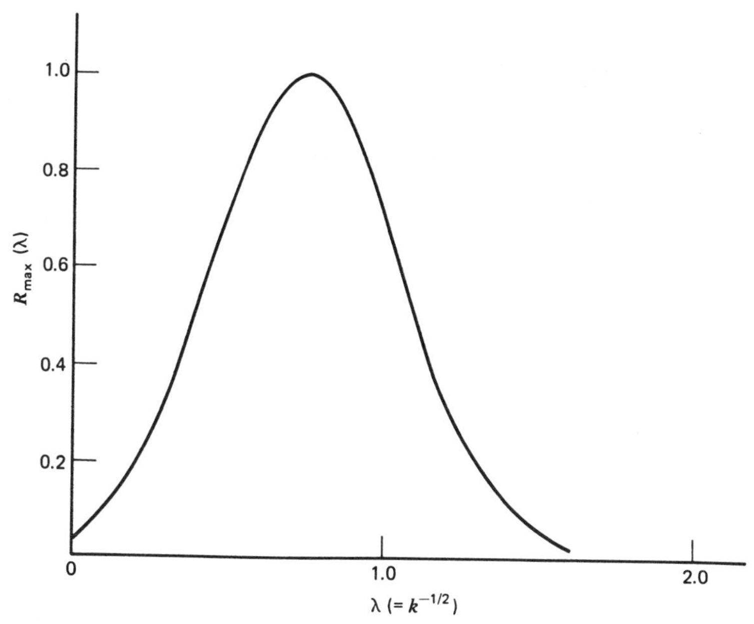

Figure 6.6.1 Maximized relative likelihood for $\lambda=k^{-1/2}$ in log-gamma regression model (Example 6.6.1)

likelihood $R_{\max}(k)=L_{\max}(k)/L_{\max}(1.83)$; this indicates plausible values for k. The value $k=1$ gives $R_{\max}(1)=.698$ and a likelihood ratio statistic value $\Lambda=-2\log(.698)=0.720$ and provides no evidence against the extreme value model. There is, on the other hand, evidence against the normal model, the value $k=\infty$ ($\lambda=0$) giving $R_{\max}(\infty)=.043$ and an associated likelihood ratio statistic value $\Lambda=6.306$, which is significant at the .01 level.

Confidence intervals for parameters and other distribution characteristics can be calculated for different values of k by treating $[\tilde{\alpha}(k), \tilde{\beta}(k), \tilde{\sigma}(k)]$ as normally distributed or by likelihood ratio methods. An effect similar to that in Example 5.3.1 is observed, namely, that estimates of quantities concerned with the center of the distribution vary moderately with k, whereas the dependence on k is much stronger for quantities concerned with the tails of the distributions. As a final comment on this example we remark that k itself appears to depend upon the voltage level. Farewell and Prentice (1977) note that when separate log-gamma models are fitted to the data with voltage levels of 32 kilovolts and less, and that with voltage levels greater than 32 kilovolts the m.l.e.'s of k are about 0.53 and 9.8, respectively, and a likelihood ratio test of equality of the k values gives a significance level of about .025.

6.7 LEAST SQUARES ESTIMATION

6.7.1 Ordinary Least Squares

Least squares methods can be used with models of the form (6.1.9); some examples with the extreme value model have already been given in Section 6.4.2 and in Example 6.3.3. Least squares estimation is not particularly efficient in many instances, but the fact that estimates are easily calculated with uncensored data makes them valuable. Let us change our notation slightly from previous sections and consider models of the form

$$Y_i=\beta_0+\mathbf{x}_i\boldsymbol{\beta}+\sigma e_i \qquad i=1,\ldots,n \tag{6.7.1}$$

where $\sigma>0$ is a scale parameter and the e_i's are i.i.d. with mean $E(e_i)=0$ and constant variance, denoted $\operatorname{Var}(e)$. In (6.7.1) $\mathbf{x}_i=(x_{i1},\ldots,x_{ip})$ and $\boldsymbol{\beta}=(\beta_1,\ldots,\beta_p)'$, and the regressor variables will be taken to be centered without loss of generality so that

$$\sum_{i=1}^{n} x_{ir}=0 \qquad r=1,\ldots,p.$$

Procedures that might be used with censored data are suggested later, but most of the initial discussion deals only with uncensored data. In this case the least squares estimates of β_0 and $\boldsymbol{\beta}$ in (6.7.1) are well known to be

$$\tilde{\beta}_0 = \bar{y} \qquad \tilde{\boldsymbol{\beta}} = (\mathbf{X}'\mathbf{X})^{-1}\mathbf{X}'\mathbf{Y} \tag{6.7.2}$$

where $X = (x_{ir})_{n \times p}$ is assumed to be of rank p ($p < n$) and $\mathbf{Y} = (y_1, \ldots, y_n)'$. These are unbiased estimators of β_0 and $\boldsymbol{\beta}$, and the covariance matrix of $\tilde{\boldsymbol{\beta}}^* = (\tilde{\beta}_0, \tilde{\boldsymbol{\beta}})$ is

$$\sigma^2 \operatorname{Var}(e) \begin{pmatrix} n^{-1} & 0 \\ 0 & (\mathbf{X}'\mathbf{X})^{-1} \end{pmatrix}. \tag{6.7.3}$$

If $\operatorname{Var}(e)$ is known, an unbiased estimate of σ^2 is

$$\tilde{\sigma}^2 = \frac{1}{(n-p-1)\operatorname{Var}(e)} \sum_{i=1}^{n} (y_i - \tilde{\beta}_0 - x_i \tilde{\boldsymbol{\beta}})^2. \tag{6.7.4}$$

The least squares estimates are easy to calculate, but for many models they are rather inefficient. Cox and Hinkley (1968) discuss their efficiency, essentially as follows: suppose that the e_i's in (6.7.1) have a distribution with p.d.f. $f(e; \boldsymbol{\lambda})$, where $\boldsymbol{\lambda} = (\lambda_1, \ldots, \lambda_q)$ represents possible additional unknown parameters. The log likelihood function based on (6.7.1) would then be

$$\log L(\boldsymbol{\lambda}, \sigma, \beta_0, \boldsymbol{\beta}) = \sum_{i=1}^{n} \log\left[\frac{1}{\sigma} f\left(\frac{y_i - \beta_0 - \mathbf{x}_i \boldsymbol{\beta}}{\sigma}\right)\right].$$

With the regressor variables centered so that $\Sigma_{i=1}^{n} x_{ir} = 0$ ($r = 1, \ldots, p$), it is easily shown (see Problem 6.12) that

$$E\left(\frac{-\partial^2 \log L}{\partial \beta_r \partial \beta_s}\right) = \frac{A_e}{\sigma^2} (\mathbf{X}'\mathbf{X})_{rs} \qquad E\left(\frac{\partial^2 \log L}{\partial \beta_r \partial \lambda_s}\right) = 0$$

$$E\left(\frac{\partial^2 \log L}{\partial \beta_r \partial \beta_0}\right) = 0 \qquad E\left(\frac{\partial^2 \log L}{\partial \beta_r \partial \sigma}\right) = 0$$

where

$$A_e = E\left(\frac{-\partial^2 \log f(e; \boldsymbol{\lambda})}{\partial e^2}\right) = \int_{-\infty}^{\infty} \frac{-\partial^2 \log f(e; \boldsymbol{\lambda})}{\partial e^2} f(e; \boldsymbol{\lambda})\, de. \tag{6.7.5}$$

The expected information matrix for the parameters $(\boldsymbol{\lambda}, \sigma, \beta_0, \boldsymbol{\beta})$ thus has the partitioned form

$$\begin{pmatrix} \mathbf{I}_1 & 0 \\ 0 & \mathbf{I}_2 \end{pmatrix} = \begin{pmatrix} \mathbf{I}_1 & 0 \\ 0 & (A_e/\sigma^2)(\mathbf{X}'\mathbf{X}) \end{pmatrix}$$

where $\mathbf{I}_2$ refers to $\boldsymbol{\beta}$ and $\mathbf{I}_1$ to $\boldsymbol{\lambda}$, σ, and β_0. The asymptotic covariance matrix of the m.l.e. $\hat{\boldsymbol{\beta}}$ of $\boldsymbol{\beta}$ is

$$\text{Asvar}(\hat{\boldsymbol{\beta}}) = A_e^{-1}\sigma^2(\mathbf{X}'\mathbf{X})^{-1}.$$

From (6.7.3) the asymptotic efficiency of the least squares estimators relative to the m.l.e.'s is

$$\frac{\text{Asvar}(\hat{\beta}_i)}{\text{Var}(\tilde{\beta}_i)} = \frac{1}{A_e \text{Var}(e)}. \tag{6.7.6}$$

This gives the asymptotic relative efficiency of the least squares estimate for an arbitrary model of the form (6.7.1). For the generalized log-gamma family given by (6.6.1) it is easily determined (see Problem 6.12) that (6.7.6) gives

$$\frac{\text{Asvar}(\hat{\beta}_i)}{\text{Var}(\tilde{\beta}_i)} = \frac{1}{k\psi'(k)} \tag{6.7.7}$$

where k is the log-gamma index parameter. For the extreme value model, $k=1$ and (6.7.7) equals .608. Hence least squares with extreme value data gives estimates of $\boldsymbol{\beta}$ with an asymptotic relative efficiency of just over 60%. For k values less than unity (6.7.7) drops off rapidly; for $k=0.5$, for example, it is just .405. On the other hand, as k increases, (6.7.7) rapidly approaches its limiting value of unity; for $k \geqslant 5$ (6.7.7) exceeds .90.

This discussion concerns the efficiency of $\tilde{\boldsymbol{\beta}}$. It should be noted that the efficiency of the estimate (6.7.4) of σ^2 may be low; this can be examined for any particular model. To illustrate this and some other points we shall examine the extreme value regression model (6.4.3) in more detail.

Example 6.7.1 (the extreme value linear model) The extreme value regression model (6.4.3) can be written as

$$Y_i = \beta_0' + \mathbf{x}_i\boldsymbol{\beta} + \sigma z_i \qquad i=1,\ldots,n$$

where z_i has a standard extreme value distribution with p.d.f. $\exp(z-e^z)$, $-\infty<z<\infty$. The mean and variance of z_i are $\psi(1)=-\gamma=-.5772\ldots$ and $\psi'(1)=\pi^2/6=1.6449\ldots$, respectively. Since z_i does not have mean 0, let us slightly rewrite the model to bring it into the general form (6.7.1). Letting $e_i=z_i+\gamma$, we have

$$Y_i=\beta_0+\mathbf{x}_i\boldsymbol{\beta}+\sigma e_i \quad i=1,\ldots,n \tag{6.7.8}$$

where $\beta_0=\beta_0'-\sigma\gamma$. In (6.7.8) the e_i's are i.i.d. with mean 0 and variance $\pi^2/6$. The x_{ir}'s are assumed to be centered so that $\sum_{i=1}^n x_{ir}=0$ $(r=1,\ldots,p)$.

The least squares estimates of β_0 and $\boldsymbol{\beta}$ are given by (6.7.2), and the unbiased estimate (6.7.4) of σ^2 is

$$\tilde{\sigma}^2=\frac{6}{\pi^2}\sum_{i=1}^n\frac{\left(y_i-\tilde{\beta}_0-\mathbf{x}_i\tilde{\boldsymbol{\beta}}\right)^2}{n-p-1}. \tag{6.7.9}$$

The asymptotic relative efficiency of $\tilde{\boldsymbol{\beta}}$ is .608, as noted earlier. The asymptotic efficiencies of $\tilde{\beta}_0$ and $\tilde{\sigma}^2$ can be determined from the full expected information matrix for $(\sigma,\beta_0,\boldsymbol{\beta})$. Proceeding in the way leading up to (6.4.10), but employing the parameterization in (6.7.8), one gets the inverse of the expected information matrix:

$$\sigma^2\begin{pmatrix} 6/\pi^2 n & -6/\pi^2 n & 0 \\ -6/\pi^2 n & 1/n+6/\pi^2 n & 0 \\ 0 & 0 & (\mathbf{X}'\mathbf{X})^{-1} \end{pmatrix}.$$

This is the asymptotic covariance matrix for the m.l.e. $(\hat{\sigma},\hat{\beta}_0,\hat{\boldsymbol{\beta}})$. The asymptotic relative efficiency of $\hat{\beta}_0$ is therefore

$$\frac{\operatorname{Asvar}(\hat{\beta}_0)}{\operatorname{Var}(\tilde{\beta}_0)}=\frac{6}{\pi^2}\left(1+\frac{6}{\pi^2}\right)$$
$$=.978.$$

The variance of $\tilde{\sigma}^2$ is required in order to assess its efficiency. This depends on the $\mathbf{x}_i$'s; from results of Atiqullah (1962) it follows that

$$\operatorname{Var}(\tilde{\sigma}^2)=\frac{\sigma^4}{n-p-1}\left(2+\frac{\gamma_2\mathbf{a}'\mathbf{a}}{n-p-1}\right)$$

where $\gamma_2=2.40$ is the kurtosis of the extreme value distribution and $\mathbf{a}$ is a vector containing the diagonal entries in the $n\times n$ matrix $\mathbf{I}_n-\mathbf{X}(\mathbf{X}'\mathbf{X})^{-1}\mathbf{X}'$. It can be shown that $\mathbf{a}'\mathbf{a}\leqslant n-p-1$ so that $\operatorname{Var}(\tilde{\sigma}^2)\leqslant 4.40\sigma^2/(n-p-1)$. The asymptotic variance of $\hat{\sigma}^2$ is $4\sigma^4(6/\pi^2 n)$. In the limit as $n\to\infty$ the ratio $\operatorname{Var}(\hat{\sigma}^2)/\operatorname{Var}(\tilde{\sigma}^2)$ is therefore greater than or equal to $(24/\pi^2)/4.40=.553$, and hence estimation of σ^2 using (6.7.9) can be quite inefficient.

We have seen that for the extreme value model, least squares produces relatively inefficient estimates in uncensored samples. Prentice and Shillington (1975) and Williams (1978) discuss the efficiency of least squares in a little more detail for some special extreme value regression models. Williams in addition, proposes a method for obtaining more efficient estimates of parameters.

Censored Data

When the data are censored, some modification to the usual procedures is necessary if least squares is to be used. These should preferably keep calculations fairly simple, or else the advantage of easy computation is lost. The most straightforward procedure is to adjust censored observations in some way and to then compute least squares estimates as though the data were uncensored. The iterations based on (6.5.10) and (6.5.11) for censored normal data are of this type. In this case the procedure, when iterated, actually produces the m.l.e.'s of the parameters. Since with uncensored data least squares is equivalent to maximum likelihood estimation under a normal model, one approach with censored data would be to use the iterative least squares procedure of Section 6.5.1 to calculate estimates. Then, as in the uncensored case, the estimates would be asymptotically fully efficient when the e_i's are normally distributed, but they become relatively inefficient as the distribution of the e_i's moves further away from normality.

With most models other relatively *ad hoc* procedures of this kind can be devised. To illustrate let us return once again to the extreme value model (6.7.8), where $z_i=(y_i-\mathbf{x}_i\boldsymbol{\beta})/\sigma$ has a standard extreme value distribution. Then $\exp z_i$ has a standard exponential distribution and it follows that

$$E\left(e^{(y_i-\mathbf{x}_i\boldsymbol{\beta})/\sigma}\mid y_i\geqslant y_i^*\right)=1+e^{(y_i^*-\mathbf{x}_i\boldsymbol{\beta})/\sigma}.$$

Since for a censored observation y_i^* we know only that the actual log lifetime Y_i was $\geqslant y_i^*$, we can consider replacing a censored observation y_i^* with a value y_i' chosen so that

$$e^{(y_i'-\mathbf{x}_i\boldsymbol{\beta})/\sigma}=1+e^{(y_i^*-\mathbf{x}_i\boldsymbol{\beta})/\sigma}.$$

This gives the adjusted value

$$y_i' = \mathbf{x}_i\boldsymbol{\beta} + \sigma \log\left(1 + \exp\frac{(y_i^* - \mathbf{x}_i\boldsymbol{\beta})}{\sigma}\right). \tag{6.7.10}$$

This depends on the unknown parameters, but an iteration procedure can be based on (6.7.10). For this, initial estimates of β_0, $\boldsymbol{\beta}$, and σ are used to obtain adjusted values y_i' for all censored observations, following which new estimates of β_0, $\boldsymbol{\beta}$, and σ are calculated using least squares as though the data were uncensored. New adjusted values y_i' can then be calculated using the new estimates, and the entire process repeated. This could be repeated, hopefully until convergence is obtained, though it seems unlikely that much is to be gained going beyond two or three iterations.

Not much is known about procedures of this kind. The efficiency of the estimates so obtained may not be particularly high when the distribution of the e_i's is not close to normal, though they should be adequate for preliminary assessment of the data and preliminary examination of models. Miller (1976) and Buckley and James (1979) discuss another approach to least squares estimation with censored data.

It is possible to go further with least squares methods and to construct test and interval estimation procedures based on these. Prentice and Shillington (1975), for example, discuss modifications to the standard normal theory F and t tests for the purpose of analyzing extreme value data. Although these methods are simple to use, they are relatively inefficient for models like the extreme value model and, in addition, do not apply to censored data. In situations where the distribution of errors is not too far from normal and there is little or no censoring, test and interval estimation procedures of this sort can be considered, but it will usually be preferable to use maximum likelihood methods.

6.7.2 Data With Several Observations at Each x

In situations where several observations are taken at each regression vector $\mathbf{x}$, let us rewrite the model (6.7.8) as

$$\begin{aligned} Y_{ij} &= \beta_0 + \mathbf{x}_i\boldsymbol{\beta} + \sigma e_{ij} \qquad i=1,\ldots,m \quad j=1,\ldots,n_i \\ &= \mu_i + \sigma e_{ij} \end{aligned} \tag{6.7.11}$$

where Y_{ij} is the jth log lifetime at $\mathbf{x}_i$. Ordinary least squares estimation applies here, of course, but other simple procedures can also be useful. For example, if one calculates unbiased estimates $\tilde{\mu}_i$ having variances σ^2/w_i directly from the sample at each $\mathbf{x}_i$, then an estimate of $\boldsymbol{\beta}$ can be obtained by minimizing $\Sigma w_i(\tilde{\mu}_i - \beta_0 - \mathbf{x}_i\boldsymbol{\beta})^2$. This gives

$$\tilde{\boldsymbol{\beta}}^* = (\mathbf{X}'\mathbf{W}\mathbf{X})^{-1}\mathbf{X}'\mathbf{W}\tilde{\boldsymbol{\mu}}$$

where $\boldsymbol{\beta}^* = (\beta_0, \boldsymbol{\beta}_1)$ and

$$\mathbf{X}_{m\times(p+1)} = \begin{pmatrix} 1 & \mathbf{x}_1 \\ \vdots & \vdots \\ 1 & \mathbf{x}_m \end{pmatrix} \qquad \mathbf{W}_{m\times m} = \text{diag}(w_1, \ldots, w_m) \qquad \boldsymbol{\mu}_{m\times 1} = \begin{pmatrix} \tilde{\mu}_1 \\ \vdots \\ \tilde{\mu}_m \end{pmatrix}.$$

This estimate is unbiased and has covariance matrix $\sigma^2(\mathbf{X}'\mathbf{W}\mathbf{X})^{-1}$.

This approach is convenient when one has simple, efficient estimates of the μ_i's. For example, for the extreme value and normal distributions simple linear estimators are available, as described in Chapters 4 and 5. The procedures require more calculation than ordinary least squares, but this is often worthwhile. In the early stages of analysis calculation of estimates $\tilde{\mu}_i$ is a useful first step, and the estimate $\tilde{\boldsymbol{\beta}}^*$ then requires little additional calculation. If sufficiently precise estimates of μ_i are used, the estimate $\tilde{\boldsymbol{\beta}}^*$ will, in addition, be more efficient than that given by ordinary least squares. An illustration of this approach with the extreme value model was given previously in Section 6.4.2, and Example 6.3.3 suggests similar procedures for censored exponential data.

It is also possible to construct best linear unbiased estimators (b.l.u.e.'s) of both $\boldsymbol{\beta}^*$ and σ in the model (6.7.11) in a way that is analogous to that used to construct b.l.u.e.'s of location and scale parameters in the nonregression case (see Problem 4.1). The procedures can be used when the data at each $\mathbf{x}_i$ are possibly Type II censored. The model can then be written as

$$y_{ij} = \beta_0 + \mathbf{x}_i\boldsymbol{\beta} + \sigma e_{ij} \qquad i = 1, \ldots, m \quad j = 1, \ldots, r_i$$

where $y_{i1} \leqslant \cdots \leqslant y_{ir_i}$ are the r_i smallest observations in a random sample of size n_i at $\mathbf{x}_i$. Note that $e_{i1}, \ldots, e_{ir_i}$ are the r_i smallest observations in a sample of n_i from the distribution of e. Denoting $E(e_{ij})$ by α_{ij}, we have

$$\begin{aligned} E(y_{ij}) &= \beta_0 + \mathbf{x}_i\boldsymbol{\beta} + \sigma\alpha_{ij} \\ &= \mathbf{x}_{ij}^+\boldsymbol{\beta}^+ \end{aligned} \tag{6.7.12}$$

where $\mathbf{x}_{ij}^+ = (1, \mathbf{x}_i, \alpha_{ij})$ and $\boldsymbol{\beta}^+ = (\beta_0, \boldsymbol{\beta}_1, \sigma)$. Let the covariance matrix of

$\mathbf{Y}=(Y_{11},\ldots,Y_{1r_1},Y_{21},\ldots,Y_{2r_2},\ldots,Y_{mr_m})$ be $\sigma^2\mathbf{V}$; that is, $\mathbf{V}$ is the covariance matrix of $(e_{11},\ldots,e_{1r_1},e_{21},\ldots,e_{2r_2},\ldots,e_{mr_m})$. $\mathbf{V}$ will be a block diagonal matrix, since y_{ij}'s with different values of i are independent. From (6.7.12) and the fact that $\sigma^2\mathbf{V}$ is the covariance matrix for $\mathbf{Y}$, the b.l.u.e. of $\boldsymbol{\beta}^+$ is given by

$$\tilde{\boldsymbol{\beta}}^+=(\mathbf{X}'\mathbf{V}^{-1}\mathbf{X})^{-1}(\mathbf{X}'\mathbf{V}^{-1}\mathbf{Y})$$

where

$$\mathbf{X}=\begin{bmatrix}\mathbf{x}_{11}^+\\ \vdots\\ \mathbf{x}_{1r_1}^+\\ \mathbf{x}_{21}^+\\ \vdots\\ \mathbf{x}_{mr_m}^+\end{bmatrix}$$

is a $(\Sigma r_i)\times(p+2)$ matrix containing the $\mathbf{x}_{ij}^+$'s. The covariance matrix of $\tilde{\boldsymbol{\beta}}^+$ is $\sigma^2(\mathbf{X}'\mathbf{V}^{-1}\mathbf{X})^{-1}$.

When the data are uncensored, this gives the ordinary least squares estimate of $\boldsymbol{\beta}^*=(\beta_0,\boldsymbol{\beta})$ and a linear estimate of σ in place of the usual quadratic one based on the residual sum of squares. To use the method it is necessary to have the means, variances, and covariances of ordered observations in samples from the distribution of the error variable e. These are available for the two important cases in which e has an extreme value or a normal distribution. The same quantities are required in order to obtain b.l.u.e.'s of location and scale parameters from the extreme value and normal distributions with no regressor variables present and they are discussed in references given in Sections 4.1.1 and 5.2.1, respectively.

Best linear unbiased estimation can be useful when the data are Type II censored at each $\mathbf{x}_i$, though the estimates obtained will, in most cases, not differ greatly from estimates computed using the simpler least squares procedure described earlier. Their use is essentially restricted to situations where e has an extreme value or normal distribution, since it is only for these that the necessary means, variances, and covariances of the standardized ordered observations are available. Nelson and Hahn (1972, 1973) consider the b.l.u.e.'s in more detail and give an example involving censored normal data.

6.8 PROBLEMS AND SUPPLEMENTS

6.1 Show that the only models in both the proportional hazards family (6.1.6) for T and the location–scale family (6.1.9) for $\log T$ are the exponential and Weibull regression models.

(Section 6.1)

6.2 For the location–scale regression model (6.1.9) with $\mu(\mathbf{x})=\mathbf{x}\boldsymbol{\beta}$, prove that the m.l.e.'s $\hat{\boldsymbol{\beta}}$ and $\hat{\sigma}$ from a complete random sample $(y_i|\mathbf{x}_i)$, $i=1,\ldots,n$, satisfy the equivariance conditions (G7) of Appendix G. Show that the least squares estimator of $\boldsymbol{\beta}$ and the corresponding residual mean square estimator of σ are also equivariant.

(Section 6.3.2, Appendix G)

6.3 Consider the exponential regression model of Section 6.3.2 when there is a single regressor variable. The exact distribution of $Z_0=\tilde{\beta}_0-\beta_0$, $Z_1=\tilde{\beta}_1-\beta_1$, given $\mathbf{a}$, is given in (6.3.11) in this case. Determine the exact p.d.f. of $\sqrt{n}(Z_0, Z_1)$, given $\mathbf{a}$. Expand its logarithm to terms of order $n^{-1/2}$ and observe that a first-order approximation to the distribution is the usual large-sample normal approximation for $\sqrt{n}(Z_0, Z_1)$, with the observed information matrix as covariance matrix. Examine the $O(n^{-1/2})$ terms in the expansion. (Hint: center the x_i's, note that we then have $\Sigma e^{a_i}=n$ and $\Sigma x_i e^{a_i}=0$, and note that, given $\mathbf{a}$, the observed information matrix is fixed.)

(Section 6.3.2; Singhal, 1978)

6.4 The data in Table 6.8.1 are from a more comprehensive set given by Krall et al. (1975). The problem is to relate survival times for multiple myeloma patients to a number of prognostic variables. The data given here show survival times, in months, for 65 patients and include measurements on each patient for the following five regressor variables:

x_1 Logarithm of a blood urea nitrogen measurement at diagnosis
x_2 Hemoglobin measurement at diagnosis
x_3 Age at diagnosis
x_4 Sex: 0, male; 1, female
x_5 Serum calcium measurement at diagnosis

Asterisks denote censoring times.

a. Examine the relationship of these variables to survival time by fitting and examining exponential regression models of the form (6.3.1) with $\theta_{\mathbf{x}}=\exp(\mathbf{x}\boldsymbol{\beta})$.

Table 6.8.1 Survival Times and Regressor Variables for Multiple Myeloma Patients

t	x_1	x_2	x_3	x_4	x_5	t	x_1	x_2	x_3	x_4	x_5
1	2.218	9.4	67	0	10	26	1.230	11.2	49	1	11
1	1.940	12.0	38	0	18	32	1.322	10.6	46	0	9
2	1.519	9.8	81	0	15	35	1.114	7.0	48	0	10
2	1.748	11.3	75	0	12	37	1.602	11.0	63	0	9
2	1.301	5.1	57	0	9	41	1.000	10.2	69	0	10
3	1.544	6.7	46	1	10	42	1.146	5.0	70	1	9
5	2.236	10.1	50	1	9	51	1.568	7.7	74	0	13
5	1.681	6.5	74	0	9	52	1.000	10.1	60	1	10
6	1.362	9.0	77	0	8	54	1.255	9.0	49	0	10
6	2.114	10.2	70	1	8	58	1.204	12.1	42	1	10
6	1.114	9.7	60	0	10	66	1.447	6.6	59	0	9
6	1.415	10.4	67	1	8	67	1.322	12.8	52	0	10
7	1.978	9.5	48	0	10	88	1.176	10.6	47	1	9
7	1.041	5.1	61	1	10	89	1.322	14.0	63	0	9
7	1.176	11.4	53	1	13	92	1.431	11.0	58	1	11
9	1.724	8.2	55	0	12	4*	1.945	10.2	59	0	10
11	1.114	14.0	61	0	10	4*	1.924	10.0	49	1	13
11	1.230	12.0	43	0	9	7*	1.114	12.4	48	1	10
11	1.301	13.2	65	0	10	7*	1.532	10.2	81	0	11
11	1.508	7.5	70	0	12	8*	1.079	9.9	57	1	8
11	1.079	9.6	51	1	9	12*	1.146	11.6	46	1	7
13	0.778	5.5	60	1	10	11*	1.613	14.0	60	0	9
14	1.398	14.6	66	0	10	12*	1.398	8.8	66	1	9
15	1.602	10.6	70	0	11	13*	1.663	4.9	71	1	9
16	1.342	9.0	48	0	10	16*	1.146	13.0	55	0	9
16	1.322	8.8	62	1	10	19*	1.322	13.0	59	1	10
17	1.230	10.0	53	0	9	19*	1.322	10.8	69	1	10
17	1.591	11.2	68	0	10	28*	1.230	7.3	82	1	9
18	1.447	7.5	65	1	8	41*	1.756	12.8	72	0	9
19	1.079	14.4	51	0	15	53*	1.114	12.0	66	0	11
19	1.255	7.5	60	1	9	57*	1.255	12.5	66	0	11
24	1.301	14.6	56	1	9	77*	1.079	14.0	60	0	12
25	1.000	12.4	67	0	10						

b. Fit exponential models with $\theta_{\mathbf{x}} = \mathbf{x}\boldsymbol{\beta}$. Does this yield any different conclusions than those obtained in part (a)? Compare the fit of this model with that of part (a).

c. Analyze the data, using a Weibull model (6.4.1) with $\log \alpha(\mathbf{x}) = \mathbf{x}\boldsymbol{\beta}$. Does this provide any evidence against the exponential model of part (a)?

(Sections 6.3, 6.4)

6.5 Zelen (1959) gives the results of life tests on glass capacitors subjected to various voltage and temperature stresses. Tests were conducted at the eight voltage–temperature combinations involving voltages of 200, 250, 300, and 350 kilovolts and temperatures of 170 and 180°C. At each combination a Type II censored life test was carried out: eight capacitors were simultaneously put on test and the test was terminated at the time of the fourth failure. Table 6.8.2 shows the four failure times, in hours, at each combination.

a. Consider a regression model in which lifetimes at voltage–temperature combination (i, j), $i=1,2,3,4$, $j=1,2$, have a two-parameter exponential distribution with threshold parameter μ_{ij} and mean $\mu_{ij}+\theta_{ij}$. Examine the possibility of a multiplicative effect for temperature and voltage by testing the model with threshold parameters μ_{ij} and

$$\log\theta_{ij}=m+\alpha_i+\beta_j \qquad i=1,2,3,4 \quad j=1,2 \tag{6.8.1}$$

where $\Sigma\alpha_i=\Sigma\beta_j=0$, against the full model.

b. If the "main effects" model (6.8.1) is found to be reasonable, examine the possibility that $\log\theta_{ij}$ can be expressed as a linear function of $x_1=\log(\text{voltage})$ and $x_2=(273.2+\text{temperature})^{-1}$, which correspond to models suggested by engineering background.

(Section 6.3.3; Zelen, 1959; Lawless and Singhal, 1980)

6.6 In Table 6.8.3 McCool (1980) gives the failure times for hardened steel specimens in a rolling contact fatigue test; 10 independent observations were taken at each of 4 values of contact stress.

Table 6.8.2 Failure Times for Glass Capacitors at Eight Voltage–Temperature Combinations

	Voltage (kV)			
Temperature (°C)	200	250	300	350
170	439	572	315	258
	904	690	315	258
	1092	904	439	347
	1105	1090	628	588
180	959	216	241	241
	1065	315	315	241
	1065	455	332	435
	1087	473	380	455

Table 6.8.3 Failure Times for Steel Specimens at Four Stress Levels

Stress ($psi^2 \div 10^6$)	Ordered Failure Times
0.87	1.67, 2.20, 2.51, 3.00, 2.90, 4.70, 7.53, 14.70, 27.8, 37.4
0.99	0.80, 1.00, 1.37, 2.25, 2.95, 3.70, 6.07, 6.65, 7.05, 7.37
1.09	0.012, 0.18, 0.20, 0.24, 0.26, 0.32, 0.32, 0.42, 0.44, 0.88
1.18	0.073, 0.098, 0.117, 0.135, 0.175, 0.262, 0.270, 0.350, 0.386, 0.456

Engineering background suggests that at stress level s failure time should have approximately a Weibull distribution with a scale parameter α related to s by a power law relationship $\alpha = cs^p$ and with a shape parameter β that is independent of s. Assess this model graphically and by a formal test of it against the alternative that failure times at the ith stress level have a Weibull distribution with parameters α_i and β ($i=1,2,3,4$). Are you satisfied that β can be considered the same for all four stresses?

(Section 6.4.2; McCool, 1980)

6.7 Matched pairs. Consider lifetimes for n pairs of individuals and suppose that for the ith pair the individuals have lifetimes T_{i1} and T_{i2} that are independent and follow a Weibull distribution with hazard function

$$h_i(t) = \lambda_i t^{\alpha-1} e^{\beta x} \qquad i=1,\ldots,n \tag{6.8.2}$$

where x is a regressor variable and the λ_i's, α, and β are unknown parameters. Suppose that the λ_i's are nuisance parameters, our main interest being in α and β.

a. Show that $W_i = T_{1i}/T_{2i}$ ($i=1,\ldots,n$) have a joint p.d.f. proportional to

$$\prod_{i=1}^{n} \frac{\alpha w_i^{\alpha-1} \exp(\beta d_i)}{[1 + w_i^{\alpha} \exp(\beta d_i)]^2} \tag{6.8.3}$$

where $d_i = x_{1i} - x_{2i}$. Note that given d_i, $\log W_i$ has a logistic distribution.

b. The data in Table 6.8.4 are an altered version of data given by Batchelor and Hackett (1970), concerning the survival times of closely matched (T_{1i}) and poorly matched (T_{2i}) skin grafts on 11 burn patients. That is,

Table 6.8.4 Survival Times in Days for Well- and Poorly-Matched Skin Grafts

Patient:	1	2	3	4	5	6	7	8	9	10	11
T_{1i} (good match)	37	19	67	93	16	23	20	18	63	29	70
T_{2i} (poor match)	29	13	15	26	11	18	26	23	43	15	42

a well-matched and a poorly matched (in terms of immunological factors related to possible rejection of the graft) graft were applied to each of the 11 persons. Analyze these data under the model (6.8.2) by letting x be a dummy variable taking values $x_{1i}=0$ and $x_{2i}=1$, for $i=1,\ldots,n$. Use (6.8.3) as a marginal likelihood for α and β to obtain m.l.e.'s $\hat{\alpha}$ and $\hat{\beta}$ and to test the hypothesis that $\beta=0$. Note that $E(T_{1i})/E(T_{2i})=\exp(\beta/\alpha)$ and obtain a confidence interval for β/α.

(Holt and Prentice, 1974; Kalbfleisch and Prentice, 1980, Sec. 8.1)

6.8 (continuation of Problem 6.7) Discuss the difference between the analysis based on (6.8.2) and the analysis you would get by assuming that T_{1i} $(i=1,\ldots,n)$ and T_{2i} $(i=1,\ldots,n)$ are respectively samples from Weibull distributions with hazard functions

$$h_1(t)=\lambda_1 t^{\alpha-1} \quad \text{and} \quad h_2(t)=\lambda_2 t^{\alpha-1}. \tag{6.8.4}$$

Describe how you would decide whether (6.8.2) or (6.8.4), if either, is appropriate.

6.9 Stone (1978) reports an experiment in which specimens of solid epoxy electrical insulation were studied in an accelerated voltage life test. In all, 20 specimens were tested at each of three voltage levels: 52.5, 55.0, and 57.5 kilovolts. Failure times, in minutes, for the insulation specimens are given in Table 6.8.5.

a. Examine whether the data at each voltage level might have arisen from a Weibull distribution. It may be necessary to consider three-parameter distributions, since with the failure process involved here there is an initiation period during which failure does not normally occur. In this case examine the possibility that each of the three parameters might depend on voltage.
b. Does a Weibull model in which there is a constant threshold and shape parameter and a scale parameter related to voltage by a power law relationship $\alpha=cv^p$ appear plausible?

(Section 6.4.2; Stone, 1978)

Table 6.8.5 Failure Times for Epoxy Insulation Specimens at Three Voltage Levels[a]

Voltage (kV)	Failure times (min)
52.5	4690, 740, 1010, 1190, 2450, 1390, 350, 6095, 3000, 1458, 6200*, 550, 1690, 745, 1225, 1480, 245, 600, 246, 1805
55.0	258, 114, 312, 772, 498, 162, 444, 1464, 132, 1740*, 1266, 300, 2440*, 520, 1240, 2600*, 222, 144, 745, 396
57.5	510, 1000*, 252, 408, 528, 690, 900*, 714, 348, 546, 174, 696, 294, 234, 288, 444, 390, 168, 558, 288

[a]Asterisk denotes a censored observation.

6.10 Analyze the data in Problem 6.6 under a generalized log-gamma model [see (6.6.1) and (6.6.2)] for log lifetimes. Assess the Weibull and log-normal lifetime models within this framework.

(Section 6.6)

6.11 The logistic linear model. Consider the linear regression model (6.1.9) with $\mu(\mathbf{x})=\mathbf{x}\boldsymbol{\beta}$ where e has a logistic distribution with p.d.f.

$$f(e)=\frac{\exp e}{(1+\exp e)^2} \qquad -\infty<e<\infty.$$

Develop maximum likelihood methods for this model, based on a possibly censored sample of log lifetimes $y_1,\ldots,y_n$, corresponding to regression vectors $\mathbf{x}_1,\ldots,\mathbf{x}_n$. Analyze the data of Example 6.5.1 under this model and compare results with the results of the normal model analysis given in the example.

6.12

a. Verify the expressions for the entries in the information matrix for the general linear model (6.7.1) [see (6.7.5) and the expressions preceding it].
b. Determine A_e for the generalized log-gamma model given by (6.6.1), and thus obtain (6.7.7).

(Section 6.7.1; Cox and Hinkley, 1968)

6.13 Consider the extreme value regression model (6.4.2) with $\mu(\mathbf{x})=\alpha+\beta x$. Let $\hat{\alpha}$, $\hat{\beta}$, and $\hat{\sigma}$ be the m.l.e.'s from a random sample $y_1,\ldots,y_n$ corresponding to fixed $x_1,\ldots,x_n$. Obtain the p.d.f. of $Z_1=(\hat{\alpha}-\alpha)/\hat{\sigma}$, $Z_2=(\hat{\beta}-\beta)/\hat{\sigma}$, and $Z_3=\hat{\sigma}/\sigma$, given the ancillary statistics $a_i=(y_i-\hat{\alpha}-\hat{\beta}x_i)/\hat{\sigma}$

from (G9) of Appendix G. Show that z_1 can be integrated out of this density, to give the p.d.f. of Z_2 and Z_3, given **a**, as

$$k(\mathbf{a},\mathbf{x})z_3^{n-2}\exp\left(z_3\sum_{i=1}^{n}a_i\right)\Big/\left(\sum_{i=1}^{n}e^{a_iz_3+x_iz_2z_3}\right)^n \qquad -\infty<z_2<\infty \quad z_3>0.$$

Note that double numerical integration is required to evaluate probabilities for Z_2 or Z_3.

(Section 6.4)

CHAPTER 7

Distribution-Free Methods for the Proportional Hazards and Related Regression Models

7.1 INTRODUCTION

Models in which factors related to lifetime have a multiplicative effect on the hazard function play an important part in the analysis of lifetime data. These so-called proportional hazards (PH) models were introduced in Chapter 6, where certain parametric regression models were examined. In this chapter we discuss an ingenious distribution-free approach to the analysis of data under proportional hazards models, first suggested by Cox (1972a).

Let us recall the proportional hazards model introduced in Section 6.1. Let T be a continuous random variable representing an individual's lifetime and let $\mathbf{x}=(x_1,\ldots,x_p)$ be a known vector of regressor variables associated with the individual. Under the proportional hazards assumption the hazard function of T, given $\mathbf{x}$, is of the form $h(t|\mathbf{x})=h_0(t)g(\mathbf{x},\boldsymbol{\beta})$, where $\boldsymbol{\beta}$ is a vector of unknown parameters. Following Cox (1972a), we will focus on the particular model that has

$$h(t|\mathbf{x})=h_0(t)\exp(\mathbf{x}\boldsymbol{\beta}) \tag{7.1.1}$$

where $\boldsymbol{\beta}=(\beta_1,\ldots,\beta_p)'$ is a vector of regression coefficients. Most of the procedures discussed in this chapter can be applied to the more general model in which $\exp(\mathbf{x}\boldsymbol{\beta})$ is replaced with $g(\mathbf{x},\boldsymbol{\beta})$, but (7.1.1) is flexible enough for many purposes. In (7.1.1) $h_0(t)$ is a hazard function and is specifically the baseline hazard function for an individual with $\mathbf{x}=\mathbf{0}$.

If one assumes a particular form for $h_0(t)$, a fully parametric proportional hazards model is obtained. The most important such model is the Weibull model, for which $h_0(t)=\delta\alpha^{-1}(t/\alpha)^{\delta-1}$; this also includes the exponential model as the special case $\delta=1$. Analysis of data under these models was discussed in Chapter 6. The approach taken in this chapter is a distribution-free one where no specific form is assumed for $h_0(t)$. The main idea for this originated with Cox (1972a), who proposed a method of estimating $\boldsymbol{\beta}$ in the absence of knowledge of $h_0(t)$ and also of estimating $h_0(t)$. Many others have contributed to this area since Cox's original paper (e.g., Kalbfleisch and Prentice, 1973; Breslow, 1974; Cox, 1975; Efron, 1977), and this body of methodology now constitutes one of the main approaches to handling lifetime data.

An advantage of the methods described in this chapter is that they are essentially distribution-free: certain properties of the procedures do not depend upon the underlying lifetime distribution or, in other words, on $h_0(t)$. This is actually true only when there is no censoring, but with many types of censoring the dependence on $h_0(t)$ is small. If the data come from a specific proportional hazards model such as the Weibull model, there will be some loss of efficiency in using the distribution-free approach rather than the one based on the correct parametric model. In certain situations, however, this loss of efficiency is only slight, as we shall see.

Procedures given here for testing and estimating $\boldsymbol{\beta}$ are closely related to certain distribution-free rank tests. The connections are noted here; distribution-free methods are discussed from a more conventional point of view in Chapter 8. We begin by considering estimation of $\boldsymbol{\beta}$ in (7.1.1) in the absence of assumptions about $h_0(t)$ and then consider estimation of $h_0(t)$. Methods of dealing with grouped data are examined in Section 7.3. Section 7.4 concludes with brief discussions of efficiency properties of the procedures, a generalization of (7.1.1) to include time-dependent regressor variables, and the assessment of departures from the proportional hazards assumption.

7.2 STATISTICAL METHODS FOR THE CONTINUOUS PH MODEL

There are two unknown components in the model (7.1.1): the regression parameter $\boldsymbol{\beta}$ and the baseline hazard function $h_0(t)$. Equivalently, one can think in terms of $\boldsymbol{\beta}$ and the baseline survivor function

$$\begin{aligned} S_0(t) &= \exp\left(-\int_0^t h_0(u)\,du\right) \\ &= \exp[-H_0(t)] \end{aligned} \tag{7.2.1}$$

where $H_0(t)$ is the baseline cumulative hazard function. The survivor function for T, given $\mathbf{x}$, can be written as

$$S(t|\mathbf{x}) = \exp\left(-\int_0^t h(u|\mathbf{x})\, du \right)$$

$$= \left[S_0(t) \right]^{\exp(\mathbf{x}\boldsymbol{\beta})}. \tag{7.2.2}$$

We wish to estimate $\boldsymbol{\beta}$ and $S_0(t)$, or $h_0(t)$, from data that are possibly censored. One approach would be to attempt to maximize the likelihood function for the observed data simultaneously with respect to $\boldsymbol{\beta}$ and $S_0(t)$. A more attractive approach is that given by Cox (1972a), in which a likelihood function that does not depend upon $h_0(t)$ is obtained for $\boldsymbol{\beta}$. This can be maximized to give an estimate of $\boldsymbol{\beta}$ and to provide tests for $\boldsymbol{\beta}$ in the absence of knowledge of $h_0(t)$. Once $\boldsymbol{\beta}$ has been estimated, $S_0(t)$ can be estimated along lines similar to those leading to the product-limit estimate discussed in Section 2.3.1. This approach is taken here.

7.2.1 Estimation and Tests for $\boldsymbol{\beta}$

Suppose that a random sample of n individuals yields a sample with k distinct observed lifetimes and $n-k$ censoring times. The k observed lifetimes will be denoted by $t_{(1)} < \cdots < t_{(k)}$, and $R_i = R(t_{(i)})$ will be used to represent the risk set at time $t_{(i)}$, that is, the set of individuals alive and uncensored just prior to $t_{(i)}$. On somewhat heuristic grounds, Cox suggested the following "likelihood function" for estimating $\boldsymbol{\beta}$ in (7.1.1) in the absence of knowledge of $h_0(t)$:

$$L(\boldsymbol{\beta}) = \prod_{i=1}^{k} \left(e^{\mathbf{x}_{(i)}\boldsymbol{\beta}} \Big/ \sum_{l \in R_i} e^{\mathbf{x}_l \boldsymbol{\beta}} \right) \tag{7.2.3}$$

where $\mathbf{x}_{(i)}$ is the regression vector associated with the individual observed to die at $t_{(i)}$. The motivation for (7.2.3) is that given $R(t)$ and given that a death occurs at t, the probability of it being individual $i[i \in R(t)]$ who dies is

$$h(t|\mathbf{x}_i) \Big/ \sum_{l \in R(t)} h(t|\mathbf{x}_l) = e^{\mathbf{x}_i \boldsymbol{\beta}} \Big/ \sum_{l \in R(t)} e^{\mathbf{x}_l \boldsymbol{\beta}}.$$

The "likelihood" (7.2.3) is formed by taking the product of all such factors over the k observed lifetimes.

The function (7.2.3) does not depend on $h_0(t)$ and can be maximized to yield an estimate $\hat{\boldsymbol{\beta}}$. We have referred to $L(\boldsymbol{\beta})$ as a "likelihood," since it is not actually a likelihood function in the usual sense. That is, it cannot be derived as the probability of some observable outcome under the stated model. The validity of (7.2.3) has been discussed by several authors, including Cox (1975), who indicates that for purposes of inference about $\boldsymbol{\beta}$, $L(\boldsymbol{\beta})$ can, however, be treated as an ordinary likelihood function. In particular, under suitable conditions maximization of $L(\boldsymbol{\beta})$ leads to an estimate $\hat{\boldsymbol{\beta}}$ that is asymptotically normally distributed with a covariance matrix which can be consistently estimated using the usual matrix of second derivatives of $\log L(\boldsymbol{\beta})$. We shall return to the question of (7.2.3)'s validity later, but in the meantime we shall treat it as an ordinary likelihood function.

Data on continuous variables frequently include ties because of rounding off or grouping. If there are many ties, one should take explicit account of this by using a discrete model, perhaps one that is based on grouped observations from the continuous model. This is discussed for the proportional hazards model in Section 7.3. If there are relatively few ties, it is usually convenient to work with the continuous model (7.2.2). This requires some modification of the likelihood (7.2.3) or else requires ties to be broken in some way. A small number of ties in a univariate sample might be broken at random, but this is usually undesirable when there are regressor variables. Another suggestion that has been made is to replace $L(\boldsymbol{\beta})$ in (7.2.3) with

$$L(\boldsymbol{\beta})=\prod_{i=1}^{k} e^{\mathbf{S}_i\boldsymbol{\beta}}\Big/\left(\sum_{l\in R_i} e^{\mathbf{x}_l\boldsymbol{\beta}}\right)^{d_i} \tag{7.2.4}$$

where d_i is the number of lifetimes equal to $t_{(i)}$ and $\mathbf{S}_i$ is the sum of the regression vectors $\mathbf{x}$ for these d_i individuals. That is, if D_i represents the set of individuals who die at $t_{(i)}$, then $d_i=|D_i|$ and $\mathbf{S}_i=\Sigma_{l\in D_i}\mathbf{x}_l$. When there are no ties, all $d_i=1$ and (7.2.4) reduces to (7.2.3). Motivation for (7.2.4) in the presence of a small number of ties and alternate expressions for handling this are discussed in Section 7.2.3. Results below will be given in terms of (7.2.4); (7.2.3) is included as a special case.

The log likelihood arising from (7.2.4) is

$$\log L(\boldsymbol{\beta})=\sum_{i=1}^{k}\mathbf{S}_i\boldsymbol{\beta}-\sum_{i=1}^{k} d_i\log\left(\sum_{l\in R_i} e^{\mathbf{x}_l\boldsymbol{\beta}}\right) \tag{7.2.5}$$

and the first derivatives of $\log L$ are

$$\frac{\partial \log L}{\partial \beta_r}=\sum_{i=1}^{k}\left(S_{ir}-d_i\sum_{l\in R_i} x_{lr}e^{\mathbf{x}_l\boldsymbol{\beta}}\Big/\sum_{l\in R_i} e^{\mathbf{x}_l\boldsymbol{\beta}}\right)\qquad r=1,\dots,p. \tag{7.2.6}$$

In (7.2.6) S_{ir} is the rth component in $\mathbf{S}_i=(S_{i1},\ldots,S_{ip})$. The matrix $\mathbf{I}$ containing minus the second partial derivatives of $\log L(\boldsymbol{\beta})$ has entries

$$I_{rs}(\boldsymbol{\beta})=\frac{-\partial^2\log L}{\partial\beta_r\,\partial\beta_s}$$

$$=\sum_{i=1}^{k} d_i\left[\sum_{l\in R_i} x_{lr}x_{ls}e^{\mathbf{x}_l\boldsymbol{\beta}}\Big/\sum_{l\in R_i} e^{\mathbf{x}_l\boldsymbol{\beta}}\right.$$

$$\left.-\left(\sum_{l\in R_i} x_{lr}e^{\mathbf{x}_l\boldsymbol{\beta}}\right)\left(\sum_{l\in R_i} x_{ls}e^{\mathbf{x}_l\boldsymbol{\beta}}\right)\Big/\left(\sum_{l\in R_i} e^{\mathbf{x}_l\boldsymbol{\beta}}\right)^2\right]$$

$$r,s=1,\ldots,p. \qquad (7.2.7)$$

The maximum likelihood equations $\partial\log L/\partial\beta_r=0$ $(r=1,\ldots,p)$ can generally be solved without difficulty by using the Newton–Raphson method.

To make inferences about $\boldsymbol{\beta}$ one mainly has to rely on large-sample procedures. Computation of the expected value of $\mathbf{I}_{rs}(\boldsymbol{\beta})$ is impossible without detailed knowledge of the censoring mechanism, and even if this is known, the required expectations will be difficult, if not impossible, to obtain. The simplest approach is to treat $\hat{\boldsymbol{\beta}}$ as normally distributed with mean $\boldsymbol{\beta}$ and covariance matrix $\mathbf{I}(\hat{\boldsymbol{\beta}})^{-1}$. Inferences can also be based on likelihood ratio methods. A third possibility that is convenient in some situations is to base tests on the $p\times 1$ score vector $\mathbf{U}(\boldsymbol{\beta})=(\partial\log L/\partial\beta_r)$, which in large samples can be considered to be normally distributed with mean $\mathbf{0}$ and covariance matrix $\mathbf{I}(\boldsymbol{\beta})$.

The examination of precise conditions under which large-sample maximum likelihood procedures are valid for the likelihood (7.2.4) is difficult. This is briefly addressed in Section 7.2.3, but henceforth it is assumed in applying these methods that conditions necessary for their validity are satisfied.

When calculating the quantities in expressions (7.2.5) to (7.2.7), it is helpful to order the lifetimes and censoring times from largest to smallest and to use an indicator variable to signify whether observations are lifetimes or censoring times. When all lifetimes are distinct, for example, the ordering of times will look like

$$\{t^*_{(k),i}\},\, t_{(k)},\, \{t^*_{(k-1),i}\},\, t_{(k-1)},\ldots,\{t^*_{(1),i}\},\, t_{(1)},\, \{t^*_{(0),i}\}$$

where $\{t^*_{(j),i}\}$ is the set of censoring times in the interval $[t_{(j)},t_{(j+1)})$ and where for convenience we define $t_{(0)}=0$ and $t_{(k+1)}=\infty$. Since the risk set R_i at $t_{(i)}$ consists of the set R_{i+1} at $t_{(i+1)}$ plus the individuals who died or were censored in $[t_{(i)}, t_{(i+1)})$, terms in (7.2.5) to (7.2.7) can be computed recursively, starting with R_k and working back to R_1.

An example demonstrating these calculations is given in the next section, where the important problem of comparing lifetime distributions is treated as a special case of (7.2.2). A more complicated example involving the estimation of both $\boldsymbol{\beta}$ and the underlying survivor function $S_0(t)$ is given in Section 7.2.5.

7.2.2 Comparison of Two or More Life Distributions

An important application of the methods introduced in the preceding section is to the comparison of life distributions. To start, let us consider the comparison of two distributions with survivor functions $S_1(t)$ and $S_2(t)$. Specifically, a test will be developed for the hypothesis

$$H_0: S_1(t) = S_2(t)$$

that the two distributions are the same. This can be handled under the model (7.2.2) by treating observations from both distributions as coming from a single population and by defining a dummy regressor variable x that takes on the value 0 or 1 according to whether an observation comes from the first or second distribution. The hazard functions for the two distributions are then, from (7.1.1), $h_1(t) = h_0(t)$ and $h_2(t) = h_0(t)e^{\beta}$, and the two distributions are identical if and only if $\beta = 0$. This is equivalent to assuming that $S_1(t)$ and $S_2(t)$ are related by

$$S_2(t) = S_1(t)^{\exp\beta} \tag{7.2.8}$$

so that by testing that $\beta = 0$, one is testing the hypothesis

$$H_0: S_2(t) = S_1(t) \quad \text{vs.} \quad H_1: S_2(t) = S_1(t)^{\delta} \qquad \delta \neq 1 \tag{7.2.9}$$

where $\delta = \exp\beta$. The family of hypotheses represented by H_1 in (7.2.9) is sometimes referred to as the Lehmann family of alternatives; it specifies that the two distributions have proportional hazard functions. The test of H_0 obtained below should consequently be powerful against alternatives in H_1. This is discussed in Sections 7.4 and 8.2.

Let us examine the test of H_0 versus H_1 in detail. Suppose that observations are taken on N_1 individuals from the first population [that with survivor function $S_1(t)$] and on N_2 individuals from the second population [that with survivor function $S_2(t)$]. Some observations may be censored. To apply the results of Section 7.2.1 we regard the two samples together as a single sample from the model (7.2.2), where the dummy variable x signifies whether an observation is from that first ($x=0$) or second ($x=1$) population. Let $t_{(1)}<\cdots<t_{(k)}$ denote the k distinct observed lifetimes in the combined sample and let d_i represent the number of deaths at $t_{(i)}$. Further, let n_{1i} and n_{2i} be the numbers of individuals in the risk set R_i at $t_{(i)}$ who are from the first and second populations, respectively, and let d_{1i} and d_{2i} denote the number of deaths at $t_{(i)}$ among individuals from populations 1 and 2. In our previous notation $n_{1i}+n_{2i}=n_i$ and $d_{1i}+d_{2i}=d_i$. Under this setup the log likelihood (7.2.5) for β becomes

$$\log L(\beta)=r_2\beta-\sum_{i=1}^{k} d_i\log\left(n_{1i}+n_{2i}e^{\beta}\right) \tag{7.2.10}$$

where $r_2=\sum d_{2i}$ is the total number of deaths among individuals from population 2. The score function and the negative of the second derivative of $\log L(\beta)$ are

$$U(\beta)=\frac{\partial \log L}{\partial\beta}$$

$$=r_2-\sum_{i=1}^{k}\frac{d_i n_{2i}e^{\beta}}{n_{1i}+n_{2i}e^{\beta}} \tag{7.2.11}$$

and

$$I(\beta)=\frac{-\partial^2 \log L}{\partial\beta^2}$$

$$=\sum_{i=1}^{k}\frac{d_i n_{1i}n_{2i}e^{\beta}}{\left(n_{1i}+n_{2i}e^{\beta}\right)^2}. \tag{7.2.12}$$

The likelihood equation $U(\beta)=0$ is readily solved by iteration to get the m.l.e. $\hat{\beta}$. Inferences for β can be obtained by the likelihood ratio method, by treating $\hat{\beta}$ as being approximately normal with mean β and variance $I(\hat{\beta})^{-1}$, or by treating $U(\beta)$ as normally distributed with mean 0 and variance $I(\beta)$.

The score function provides a particularly simple test of the equality of $S_1(t)$ and $S_2(t)$ without the need to compute $\hat{\beta}$. In particular, under $H_0: \beta=0$ the statistic

$$Z=\frac{U(0)}{\left[I(0)\right]^{1/2}} \tag{7.2.13}$$

is approximately normally distributed with mean 0 and variance 1. Large absolute values of Z provide evidence against equality of the two distributions. From (7.2.11) and (7.2.12) the expressions for $U(0)$ and $I(0)$ are

$$U(0)=r_2-\sum_{i=1}^{k}\frac{d_i n_{2i}}{n_i}$$

$$I(0)=\sum_{i=1}^{k}\frac{d_i n_{1i} n_{2i}}{n_i^2}. \qquad (7.2.14)$$

Several comments about the test based on (7.2.13) are in order. First, one should remember that the likelihood function (7.2.4) on which the test is based is suitable if the data contain relatively few ties. When there are a substantial number of ties, a test that takes into account the discrete nature of the data should be used. This is discussed in Section 7.3.2; there it is shown that a test of the equality of two life distributions can be based on (7.2.13), with $U(0)$ still given by (7.2.14) but $I(0)$ in (7.2.14) replaced by

$$\sum_{i=1}^{k}\frac{d_i(n_i-d_i)n_{1i}n_{2i}}{n_i^2(n_i-1)}. \qquad (7.2.15)$$

The two expressions for $I(0)$ are the same when all d_i's equal unity. When only a very few d_i's are greater than unity, they also agree closely and either expression can reasonably be used in the denominator of (7.2.13). When there are many ties, (7.2.15) should be used.

A second comment is that $U(0)$ in (7.2.14) can be written as

$$U(0)=\sum_{i=1}^{k} d_{2i}-\frac{d_i n_{2i}}{n_i}.$$

Thus, if $d_i n_{2i}/n_i$ is thought of as the expected number of deaths in population 2 at $t_{(i)}$, $U(0)$ is just the sum of the differences in the observed and expected number of deaths over the k observed lifetimes. The test that uses (7.2.15) for $I(0)$ has been discussed from this point of view by Mantel (1966), who was the first to propose this procedure.

A third remark is that this test can also be viewed as a generalization to the case of censored data of a two-sample rank test proposed by Savage (1956). The test, sometimes called the exponential ordered scores test (Cox, 1964), is reexamined from this angle in Section 8.2. For now, we note that when there are no censoring and no ties in the data, $U(0)$ can be written in the form (see Problem 7.2 and Section 8.2)

$$U(0) = N_2 - \sum_{j \in S_2} e_{(j),N} \tag{7.2.16}$$

where the sum in (7.2.16) is over all observations j from the second population ($S_2 \equiv$ "sample two"), (j) is the rank of observation j in the combined sample of size $N = N_1 + N_2$, and

$$e_{l,N} = \sum_{i=1}^{l} \frac{1}{N-i+1}$$

is the expected value of the lth ordered observation in a random sample of size N from the standard exponential distribution. When there are no censoring and no ties, the exact variance of $U(0)$ can be calculated, and it is (see Problem 7.2 and Section 8.2)

$$\begin{aligned}\text{Var}[U(0)] &= E[I(0)] \\ &= \frac{N_1 N_2}{N(N-1)}(N - e_{N,N}).\end{aligned} \tag{7.2.17}$$

Example 7.2.1 The data below show remission times, in weeks, for leukemia patients given two types of treatments. In the study 20 patients were given treatment A and 20 treatment B. Starred observations are censoring times.

Treatment A	1, 3, 3, 6, 7, 7, 10, 12, 14, 15, 18, 19, 22, 26, 28*, 29, 34, 40, 48*, 49*
Treatment B	1, 1, 2, 2, 3, 4, 5, 8, 8, 9, 11, 12, 14, 16, 18, 21, 27*, 31, 38*, 44

Let us test the hypothesis that the remission time distributions are the same for patients on the two treatments. To carry out the test it is simply necessary to combine the samples and calculate the d_i's, n_{1i}'s, n_{2i}'s, and r_2. We note that there are 25 distinct remission times observed; the required quantities are given in Table 7.2.1 for each of these times.

Noting that $r_2 = 18$, we easily calculate that $U(0) = 3.323$ and $I(0) = 8.4087$ and thus that $U(0)^2/I(0) = 1.31$. The $\chi^2_{(1)}$ distribution gives a significance level of about .25, and so there is no evidence of a difference in distributions.

Table 7.2.1 Calculation of Two-Sample Test

$t_{(i)}$	d_i	n_{1i}	n_{2i}	$t_{(i)}$	d_i	n_{1i}	n_{2i}	$t_{(i)}$	d_i	n_{1i}	n_{2i}
1	3	20	20	10	1	14	10	21	1	8	5
2	2	19	18	11	1	13	10	22	1	8	4
3	3	19	16	12	2	13	9	26	1	7	4
4	1	17	15	14	2	12	8	29	1	5	3
5	1	17	14	15	1	11	7	31	1	4	3
6	1	17	13	16	1	10	7	34	1	4	2
7	2	16	13	18	2	10	6	40	1	3	1
8	2	14	13	19	1	9	5	44	1	2	1
9	1	14	11								

There are several ties in the data, and one might use (7.2.15) in place of $I(0)$; this gives 8.1962, which is close to $I(0)$. The test could also be carried out by using either the statistic $Z^2=\hat{\beta}^2/I^{-1}(\hat{\beta})$ or $\Lambda=-2\log[L(0)/L(\hat{\beta})]$, where $\log L(\beta)$ is given by (7.2.10). We find here that $\hat{\beta}=0.388$ and $I(\hat{\beta})=8.6191$, giving $Z^2=1.30$; we also find that $\log L(\hat{\beta})=-103.298$ and $\log L(0)=-103.945$, which gives $\Lambda=1.29$. In this case the three statistics are almost equal.

An alternate procedure with these data would be to assume that the two distributions are Weibull distributions with equal shape parameters. A plot of $\log[-\log \hat{S}(t)]$ versus $\log t$ for each sample, where $\hat{S}(t)$ is the product-limit estimate, gives roughly linear and parallel graphs and suggests this model. Equality of the two distributions then amounts to equality of the two Weibull scale parameters (see Section 4.3.2).

An m Sample Test

Tests of equality of three or more lifetime distributions can be devised in a similar manner to the two-sample test. To compare m distributions we define a vector of $m-1$ dummy regressor variables $\mathbf{x}=(x_1,\ldots,x_{m-1})$ as follows: individuals in populations $1,\ldots,m-1$ have vectors $\mathbf{x}=(1,0,\ldots,0),\ldots,(0,\ldots,0,1)$, respectively, and individuals in population m have $\mathbf{x}=(0,\ldots,0)$. We then assume that the survivor functions $S_i(t)$, $i=1,\ldots,m$, in the m populations are given by (7.2.2), so that

$$S_1(t)=S_0(t)^{\delta_1},\ldots,S_{m-1}(t)=S_0(t)^{\delta_{m-1}},S_m(t)=S_0(t) \quad (7.2.18)$$

where $\delta_i=\exp\beta_i$. To test the equality of the m distributions one tests that $\boldsymbol{\beta}=(\beta_1,\ldots,\beta_{m-1})=\mathbf{0}$. This test will be good at detecting departures from equality in which the m distributions have proportional hazard functions.

A scores test of $\boldsymbol{\beta}=\mathbf{0}$ gives simple results, as in the two-sample problem. Assume that observations are taken on N individuals, of whom N_r are from population r $(r=1,\dots,m)$. Some observations may be censored; in the combined sample of N let there be k distinct lifetimes denoted by $t_{(1)} < \cdots < t_{(k)}$. The total numer of individuals at risk at $t_{(i)}$ is n_i and the number of deaths is d_i; of the n_i at risk, n_{ri} are from population r, and of the d_i deaths, d_{ri} are from population r. The elements in the score vector and information matrix with $\boldsymbol{\beta}=\mathbf{0}$ can be written down from (7.2.6) and (7.2.7). These give

$$U_r(\mathbf{0}) = \left(\frac{\partial \log L}{\partial \beta_r} \right)_{\boldsymbol{\beta}=\mathbf{0}}$$

$$= \sum_{i=1}^{k} \left(d_{ri} - \frac{d_i n_{ri}}{n_i} \right) \qquad r=1,\dots,m-1 \tag{7.2.19}$$

$$I_{rs}(\mathbf{0}) = \sum_{i=1}^{k} d_i \frac{n_{ri}}{n_i} \left(\delta_{rs} - \frac{n_{si}}{n_i} \right) \qquad r,s=1,\dots,m-1 \tag{7.2.20}$$

where $\delta_{rs}=1$ $(r=s)$ or 0 $(r\neq s)$ is the Kronecker δ.

Under the hypothesis $\boldsymbol{\beta}=\mathbf{0}$, $\mathbf{U}=[U_1(\mathbf{0}),\dots,U_{m-1}(\mathbf{0})]'$ can be treated as approximately normal with mean $\mathbf{0}$ and covariance matrix $\mathbf{I}(\mathbf{0})$. A test of $\boldsymbol{\beta}=\mathbf{0}$ can be based on the statistic

$$X^2 = \mathbf{U}'\mathbf{I}(\mathbf{0})^{-1}\mathbf{U}. \tag{7.2.21}$$

Large values of X^2 provide evidence against equality of the m distributions; under H_0 it is approximately $\chi^2_{(m-1)}$.

When $m=2$, this test is that based on (7.2.13), except that the roles of populations 1 and 2 have been interchanged. As with (7.2.13), (7.2.21) is applicable when there are not too many ties in the data. If the number of ties is substantial, the methods of Section 7.3.2 should be used. There it is shown that a test can be based on a statistic of the form (7.2.21), with $U_r(\mathbf{0})$ given by (7.2.19) but with (7.2.20) replaced by

$$I_{rs}(\mathbf{0}) = \sum_{i=1}^{k} \frac{d_i(n_i-d_i)n_{ri}}{(n_i-1)n_i} \left(\delta_{rs} - \frac{n_{si}}{n_i} \right) \qquad r,s=1,\dots,m-1. \tag{7.2.22}$$

Like the two-sample test, the m-sample test has been discussed by several authors, from several points of view. The treatment here is closest to that of Cox (1972a). Mantel (1966) first proposed the tests, based on work of

Mantel and Haenszel (1959), and they are sometimes referred to as Mantel–Haenszel tests. The tests have also been considered by Peto and others under the name of "log rank tests" (e.g., Peto and Peto, 1972; Peto et al., 1977) and, when there is no censoring, as "exponential ordered scores tests" (Cox, 1964). In Section 8.2 these tests are reconsiderd in the framework of rank test procedures.

7.2.3 Justification of the Likelihood Function (7.2.3)*

It was noted in Section 7.2.1 that the "likelihood" function (7.2.3) used to estimate $\boldsymbol{\beta}$ is not a likelihood function in the ordinary sense, but justifications of it have been presented by several authors. The legitimacy of (7.2.3) can only be discussed, of course, under specific assumptions about the lifetime and censoring time processes and the structure of the regressor variables, but it appears that under fairly general conditions it is reasonable to treat it as an "ordinary" likelihood function. In particular, the estimate $\hat{\boldsymbol{\beta}}$ obtained by maximizing (7.2.3) will be consistent and in large samples can be treated as approximately normally distributed with mean $\boldsymbol{\beta}$ and covariance matrix $\mathbf{I}(\hat{\boldsymbol{\beta}})^{-1}$. Concomitantly, likelihood ratio statistics based on $\log L(\boldsymbol{\beta})$ can be treated as approximately χ^2 in large samples. A detailed examination of these problems is beyond the scope of this book and, indeed, many questions are far from answered at present. We shall sketch some arguments leading to (7.2.3) and justifying its use. More detailed and rigorous treatments of these problems can be found in Aalen (1978), Kalbfleisch and Prentice (1980, Chs. 4 and 5), and in other references mentioned below.

Justification as a Marginal Likelihood

When there is no censoring, (7.2.3) can be derived as a marginal likelihood function based on the rank statistic for the data, as shown by Kalbfleisch and Prentice (1973). Specifically, suppose that lifetimes $t_1, \dots, t_n$ of n individuals with regression vectors $\mathbf{x}_1, \dots, \mathbf{x}_n$ are observed, with no censoring present. The hazard and survivor functions of t_i, given $\mathbf{x}_i$, are given by (7.1.1) and (7.2.2), and the p.d.f. of t_i, given $\mathbf{x}_i$, is correspondingly

$$f(t_i|\mathbf{x}_i) = h_0(t)e^{\mathbf{x}_i\boldsymbol{\beta}}\exp\left[-H_0(t_i)e^{\mathbf{x}_i\boldsymbol{\beta}}\right] \qquad i=1,\dots,n \qquad (7.2.23)$$

where $H_0(t) = \int_0^t h_0(u)\,du$ is the underlying cumulative hazard function. Let $\mathbf{r} = [(1), \dots, (n)]$ denote the rank statistic for the data; that is, (i) is the label of the individual with the ith smallest lifetime. The distribution of $\mathbf{r}$ is discrete: note that there are $n!$ different rank vectors possible. We may ignore the possibility of ties, since they have probability 0 of occurring under a continuous model. The probability function for $\mathbf{r}$ is found as

$$\Pr\{\mathbf{r}=[(1),\ldots,(n)]\}=\Pr[t_{(1)}<t_{(2)}<\cdots<t_{(n)}]$$

$$=\int_0^\infty\int_{t_{(1)}}^\infty\cdots\int_{t_{(n-1)}}^\infty f(t_{(1)}|\mathbf{x}_{(1)})\cdots f(t_{(n)}|\mathbf{x}_{(n)})\,dt_{(n)}\cdots dt_{(1)}$$

$$=\int_0^\infty\int_{t_{(1)}}^\infty\cdots\int_{t_{(n-1)}}^\infty\prod_{i=1}^n\left\{h_0(t_{(i)})e^{\mathbf{x}_{(i)}\boldsymbol{\beta}}\exp\left[-H_0(t_{(i)})e^{\mathbf{x}_{(i)}\boldsymbol{\beta}}\right]\right\}dt_{(n)}\cdots dt_{(1)}.$$

Straightforward integration of this expression yields

$$\Pr\{\mathbf{r}=[(1),\ldots,(n)]\}=\prod_{i=1}^n\left(e^{\mathbf{x}_{(i)}\boldsymbol{\beta}}\Big/\sum_{l\in R(t_{(i)})}e^{\mathbf{x}_l\boldsymbol{\beta}}\right)\qquad(7.2.24)$$

which is the likelihood function (7.2.3). In obtaining this result we have made use of the fact that $R(t_{(i)})=[(i),(i+1),\ldots,(n)]$, since there is no censoring.

In the simple noncensored case, therefore, (7.2.3) is a legitimate likelihood function arising from the probability distribution of the rank statistic. Under suitable assumptions concerning the $\mathbf{x}_i$'s such as those given in Hajek and Sidak (1967), $L(\boldsymbol{\beta})$ behaves in the usual way, with $\hat{\boldsymbol{\beta}}$ being asymptotically normally distributed with mean $\boldsymbol{\beta}$ and covariance matrix $\mathbf{I}^{-1}$, where $\mathbf{I}$ has entries $I_{rs}=E(-\partial^2\log L/\partial\beta_r\,\partial\beta_s)$. $\mathbf{I}$ is consistently estimated by $\mathbf{I}(\hat{\boldsymbol{\beta}})$ as given by (7.2.7). It can also be noted that $L(\boldsymbol{\beta})$ is a marginal likelihood in the sense of Fraser (1968) or Kalbfleisch and Sprott (1970). This follows from the fact that under the group of differentiable and strictly monotone increasing transformations on t, $\mathbf{r}$ and $\boldsymbol{\beta}$ are invariant, whereas the transformation is transitive on $t_1,\ldots,t_n$ and $h_0(t)$ (Kalbfleisch and Prentice, 1973). From this point of view $L(\boldsymbol{\beta})$ is uniquely appropriate for making inferences about $\boldsymbol{\beta}$ with $h_0(t)$ unknown and completely arbitrary.

A Partial Likelihood Justification

If the data are subject to Type II censoring, an extension of the preceding argument shows that $L(\boldsymbol{\beta})$ is once again a legitimate marginal likelihood function based on a rank statistic. For more general types of censoring the argument breaks down, however. One approach (Kalbfleisch and Prentice, 1973) to justifying (7.2.3) in general situations is to consider the data as supplying only partial information on the rank statistic, and to consider the probability of the set of possible rank vectors. A more flexible approach is one based on the concept of partial likelihood (see Appendix E) as presented by Cox (1975). Partial likelihood applied to censored data problems has been discussed by Cox (1975), Efron (1977), Kalbfleisch and MacKay (1978), Kalbfleisch and Prentice (1980), and others. We shall merely indicate the nature of this approach; for a more detailed and rigorous treatment see the references just cited and Aalen (1978).

Consider the full likelihood function for a set of censored data on n individuals when the lifetime distribution of an individual with regression vector $\mathbf{x}_i$ is given by (7.2.23). Under conditions like those stated in Section 1.4.1, the likelihood function can be taken to be

$$L[\boldsymbol{\beta}, h_0(t)] = \prod_{i=1}^{n} \left[h_0(t_i) e^{\mathbf{x}_i\boldsymbol{\beta}} S_0(t_i)^{\exp(\mathbf{x}_i\boldsymbol{\beta})}\right]^{\delta_i} \left[S_0(t_i)^{\exp(\mathbf{x}_i\boldsymbol{\beta})}\right]^{1-\delta_i}$$

where t_i is an observed lifetime or censoring time for the ith individual and δ_i is the usual indicator variable taking on the value 1 if t_i is a lifetime, and 0 if it is a censoring time. As earlier, the individuals who die are labeled $(1),\ldots,(k)$, with $t_{(1)} < \cdots < t_{(k)}$ being the k distinct observed lifetimes. This likelihood can be rewritten as

$$\begin{aligned} L[\boldsymbol{\beta}, h_0(t)] &= \prod_{i=1}^{n} S_0(t_i)^{e^{\mathbf{x}_i\boldsymbol{\beta}}} \prod_{i\in D} h_0(t_i) e^{\mathbf{x}_i\boldsymbol{\beta}} \\ &= \prod_{i\in D} \left(e^{\mathbf{x}_i\boldsymbol{\beta}} \Big/ \sum_{l\in R_i} e^{\mathbf{x}_l\boldsymbol{\beta}} \right) \prod_{i\in D} \left(h_0(t_i) \sum_{l\in R_i} e^{\mathbf{x}_l\boldsymbol{\beta}} \right) \prod_{i=1}^{n} S_0(t_i)^{e^{\mathbf{x}_i\boldsymbol{\beta}}} \end{aligned} \tag{7.2.25}$$

where D denotes the set of individuals who are observed to die and R_i is the set of individuals alive just prior to the observed lifetime t_i $(i\in D)$. It is assumed, in conjunction with the continuous model, that all lifetimes are distinct and that they are distinct from any censoring times.

The likelihood (7.2.3) suggested by Cox (1972a) is the first term in (7.2.25), the remaining portion of the likelihood function being ignored. To better understand the information being disregarded we can rewrite (7.2.25) using an approach taken by Efron (1977). This involves considering the process of death and censoring, starting from time 0. Let $R(t)$ denote the risk set at time t, that is, the set of individuals alive and uncensored just prior to t. Define the "overall" hazard function at time t as

$$h(t) = \sum_{l\in R(t)} h_l(t)$$

where $h_l(t)=h(t|\mathbf{x}_l)=h_0(t)e^{\mathbf{x}_l\boldsymbol{\beta}}$ is the hazard function for individual l. Note that $h(t)$ is a random variable, since $R(t)$ is. With this notation (7.2.25) can be rewritten as

$$L[\boldsymbol{\beta}, h_0(t)]=\left[\prod_{i=1}^{k}\left(e^{\mathbf{x}_{(i)}\boldsymbol{\beta}}\Big/\sum_{l\in R(t_{(i)})} e^{\mathbf{x}_l\boldsymbol{\beta}}\right)h(t_{(i)})\exp\left(-\int_{t_{(i-1)}}^{t_{(i)}} h(u)\,du\right)\right]$$

$$\times\exp\left(-\int_{t_{(k)}}^{\infty} h(u)\,du\right). \tag{7.2.26}$$

To write down (7.2.26) we have noted that

$$\prod_{i=1}^{n} S_0(t_i)^{e^{\mathbf{x}_i\boldsymbol{\beta}}}=\exp\left(-\sum_{i=1}^{n}\int_0^{t_i} e^{\mathbf{x}_i\boldsymbol{\beta}}h_0(u)\,du\right)$$

$$=\exp\left(-\sum_{i=1}^{k}\int_{t_{(i-1)}}^{t_{(i)}}\sum_{l\in R(t)} h_0(u)e^{\mathbf{x}_l\boldsymbol{\beta}}\,du-\int_{t_{(k)}}^{\infty} h(u)\,du\right)$$

$$=\exp\left(-\sum_{i=1}^{k}\int_{t_{(i-1)}}^{t_{(i)}} h(u)\,du-\int_{t_{(k)}}^{\infty} h(u)\,du\right)$$

where $t_{(0)}=0$. The first terms in the product in (7.2.26) give the partial likelihood (7.2.3), and the individual terms are recognizable as the probabilities $\Pr[(i)|R(t_{(i)}), t_{(i)}]$. The second term gives the conditional probabilities of a death at $t_{(i)}$, with no deaths in $(t_{(i-1)}, t_{(i)})$, by standard Poisson process arguments and incorporates information involving the censoring times. The last term gives the probability of there being no further deaths beyond $t_{(k)}$. The partial likelihood

$$L_1(\boldsymbol{\beta})=\prod_{i=1}^{k}\left(e^{\mathbf{x}_{(i)}\boldsymbol{\beta}}\Big/\sum_{l\in R(t_{(i)})} e^{\mathbf{x}_l\boldsymbol{\beta}}\right)$$

is not a likelihood function in the usual sense in that it is not determined from the probability of any observable event but is the product of terms in the factored likelihood (7.2.26).

It appears that maximum partial likelihood estimates (m.p.l.e.'s) obtained by maximizing $L_1(\boldsymbol{\beta})$ possess the usual asymptotic properties of ordinary m.l.e.'s under quite broad conditions. Cox (1975) and Kalbfleisch and MacKay (1978) give heuristic treatments that attempt to place only mild conditions on the censoring and lifetime processes. Aalen (1978) gives some relevant results as part of a general treatment of counting processes. Tsiatis (1978a) and Liu and Crowley (1978) demonstrate under models involving random independent censoring mechanisms that the m.p.l.e. is consistent and asymptotically normal and that likelihood ratio tests based on $L_1(\boldsymbol{\beta})$ are valid. Earlier, Crowley (1974) studied the partial likelihood for the special case of the m-sample problem under the random independent censoring model. The main problem that needs further study is the precise conditions on the regressor variables and the censoring mechanism necessary for various asymptotic results to hold. For example, Crowley (1974) shows under the random independent censorship model that the score statistic (7.2.21) is asymptotically χ^2 under the null hypothesis of equal distributions, provided, however, that the censoring time distributions are the same in the different populations. On the other hand, certain results appear to be valid under fairly weak conditions (e.g., Kalbfleisch and MacKay, 1978). Further work is needed to clarify the picture.

Other Approaches

Several authors have provided other justifications for the Cox likelihood function (e.g., Breslow, 1974; Holford, 1976). These are mostly heuristic and will not be discussed here, but they do provide motivation for the likelihood. It should also be mentioned that $\boldsymbol{\beta}$ can be estimated by direct maximum likelihood instead of through the partial likelihood. This is less convenient than the partial likelihood approach, since it requires simultaneous consideration of $\boldsymbol{\beta}$ and $h_0(t)$ (see Problem 7.4).

Adjustments for Ties

In the discussion above the possibility of ties in the data was disregarded, since these have probability zero under a continuous model. In practice ties frequently occur, however. If there is a substantial number of ties, the discrete nature of the lifetimes should be explicitly recognized, and methods for handling discrete or grouped data should be used. These are discussed in Section 7.3. On the other hand, if there are only a few ties, it is often convenient to work with the continuous model, making some adjustment to handle the ties.

Cox (1972a) suggested replacing $e^{\mathbf{x}_{(i)}\boldsymbol{\beta}}/\sum_{l\in R_i} e^{\mathbf{x}_l\boldsymbol{\beta}}$ in (7.2.3) with the expression

$$e^{\mathbf{S}_i\boldsymbol{\beta}}\Big/\sum_{\text{all } D_j} e^{\mathbf{S}_j\boldsymbol{\beta}} \tag{7.2.27}$$

when there are $d_i > 1$ deaths at $t_{(i)}$. Here $\mathbf{S}_i$ is the sum of the regression vectors $\mathbf{x}$ for the d_i individuals observed to die at $t_{(i)}$, the sum in the denominator of (7.2.27) is over all possible d_i-subsets D_j of R_i, and $\mathbf{S}_j = \Sigma_{l \in D_j} \mathbf{x}_l$ is the sum of the regression vectors for individuals in D_j. This expression is intuitively appealing and arises formally in a grouped data model introduced by Cox and discussed in Section 7.3. Another reasonable modification is one suggested by Peto (1972a); in this case the position taken is that although d_i deaths are observed at $t_{(i)}$, the times of death are actually distinct, and thus the general term in (7.2.3) should be replaced with a sum of probabilities

$$\sum \Pr(l_1 \text{ died, then } l_2 \text{ died}, \ldots, \text{then } l_{d_i} \text{ died})$$

$$= \sum \left[\left(e^{\mathbf{x}_{l_1}\boldsymbol{\beta}} \Big/ \sum_{l \in R_i} e^{\mathbf{x}_l \boldsymbol{\beta}} \right) \times \left(e^{\mathbf{x}_{l_2}\boldsymbol{\beta}} \Big/ \sum_{l \in R_i - \{l_1\}} e^{\mathbf{x}_l \boldsymbol{\beta}} \right) \cdots \left(e^{\mathbf{x}_{l_{d_i}}\boldsymbol{\beta}} \Big/ \sum_{j \in R_i - \{l_1, \ldots, l_{d_i - 1}\}} e^{\mathbf{x}_l \boldsymbol{\beta}} \right) \right] \tag{7.2.28}$$

where the sum ranges over all of the ordered d_i-subsets of R_i.

If d_i is even as large as 2, both (7.2.27) and (7.2.28) are laborious to compute when $|R_i|$ is large. When $d_i/|R_i|$ is small, however, (7.2.27) and (7.2.28) do not differ much from each other nor from

$$e^{\mathbf{S}_i \boldsymbol{\beta}} \Big/ \binom{n_i}{d_i} \left(\frac{1}{n_i} \sum_{l \in R_i} e^{\mathbf{x}_l \boldsymbol{\beta}} \right)^{d_i}. \tag{7.2.29}$$

If the continuous model is retained only when there are relatively few ties in the data, this expression is the most convenient to use, since it requires little calculation; (7.2.4) is based on this (with the proportionality constant $\binom{n_i}{d_i} / n_i^{d_i}$ dropped). This likelihood has also been proposed by Breslow (1974) on somewhat different grounds. Farewell and Prentice (1980) discuss its accuracy.

7.2.4 Estimation of the Survivor Function

By (7.2.2), the survivor function for an individual with covariate vector $\mathbf{x}$ is

$$S(t|\mathbf{x}) = S_0(t)^{\exp(\mathbf{x}\boldsymbol{\beta})}$$

where $S_0(t)$ is the baseline survivor function of an individual with $\mathbf{x}=\mathbf{0}$. It is clearly of interest to estimate $S_0(t)$, since this would give estimates of $S(t|\mathbf{x})$ for any $\mathbf{x}$. One approach is to attempt to jointly maximize the full likelihood function (7.2.25) for $\boldsymbol{\beta}$ and $S_0(t)$ (see Problem 7.4). A slightly simpler approach, and the one taken here, is to estimate $\boldsymbol{\beta}$ from the partial likelihood function (7.2.4) and then to maximize the full likelihood for $S_0(t)$, assuming that $\boldsymbol{\beta}$ is equal to the m.p.l.e. $\hat{\boldsymbol{\beta}}$ obtained from (7.2.4). This involves arguments similar to those used in Section 2.3.1 to derive the product-limit estimate and produces what is essentially a nonparametric m.l.e. of $S_0(t)$. The approach was introduced by Kalbfleisch and Prentice (1973).

Define, for convenience, $t_{(0)}=0$, $t_{(k+1)}=\infty$ and assume, as usual, that $t_{(1)}<\cdots<t_{(k)}$ are the observed lifetimes in the sample. There are n_i at risk and d_i deaths at $t_{(i)}$, and in the interval $[t_{(i-1)}, t_{(i)})$ there are, say, λ_i censoring times, which we will denote by $L_j^{(i)}$ ($j=1,\ldots,\lambda_i$), in keeping with the notation of Section 2.3.1. According to the same argument used for (2.3.5), the observed likelihood function is of the form

$$L=\prod_{i=1}^{k}\left[\prod_{j=1}^{\lambda_i} S\left(L_j^{(i)}|\mathbf{x}_j^{(i)}\right)\prod_{j\in D_i}\left[S\left(t_{(j)}|\mathbf{x}_{(j)}\right)-S\left(t_{(j)}+0|\mathbf{x}_{(j)}\right)\right]\right]$$

$$\times\prod_{j=1}^{\lambda_{k+1}} S\left(L_j^{(k+1)}|\mathbf{x}_j^{(k+1)}\right)$$

where $\mathbf{x}_j^{(i)}$ denotes the regressor associated with the individual censored at $L_j^{(i)}$. We want to maximize L with respect to $S_0(t)$, where $S(t|\mathbf{x})=S_0(t)^{\exp(\mathbf{x}\boldsymbol{\beta})}$ and $\boldsymbol{\beta}$ is assumed to be known. Since $S_0(t)$ is a survivor function, it is nonincreasing and left continuous, and thus, exactly as in Section 2.3.1, it follows that $\hat{S}_0(t)$ must be constant except for jumps at the observed lifetimes $t_{(1)},\ldots,t_{(k)}$. Further, it follows that

$$\hat{S}_0\left(t_{(1)}\right)=\hat{S}_0\left(L_j^{(1)}\right)=1 \qquad j=1,\ldots,\lambda_1$$

and

$$\hat{S}_0\left(t_{(i)}+0\right)=\hat{S}_0\left(t_{(i+1)}\right)=\hat{S}_0\left(L_j^{(i+1)}\right) \qquad i=1,\ldots,k \quad j=1,\ldots,\lambda_{i+1}.$$

Writing $S_0(t_{(i)}+0)=P_i$ ($i=1,\ldots,k$), we are required to maximize

$$L_1=\prod_{i=1}^{k}\left(\prod_{l\in D_i}\left(P_{i-1}^{\exp(\mathbf{x}_l\boldsymbol{\beta})}-P_i^{\exp(\mathbf{x}_l\boldsymbol{\beta})}\right)\prod_{l\in C_i}P_{i-1}^{\exp(\mathbf{x}_l\boldsymbol{\beta})}\right)\prod_{l\in C_{k+1}}P_k^{\exp(\mathbf{x}_l\boldsymbol{\beta})}$$

where D_i is the set of individuals dying at $t_{(i)}$ and C_i is the set of individuals with censoring times in $[t_{(i-1)}, t_{(i)})$. If we let $\alpha_i = P_i / P_{i-1}$ $(i=1,\ldots,k)$, L_1 can be written as

$$L_1 = \prod_{i=1}^{k} \left(\prod_{l \in D_i} \left(1 - \alpha_i^{\exp(\mathbf{x}_l \boldsymbol{\beta})}\right) \right) \prod_{i=1}^{k+1} \left(\prod_{l \in D_i \cup C_i} (\alpha_1 \cdots \alpha_{i-1})^{\exp(\mathbf{x}_l \boldsymbol{\beta})} \right)$$

$$= \prod_{i=1}^{k} \left(\prod_{l \in D_i} \left(1 - \alpha_i^{\exp(\mathbf{x}_l \boldsymbol{\beta})}\right) \prod_{l \in R_i - D_i} \alpha_i^{\exp(\mathbf{x}_l \boldsymbol{\beta})} \right).$$

Differentiating $\log L_1$ with respect to $\alpha_1, \ldots, \alpha_k$, we get equations

$$\frac{\partial \log L_1}{\partial \alpha_i} = - \sum_{l \in D_i} \frac{\exp(\mathbf{x}_l \boldsymbol{\beta}) \alpha_i^{\exp(\mathbf{x}_l \boldsymbol{\beta}) - 1}}{1 - \alpha_i^{\exp(\mathbf{x}_l \boldsymbol{\beta})}} + \sum_{l \in R_i - D_i} \frac{\exp(\mathbf{x}_l \boldsymbol{\beta})}{\alpha_i} = 0, \; i = 1, \ldots, k.$$

These can be rearranged to give

$$\sum_{l \in D_i} \frac{\exp(\mathbf{x}_l \boldsymbol{\beta})}{1 - \alpha_i^{\exp(\mathbf{x}_l \boldsymbol{\beta})}} = \sum_{l \in R_i} \exp(\mathbf{x}_l \boldsymbol{\beta}) \qquad i = 1, \ldots, k. \tag{7.2.30}$$

When $d_i = |D_i| = 1$, (7.2.30) has a solution given by

$$\hat{\alpha}_i^{\exp(\mathbf{x}_{(i)} \boldsymbol{\beta})} = 1 - \exp(\mathbf{x}_{(i)} \boldsymbol{\beta}) \Big/ \sum_{l \in R_i} \exp(\mathbf{x}_l \boldsymbol{\beta}). \tag{7.2.31}$$

When $d_i > 1$, (7.2.30) must be solved iteratively for $\hat{\alpha}_i$. A good initial approximation to $\hat{\alpha}_i$ can be obtained by noting that when α_i is close to unity, $\alpha_i^{\exp(\mathbf{x}_l \boldsymbol{\beta})} \doteq 1 + (\log \alpha_i) \exp(\mathbf{x}_l \boldsymbol{\beta})$. Substituting this into (7.2.30), we get

$$\alpha_i = \exp\left(\frac{-d_i}{\sum_{l \in R_i} \exp(\mathbf{x}_l \boldsymbol{\beta})} \right). \tag{7.2.32}$$

The m.l.e. of P_i is given by $\hat{P}_i = \hat{\alpha}_1 \cdots \hat{\alpha}_i$.

The procedure to be used in estimating the underlying survivor function $S_0(t)$ is therefore to determine $\hat{\boldsymbol{\beta}}$ from the partial likelihood (7.2.4) and to then estimate $S_0(t)$ via (7.2.30), this being taken as the given value of $\boldsymbol{\beta}$. From the results above, the estimate of $S_0(t)$ is then

$$\hat{S}_0(t) = \prod_{i : t_{(i)} < t} \hat{\alpha}_i. \tag{7.2.33}$$

The estimated survivor function for an individual with covariate vector $\mathbf{x}$ is $\hat{S}(t|\mathbf{x})=\hat{S}_0(t)^{\exp(\mathbf{x}\hat{\boldsymbol{\beta}})}$. A minor qualification is that, as in the case of the product-limit estimate, if the largest observation is a censoring time L^*, (i.e., $\lambda_{k+1}>0$), then $\hat{S}_0(t)$ is undefined past L^*, since L does not depend on $S_0(t)$ for $t>L^*$, whereas $\hat{S}_0(L^*)>0$.

The estimate of $S_0(t)$ presented is easily calculated when there are no ties in the data; if there are ties, some iterative calculation is required. Breslow (1974) suggests an estimate that is not much different numerically in most cases from (7.2.33) but that does not require iteration when d_i's are greater than unity. His estimate is

$$\tilde{H}_0(t)=-\log \tilde{S}_0(t)$$

$$=\sum_{i:\,t_{(i)}<t}\left(d_i \Big/ \sum_{l\in R_i} e^{\mathbf{x}_l\hat{\boldsymbol{\beta}}}\right). \tag{7.2.34}$$

Since $-\log(1-u)\doteq u$ for small u, the estimates of $S_0(t)$ given by (7.2.33) and (7.2.34) will not differ much when all d_i's are unity, except in the right-hand tail of the distribution. They will also not differ greatly when only a few d_i's exceed unity; note that if $\hat{\alpha}_i$ were given exactly by (7.2.32), then $\hat{S}_0(t)$ would be the same as $\tilde{S}_0(t)$.

$\hat{S}_0(t)$ and $\tilde{S}_0(t)$ correspond to the product-limit and empirical cumulative hazard function estimates of the survivor function (see Section 2.3). In particular, when $\hat{\boldsymbol{\beta}}=\mathbf{0}$, $\hat{S}_0(t)$ reduces to the PL estimate (2.3.2) and $\tilde{H}_0(t)$ reduces to the empirical cumulative hazard function (2.3.13).

Variance estimation for $\hat{S}_0(t)$ or $\tilde{S}_0(t)$ will not be examined here. Kalbfleisch and Prentice (1980, Ch. 4) obtain a few results about the asymptotic distribution of $\hat{S}_0(t)$ when $\boldsymbol{\beta}$ is known; these are analogous to the results in Section 2.3.2, concerning asymptotic properties of the PL estimate. Tsiatis (1978a, 1978b) gives a more detailed discussion of the asymptotic properties of (7.2.34) under a random censorship model. He suggests an estimate of the variance of $\tilde{H}_0(t)$, which then yields variance estimates for $\tilde{S}_0(t)$ and also for $\hat{S}_0(t)$, which is asymptotically equivalent.

Example 7.2.2 (Example 7.2.1 revisited) Let us consider the situation in Example 7.2.1, where $\mathbf{x}$ was a single dummy variable ($x=0,1$), so that the survivor function of T, given $\mathbf{x}$, was

$$S(t|x)=\begin{cases} S_0(t) & x=0 \\ S_0(t)^{e^{\beta}} & x=1. \end{cases}$$

Table 7.2.2 Estimation of the Baseline Survivor Function in Example 7.2.2

$t_{(i)}$	$\hat{\alpha}_i$	$\hat{S}_0(t_{(i)}+0)$	$t_{(i)}$	$\hat{\alpha}_i$	$\hat{S}_0(t_{(i)}+0)$	$t_{(i)}$	$\hat{\alpha}_i$	$\hat{S}_0(t_{(i)}+0)$
1	.9388	.939	10	.9652	.637	21	.9339	.368
2	.9556	.898	11	.9636	.614	22	.9280	.341
3	.9291	.834	12	.9231	.567	26	.9225	.315
4	.9743	.813	14	.9151	.519	29	.8939	.281
5	.9733	.791	15	.9531	.495	31	.8776	.247
6	.9723	.770	16	.9502	.470	34	.8561	.211
7	.9431	.726	18	.8924	.419	40	.7765	.164
8	.9388	.682	19	.9389	.394	44	.6876	.113
9	.9666	.659						

One can estimate $S_0(t)$ using either of the procedures outlined above. For (7.2.33), for example, one needs to obtain the $\hat{\alpha}_i$'s. This is easily done: note that when $d_i > 1$, (7.2.30) becomes

$$\frac{d_{1i}}{1-\alpha_i} + \frac{d_{2i}e^{\beta}}{1-\alpha_i^{e^{\beta}}} = n_{1i} + n_{2i}e^{\beta}$$

and when $d_i = 1$, (7.2.31) gives

$$\alpha_i^{\exp(x_i\beta)} = 1 - \frac{e^{x_i\beta}}{n_{1i} + n_{2i}e^{x_i\beta}}$$

where $x_i = 0$ or 1 according to whether the death at t_i is from population 1 or population 2. With $\beta = 0.388$, the estimate obtained from the partial likelihood, the $\hat{\alpha}_i$'s and $\hat{S}_0(t)$ are as given in Table 7.2.2.

The estimate $\hat{S}_0(t)$ gives estimates of the survivor functions for the two life distributions, under the assumption that they have proportional hazard functions. The estimates for populations 1 and 2 are $\hat{S}_0(t)$ and $\hat{S}_0(t)^{\exp\hat{\beta}} = \hat{S}_0(t)^{1.474}$, respectively.

7.2.5 Data Analysis Using the Proportional Hazards Model

Distribution-free methods for the proportional hazards model provide a flexible approach to the investigation of lifetime data when regressor variables are thought to affect the hazard function in a multiplicative manner. The methods described in this section are useful both in the preliminary assessment of various factors and in testing and estimation, once a specific model has been adopted. If the data suggest serious departures from proportional hazards assumptions, preliminary analysis as suggested here may indicate the nature of these departures. A distribution-free analysis may also suggest that a particular parametric family such as the Weibull family might be adopted.

We shall consider some points related to the use of the model (7.1.1) in situations where there may be several regressor variables. Other good discussions of various aspects of regression analysis under (7.1.1) are given, for example, by Breslow (1975), Kay (1977), and Kalbfleisch and Prentice (1980, Ch. 4). Armitage and Gehan (1974) make a number of useful general comments on regression methods.

In the initial stages of analysis one has to decide what regressor variables are to be considered further, and roughly how these affect lifetimes. The same ideas apply here as were discussed in Section 6.2.1, which dealt with parametric regression models. Once variables have been labeled for investigation, a main problem is to assess whether particular variables act upon the hazard function in a manner consonant with (7.1.1). If a regressor variable x_1 takes on only two values x_{11} and x_{12}, a convenient way to examine its effect is to split the data into two parts, according to the values of the variable, and to fit separate proportional hazards models to each. That is, models with hazard functions of the form

$$h_i(t|\mathbf{x}^-)=h_{0i}(t)e^{\mathbf{x}^-\boldsymbol{\beta}^-} \qquad i=1,2 \tag{7.2.35}$$

are fitted to the parts of the data with $x_1=x_{11}$ and $x_1=x_{12}$, respectively. Here, $\mathbf{x}^-$ denotes the regression vector $(x_2,\dots,x_p)$ with x_1 missing, and $\boldsymbol{\beta}^-=(\beta_2,\dots,\beta_p)'$. Under the model (7.1.1) we have

$$\begin{aligned} h_i(t|\mathbf{x}^-)&=h(t|x_{1i},\mathbf{x}^-)\\ &=h_0(t)e^{x_1\beta_1}e^{\mathbf{x}^-\boldsymbol{\beta}^-} \end{aligned}$$

and thus if (7.1.1) is appropriate, the $h_{0i}(t)$'s in (7.2.35) are proportional. Thus, when the model (7.2.35) is fitted, plots of

$$\log\left[-\log\hat{S}_{0i}(t)\right] \quad \text{vs.} \quad t \qquad i=1,2$$

should be roughly parallel, where

$$S_{0i}(t)=\exp\left(-\int_0^t h_{0i}(u)\,du\right)$$

is the baseline survivor function for the individuals with $x_1 = x_{1i}$. Departures from parallelism suggest a possible nonmultiplicative effect for x_1 on the hazard function. In addition, if $\boldsymbol{\beta}^-$ is estimated separately for each of $i=1,2$, the estimates should not differ substantially unless there is an interaction effect involving x_1 and other variables.

Variables that take on several different values can be examined in a similar way by splitting the data into more groups. For regressor variables taking on many values, one can form a few groups by a sensible partitioning of the values of the variable. It is also sometimes useful to partition the data according to joint values of two or more variables.

Stratification

When a factor does not affect the hazard multiplicatively, stratification may be useful in model building. Suppose that individuals can be assigned to one of s different strata, defined in terms of one or more factors. Sometimes it is reasonable to assume that the hazard function for an individual in stratum j with regressor variable $\mathbf{x}$ is

$$h_j(t|\mathbf{x}) = h_{0j}(t)e^{\mathbf{x}\boldsymbol{\beta}}. \tag{7.2.36}$$

That is, individuals in the same stratum have proportional hazard functions, but this is not necessarily the case for individuals in different strata. In (7.2.36) it is also assumed that the relative effect of the regressor variables is the same in each stratum; this condition sometimes needs to be relaxed, with $\boldsymbol{\beta}$ varying from stratum to stratum.

There is no difficulty in handling the model (7.2.36): a partial likelihood function $L_j(\boldsymbol{\beta})$ of the form (7.2.3) is obtained for each stratum, and then the overall partial likelihood function for $\boldsymbol{\beta}$ is $L(\boldsymbol{\beta}) = L_1(\boldsymbol{\beta}) \dots L_s(\boldsymbol{\beta})$. After $\boldsymbol{\beta}$ is estimated by maximizing this, the underlying survivor functions

$$S_{0j}(t) = \exp\left(-\int_0^t h_{0j}(u)\,du\right) \qquad j=1,\dots,s$$

can be estimated as described in Section 7.2.4. Tests and interval estimates for $\boldsymbol{\beta}$ can be based on the usual large-sample procedures for $L(\boldsymbol{\beta})$ and $\hat{\boldsymbol{\beta}}$.

Residual Analysis

One of the most useful ways of assessing models that have been fitted to the data is by examining residuals. The formation and use of residuals in parametric models was discussed in Section 6.2.2, and similar procedures can be followed here. The simplest way to define residuals for the model

(7.1.1) is to use (6.2.11). The residual corresponding to an uncensored lifetime is then

$$\begin{aligned} \hat{e}_i &= \hat{H}(t_i|\mathbf{x}_i) \\ &= \hat{H}_0(t_i)e^{\mathbf{x}_i\hat{\boldsymbol{\beta}}} \\ &= \left[-\log \hat{S}_0(t_i)\right]e^{\mathbf{x}_i\hat{\boldsymbol{\beta}}} \end{aligned} \tag{7.2.37}$$

where $H_0(t)$ is the baseline cumulative hazard function and $\hat{H}_0(t) = -\log \hat{S}_0(t)$ is an estimate of it, obtained as described in Section 7.2.4. The estimate $\hat{\boldsymbol{\beta}}$ is the m.p.l.e. of Section 7.2.1. Since quantities $H(t_i|\mathbf{x}_i)$ are independent and have standard exponential distributions, the $\hat{e}_i$'s, if there is no censoring, should look roughly like a random sample from the standard exponential distribution. When there are censored observations, one of the approaches discussed in Section 6.2.2 can be used. For example, residuals $\hat{e}_i$ for censoring times t_i, if defined as in (7.2.37), can be treated as censored standard exponential observations. One can then form a product-limit survivor function estimate from the set of censored and uncensored residuals. The resulting estimate $\hat{S}(e)$ should be consonant with an underlying standard exponential distribution; for example, a plot of $\log \hat{S}(e)$ versus e should be roughly linear with slope -1.

Alternately, residuals for a censored observation t_i^* can be defined as 1 plus the value given by (7.2.37),

$$\hat{e}_i = \left[-\log \hat{S}_0(t_i^*)\right]e^{\mathbf{x}_i\hat{\boldsymbol{\beta}}} + 1.$$

Then the full set of residuals corresponding to censored and uncensored observations can be treated roughly as a random sample from the standard exponential distribution. This approach is especially convenient if one wishes to plot residuals against specific regressor variables or other factors of some kind.

If a satisfactory proportional hazards model (7.1.1) is established, one can also investigate whether some simple parametric proportional hazards model might be appropriate by examining the estimated baseline survivor function $\hat{S}_0(t)$. For example, a roughly linear plot of $\log[-\log \hat{S}_0(t)]$ versus $\log t$ suggests a Weibull model with $h_0(t)$ of the form $\lambda\alpha t^{\alpha-1}$. If a particular parametric model is adopted, inferences and model checks can be carried out as described in Chapter 6.

In many instances a parametric model may be found to fit the data adequately. In this case one has the option of carrying out tests and other inferences about $\boldsymbol{\beta}$ using either the distribution-free procedures of this chapter or procedures based on the parametric model. Each approach has its advantages. One useful feature of a fully parametric model, when it can safely be assumed, is that continuous estimates of the underlying survivor function $S_0(t)$ and hazard function $h_0(t)$ automatically result. On the other hand, distribution-free procedures enjoy superior robustness properties, in many instances without sacrificing much in terms of efficiency. The efficiency of the methods presented in this section is considered in Section 7.4.1.

Example 7.2.3 (Example 6.3.1 revisited) In Example 6.3.1 some survival data for lung cancer patients were examined under an exponential model; the data were also examined under a Weibull model in Example 6.4.2. The data, given in Table 6.3.1, involve 40 patients, 21 of whom were given one chemotherapy treatment (standard), and 19 another (test). Concomitant variables included performance status, age, the number of months since diagnosis of cancer, and, in addition, tumors were classified into four types.

Let us analyze these data under the proportional hazards model (7.1.1). Employing the regressor variables $x_1,\ldots,x_7$ defined in Example 6.3.1, we consider the model for which the hazard function for survival time, given $\mathbf{x}=(x_1,\ldots,x_7)$, is

$$h(t|\mathbf{x})=h_0(t)\exp\left(\beta_1(x_1-\bar{x}_1)+\beta_2(x_2-\bar{x}_2)+\beta_3(x_3-\bar{x}_3)+\sum_{i=4}^{7}\beta_i x_i\right). \tag{7.2.38}$$

Following the treatment in Example 6.3.1, we have centered the regressor variables x_1, x_2, and x_3 about their means.

There are two ties in the data and, for convenience, these were randomly broken, with the result that $t=8$ in the "standard, squamous" group was treated as infinitesimally smaller than $t=8$ in the "standard, adeno" group and $t=12$ in the "standard, adeno" group was considered infinitesimally smaller than $t=12$ in the "standard, large" group. The maximum likelihood estimates from (7.2.4) are readily found by solving $\partial\log L/\partial\beta_r=0$, $r=1,\ldots,7$ [see (7.2.6)] by Newton–Raphson iteration. It is desirable to have good initial estimates; these can be obtained from the exponential model fitted in Example 6.3.1, noting that $\boldsymbol{\beta}$ there corresponds to $-\boldsymbol{\beta}$ in (7.2.38). Satisfactory initial estimates can also be found by estimating $\boldsymbol{\beta}$ in Example 6.3.1 by least squares and by using this as an estimate of $-\boldsymbol{\beta}$ in (7.2.38).

The m.l.e.'s from the partial likelihood (7.2.4), along with their estimated standard deviations obtained from the observed information matrix $\mathbf{I}(\hat{\boldsymbol{\beta}})^{-1}$

[see (7.2.7) for the entries in $\mathbf{I}(\boldsymbol{\beta})$], are as follows:

$$\hat{\beta}_1 = -0.0596\ (0.014) \qquad \hat{\beta}_2 = -0.0115\ (0.021) \qquad \hat{\beta}_3 = 0.0008\ (0.012)$$

$$\hat{\beta}_4 = -0.3259\ (0.485) \qquad \hat{\beta}_5 = -0.0290\ (0.506) \qquad \hat{\beta}_6 = 1.068\ (0.630)$$

$$\hat{\beta}_7 = 0.4118\ (0.408)$$

These results are in close agreement with those obtained under the exponential model in Example 6.3.1. Performance status (x_1) appears to be an important factor, but age (x_2) and the number of months since diagnosis (x_3) do not. Treatment and tumor type do not appear to have sizeable effects, though there is a suggestion that the adeno tumor type is important.

Formal significance tests can be based on likelihood ratio methods or the large-sample normal approximation $\hat{\boldsymbol{\beta}} \sim N(\boldsymbol{\beta}, \mathbf{I}(\hat{\boldsymbol{\beta}})^{-1})$. The latter alternative requires much less computation and tends to give results that are in good agreement with the likelihood ratio method when there are moderately many observations. Table 7.2.3 shows likelihood ratio statistic values Λ for tests of various submodels against the full model (7.2.38). In each case $\Lambda = -2\log[L(\tilde{\boldsymbol{\beta}})/L(\hat{\boldsymbol{\beta}})]$, where $L(\boldsymbol{\beta})$ is the partial likelihood (7.2.4), $\hat{\boldsymbol{\beta}}$ is the m.p.l.e. under the full model and $\tilde{\boldsymbol{\beta}}$ is the m.p.l.e. under the submodel being tested. The table also shows statistic values Λ_1 obtained from the normal approximation $\hat{\boldsymbol{\beta}} \sim N(\boldsymbol{\beta}, \mathbf{I}(\hat{\boldsymbol{\beta}})^{-1})$. The values of Λ_1 for β_1, β_2, β_3, and β_7 are $\hat{\beta}_i^2 / I(\hat{\boldsymbol{\beta}})^{ii}$ $(i = 1,2,3,7)$, where the (i,i) entry $I(\hat{\boldsymbol{\beta}})^{ii}$ in $\mathbf{I}(\hat{\boldsymbol{\beta}})^{-1}$ is the estimated variance of $\hat{\beta}_i$. In line 5, Λ_1 is $(\hat{\beta}_4 \hat{\beta}_5 \hat{\beta}_6)\mathbf{C}^{-1}(\hat{\beta}_4 \hat{\beta}_5 \hat{\beta}_6)'$, where $\mathbf{C}$ is the estimated covariance matrix for $(\hat{\beta}_4 \hat{\beta}_5 \hat{\beta}_6)$, taken from $\mathbf{I}(\hat{\boldsymbol{\beta}})^{-1}$. In each case the statistic has an approximate χ^2 distribution under the given submodel.

There is close agreement between the values of Λ and Λ_1. There is also close agreement between the values of the test statistics here and those obtained under the analogous exponential model in Example 6.3.1. This is

Table 7.2.3 Tests of Submodels for Lung Cancer Data

Model	$\log L$	Λ	d.f.	Λ_1
Full model $(\beta_1, \ldots, \beta_7)$	-87.677			
$\beta_1 = 0$	-97.549	19.74	1	18.65
$\beta_2 = 0$	-87.831	0.308	1	0.300
$\beta_3 = 0$	-87.679	0.004	1	0.004
$\beta_4 = \beta_5 = \beta_6 = 0$	-89.899	4.444	3	5.087
$\beta_7 = 0$	-88.197	1.040	1	1.019

expected, since the exponential model appears to give a satisfactory fit to the data, and is a special case of the general proportional hazards model that we are employing here. Other questions, such as the possibility of interaction effects, can be examined in a manner analogous to that in Example 6.3.1. Results are once again in close agreement with those obtained under the exponential model.

To conclude this example let us consider how we can use the proportional hazards model to examine the possibility of a fully parametric model. Once $\hat{\boldsymbol{\beta}}$ is obtained we can estimate the underlying survivor function $S_0(t)=S(t|\mathbf{x}=\mathbf{0})$ as described in Section 7.2.4. This involves solving (7.2.31) for the 37 uncensored lifetimes and getting $\hat{S}_0(t)$ from (7.2.33). The resulting estimate is given in Table 7.2.4. Specific parametric forms for $h_0(t)$ can be examined by looking at $\hat{S}_0(t)$. For example, a plot of $\log[-\log \hat{S}_0(t)]$ versus $\log t$ can be used to ascertain whether a Weibull model would be appropriate: the plot should be roughly linear if this is the case. Figure 7.2.1 shows such a plot and there is indeed a suggestion that a Weibull model might be reasonable. The slope of a straight line through the points in the plot is in the neighborhood of 1, and this suggests that an exponential model may actually be appropriate. Alternately, one can plot $\log[\hat{S}_0(t)]$ versus t to assess whether an exponential model is feasible, expecting a roughly linear plot if this is the case.

Example 7.2.4 ("quick" analysis using stratification to adjust for concomitant variables) Sometimes it is possible to carry out a "quick" analysis of

Table 7.2.4 Estimate of Baseline Survivor Function $S_0(t)$

$t_{(i)}$	$\hat{S}_0(t_{(i)}+0)$	$t_{(i)}$	$\hat{S}_0(t_{(i)}+0)$	$t_{(i)}$	$\hat{S}_0(t_{(i)}+0)$
1	.987	21	.781	164	.465
2	.974	43	.763	177	.431
8−	.961	44	.744	200	.395
8+	.947	51	.725	201	.358
10	.933	54	.705	231	.320
11	.918	56	.684	250	.278
12−	.902	82	.663	287	.225
12+	.885	84	.640	340	.162
15	.869	90	.616	411	.097
16	.852	100	.589	991	.037
18	.835	118	.559	999	0
19	.817	126	.529		
20	.799	153	.498		

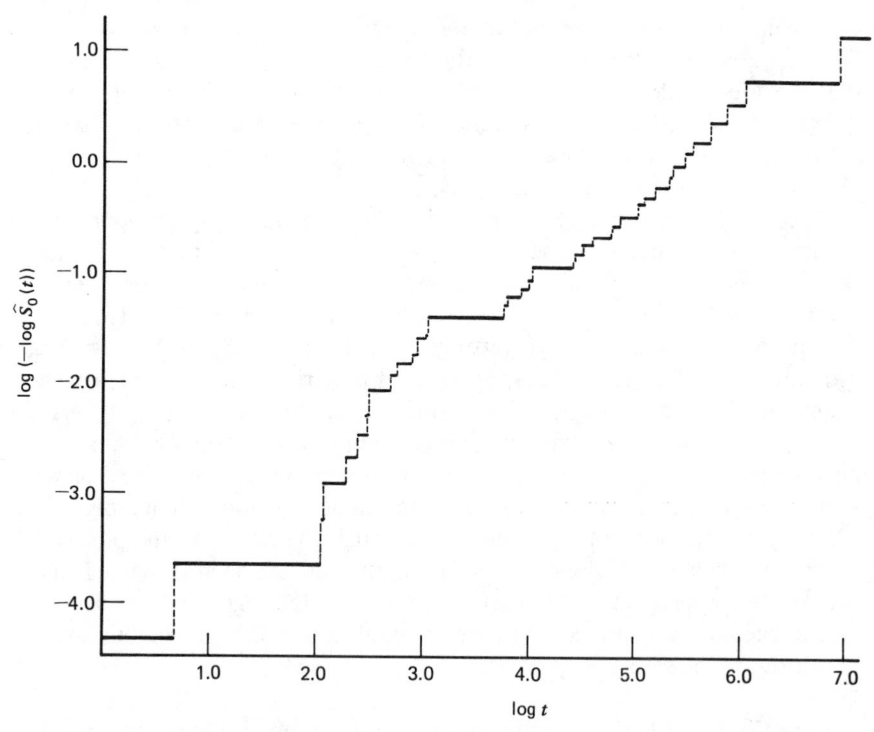

Figure 7.2.1 Plot of transformed baseline survivor function estimate in PH model (Example 7.2.3)

data like that in Example 7.2.3 by using stratification to adjust for concomitant variables. This can also be useful in presenting results following a more complicated analysis. Suppose that individuals are grouped into s strata on the basis of certain concomitant variables and that, in addition, individuals can come from one of m populations. Aside from this there are no other concomitant variables under consideration. Assume that the model (7.2.36) applies, where individuals in stratum j have hazard function

$$h_j(t|\mathbf{x})=h_{0j}(t)\exp(\mathbf{x}\boldsymbol{\beta}) \qquad j=1,\ldots,s \tag{7.2.39}$$

where, as in Section 7.2.2, $\mathbf{x}$ is a $1\times(m-1)$ regression vector used to indicate which of the m populations an individual is from. To test for absence of a "population" effect, we consider $H_0:\boldsymbol{\beta}=\mathbf{0}$.

Let $\mathbf{U}_j$ and $\mathbf{I}_j$ be the score vector and information matrix used for testing $\boldsymbol{\beta}=\mathbf{0}$ within stratum j; $\mathbf{U}_j$ and $\mathbf{I}_j$ are given by (7.2.19) and (7.2.20). Under

(7.2.39) it is easily shown that the score statistic for testing $H_0: \boldsymbol{\beta}=\mathbf{0}$ is

$$X^2 = \left(\sum_{j=1}^{s} \mathbf{U}_j \right)' \left(\sum_{j=1}^{s} \mathbf{I}_j \right)^{-1} \left(\sum_{j=1}^{s} \mathbf{U}_j \right). \qquad (7.2.40)$$

As an example of this approach let us consider the lung cancer data of Example 7.2.3 and test for a treatment effect after stratifying for concomitant variables. Among the concomitant variables it appears that performance status is the only one with a sizeable effect, and to simplify the illustration we shall stratify only on this variable. We arbitrarily form three strata consisting of individuals whose performance status rating is in the range 10 to 30, 40 to 60, and 70 to 90, respectively. There are two populations of interest, consisting of the individuals given the standard and test treatments, respectively; $\mathbf{x}$ is a single indicator variable taking on the value 0 for individuals given the standard treatment and 1 for individuals given the test treatment. For each stratum j we compute $U_j = U_j(0)$ and $I_j = I_j(0)$; U_j values calculated according to (7.2.14) and I_j values calculated according to (7.2.15) are given in Table 7.2.5. To test for absence of a treatment effect we compute $U = \Sigma U_j = 1.71$ and $I = \Sigma I_j = 5.8468$, which by (7.2.40) gives the value $X^2 = U^2/I = .50$. This is clearly nonsignificant on $\chi^2_{(1)}$ and provides no evidence of a difference in the two treatments. This test supposes that treatment effects are as given by (7.2.39), wherein the treatments have the same relative effect on the hazard within each stratum. If we suspect a possibly different effect within different strata, we can examine the individual score statistics U_j^2/I_j for each stratum. In this case these have observed values 1.59, 1.54, and 2.56, respectively, none of which is significant at the .10 level on $\chi^2_{(1)}$.

This type of analysis provides an easy method of testing for a treatment effect (or, in general, for a "population" effect) while adjusting for important concomitant variables. It is, however, directed at testing and does not permit the flexibility in modeling and estimation that is obtainable with a full regression analysis. There is also the problem of deciding what variables to stratify on in any given situation. Nevertheless, the approach is

Table 7.2.5 Score Statistics From Stratified Sample

Stratum (Performance Status)		
10–30	40–60	70–90
$U_1 = 1.02$	$U_2 = -1.95$	$U_3 = 2.64$
$I_1 = 0.6497$	$I_2 = 2.4707$	$I_3 = 2.7264$

a handy one in many instances. An extended discussion of it, with examples, is given by Peto et al. (1976, 1977).

7.3 REGRESSION METHODS FOR GROUPED DATA

The life table methods presented in Chapter 2 are extremely useful for displaying univariate lifetime data and estimating survival probabilities. In this section similar distribution-free methods are considered for problems in which regressor variables are present. These greatly broaden the scope of life table analysis.

7.3.1 Life Table Models with Regressor Variables

The situation we consider here is basically that of Section 2.2, except that regressor variables are present. Observations are taken on n individuals, with a lifetime t and a $1 \times p$ vector $\mathbf{x}$ of regressor variables associated with each individual. The exact regression vector $\mathbf{x}$ is known for each individual, but lifetimes are grouped. In particular, it is assumed, as in Section 2.2, that time is partitioned into $k+1$ intervals $I_j = [a_{j-1}, a_j)$, $j = 1, \ldots, k+1$, where $a_0 = 0$, $a_k = T$, and $a_{k+1} = \infty$, and that we know only in what interval an individual died or was censored.

Define the quantities

$$P_i(\mathbf{x}) = \Pr(\text{an individual survives past } a_i | \mathbf{x})$$

$$p_i(\mathbf{x}) = \frac{P_i(\mathbf{x})}{P_{i-1}(\mathbf{x})} \tag{7.3.1}$$

$$= \Pr(\text{an individual survives past } a_i | \text{he survives past } a_{i-1}, \mathbf{x}).$$

In addition, $P_0(\mathbf{x})$ is assumed equal to unity for all $\mathbf{x}$. Just as in ordinary life table work, there is a need for explicit assumptions regarding censoring times. For the time being we will assume that all censoring takes place at the ends of the intervals; modifications will be introduced later for occasions where this assumption is unreasonable. Let R_i be the risk set at time a_{i-1}, D_i the set of individuals observed to die in $I_i = [a_{i-1}, a_i)$, and C_i the set of individuals censored in I_i. Under assumptions like those in Section 1.4.1, the likelihood function is then

$$\prod_{i=1}^{k} \left(\prod_{l \in D_i} [1 - p_i(\mathbf{x}_l)] \prod_{l \in R_i - D_i} p_i(\mathbf{x}_l) \right) \tag{7.3.2}$$

remembering that all censoring in I_i is assumed to take place just prior to a_i.

We want to develop distribution-free procedures like those presented in the previous two sections for ungrouped data, and two approaches will be discussed. A third approach is discussed by Pierce et al. (1979). The models to be considered are as follows.

A Model Based on Grouping Data From the Model (7.2.2)
In many instances the lifetime T of an individual with regressor variable $\mathbf{x}$ might be assumed to come from a continuous proportional hazards model (7.2.2); this was the basis for the methods of Section 7.2. If lifetimes are grouped into classes in the way described here, then a "grouped" regression model is obtained for which

$$\begin{aligned} P_i(\mathbf{x}) &= S(a_i|\mathbf{x}) \\ &= S_0(a_i)^{\exp(\mathbf{x}\boldsymbol{\beta})} = P_i^{\exp(\mathbf{x}\boldsymbol{\beta})} \end{aligned} \qquad (7.3.3)$$

where $S(t|\mathbf{x}) = S_0(t)^{\exp(\mathbf{x}\boldsymbol{\beta})}$ is the survivor function of T, given $\mathbf{x}$, and

$$P_i = P_i(0) \qquad i = 1, \ldots, k$$

with $P_0 = 0$. This gives

$$\begin{aligned} p_i(\mathbf{x}) &= \frac{P_i(\mathbf{x})}{P_{i-1}(\mathbf{x})} \\ &= p_i^{\exp(\mathbf{x}\boldsymbol{\beta})} \end{aligned} \qquad (7.3.4)$$

where

$$p_i = \frac{P_i}{P_{i-1}} \qquad i = 1, \ldots, k.$$

Use of this model produces a likelihood (7.3.2) that can be used to estimate $\boldsymbol{\beta}$ and $p_1, \ldots, p_k$. This approach was introduced by Kalbfleisch and Prentice (1973).

A Logistic Model
Model (7.3.3) is appealing because it is based on the continuous proportional hazards model, which is the main topic of discussion in this chapter. We consider here a second model that is useful for analyzing grouped data, though it cannot be obtained by grouping from a continuous proportional hazards model. This model takes $p_i(\mathbf{x})$ of the form

$$p_i(\mathbf{x}) = \left(1 + \gamma_i e^{\mathbf{x}\boldsymbol{\beta}}\right)^{-1} \qquad i = 1, \ldots, k \qquad (7.3.5)$$

where $\gamma_1, \ldots, \gamma_k$ are given by

$$\gamma_i = \frac{1-p_i(\mathbf{0})}{p_i(\mathbf{0})}$$

$$= \frac{1-p_i}{p_i}.$$

This is a logistic model; note that

$$\log\left(\frac{1-p_i(\mathbf{x})}{p_i(\mathbf{x})}\right) = \log \gamma_i + \mathbf{x}\boldsymbol{\beta}.$$

This model is flexible and convenient, as logistic models are in many other discrete data situations (e.g., Cox, 1970). It can be noted that as the number of intervals increases and interval lengths decrease, (7.3.5) and (7.3.4) agree more and more closely. In the limit as interval lengths approach 0 it is easily seen that

$$\frac{1-p_i^{\exp(\mathbf{x}\boldsymbol{\beta})}}{p_i^{\exp(\mathbf{x}\boldsymbol{\beta})}} \sim \frac{1-p_i}{p_i} e^{\mathbf{x}\boldsymbol{\beta}}$$

provided that the p_i's approach unity as interval lengths decrease. This model was proposed by Cox (1972a) and developed further by Thompson (1977).

Maximum likelihood estimation under either (7.3.4) or (7.3.5) is basically straightforward, and will be examined for the two models in turn. A major application of the models is in providing tests for the equality of life distributions based on grouped data; this is discussed in Section 7.3.2.

The Grouped PH Model (7.3.4)

Maximum likelihood estimation for (7.3.4) is discussed by Prentice and Gloeckler (1978). For (7.3.2), with $p_i(\mathbf{x})$ given by (7.3.4), the log likelihood function for the parameters $\boldsymbol{\beta}=(\beta_1, \ldots, \beta_p)$ and $\mathbf{p}=(p_1, \ldots, p_k)$ can be written as

$$\log L(\boldsymbol{\beta}, \mathbf{p}) = \sum_{i=1}^{k} \left[\sum_{l \in D_i} \log\left(\frac{1-p_i^{\exp(\mathbf{x}_l\boldsymbol{\beta})}}{p_i^{\exp(\mathbf{x}_l\boldsymbol{\beta})}}\right) + \sum_{l \in R_i} \log\left(p_i^{\exp(\mathbf{x}_l\boldsymbol{\beta})}\right) \right].$$

The p_i's are restricted to lie between 0 and 1 and it is advantageous to reparameterize; Prentice and Gloeckler suggest using the parameters $\gamma_i = \log(-\log p_i)$, $i=1,\ldots,k$. The γ_i's are unrestricted, and Prentice and Gloeckler found that convergence in a Newton–Raphson iteration procedure was improved when $\boldsymbol{\gamma}=(\gamma_1,\ldots,\gamma_k)$ was used in place of $\mathbf{p}$. If the approximate normality of the m.l.e.'s in large samples is used to obtain approximate confidence intervals or tests, it is preferable to develop these in terms of $\boldsymbol{\gamma}$ too.

In terms of $\boldsymbol{\beta}$ and $\boldsymbol{\gamma}$ the log likelihood is

$$\log L(\boldsymbol{\beta},\boldsymbol{\gamma}) = \sum_{i=1}^{k}\left(\sum_{l\in D_i}\log\left[e^{\exp(\gamma_i+\mathbf{x}_l\boldsymbol{\beta})}-1\right] - \sum_{l\in R_i} e^{\gamma_i+\mathbf{x}_l\boldsymbol{\beta}}\right). \quad (7.3.6)$$

To write down the required derivatives of $\log L$ we define

$$z_{il} = e^{\gamma_i+\mathbf{x}_l\boldsymbol{\beta}} \qquad i=1,\ldots,k \quad l=1,\ldots,n.$$

Then

$$\frac{\partial \log L}{\partial \beta_r} = \sum_{i=1}^{k}\left(\sum_{l\in D_i}\frac{x_{lr}z_{il}}{1-e^{-z_{il}}} - \sum_{l\in R_i} x_{lr}z_{il}\right) \qquad r=1,\ldots,p \quad (7.3.7)$$

$$\frac{\partial \log L}{\partial \gamma_i} = \sum_{l\in D_i}\frac{z_{il}}{1-e^{-z_{il}}} - \sum_{l\in R_i} z_{il} \qquad i=1,\ldots,k. \quad (7.3.8)$$

The second derivatives of $\log L$ are

$$\frac{\partial^2 \log L}{\partial \beta_r\,\partial \beta_s} = \sum_{i=1}^{k}\left(\sum_{l\in D_i}\frac{x_{lr}x_{ls}z_{il}(1-e^{-z_{il}}-z_{il}e^{-z_{il}})}{(1-e^{-z_{il}})^2} - \sum_{l\in R_i} x_{lr}x_{ls}z_{il}\right)$$

$$r,s=1,\ldots,p \quad (7.3.9)$$

$$\frac{\partial^2 \log L}{\partial \beta_r\,\partial \gamma_i} = \sum_{l\in D_i}\frac{x_{lr}z_{il}(1-e^{-z_{il}}-z_{il}e^{-z_{il}})}{(1-e^{-z_{il}})^2} - \sum_{l\in R_i} x_{lr}z_{il}$$

$$r=1,\ldots,p \quad i=1,\ldots,k \quad (7.3.10)$$

$$\frac{\partial^2 \log L}{\partial \gamma_i^2} = \sum_{l\in D_i}\frac{z_{il}(1-e^{-z_{il}}-z_{il}e^{-z_{il}})}{(1-e^{-z_{il}})^2} - \sum_{l\in R_i} z_{il} \qquad i=1,\ldots,k \quad (7.3.11)$$

$$\frac{\partial^2 \log L}{\partial \gamma_i\,\partial \gamma_j} = 0 \qquad i\neq j.$$

The maximum likelihood equations $\partial \log L/\partial\beta_r = 0$ $(r=1,\ldots,p)$, $\partial \log L/\partial\gamma_i = 0$ $(i=1,\ldots,k)$ can usually be solved without any trouble by Newton–Raphson iteration. The dimension of the matrix of second partial derivatives is $(p+k)\times(p+k)$, which will often be large. However, the fact that $\partial^2 \log L/\partial\gamma_i\,\partial\gamma_j = 0$ $(i\neq j)$ enables a reduction in the computation required to invert this: well-known expressions for the inverse of a partitioned matrix (e.g., Rao, 1965, p. 29) allow the inverse of the second derivative matrix for $(\boldsymbol{\beta},\boldsymbol{\gamma})$ to be determined by inverting only the $p\times p$ submatrix corresponding to $\boldsymbol{\beta}$ (see Prentice and Gloeckler, 1978, p. 60). Another way to iterate is to work alternately with (7.3.7) and (7.3.8): start with initial estimates of $\boldsymbol{\gamma}$ and $\boldsymbol{\beta}$ and then iterate using (7.3.7) to obtain a new $\boldsymbol{\beta}$ value. Then, with this $\boldsymbol{\beta}$, iterate using (7.3.8) to obtain a new $\boldsymbol{\gamma}$, and so on. Iteration using the Newton–Raphson method will require matrix inversion only when working with equations (7.3.7), since (7.3.8) gives separate equations for $\gamma_1,\ldots,\gamma_k$.

When $\boldsymbol{\beta}=\mathbf{0}$, equations (7.3.8) give estimates

$$\hat{p}_i = e^{-\exp(\hat{\gamma}_i)} = \frac{n_i - d_i}{n_i}$$

where $d_i = |D_i|$ and $n_i = |R_i|$ are the number of deaths and the number at risk in the ith interval, respectively. These are the life table estimates of p_i discussed in Section 2.2 when censoring is assumed to be at the ends of intervals. In the present situation these can be used to give initial estimates of $\gamma_1,\ldots,\gamma_p$ for the maximum likelihood iteration procedure.

Prentice and Gloeckler (1978) give a detailed account of the use of these procedures in a large set of breast cancer survival data.

The Logistic Model (7.3.5)

With $p_i(\mathbf{x})$ given by (7.3.5), the likelihood function (7.3.2) can be written as

$$L(\boldsymbol{\beta},\boldsymbol{\gamma}) = \prod_{i=1}^{k}\left(\prod_{l\in D_i}\gamma_i e^{\mathbf{x}_l\boldsymbol{\beta}} \prod_{l\in R_i}\left(1+\gamma_i e^{\mathbf{x}_l\boldsymbol{\beta}}\right)^{-1}\right)$$

where $\gamma_i = (1-p_i)/p_i$. To improve convergence and large-sample normal approximations to distributions of the m.l.e.'s, the reparameterization

$$\alpha_i = \log\gamma_i \qquad i=1,\ldots,k$$

is useful. The log likelihood function for $\boldsymbol{\beta}$ and $\boldsymbol{\alpha}=(\alpha_1,\ldots,\alpha_k)$ is

$$\log L(\boldsymbol{\beta},\boldsymbol{\alpha}) = \sum_{i=1}^{k}\left(\sum_{l\in D_i}(\alpha_i + \mathbf{x}_l\boldsymbol{\beta}) - \sum_{l\in R_i}\log\left(1+e^{\alpha_i+\mathbf{x}_l\boldsymbol{\beta}}\right)\right). \tag{7.3.12}$$

First partial derivatives of $\log L$ are

$$\frac{\partial \log L}{\partial \beta_r} = \sum_{i=1}^{k} \left(\sum_{l \in D_i} x_{lr} - \sum_{l \in R_i} \frac{x_{lr} e^{\alpha_i + \mathbf{x}_l \boldsymbol{\beta}}}{1 + e^{\alpha_i + \mathbf{x}_l \boldsymbol{\beta}}} \right) \qquad r = 1, \ldots, p \tag{7.3.13}$$

$$\frac{\partial \log L}{\partial \alpha_i} = d_i - \sum_{l \in R_i} \frac{e^{\alpha_i + \mathbf{x}_l \boldsymbol{\beta}}}{1 + e^{\alpha_i + \mathbf{x}_l \boldsymbol{\beta}}} \qquad i = 1, \ldots, k. \tag{7.3.14}$$

Second partial derivatives are

$$\frac{\partial^2 \log L}{\partial \beta_r \partial \beta_s} = -\sum_{i=1}^{k} \sum_{l \in R_i} \frac{x_{lr} x_{ls} e^{\alpha_i + \mathbf{x}_l \boldsymbol{\beta}}}{(1 + e^{\alpha_i + \mathbf{x}_l \boldsymbol{\beta}})^2} \qquad r, s = 1, \ldots, p \tag{7.3.15}$$

$$\frac{\partial^2 \log L}{\partial \beta_r \partial \alpha_i} = -\sum_{l \in R_i} \frac{x_{lr} e^{\alpha_i + \mathbf{x}_l \boldsymbol{\beta}}}{(1 + e^{\alpha_i + \mathbf{x}_l \boldsymbol{\beta}})^2} \qquad r = 1, \ldots, p \quad i = 1, \ldots, k$$

$$\tag{7.3.16}$$

$$\frac{\partial^2 \log L}{\partial \alpha_i^2} = -\sum_{l \in R_i} \frac{e^{\alpha_i + \mathbf{x}_l \boldsymbol{\beta}}}{(1 + e^{\alpha_i + \mathbf{x}_l \boldsymbol{\beta}})^2} \qquad i = 1, \ldots, k$$

$$\frac{\partial^2 \log L}{\partial \alpha_i \partial \alpha_j} = 0 \qquad i \neq j. \tag{7.3.17}$$

The maximum likelihood equations $\partial \log L / \partial \beta_r = 0$ $(r = 1, \ldots, p)$, $\partial \log L / \partial \alpha_i = 0$ $(i = 1, \ldots, k)$ can be solved using Newton–Raphson iteration or some other iterative procedure. Remarks similar to those made for the grouped proportional hazards model also apply here. In particular, if Newton–Raphson iteration is used, it is only necessary to invert matrices of dimension $p \times p$. In addition, when $\boldsymbol{\beta} = \mathbf{0}$, (7.3.14) yields the life table estimates $\hat{p}_i = (1 + e^{\hat{\alpha}_i})^{-1} = (n_i - d_i)/n_i$, which can be used as initial estimates in an iteration procedure.

Tests and Interval Estimates

Tests and interval estimates for $\boldsymbol{\beta}$ or the p_i's can be obtained via the usual large-sample theory approximations. If the approximate normality of the m.l.e.'s is used, it is desirable to work with the parameterizations $(\boldsymbol{\beta}, \boldsymbol{\gamma})$ and

$(\boldsymbol{\beta}, \boldsymbol{\alpha})$, respectively, for the grouped proportional hazards and the logistic model, since the normal approximations for $(\hat{\boldsymbol{\beta}}, \hat{\boldsymbol{\gamma}})$ and $(\hat{\boldsymbol{\beta}}, \hat{\boldsymbol{\alpha}})$ are likely to be somewhat better than those for $(\hat{\boldsymbol{\beta}}, \hat{\mathbf{p}})$. Alternately, one can use likelihood ratio methods for tests and interval estimation.

Tests about $\boldsymbol{\beta}$ are of particular interest in many situations. In the case of the logistic model a partial likelihood for $\boldsymbol{\beta}$ is available and can be used for inferences about $\boldsymbol{\beta}$ in the absence of knowlege of $p_1, \ldots, p_k$, though this requires excessive computation if the number of deaths per interval is substantial. This approach is discussed below.

One may also want estimates or tests for

$$P_i(\mathbf{x}) = \prod_{j=1}^{i} p_j(\mathbf{x}).$$

For the grouped proportional hazards model (7.3.4),

$$\hat{P}_i(\mathbf{x}) = \exp\left(-\sum_{j=1}^{i} e^{\hat{\gamma}_j + \mathbf{x}\hat{\boldsymbol{\beta}}} \right)$$

is the m.l.e. for $P_i(\mathbf{x})$ and inferences can be obtained by treating $\hat{P}_i(\mathbf{x})$ or some function of it as normally distributed. A satisfactory procedure in many instances is to treat $\log \hat{P}_i(\mathbf{x})$ as normally distributed, with mean $\log P_i(\mathbf{x})$ and variance

$$\left[\log \hat{P}_i(\mathbf{x})\right]^2 \mathbf{a}_i \mathbf{I}_0^{-1} \mathbf{a}_i'$$

where $\mathbf{I}_0$ is the observed information matrix for $(\boldsymbol{\beta}, \boldsymbol{\gamma})$ and $\mathbf{a}_i = -(\mathbf{x}, \mathbf{m}_i)$, where m_i is a $1 \times k$ vector with entries

$$m_{ij} = -\frac{e^{\hat{\gamma}_j + \mathbf{x}\hat{\boldsymbol{\beta}}}}{\sum_{j=1}^{i} e^{\hat{\gamma}_j + \mathbf{x}\hat{\boldsymbol{\beta}}}} \qquad j = 1, \ldots, i$$

$$m_{ij} = 0 \qquad j > i.$$

This follows from Theorem C1 in Appendix C and the fact that $(\hat{\boldsymbol{\beta}}, \hat{\boldsymbol{\gamma}})$ is asymptotically $N((\boldsymbol{\beta}, \boldsymbol{\gamma}), \mathbf{I}_0^{-1})$. Estimation of $P_i(\mathbf{x})$ under the logistic model (7.3.5) can be handled similarly, the analogous result being that

$$\log \hat{P}_i(\mathbf{x}) = -\sum_{j=1}^{i} \log\left(1 + e^{\hat{\alpha}_j + \mathbf{x}\hat{\boldsymbol{\beta}}}\right)$$

is approximately normal, with mean $\log P_i(\mathbf{x})$ and variance

$$\left[\log \hat{P}_i(\mathbf{x})\right]^2 \mathbf{a}_i \mathbf{I}_0^{-1} \mathbf{a}_i'$$

where $\mathbf{a}_i = (-\mathbf{x}, \mathbf{q}_i)$ and $\mathbf{q}_i$ is a $1 \times k$ vector with entries

$$q_{ij} = \frac{-\left(1+e^{-\hat{\alpha}_j - \mathbf{x}\hat{\boldsymbol{\beta}}}\right)^{-1}}{\sum_{j=1}^{i} \log\left(1+e^{\hat{\alpha}_j + \mathbf{x}\hat{\boldsymbol{\beta}}}\right)} \qquad j=1,\ldots,i$$

$$q_{ij} = 0 \qquad j > i.$$

A Partial Likelihood for the Logistic Model

Another question of interest concerns the possibility of obtaining a partial likelihood function for $\boldsymbol{\beta}$ analogous to that obtained in Section 7.2.1 for the continuous proportional hazards model. This is not possible for the grouped proportional hazards model (7.3.4), though an approximation can be based on an argument concerning the "true" ranks of observations within intervals (Kalbfleisch and Prentice, 1973). With the logistic model (7.2.5), a partial likelihood for $\boldsymbol{\beta}$ can be developed by considering for each interval the probability of the observed set of individuals who die, conditional on the observed number of deaths. For $d_i > 0$, we have

$$\Pr(\text{the individuals in } D_i \text{ die in } I_i \mid d_i \text{ individuals die in } I_i)$$

$$= \left(\prod_{l \in D_i} e^{\mathbf{x}_l \boldsymbol{\beta}}\right) \Big/ \sum_{\text{all } D_j} \left(\prod_{l \in D_j} e^{\mathbf{x}_l \boldsymbol{\beta}}\right)$$

$$= e^{\mathbf{s}_i \boldsymbol{\beta}} \Big/ \sum_{\text{all } D_j} e^{\mathbf{s}_j \boldsymbol{\beta}}. \qquad (7.3.18)$$

In (7.3.18) the sum in the denominator is over all possible d_i-subsets D_j of R_i, and $\mathbf{s}_j = \Sigma_{l \in D_j} \mathbf{x}_l$ is the sum of the regression vectors for individuals in D_j. Multiplying the factors for different intervals gives the partial likelihood

$$L_1(\boldsymbol{\beta}) = \prod_{i=1}^{k} \left(e^{\mathbf{s}_i \boldsymbol{\beta}} \Big/ \sum_{\text{all } D_j} e^{\mathbf{s}_j \boldsymbol{\beta}} \right). \qquad (7.3.19)$$

This likelihood was mentioned previously in Section 7.2.3, in connection with ties in the continuous proportional hazards model. As noted there, (7.3.19) usually requires excessive computation unless the d_i's are small, because of the large number of d_i-subsets of R_i. One problem for which this likelihood is very useful, however, is in testing that $\boldsymbol{\beta}=\mathbf{0}$, where a statistic based on the score function from (7.3.19) is easily computed. This is discussed in Section 7.3.2 and is used to test the equality of two or more lifetime distributions on the basis of grouped data. Farewell and Prentice (1980) discuss approximations to (7.3.19); see also Problem 7.3.

Assumptions About Censoring

It has been assumed thus far that all censoring takes place at the ends of intervals. If this assumption is unreasonable, modifications can be made to the arguments above, based on other suppositions about censoring, but these will usually have to be somewhat *ad hoc*. If censoring times are more or less randomly distributed across intervals, for example, Thompson (1977) suggests the following procedure. Partition R_i into three groups: D_i is the set of individuals who die in I_i, C_i is the set of individuals censored in I_i, and $G_i=R_i-D_i-C_i$ is the set of individuals surviving beyond I_i. The likelihood (7.3.2) is then replaced by

$$\prod_{i=1}^{k}\left(\prod_{l\in D_i}[1-p_i(\mathbf{x}_l)]\prod_{l\in G_i}p_i(\mathbf{x}_l)\prod_{l\in C_i}p_i(\mathbf{x}_l)^{1/2}\right). \tag{7.3.20}$$

With (7.3.20), the only difference in the formulas given above is that sums $\Sigma_{l\in R_i}$ are replaced by

$$\sum_{l\in R_i-C_i}+\frac{1}{2}\sum_{l\in C_i}.$$

That is, the contribution to the log likelihood of individuals censored in I_i is halved. This modification is relatively *ad hoc* but often sensible. It can be noted that when $\boldsymbol{\beta}=\mathbf{0}$, maximization of (7.3.20) under either the grouped proportional hazards or logistic model gives the estimates $\hat{p}_i=1-d_i/(n_i-\frac{1}{2}w_i)$, where $w_i=|C_i|$; these are the standard life table estimates discussed in Section 2.2.

7.3.2 Testing the Equality of Distributions With Grouped Data

The procedures described in the preceding section can be specialized to provide tests for the equality of two or more lifetime distributions when the data are grouped. Tests can be based on either of the two models; one or the other of the tests may be preferred in a particular situation, according to the types of departures from the null hypothesis that are envisaged. When intervals are short, the two models are nearly equivalent and the tests are virtually identical. A test based on the logistic model (7.3.5) is derived here, and similar results for the grouped proportional hazards model are noted.

Tests of equality of two or more distributions can be formulated as tests of $\boldsymbol{\beta}=\mathbf{0}$ in a suitably defined regression model. Results for such tests are slightly more involved than in the case of the continuous proportional hazards model (see Section 7.2.2), so we shall derive the test of $\boldsymbol{\beta}=\mathbf{0}$ first and then write down results for the tests of equality. We shall employ the scores test of $\boldsymbol{\beta}=\mathbf{0}$ based on the partial likelihood function (7.3.19). This turns out to have a simple form and does not require the calculation of any maximum likelihood estimates.

The log partial likelihood for $\boldsymbol{\beta}$ is, from (7.2.19),

$$\log L_1(\boldsymbol{\beta})=\sum_{i=1}^{k}\mathbf{s}_i\boldsymbol{\beta}-\sum_{i=1}^{k}\log\left(\sum_{\text{all } D_j} e^{\mathbf{s}_j\boldsymbol{\beta}}\right)$$

where $\mathbf{s}_j=\Sigma_{l\in D_j}\mathbf{x}_l$ is the sum of the regression vectors for all individuals in a particular set D_j and the sum $\Sigma_{\text{all } D_j}$ is taken over all d_i-subsets of R_i. The first and second derivatives of $\log L_1$ are

$$\frac{\partial \log L_1}{\partial \beta_r}=\sum_{i=1}^{k} s_{ir}-\sum_{i=1}^{k}\left(\frac{\sum_{\text{all } D_j} s_{jr}e^{\mathbf{s}_j\boldsymbol{\beta}}}{\sum_{\text{all } D_j} e^{\mathbf{s}_j\boldsymbol{\beta}}}\right) \qquad r=1,\ldots,p \tag{7.3.21}$$

$$\frac{\partial^2 \log L_1}{\partial \beta_r\,\partial \beta_t}$$

$$=-\sum_{i=1}^{k}\frac{\left(\sum_{\text{all } D_j} s_{jr}s_{jt}e^{\mathbf{s}_j\boldsymbol{\beta}}\right)\left(\sum_{\text{all } D_j} e^{\mathbf{s}_j\boldsymbol{\beta}}\right)-\left(\sum_{\text{all } D_j} s_{jr}e^{\mathbf{s}_j\boldsymbol{\beta}}\right)\left(\sum_{\text{all } D_j} s_{jt}e^{\mathbf{s}_j\boldsymbol{\beta}}\right)}{\left(\sum_{\text{all } D_j} e^{\mathbf{s}_j\boldsymbol{\beta}}\right)^2}$$

$$r,t=1,\ldots,p. \tag{7.3.22}$$

The $p\times 1$ score function is $\mathbf{U}(\boldsymbol{\beta})=\partial \log L_1/\partial \beta_r$ and the $p\times p$ second derivative matrix is $\mathbf{I}(\boldsymbol{\beta})=(-\partial^2 \log L/\partial\beta_r\,\partial\beta_t)$. Under $H_0:\boldsymbol{\beta}=\mathbf{0}$, $\mathbf{U}(\mathbf{0})$ is approximately normally distributed with mean $\mathbf{0}$ and covariance matrix $\mathbf{I}(\mathbf{0})$, and to test H_0 the statistic

$$\mathbf{U}(\mathbf{0})'\mathbf{I}(\mathbf{0})^{-1}\mathbf{U}(\mathbf{0})$$

can be used. Under H_0 this is distributed approximately as $\chi^2_{(p)}$, and large values of the statistic provide evidence against H_0. Though $\mathbf{U}(\boldsymbol{\beta})$ and $\mathbf{I}(\boldsymbol{\beta})$ typically require a lot of computation, $\mathbf{U}(\mathbf{0})$ and $\mathbf{I}(\mathbf{0})$ require relatively little, as we now show.

The components of $\mathbf{U}(\mathbf{0})$ and $\mathbf{I}(\mathbf{0})$ can be written as

$$U_r(\mathbf{0}) = \sum_{i=1}^{k} \left(s_{ir} - \frac{A_{ir}}{B_i} \right) \qquad r = 1, \ldots, p \tag{7.3.23}$$

and

$$I_{rt}(\mathbf{0}) = \sum_{i=1}^{k} \frac{B_i C_{irt} - A_{ir} A_{it}}{B_i^2} \qquad r, t = 1, \ldots, p \tag{7.3.24}$$

where

$$A_{ir} = \sum_{\text{all } D_j} s_{jr}$$

$$C_{irt} = \sum_{\text{all } D_j} s_{jr} s_{jt}$$

and

$$B_i = \sum_{\text{all } D_j} 1 = \binom{n_i}{d_i}.$$

To simplify calculations we note that

$$A_{ir} = \sum_{\text{all } D_j} \sum_{l \in R_i} x_{lr}$$

$$= \binom{n_i - 1}{d_i - 1} \sum_{l \in R_i} x_{lr} \tag{7.3.25}$$

since each $l \in R_i$ appears in exactly $\binom{n_i - 1}{d_i - 1}$ of the d_i-subsets of R_i. Also

$$C_{irt} = \sum_{\text{all } D_j} \left(\sum_{l \in D_j} x_{lr} \right) \left(\sum_{m \in D_j} x_{mt} \right)$$

$$= \sum_{\text{all } D_j} \sum_{l \in D_j} x_{lr} x_{lt} + \sum_{\text{all } D_j} \sum_{\substack{l, m \in D_j \\ l \neq m}} x_{lr} x_{mt}$$

$$= \binom{n_i - 1}{d_i - 1} \sum_{l \in R_i} x_{lr} x_{lt} + \binom{n_i - 2}{d_i - 2} \sum_{\substack{l, m \in R_i \\ l \neq m}} x_{lr} x_{mt}. \tag{7.3.26}$$

Using (7.3.25) and (7.3.26) in (7.3.23) and (7.3.24) and simplifying a little more, we get

$$U_r(\mathbf{0}) = \sum_{i=1}^{k} \left(\sum_{l \in D_i} x_{lr} - \frac{d_i}{n_i} \sum_{l \in R_i} x_{lr} \right) \qquad r = 1, \ldots, p \tag{7.3.27}$$

$$I_{rt}(\mathbf{0}) = \sum_{i=1}^{k} \frac{(n_i - d_i) d_i}{n_i (n_i - 1)} \left[\sum_{l \in R_i} x_{lr} x_{lt} - \frac{1}{n_i} \left(\sum_{l \in R_i} x_{lr} \right) \left(\sum_{l \in R_i} x_{lt} \right) \right]$$

$$r, t = 1, \ldots, p. \tag{7.3.28}$$

Tests for the equality of two or more distributions can be obtained from these results.

Testing the Equality of Two Lifetime Distributions

Suppose that grouped lifetimes and censoring times are available from each of two populations, the grouping intervals $I_i = [a_{i-1}, a_i)$ being the same in both cases. We define a dummy regressor variable x that takes on the values 0 or 1 according to whether an individual comes from the first or second population, and we suppose that conditional probabilities of surviving past an interval for individuals in the two populations are given by the logistic model (7.3.5). This implies that

$$p_{1i} = p_i(0) = (1 + e^{\gamma_i})^{-1}$$

and

$$p_{2i} = p_i(1) = (1 + e^{\gamma_i + \beta})^{-1} \tag{7.3.29}$$

where p_{1i} is the probability of an individual from population 1 surviving to a_i, given that he survives to a_{i-1}, and p_{2i} is the similar probability for population 2. That $\beta=0$ in the model $p_i(x) = (1 + e^{\gamma_i + x\beta})^{-1}$ then tests that $p_{1i} = p_{2i}$ $(i=1,\ldots,k)$ or that survival probabilities are the same for the two populations.

Let n_i be the total number of individuals at risk at the start of I_i from the two samples combined and let n_{1i} and n_{2i} be the numbers among these who are from populations 1 and 2, respectively. Similarly, let d_{1i} and d_{2i} be the number of deaths in I_i among individuals from populations 1 and 2, with $d_i = d_{1i} + d_{2i}$ being the total number of deaths in I_i. With the combined sample from the two populations and the dummy regressor variable x, (7.3.27) and (7.3.28) reduce to

$$U(0) = \sum_{i=1}^{k} \left(d_{2i} - \frac{d_i n_{2i}}{n_i} \right)$$

$$I(0) = \sum_{i=1}^{k} \frac{d_i(n_i - d_i) n_{1i} n_{2i}}{n_i^2 (n_i - 1)}. \tag{7.3.30}$$

Tests of $H_0: \beta = 0$ (i.e., $p_{1i} = p_{2i}$; $i=1,\ldots,k$) can be based on the statistic

$$Z^2 = \frac{U(0)^2}{I(0)}.$$

Large values of Z^2 provide evidence against H_0; under H_0, Z^2 is approximately $\chi^2_{(1)}$.

The statistic is similar to that obtained in the two-sample problem for the continuous proportional hazards model [see (7.2.13)]. As noted there, the test is sometimes referred to as the Mantel–Haenszel test. It was derived by Mantel and Haenszel (1959) from a somewhat different point of view, involving the combination of statistics in contingency tables. In this respect note that the ith term in $I(0)$ is the variance of $(d_{2i} - d_i n_{2i}/n_i)$, given d_i, n_{2i}, and n_i, so that $I(0)$ is a sum of conditional variance terms.

This test will be sensitive at detecting differences between distributions of the type embodied in the logistic model (7.3.5), which stipulates that the conditional probabilities of survival for the two distributions are given by (7.3.29). This implies that the ordering between the p_{1i}'s and p_{2i}'s is uniform across the k time intervals. That is if $p_{1i} > p_{2i}$ for any i, then $p_{1i} > p_{2i}$ for all

$i=1,\ldots,k$. Equivalently, the probabilities P_{1i} and P_{2i} of surviving past the ends of intervals are similarly ordered, and P_{1i}/P_{2i} increases or decreases monotonically. To assess in a particular situation whether the model (7.3.29) is in fact reasonable, one can calculate for each sample the usual life table estimates of p_{1i} and p_{2i}, namely, $\hat{p}_{1i}=(n_{1i}-d_{1i})/n_{1i}$ and $\hat{p}_{2i}=(n_{2i}-d_{2i})/n_{2i}, i=1,\ldots,k$. Differences in the quantities $\log[(1-\hat{p}_{1i})/\hat{p}_{1i}]$ and $\log[(1-\hat{p}_{2i})/\hat{p}_{2i}]$ should be roughly constant for $i=1,\ldots,k$, the approximate difference providing an estimate of β in (7.3.29).

The logistic model approaches the continuous proportional hazards model of Section 7.2 as the number of intervals becomes arbitrarily large and the lengths of intervals approach 0. In the continuous situation $U(0)$ and $I(0)$ of (7.3.30) approach $U(0)$ and $I(0)$ of (7.2.15), since in the limit all d_i's will equal 1 or 0. For data that are essentially continuous but contain some ties, the statistic Z^2 based on (7.3.30) was suggested in Section 7.2.2 for use with the continuous proportional hazards model.

Testing the Equality of Three or More Distributions

A test of the equality of three or more distributions is similarly easily obtained from (7.3.27) and (7.3.28). If there are m distributions to be compared, define a regression vector $\mathbf{x}=(x_1,\ldots,x_{m-1})$ such that individuals from the 1st, 2nd,...,$(m-1)$th and mth populations, respectively, have $\mathbf{x}$ vectors $(1,0,\ldots,0)$, $(0,1,0,\ldots,0),\ldots,(0,\ldots,0,1)$ and $(0,\ldots,0)$. The model (7.3.5) then specifies that the probabilities p_{ri} $(r=1,\ldots,m)$ for the m distributions are given by

$$p_{1i}=\frac{1}{1+e^{\gamma_i+\beta_1}},\ldots,p_{m-1,i}=\frac{1}{1+e^{\gamma_i+\beta_{m-1}}},\ p_{mi}=\frac{1}{1+e^{\gamma_i}} \qquad i=1,\ldots,k$$

where p_{ri} is the probability of an individual from population r surviving to a_i, given that she survives to a_{i-1}. Equality of the conditional probabilities for all distributions is tested by testing that $\boldsymbol{\beta}=(\beta_1,\ldots,\beta_{m-1})=\mathbf{0}$.

Consider the combined sample of observations from all m populations. Letting n_{ri} represent the number of individuals at risk in the ith time interval who are from the rth population $(n_{1i}+\cdots+n_{mi}=n_i)$ and letting d_{ri} represent the number of deaths in the ith interval among these individuals $(d_{1i}+\cdots+d_{mi}=d_i)$, we find that (7.3.27) and (7.3.28) give

$$U_r(\mathbf{0})=\sum_{i=1}^{k}\left(d_{ri}-\frac{d_i n_{ri}}{n_i}\right) \qquad r=1,\ldots,m-1 \qquad (7.3.31)$$

$$I_{rt}(\mathbf{0})=\sum_{i=1}^{k}\frac{(n_i-d_i)d_i}{n_i(n_i-1)}\left(n_{ri}\delta_{rt}-\frac{n_{ri}n_{ti}}{n_i}\right) \qquad r,t=1,\ldots,m-1 \qquad (7.3.32)$$

where δ_{rt} is the Kronecker delta ($\delta_{rr}=1, \delta_{rt}=0$ if $r \neq t$). The appropriate test statistic for testing the equality of the m distributions is then

$$\mathbf{U}'(\mathbf{0})\mathbf{I}(\mathbf{0})^{-1}\mathbf{U}(\mathbf{0}) \tag{7.3.33}$$

where $\mathbf{U}'(\mathbf{0})=[U_1(\mathbf{0}),\ldots,U_{m-1}(\mathbf{0})]$. Large values of this statistic provide evidence against equality; under H_0 the statistic is approximately distributed as $\chi^2_{(m-1)}$.

The statistic $U_r(\mathbf{0})$ in (7.3.31) can be written as

$$\sum_{i=1}^{k} d_{ri} - \sum_{i=1}^{k} \frac{d_i n_{ri}}{n_i} = O_r - E_r$$

where O_r is the total number of deaths observed for individuals in population r and E_r can be thought of as the expected number of deaths under the hypothesis that the m distributions are identical. The statistic $\sum_{r=1}^{m}(O_r - E_r)^2/E_r$ is sometimes suggested as a simple approximation to the score statistic (7.3.33) (e.g., Peto and Pike, 1973; Crowley and Breslow, 1975), but should only be used in certain situations. This statistic can be shown to be conservative, in the sense that it is always less than or equal to (7.3.33), but unless the m distributions are not much different from each other and there is roughly the same censoring pattern in each sample, it can greatly underestimate (7.3.33).

An example illustrating these tests is given later.

Stratification

The general logistic model (7.3.5) allows adjustments to be made for important concomitant variables when comparing distributions by including these variables in the regression model. However, the simplicity of the m-sample tests is then lost because hypotheses of equality of distributions are no longer of the form $\boldsymbol{\beta}=\mathbf{0}$. In some situations a simple way to adjust for concomitant variables is to stratify the population according to them. Suppose, for example, that individuals within each of the m populations of interest can be put into strata $j=1,\ldots,s$ and that the model

$$\log\left(\frac{1-p_{rij}}{p_{rij}}\right)=\gamma_{ij}+\beta_r \qquad r=1,\ldots,m \quad i=1,\ldots,k \quad j=1,\ldots,s \tag{7.3.34}$$

is reasonable. Here, as earlier, $\beta_m=0$. The score statistic for testing that

$\boldsymbol{\beta}=(\beta_1,\ldots,\beta_{m-1})=\mathbf{0}$ is easily seen to be

$$X^2=\left(\sum_{j=1}^{s}\mathbf{U}_j\right)'\left(\sum_{j=1}^{s}\mathbf{I}_j\right)^{-1}\left(\sum_{j=1}^{s}\mathbf{U}_j\right)$$

where $\mathbf{U}_j$ and $\mathbf{I}_j$ are the $(m-1)\times 1$ score statistic $\mathbf{U}(\mathbf{0})$ and $(m-1)\times(m-1)$ information matrix $\mathbf{I}(\mathbf{0})$ obtained from the observations in the jth stratum. Under the hypothesis that $\boldsymbol{\beta}=\mathbf{0}$, X^2 is approximately $\chi^2_{(m-1)}$.

Tests Based on the Grouped PH Model

Tests for the equality of two or more life distributions can also be developed from the grouped proportional hazards model (7.3.4). In this case, however, there is no partial likelihood for $\boldsymbol{\beta}$ alone, so that procedures must be based on the full likelihood function (7.3.6) for $\boldsymbol{\beta}$ and $\boldsymbol{\gamma}$. Prentice and Gloeckler (1978) propose a partial scores test: let

$$\mathbf{U}=\begin{bmatrix}\mathbf{U}_1\\ \mathbf{U}_2\end{bmatrix}=\begin{bmatrix}\partial\log L/\partial\boldsymbol{\beta}\\ \partial\log L/\partial\boldsymbol{\gamma}\end{bmatrix}\quad\text{and}\quad\mathbf{I}=\begin{bmatrix}\mathbf{I}_{11} & \mathbf{I}_{12}\\ \mathbf{I}'_{12} & \mathbf{I}_{22}\end{bmatrix}$$

where $\mathbf{I}$ is evaluated at $\boldsymbol{\beta}=\mathbf{0}, \boldsymbol{\gamma}=\hat{\boldsymbol{\gamma}}(\mathbf{0})$, and is partitioned so that $\mathbf{I}_{11}$ contains second partial derivatives of $\log L$ with respect to $\boldsymbol{\beta}$ and $\mathbf{I}_{22}$ with respect to $\boldsymbol{\gamma}$. A test for $H_0: \boldsymbol{\beta}=\mathbf{0}$ can be based on

$$\mathbf{U}_1[\mathbf{0},\hat{\boldsymbol{\gamma}}(\mathbf{0})]=\left(\frac{\partial\log L}{\partial\boldsymbol{\beta}}\right)_{[\mathbf{0},\hat{\boldsymbol{\gamma}}(\mathbf{0})]}$$

where $\hat{\boldsymbol{\gamma}}(\mathbf{0})$ is the m.l.e. of $\boldsymbol{\gamma}$ when $\boldsymbol{\beta}=\mathbf{0}$. Under H_0, $\mathbf{U}_1[\mathbf{0},\hat{\boldsymbol{\gamma}}(\mathbf{0})]$ is asymptotically normal with mean $\mathbf{0}$ and covariance matrix $\mathbf{V}_1=\mathbf{I}_{11}-\mathbf{I}_{12}\mathbf{I}_{22}^{-1}\mathbf{I}'_{12}$. It is readily found from (7.3.7) to (7.3.11) that

$$U_{1r}=\left.\frac{\partial\log L}{\partial\beta_r}\right|_{[\mathbf{0},\hat{\boldsymbol{\gamma}}(\mathbf{0})]}$$

$$=\sum_{i=1}^{k}\left[\left(-\sum_{l\in D_i}x_{lr}\right)\frac{n_i}{d_i}\log\left(\frac{n_i-d_i}{n_i}\right)+\left(\sum_{l\in R_i}x_{lr}\right)\log\left(\frac{n_i-d_i}{n_i}\right)\right]$$

$$r=1,\ldots,p. \qquad (7.3.35)$$

Actually, Prentice and Gloeckler replace $\mathbf{V}_1$ with the matrix $\mathbf{V}$, which has entries

$$V_{rt}=\sum_{i=1}^{k} q_i\left[\frac{1}{n_i}\sum_{l\in R_i} x_{lr}x_{lt}-\frac{1}{n_i^2}\left(\sum_{l\in R_i} x_{lr}\right)\left(\sum_{l\in R_i} x_{lt}\right)\right] \qquad r,t=1,\ldots,p$$

(7.3.36)

where $q_i=[n_i(n_i-d_i)/d_i]\log^2[(n_i-d_i)/n_i]$. This matrix is obtained by taking some conditional expectations in $\mathbf{I}$ before evaluating $\mathbf{I}_{11}-\mathbf{I}_{12}\mathbf{I}_{22}^{-1}\mathbf{I}'_{12}$. (Specifically, for interval i expectations are taken conditionally on d_i and n_{ri}, $r=1,\ldots,p$.)

A test of $\boldsymbol{\beta}=\mathbf{0}$ can be based on the statistic $\mathbf{U}_1'\mathbf{V}^{-1}\mathbf{U}_1$, which is asymptotically distributed as $\chi^2_{(p)}$ under H_0. Tests for the equality of two or more lifetime distributions are obtained in exactly the same manner as for the logistic model. To test for the equality of m distributions, we let $\mathbf{x}$ be a $1\times(m-1)$ vector taking on the values $(1,0,\ldots,0)$, $(0,1,\ldots,0),\ldots,(0,\ldots,0,1)$, and $(0,\ldots,0)$ for the 1st, 2nd,..., mth populations, respectively. The statistics $\mathbf{U}$ and $\mathbf{V}$ given by (7.3.35) and (7.3.36) then have entries

$$U_r=\sum_{i=1}^{k}\left(n_{ri}-\frac{n_i d_{ri}}{d_i}\right)\log\left(\frac{n_i-d_i}{n_i}\right) \qquad r=1,\ldots,m-1$$

$$V_{rt}=\sum_{i=1}^{k}\frac{n_i-d_i}{d_i}\left[\log\left(\frac{n_i-d_i}{n_i}\right)\right]^2\left(n_{ri}\delta_{rt}-\frac{n_{ri}n_{ti}}{n_i}\right) \qquad r,t=1,\ldots,m-1.$$

(7.3.37)

The expressions in (7.3.37) and those in (7.3.30) for the logistic model do not differ greatly when interval lengths are short and the d_i/n_i's are small. Differences in the two tests are more substantial when intervals are longer.

Example 7.3.1 To illustrate the tests for the equality of distributions, let us consider the data given in Table 7.3.1. The data are fictitious, but we will suppose that they represent failure times (in weeks) for three types of electrical components subjected to constant use. A total of 140 components are involved, with 42, 50, and 48 components of Types A, B, and C, respectively.

There are 128 failures and 12 censoring times; the number of withdrawals in the different time intervals is not shown in the table but can be deduced from the numbers at risk and the numbers of deaths for the intervals. To test for the equality of failure time distributions, using the logistic score

Table 7.3.1 Life Table Data on Component Failures[a]

	Total		Type A		Type B		Type C	
Interval	n_i	d_i	n_{1i}	d_{1i}	n_{2i}	d_{2i}	n_{3i}	d_{3i}
[0, 10)	140	21	42	4	50	6	48	11
[10, 20)	119	24	38	3	44	11	37	10
[20, 30)	94	25	35	3	32	10	27	12
[30, 40)	68	21	31	5	22	8	15	8
[40, 50)	44	16	26	6	12	6	6	4
[50, 60)	26	7	20	4	5	3	1	0
[60, 70)	17	5	15	3	1	1	1	1
[70, 80)	11	4	11	4	0	0	0	0
[80, ∞)	5	5	5	5	0	0	0	0

[a] $E_1=61.10, E_2=37.68, E_3=29.22, 0_1=37, 0_2=45, 0_3=46$.

statistic (7.3.33), it is necessary to calculate $\mathbf{U}=[U_1(\mathbf{0}), U_2(\mathbf{0})]'$ and $\mathbf{I}(\mathbf{0})$, given in (7.3.31) and (7.3.32). The values O_r and E_r ($r=1,2,3$) are easily calculated and are shown in Table 7.3.1; this gives $\mathbf{U}=(O_1-E_1, O_2-E_2)'=(-24.10, 8.32)'$. Note that as a check, $O_1+O_2+O_3$ must equal $E_1+E_2+E_3$. Note also that in the expressions involved in (7.3.33), $k=8$, the ninth interval being $[80,\infty)$, wherein all remaining individuals must die. It is also easy to calculate

$$\mathbf{I}(\mathbf{0})=\begin{bmatrix} 19.8157 & -11.3917 \\ -11.3917 & 19.1190 \end{bmatrix}.$$

This gives the observed score statistic value $\mathbf{U}'\mathbf{I}(\mathbf{0})^{-1}\mathbf{U}=32.74$, which gives a significance level much lower than .001 on $\chi^2_{(2)}$. As is pretty clear from a look at the data, there is strong evidence of a difference in the failure time distributions.

When intervals are long, there will be some difference in the results of tests based on (7.3.33) and those based on the grouped proportional hazards model, though the broad conclusions emerging from the two tests will usually be similar. Which test is more appropriate depends upon the presumed nature of possible differences in the life distributions. In the present situation we find from (7.3.37) that $\mathbf{U}=(U_1, U_2)'=(-28.08, 8.72)'$ and

$$\mathbf{V}=\begin{bmatrix} 26.0903 & -15.0919 \\ -15.0919 & 24.9656 \end{bmatrix}$$

which gives a score statistic value $\mathbf{U}'\mathbf{V}^{-1}\mathbf{U}=33.70$. This is close to the value obtained from the logistic model.

Finally, if intervals in the life table are long, one may want to make an adjustment to allow for the way withdrawals occur. The calculations above assume that all censoring occurs at the ends of intervals, but this can be modified, as indicated earlier.

7.4 SOME OTHER PROBLEMS

7.4.1 Efficiency of Procedures Based on Partial Likelihoods*

An important question that has not been discussed so far concerns the efficiency of the partial likelihood procedures in this chapter. If, for example, the data arise from a Weibull distribution with hazard function for T, given $\mathbf{x}$, of the form $h(t|\mathbf{x})=\lambda\alpha t^{\alpha-1}e^{\mathbf{x}\boldsymbol{\beta}}$, then $\boldsymbol{\beta}$ can be estimated by standard parametric methods, as can the underlying hazard function $h_0(t)=\lambda\alpha t^{\alpha-1}$. If the partial likelihood (7.2.3) is used to estimate $\boldsymbol{\beta}$, some loss of information results from neglecting a portion of the available data. The crucial question is whether this loss of information is substantial. The efficiency of the partial likelihood procedures has been examined by several authors. Broadly speaking, it appears that the efficiency of these methods is high when the underlying model is truly of the proportional hazards type and $|\boldsymbol{\beta}|$ is not too large. This is an important result, because the partial likelihood methods are more robust than methods based on a specific parametric model.

Investigation of the efficiency of partial likelihood methods for $\boldsymbol{\beta}$ in anything but simple situations is difficult and must be done by simulation to a large extent. The nature of the efficiency problem will only be sketched here; detailed results can be found in references cited below.

Suppose that the data arise from the model with continuous hazard function

$$h(t|\mathbf{x})=h_0(t;\boldsymbol{\theta})e^{\mathbf{x}\boldsymbol{\beta}} \tag{7.4.1}$$

where $h_0(t;\boldsymbol{\theta})$ is some specified underlying hazard function involving a vector $\boldsymbol{\theta}$ of unknown parameters. Given a censored sample involving observations on n individuals, assume that the likelihood function is of the usual form

$$L_1(\boldsymbol{\beta},\boldsymbol{\theta})=\prod_{i=1}^{n}\left[h_0(t_i;\boldsymbol{\theta})e^{\mathbf{x}_i\boldsymbol{\beta}}\exp\left(-\int_0^{t_i}h_0(u;\boldsymbol{\theta})e^{\mathbf{x}_i\boldsymbol{\beta}}\,du\right)\right]^{\delta_i}$$
$$\times\left[\exp\left(-\int_0^{t_i}h_0(u;\boldsymbol{\theta})e^{\mathbf{x}_i\boldsymbol{\beta}}\,du\right)\right]^{1-\delta_i}$$

where t_i is a lifetime or censoring time and $\delta_i = 1$ or 0 according to whether t_i is the former or the latter. Let us denote the expected information matrix arising from $L_1(\boldsymbol{\beta}, \boldsymbol{\theta})$ as

$$\mathbf{I} = \begin{bmatrix} E\{-\partial^2 \log L_1/\partial\boldsymbol{\beta}\,\partial\boldsymbol{\beta}\} & E\{-\partial^2 \log L_1/\partial\boldsymbol{\beta}\,\partial\boldsymbol{\theta}\} \\ E\{-\partial^2 \log L_1/\partial\boldsymbol{\theta}\,\partial\boldsymbol{\beta}\} & E\{-\partial^2 \log L_1/\partial\boldsymbol{\theta}\,\partial\boldsymbol{\theta}\} \end{bmatrix}$$

$$= \begin{bmatrix} \mathbf{I}_{11} & \mathbf{I}_{12} \\ \mathbf{I}_{21} & \mathbf{I}_{22} \end{bmatrix}.$$

Let $\hat{\boldsymbol{\beta}}$ be the m.l.e. of $\boldsymbol{\beta}$ obtained by maximizing $L_1(\boldsymbol{\beta}, \boldsymbol{\theta})$; the asymptotic covariance matrix for $\hat{\boldsymbol{\beta}}$ is the upper $p \times p$ principal submatrix of $\mathbf{I}^{-1}$, which is $\mathbf{V}_1 = (\mathbf{I}_{11} - \mathbf{I}_{12}\mathbf{I}_{22}^{-1}\mathbf{I}_{21})^{-1}$.

On the other hand, the partial likelihood in this situation is

$$L_2(\boldsymbol{\beta}) = \prod_{i=1}^{k} \left(e^{\mathbf{x}_{(i)}\boldsymbol{\beta}} \Big/ \sum_{l \in R_i} e^{\mathbf{x}_l\boldsymbol{\beta}} \right)$$

where $t_{(1)} < \cdots < t_{(k)}$ are the distinct observed lifetimes in the sample and R_i is the risk set at $t_{(i)}$. The expected information matrix based on $L_2(\boldsymbol{\beta})$ is the $p \times p$ matrix

$$\mathbf{I}_2 = \left[E\left(\frac{-\partial^2 \log L_2}{\partial\beta_r\,\partial\beta_t} \right) \right].$$

Under suitable conditions the estimate $\tilde{\boldsymbol{\beta}}$ of $\boldsymbol{\beta}$ obtained by maximizing $L_2(\boldsymbol{\beta})$ is asymptotically normal, with covariance matrix $\mathbf{V}_2 = \mathbf{I}_2^{-1}$. The asymptotic efficiency of $\tilde{\boldsymbol{\beta}}$ relative to $\hat{\boldsymbol{\beta}}$ is examined by comparing $\mathbf{V}_1$ and $\mathbf{V}_2$. Computation of $\mathbf{V}_2$ in anything but quite simple situations is, however, very difficult, and calculation of $\mathbf{V}_1$ can also be difficult unless the censoring mechanism is a simple one. Thus even asymptotic results are hard to come by, and to examine specific situations it is often necessary to resort to simulation. Efron (1977) manages to obtain some general results by comparing $\mathbf{V}_1$ and $\mathbf{V}_2$ in a large class of models and reaches some fairly general conclusions. Along with results for some specific models mentioned below, this gives at least a broad impression of the efficiency properties of the partial likelihood methods.

The main general result (Efron, 1977) is that for certain types of models the partial likelihood m.l.e. $\tilde{\boldsymbol{\beta}}$ is asymptotically fully efficient. This result is of limited direct use, since most realistic models are only approximated by a model of the type Efron considers, though it does suggest that the efficiency of the partial likelihood procedures will be high in many instances. Other results indicate this is true in a few fairly simple models. Kalbfleisch (1974), Hensler et al. (1977), Efron (1977), Oakes (1977), and Kay (1979) all consider the case of an underlying exponential distribution, where $h_0(t)=\lambda$. When there is a single regressor variable and no censoring, the partial likelihood procedures are asymptotically fully efficient at $\beta=0$, and for β not too far from 0 they have asymptotic efficiency in excess of 80%; this is particularly important in connection with tests of the equality of two distributions. When there is censoring, the efficiency falls off somewhat, but the loss does not appear to be great. Crowley and Thomas (1975) have investigated the two-sample problem and find that the partial likelihood method is still asymptotically fully efficient at $\beta=0$ if there is a common censoring time distribution in the two populations but that there is a loss of efficiency when censoring time distributions are different.

Similar efficiencies appear to be obtained with more than one regressor variable and under other common proportional hazards models, though this has not been very thoroughly investigated. Peace and Flora (1978) report results of a small simulation study involving four regressor variables and underlying exponential, Weibull, and Gompertz distributions, with some censoring. They found the efficiency of the partial likelihood methods to be fairly high in all cases. A few additional simulation results for the two-sample problem (Gehan and Thomas, 1969; Lee et al., 1975) indicate that under a Weibull model the situation is similar. The efficiency of two-sample tests is also discussed in Chapter 8.

The evidence accumulated so far thus suggests that the partial likelihood methods for $\boldsymbol{\beta}$ have good efficiency in a range of situations. Two points that should be remembered, however, are that these results assume that a proportional hazards model is appropriate and that efficiency can drop off rapidly if $|\boldsymbol{\beta}|$ is large.

7.4.2 Time-Dependent Regressor Variables

In the development of the model (7.1.1) it was assumed that the regression vector **x** was fixed and known for each individual; this yields a proportional hazards model. Sometimes it is useful to allow the **x**'s to depend on time. In this case we write $\mathbf{x}=\mathbf{x}(t)$ and the hazard function of an individual at time t is of the form

$$h[t|\mathbf{x}(t)]=h_0(t)e^{\mathbf{x}(t)\boldsymbol{\beta}}. \tag{7.4.2}$$

The variable $\mathbf{x}(t)$ can be either stochastic or deterministic. The idea of time-dependent regressor variables was introduced by Cox (1972a), and although there are still questions about the application of partial likelihood methods in this case, work by Cox (1975), Aalen (1978), Efron (1977), Kalbfleisch and MacKay (1978, 1980), Kalbfleisch and Prentice (1980), and others suggests that the partial likelihood approach is still valid. If this is so, the results given earlier in this chapter can be used in their stated form: for example, the partial likelihood (7.2.3) for $\boldsymbol{\beta}$ now becomes

$$L(\boldsymbol{\beta}) = \prod_{i=1}^{k} \left(e^{\mathbf{x}_i(t_{(i)})\boldsymbol{\beta}} \Big/ \sum_{l \in R_i} e^{\mathbf{x}_l(t_{(i)})\boldsymbol{\beta}} \right)$$

with corresponding alterations to other formulas. Let us consider two situations where time-dependent regressor variables would be helpful.

Example 7.4.1 In accelerated life tests involving a stress covariable x, the stress is sometimes changed during the course of the experiment. For example, if the lifetime of a type of electrical insulation depends on the voltage the insulation experiences, tests are often carried out in which insulation specimens are subjected to a voltage that increases with time. The voltage may be increased more or less continuously, or in stages, until the item fails. In this situation voltage can be considered as a time-dependent regressor variable. Note that in situations like this the hazard function at time t_0 may not depend just on $x(t_0)$, but on the entire history $x(t), t \leqslant t_0$.

Example 7.4.2 One useful way in which model (7.4.2) can be exploited is in testing for departures from a proportional hazards model (see Section 7.4.3). For example, consider the two-sample problem in the form discussed in Section 7.2.2, except now define two regressor variables x_1 and x_2: x_1 is a dummy variable taking on the value 0 or 1 according to whether the individual is from the first or second population, and x_2 is a time-dependent covariable defined by $x_2 = x_1 \log t$. By (7.4.2), the hazard functions for individuals from population 1 and 2, respectively, are

$$\begin{aligned} h_1(t) &= h(t|x_1=0, x_2=0) \\ &= h_0(t) \end{aligned}$$

and

$$\begin{aligned} h_2(t) &= h(t|x_1=1, x_2=\log t) \\ &= e^{\beta_1} t^{\beta_2} h_0(t). \end{aligned}$$

The hazard functions $h_1(t)$ and $h_2(t)$ are proportional if and only if $\beta_2=0$; they are equal if and only if $\beta_1=\beta_2=0$. A formal test of the proportionality of $h_1(t)$ and $h_2(t)$ can be made by testing that $\beta_2=0$.

Time-dependent regressor variables have not been really thoroughly investigated. Efficiency results of the type discussed by Efron (1977) appear to apply to the time-dependent case. In addition, Kalbfleisch and McIntosh (1977) have investigated the efficiency of the partial likelihood in testing the proportional hazards assumption $\beta_2=0$, described in Example 7.4.2, in the case in which the two distributions involved are Weibull distributions. This amounts to comparing the efficiency of the partial likelihood test of $\beta_2=0$ with the parametric test that the two Weibull distributions have equal shape parameters (see Section 4.3.1). They find that the relative efficiency of the partial likelihood procedure exceeds 70% in a variety of practical situations.

There appears to be considerable scope for the application of time-dependent regressor variables, though it is not always clear how to incorporate these into a model An interesting application to the analysis of heart transplant data has been given by Kalbfleisch and Prentice (1980, Sec. 5.4), Crowley and Hu (1977), and others. Further experience is needed, however, before the usefulness of time-dependent regressor variables can be fully ascertained.

7.4.3 Assessment of Departures From Proportional Hazards

Since this chapter deals with the analysis of data under proportional hazards (PH) models, it is fitting to conclude it with a few comments on methods of assessing the adequacy of a PH model.

Examination of residuals, as described in Section 7.2.5, provides checks on a model, though the effects of different types of departures from the model are inevitably compounded to some degree. For example, suppose we have two populations, with the same regressor variables $\mathbf{x}$ affecting individuals in each. Suppose that we entertain the model wherein the hazard functions for individuals in populations 1 and 2 are

$$h_i(t|\mathbf{x})=h_0(t)\exp(\alpha_i+\mathbf{x}\boldsymbol{\beta}) \qquad i=1,2. \tag{7.4.3}$$

That is, after adjustment for covariables, individuals in the two populations have proportional hazards. Two examples of departures from this model are (1) $h_i(t|\mathbf{x})=h_{0i}(t)\exp(\mathbf{x}\boldsymbol{\beta})$, where $h_{01}(t)$ and $h_{02}(t)$ are not necessarily proportional, and (2) $h_i(t|\mathbf{x})=h_0(t)e^{\alpha_i}+g(\mathbf{x})$. A fit of the original model and a cursory examination of residuals may suggest model departures but give little idea of their possible source. However, by fitting a variety of models and carefully examining the results, we can often build confidence in a particular model. For example, one could assess model (1) versus model (2) by fitting separate models to the data from each population. The discussion in Section 7.2.5 is relevant to model assessment of this type.

Formal tests of proportional hazards assumptions can be derived as parametric tests in more general models, though one then has the task of assessing the adequacy of the general model. For example, to assess the model (7.4.3) we could use a time-dependent regressor variable (e.g., Cox, 1972a; Kalbfleisch and McIntosh, 1977). By defining a dummy regressor variable as in Example 7.4.2, we create models

$$h_1(t|\mathbf{x})=h_0(t)e^{\alpha_1+\mathbf{x}\boldsymbol{\beta}} \qquad h_2(t|\mathbf{x})=h_0(t)t^{\gamma}e^{\alpha_2+\mathbf{x}\boldsymbol{\beta}}$$

The partial likelihood given in Section 7.4.2 can be used to test $\gamma=0$, providing a check on the proportionality of hazards (see Problem 7.11).

It is possible to model departures from the proportional hazards model in other ways. For example, Aranda-Ordaz (1980) considers models for grouped survival data that do this: consider the framework of Section 7.3.1, where

$$p_i(\mathbf{x})=\Pr(\text{an individual survives past } a_i|\text{ she survives past } a_{i-1},\mathbf{x})$$

where $\mathbf{x}$ is a (fixed) regressor variable vector and lifetimes are grouped into intervals $[a_{j-1},a_j)$, $j=1,\ldots,k+1$, where $a_0=0$ and $a_k=T$. Aranda-Ordaz considers the transformation family of models

$$V_\lambda[p_i(\mathbf{x})]=\alpha_i+\mathbf{x}\boldsymbol{\beta} \qquad i=1,\ldots,k \tag{7.4.4}$$

where $V_\lambda(p)$ is a family of transformations indexed by the parameter λ. One convenient and useful class is the power transformation family with

$$V_\lambda(p)=\frac{(-\log p)^\lambda-1}{\lambda} \qquad \lambda\neq 0.$$

As $\lambda\to 0$ this approaches $V_0(p)=\log(-\log p)$, and (7.4.4) then gives $\log[-\log p_i(\mathbf{x})]=\alpha_i+\mathbf{x}\boldsymbol{\beta}$, which is the grouped PH model (7.3.4) with $\alpha_i=\log(-\log p_i)=\log[-\log p_i(\mathbf{0})]$. Proportional hazards can be assessed by fitting models with various λ values and assessing $\lambda=0$ (see Problem 7.12).

Finally, it would be desirable to have tests of proportional hazards that were not too closely tied to parametric assumptions. Schoenfeld (1980) gives a χ^2 test of this kind, which is unfortunately rather complicated. Further work in this area would be useful.

7.5 PROBLEMS AND SUPPLEMENTS

7.1 Consider the matrix $\mathbf{I}(\boldsymbol{\beta})$ defined by (7.2.7) in the case in which there is no censoring and no ties.

a. Prove that the exact covariance matrix for $\mathbf{U}(\mathbf{0})$ is

$$E[\mathbf{I}(\mathbf{0})] = \left(\sum_{l=1}^{n} (e_{l,n} - 1)^2/(n-1) \right) \mathbf{X}_c'\mathbf{X}_c \tag{7.5.1}$$

where $\mathbf{X}_c'\mathbf{X}_c$ is the corrected sum of squares matrix with (r, s) entry

$$\sum_{i=1}^{n} (x_{ir} - \bar{x}_r)(x_{is} - \bar{x}_s) \qquad r, s = 1, \ldots, p$$

and

$$e_{l,n} = \sum_{i=1}^{l} (n - i + 1)^{-1}.$$

b. Specialize (7.5.1) to the two-sample problem discussed in Section 7.2.2 to obtain Var[$\mathbf{U}(\mathbf{0})$].
c. Specialize (7.5.1) to the m-sample problem to obtain an alternate to (7.2.20) as the covariance matrix for $\mathbf{U}(\mathbf{0})$ in the noncensored no-ties situation.

(Section 7.2.2)

7.2 (Continuation)

a. Prove that $\sum_{l=1}^{N} e_{l,N} = N$ and $\sum_{l=1}^{N} e_{l,N}^2 = 2N - e_{N,N}$.
b. For the two-sample problem show that $\mathbf{U}(\mathbf{0})$ can be expressed in the form (7.2.16) and that Var[$\mathbf{U}(\mathbf{0})$] can be given as (7.2.17).

(Section 7.2.2)

7.3

a. Verify that (7.2.29) closely approximates (7.2.27) and (7.2.28) when $d_i/|R_i|$ is very small.
b. Examine the adequacy of (7.2.29) as an approximation to (7.2.27) when $\boldsymbol{\beta} = \mathbf{0}$ and $d_i/|R_i|$ is not necessarily close to zero. Specifically examine

the approximation that (7.2.29) provides to the score vector $\mathbf{U}(\mathbf{0})$ and the log likelihood negative second derivative matrix $\mathbf{I}(\mathbf{0})$.

(Sections 7.2.3, 7.3; Farewell and Prentice, 1980.)

7.4 Consider the observed "likelihood function" $L=L[\boldsymbol{\beta}, S_0(t)]$ in Section 7.2.4 in the case in which there are no ties (i.e., all $d_i=1$). Use the fact that for a given $\boldsymbol{\beta}$, L is maximized by $\hat{S}_0(t)$ of (7.2.33), to obtain the maximized likelihood function

$$L_{\max}(\boldsymbol{\beta})=L\left[\boldsymbol{\beta}, \hat{S}_0(t)\right].$$

Compare $L_{\max}(\boldsymbol{\beta})$ with the partial likelihood function (7.2.3). Obtain the value of β that maximizes $L_{\max}(\beta)$ for the model and data in Example 7.2.1 and compare this with the estimate obtained there.

(Section 7.2; Bailey, 1979)

7.5 Consider the leukemia survival time data given in Example 6.3.2. Fit a proportional hazards model (7.1.1) by the methods of Section 7.2 and compare the estimates of the regression coefficient β and baseline survivor function $S_0(t)$ obtained with those obtained under the exponential model of Example 6.3.2. Assess the exponential model by examining $\hat{S}_0(t)$ from the distribution-free analysis.

(Section 7.2)

7.6

a. Consider the electrical insulation failure time data of Example 4.3.3. Assuming that the failure voltages for the two types of insulation discussed there have distributions with proportional hazards, use the methods of Section 7.2.2 to test that the two distributions are identical.
b. Obtain a confidence interval for $\delta=\exp\beta$ of (7.2.8) for the problem in part (a). Compare this with the estimate of δ one obtains under a Weibull model (see Problem 4.10).

(Section 7.2.2)

7.7 ***Examination of trend across several distributions.*** Suppose there are several lifetime distributions and that it is suspected that there is a trend among the hazard functions for the distributions. Certain types of trends can be examined with the methods of this chapter. For example, suppose there are m distributions, corresponding to levels $0, d_1, \ldots, d_{m-1}$ of a covariate d, and that it is suspected that hazard functions for the m distributions are proportional and vary monotonically with d.

a. Consider the model (7.1.1) in which population j has hazard function

$$h_j(t)=e^{\beta d_j}h_0(t) \qquad j=0,1,\ldots,m-1. \tag{7.5.2}$$

Derive the score function statistic $U(0)^2I(0)^{-1}$ for testing $\beta=0$ from (7.2.6) and (7.2.7). Show that $U(0)$ and $I(0)$ can be expressed as $\mathbf{d}'\mathbf{U}^*(\mathbf{0})$ and $\mathbf{d}'\mathbf{I}^*(\mathbf{0})\mathbf{d}$, respectively, where $\mathbf{d}'=(d_1,\ldots,d_{m-1})$ and $\mathbf{U}^*(\mathbf{0})$ and $\mathbf{I}^*(\mathbf{0})$ are the score vector (7.2.19) and second derivative matrix (7.2.20) used in the m-sample test of Section 7.2.2. Tarone (1975) has shown that $X_D^2=\mathbf{U}^*(\mathbf{0})'\mathbf{I}^*(\mathbf{0})^{-1}\mathbf{U}^*(\mathbf{0})-U(0)^2I(0)^{-1}$ is asymptotically $\chi^2_{(m-2)}$ under the hypothesis that the m distributions are equal. This can be used to test for departures from trend.

b. Consider the insulation data of Examples 4.3.2 and 6.4.3 for the voltage levels 30, 32, 34, and 36 kilovolts. Let $d_j=\log(v_j/30)$. Carry out a test for equality of the lifetime distributions at the four voltage levels, using (7.2.21). Also carry out a test for trend by considering the model (7.5.1) and testing $\beta=0$ as suggested in part (a). Finally, use the statistic X_D^2 of part (a) to examine departures from the trend represented by (7.5.2).

(Section 7.2; Tarone, 1975)

7.8 Holford (1976) has proposed a piecewise exponential model for analyzing data in situations where (7.1.1) might be used. Suppose that the hazard function for lifetime T, given the vector $\mathbf{x}$ of regressor variables, is

$$h(t|\mathbf{x})=e^{\alpha_j}e^{\mathbf{x}\boldsymbol{\beta}} \qquad a_{j-1}\leqslant t<a_j \tag{7.5.3}$$

where $0=a_0<a_1<\cdots<a_k<a_{k+1}=\infty$ partition the time axis into intervals. The α_j's in (7.5.3) are unknown parameters; note that this model stipulates an exponential survival distribution within each interval for any individual. Obtain the likelihood function for (7.5.3) based on an arbitrarily censored sample of lifetimes. Develop methods of testing hypotheses about $\boldsymbol{\beta}$; pay special attention to the hypothesis $\boldsymbol{\beta}=\mathbf{0}$ and give a test for the equality of m lifetime distributions.

(Section 7.2; Holford, 1976)

7.9 (continuation of Problem 7.8) Use the model (7.5.3) to reexamine the sets of data discussed in Problem 7.5 and in Problem 7.6, respectively.

7.10 Efficiency of partial likelihood in a simple situation. Suppose in the model (7.1.1) that $h_0(t)=\lambda$; that is, the distribution is actually exponential. Consider the case in which there is a single regressor variable x. Suppose that $t_1,\ldots,t_n$ are observed lifetimes in a random sample of n, corresponding to covariates $x_1,\ldots,x_n$; assume that the x_i's are centered so that $\Sigma x_i=0$.

a. Show that the joint p.d.f. of $a_2,\ldots,a_n$, where $a_i=t_i/t_1$, is

$$(n-1)!\left(\sum_{i=1}^{n} a_i e^{\beta x_i}\right)^{-n} \qquad a_i>0$$

where $a_1=1$. This can be used for inference about β when λ is unknown. Determine the expected information $I_1(\beta)$ based on this distribution and show that

$$I_1(0)=\frac{n}{n+1}\sum_{i=1}^{n} x_i^2=\frac{n\mu_2}{n+1}.$$

b. Consider the partial likelihood (7.2.3). Determine the expected information $I_2(0)$, noting that $t_1,\ldots,t_n$ are i.i.d. when $\beta=0$, whereupon each of the $n!$ possible rank vectors has probability $(n!)^{-1}$. Show that

$$I_2(0)=\frac{n\mu_2}{n-1}\sum_{i=1}^{n}\frac{n-i}{n-i+1}.$$

c. Examine $I_2(0)/I_1(0)$ for various values of n; this represents the efficiency of the partial likelihood method at $\beta=0$.

(Section 7.4.1; Kalbfleisch, 1974)

7.11 Checking the PH assumption with a time-dependent covariate. Consider the method suggested in Example 7.4.2 for testing departures from the proportional hazards model in the two-sample problem. Apply this method to the data of Example 7.2.1. [Remark: to reduce high asymptotic correlation between the two parameters in the model, it is preferable to work with the covariable x_2 defined as $x_2=x_1(\log t-c)$, where c is a constant close to the average of the $\log t_{(i)}$'s.]

(Section 7.4.2; Kalbfleisch and McIntosh, 1977)

7.12 Examine possible departures from proportional hazards for the two distributions discussed in Example 7.3.1 by working with the transformation family of models (7.4.4).

(Section 7.4.2; Aranda-Ordaz, 1980)

7.13 The data in Table 7.5.1 show the survival experience of a certain type of cancer patient in life table form. Time is given in months from treatment. The 285 patients fall into four categories, which correspond to two levels for each of two factors A and B. The data give the survival experience for each of the four patient groups A_1B_1, A_1B_2, A_2B_1, and A_2B_2.

Examine whether the survival distributions for the four patient groups might be the same. Attempt to assess the effect of factors A and B on the distributions.

(Sections 7.3, 7.4)

Table 7.5.1 Survival Data for Cancer Patients Classified by Two Factors

	A_1B_1		A_1B_2		A_2B_1		A_2B_2	
Interval	n_{1i}	d_{1i}	n_{2i}	d_{2i}	n_{3i}	d_{3i}	n_{4i}	d_{4i}
[0,3)	75	15	87	18	59	12	64	9
[3,6)	60	3	69	14	46	7	55	8
[6,9)	56	14	55	8	38	8	47	11
[9,12)	41	17	47	16	30	7	34	8
[12,15)	22	7	30	11	21	9	25	5
[15,18)	13	6	19	12	10	5	18	9
[18,21)	5	4	7	4	4	2	8	3
$[21,\infty)$	1	1	2	2	2	2	4	4

CHAPTER 8

Nonparametric and Distribution-Free Methods

When circumstances do not support the adoption of a fully parametric lifetime distribution model, one often turns to nonparametric or distribution-free methods. Some such methods have already been discussed in Chapters 2 and 7, where it was seen that certain procedures are also useful in preliminary investigations of a situation. In this chapter we examine two specific problems in a little more detail: these are the estimation of a distribution's quantiles and survivor function by distribution-free methods and the comparison of distributions through the use of rank tests.

8.1 NONPARAMETRIC ESTIMATION OF SURVIVOR FUNCTIONS AND QUANTILES

If one assumes a specific parametric model, estimation of characteristics of the distribution is, at least in principle, straightforward. Estimation for the exponential, Weibull, gamma, log-normal, and other distributions has been discussed at length in previous chapters. We now consider nonparametric interval estimation of the survivor function and quantiles of a distribution. These methods can be used when it is not feasible or desirable to assume a specific parametric model.

8.1.1 Interval Estimates of the Survivor Function

Uncensored Data

Interval estimates of the survivor function can be obtained from uncensored data by using well-known procedures based on the binomial distribution. Suppose that $t_1, \dots, t_n$ is a random sample from a distribution with survivor function $S(t)$, and suppose that it is desired to estimate $S(a)$, for some specified time a. Define the random variable X as the number of t_i's in the sample that are greater than or equal to a. Then X has a binomial distribution with probability function

$$\Pr(X=x;\,p)=\binom{n}{x}p^x(1-p)^{n-x}$$

where $p=S(a)$. Confidence intervals for p are obtained in well-known ways that are described in most elementary statistics books.

For example, to obtain a lower α confidence limit on p it is necessary to determine all p values such that

$$\Pr(X\geqslant x_0;\,p)\geqslant 1-\alpha \tag{8.1.1}$$

where x_0 is the observed value of X. The set of p values satisfying (8.1.1.) is of the form $(p_L, 1)$ and p_L is the desired lower confidence limit. It can be shown that

$$p_L=\frac{x_0}{x_0+(n-x_0+1)F_{(2(n-x_0+1)),2x_0),\alpha}}$$

(e.g., Kempthorne and Folks, 1971, pp. 367–368). Upper confidence limits for p can be found in a similar way; the upper α confidence limit is

$$p_u=\frac{x_0+1}{(x_0+1)+(n-x_0)F_{(2(n-x_0),2x_0+2),1-\alpha}}.$$

For larger values of n upper and lower confidence limits can be obtained using one of several approximations. Some approaches are

1. Treat $Z_1=(\hat{p}-p)/[\hat{p}(1-\hat{p})/n]^{1/2}$ as a standard normal pivotal quantity, where $\hat{p}=X/n$ is the m.l.e. of p. A continuity correction can be used with Z_1 and with Z_2, below.
2. Treat $Z_2=(\hat{p}-p)/[p(1-p)/n]^{1/2}$ as standard normal.
3. Let $\psi=\psi(p)$ be a one-to-one function of p and treat

$$Z_3=\frac{\hat{\psi}-\psi}{S_{\hat{\psi}}}$$

as a standard normal pivotal, where $\hat{\psi}=\psi(\hat{p})$ and the variance of $\hat{\psi}$ is estimated by

$$S_{\hat{\psi}}^2 = [\psi'(\hat{p})]^2 \frac{\hat{p}(1-\hat{p})}{n}. \tag{8.1.2}$$

The idea is to choose ψ so that the distribution of $\hat{\psi}$ is closer to normal than is that of $\hat{p}$; Z_3 will then give confidence intervals with closer to the nominal confidence coefficient, than Z_1 or Z_2 will. One good choice of ψ is (Anscome, 1964; Cox and Snell, 1968)

$$\psi = \int_0^p t^{-1/3}(1-t)^{-1/3}\, dt. \tag{8.1.3}$$

Cox and Snell (1968, Sec. 8) give a table from which to calculate ψ from p. Another choice that is only slightly less satisfactory is the logistic transform

$$\psi = \log\left(\frac{p}{1-p}\right). \tag{8.1.4}$$

4. A fourth procedure for obtaining confidence intervals for p is the likelihood ratio method. The likelihood ratio statistic for testing $H_0: p = p_0$ versus $H_1: p \neq p_0$ is

$$\Lambda = 2n\hat{p}\log\left(\frac{\hat{p}}{p_0}\right) + 2n(1-\hat{p})\log\left(\frac{1-\hat{p}}{1-p_0}\right). \tag{8.1.5}$$

Under H_0, Λ is approximately distributed as $\chi^2_{(1)}$ in large samples, and an approximate α confidence interval for p consists of all values p_0 such that $\Lambda \leqslant \chi^2_{(1),\alpha}$.

These well-known results have been reviewed because with arbitrarily censored data methods can be based on procedures similar to these. It should be remarked that the approximations in (1) to (4) agree quite well for sufficiently large n, but for moderate n differences can be expected, particularly when p is close to 0 or 1. A sensible procedure is to use the exact method where possible, reserving the approximations for situations where n is quite large. Thomas and Grunkemeier (1975) discuss the accuracy of (1), (2), and (4) in several situations.

We now turn to the case of censored data.

Censored Data

If the data are singly Type I or Type II censored, $S(a)$ can be estimated in essentially the same way as when there is no censoring. If the data are Type I censored so that only lifetimes that are less than or equal to some preassigned value T_0 are known, then for $a \leqslant T_0$, $S(a)$ can be estimated in exactly the same way as for uncensored data. Strictly speaking, $S(a)$ cannot be estimated directly for $a > T_0$, though an upper confidence limit on $S(T_0)$ is a conservative upper confidence limit on $S(a)$, $a > T_0$.

Type II censoring is slightly more complicated, since the termination time $t_{(r)}$ of the experiment is a random variable and can be either greater than or less than a specified value a. If estimation of $S(a)$ is important, the experiment can be designed so that the probability of $t_{(r)}$ being less than a is small, in which case one can proceed to estimate $S(a)$ as if the censoring were Type I, ignoring the small probability of $t_{(r)} < a$. If the probability of $t_{(r)}$ being less than a is not negligible, the procedure can be modified, but then the inferences will no longer be strictly distribution-free.

If the data are arbitrarily censored, one can extend procedures (1) through (4) for uncensored data, employing the product-limit estimate $\hat{S}(a)$ of $S(a)$. Suppose that from a sample involving n individuals, k distinct lifetimes $t_{(1)} < \cdots < t_{(k)}$ are observed. Let d_j represent the number of deaths at $t_{(j)}$, and n_j the number of individuals alive just prior to $t_{(j)}$. The product-limit estimate (2.3.2) of $S(a)$ is

$$\hat{S}(a) = \prod_{j:\, t_{(j)} < a} \frac{n_j - d_j}{n_j}$$

and an estimate of the variance of $\hat{S}(a)$ is, from (2.3.3),

$$V_a = \hat{S}(a)^2 \sum_{j:\, t_{(j)} < a} \frac{d_j}{n_j(n_j - d_j)}. \tag{8.1.6}$$

For convenience let $P = S(a)$. One obvious way to obtain confidence intervals for P is to use the large-sample approximation

$$\frac{\hat{P} - P}{\sqrt{V_a}} \sim N(0,1). \tag{8.1.7}$$

This is analogous to **1** for uncensored data and in fact it is readily seen that (8.1.7) reduces to Z_1 of **1** when the data are uncensored.

A second approach for arbitrarily censored data would be to use a transformation $\psi = \psi(P)$ to obtain confidence intervals, as described in **3**.

In this case one would use the approximation

$$\frac{\hat{\psi}-\psi}{\sqrt{V_{\hat{\psi}}}} \sim N(0,1) \tag{8.1.8}$$

where $\hat{\psi}=\psi(\hat{P})$ and $V_{\hat{\psi}}=[\psi'(\hat{P})]^2 V_a$. The transformations mentioned in **3** are useful in censored data situations too.

It is also possible to generalize **2** and **4** for uncensored data to the case of arbitrarily censored data. This has been discussed by Thomas and Grunkemeier (1975). These two methods tend to give results that are in good agreement with each other and with (8.1.8), used in conjunction with (8.1.3) or (8.1.4); we shall describe only the likelihood ratio method. For this one considers the ratio of the likelihood functions maximized unconditionally and under $H_0: S(a)=P_0$; this is appropriate for testing H_0 versus $H_1: S(a) \neq P_0$. From the arguments that follow (2.3.5), the overall maximized likelihood function is

$$L_1 = \prod_{j=1}^{k} (1-\hat{p}_j)^{d_j} \hat{p}_j^{n_j-d_j}$$

where $\hat{p}_j = 1 - d_j/n_j$. To consider the maximum of the likelihood under H_0, note that $S(a)=P_0$ in (2.3.5) implies that we must have $\hat{S}(t_l+0)=\hat{S}(t_{l+1})=P_0$, where l is such that $a \in (t_l, t_{l+1}]$. This implies that to maximize the likelihood subject to $S(a)=P_0$ it is necessary to maximize

$$L = \prod_{j=1}^{k} (1-p_j)^{d_j} p_j^{n_j-d_j}$$

subject to the restriction $p_1 \cdots p_l = P_0$. To do this we use a Lagrange multiplier and consider

$$\log L + \lambda\left(\sum_{i=1}^{l} \log p_i - \log P_0\right) = \sum_{i=1}^{k} d_j \log(1-p_j) + (n_j - d_j)\log p_j$$
$$+\lambda\left(\sum_{i=1}^{l} \log p_i - \log P_0\right).$$

Setting derivatives with respect to each of $p_1, \ldots, p_k$ equal to zero, we find the constrained m.l.e.'s to be

$$\tilde{p}_j = 1 - \frac{d_j}{n_j + \lambda} \qquad j = 1, \ldots, l$$

$$\tilde{p}_j = 1 - \frac{d_j}{n_j} \qquad j = l+1, \ldots, k$$

where $\lambda = \lambda(P_0)$ satisfies

$$\prod_{i=1}^{l} \tilde{p}_i = \prod_{i=1}^{l} \left(1 - \frac{d_i}{n_i + \lambda}\right) = P_0. \tag{8.1.9}$$

The maximum of the likelihood under $H_0: S(a) = P_0$ is thus

$$L_2 = \prod_{j=1}^{k} (1 - \tilde{p}_j)^{d_j} \tilde{p}_j^{\,n_j - d_j}$$

and the likelihood ratio statistic for testing H_0 is

$$\begin{aligned} \Lambda &= -2 \log\left(\frac{L_2}{L_1}\right) \\ &= -2\left[\sum_{j=1}^{k} (n_j - d_j) \log\left(\frac{\tilde{p}_j}{\hat{p}_j}\right) + d_j \log\left(\frac{1 - \tilde{p}_j}{1 - \hat{p}_j}\right)\right] \\ &= 2 \sum_{j=1}^{k} \left[n_j \log\left(1 + \frac{\lambda}{n_j}\right) - (n_j - d_j) \log\left(1 + \frac{\lambda}{n_j - d_j}\right)\right]. \end{aligned} \tag{8.1.10}$$

To get an α confidence interval for $S(a)$ we need to determine the set of all P_0 such that $\Lambda \leqslant \chi^2_{(1),\alpha}$. This can be done by finding the set of all λ in (8.1.10) that make $\Lambda \leqslant \chi^2_{(1),\alpha}$ and then obtaining from this the corresponding set of P_0 values. Thomas and Grunkemeier (1975) show that the sets of λ values making $\Lambda \leqslant \chi^2_{(1),\alpha}$ are closed intervals $[\lambda_L, \lambda_u]$ such that $\lambda_L < 0 < \lambda_u$, unless $\hat{S}(a) = 0$, in which case $0 = \lambda_L < \lambda_u$. Because P in (8.1.9) is an increasing function of λ, the confidence intervals for $P = S(a)$ are thus of the form $[P_L, P_u]$, with $0 < P_L < P_u < 1$, unless $\hat{S}(a) = 0$, in which case $P_L = 0 < P_u < 1$. Note that

$$P_L = \prod_{i=1}^{l} \left(1 - \frac{d_i}{n_i + \lambda_L}\right) \quad \text{and} \quad P_u = \prod_{i=1}^{l} \left(1 - \frac{d_i}{n_i + \lambda_u}\right).$$

Thomas and Grunkemeier investigate the adequacy of the likelihood ratio method and the method based on (8.1.7) in several censored sampling situations involving exponential or Weibull lifetimes and uniform censoring distributions. It appears that the likelihood ratio method performs substantially better than the other method and can be recommended for use in most situations with moderate to large sample sizes. The use of pivotals (8.1.8) was not investigated, but it seems likely that this procedure would compare favorably with the likelihood ratio method if (8.1.3) or (8.1.4) were used.

Example 8.1.1 (Example 2.3.1 revisited) The data below are remission times for two groups of leukemia patients, and they have been discussed earlier, in Example 2.3.1 and elsewhere. Times are in weeks and starred observations are censoring times.

Sample 1(control)	1, 1, 2, 2, 3, 4, 4, 5, 5, 8, 8, 8, 8, 11, 11, 12, 12, 15, 17, 22, 23
Sample 2 (drug 6 − MP)	6, 6, 6, 6*, 7, 9*, 10, 10*, 11*, 13, 16, 17*, 19*, 20*, 22, 23, 25*, 32*, 32*, 34*, 35*

Table 8.1.1 shows .95 confidence intervals for $S(10+0)$ and $S(20+0)$, obtained with (1) approximation (8.1.7), (2) approximation (8.1.8) with $\psi(p)=\log[p/(1-p)]$, and (3) the likelihood ratio (LR) procedure. In addition, since there is no censoring in sample 1, the exact confidence interval obtained from the binomial distribution is shown as (4).

To illustrate the calculations involved in the various methods, consider sample 1 and confidence limits for $S(10+0)=p$. It was determined in Example 2.3.1 that $\hat{p}=\hat{S}(10+0)=.381$; from (8.1.6) we also find $V_a^{1/2}=.106$. Hence using (8.1.7) we easily find from the approximation $(\hat{p}-p)/.106\sim N(0,1)$ the .95 confidence interval (.17,.59). To use (8.1.8) with $\psi(p)$ given as the logit transformation $\log[p/(1-p)]$ we calculate $\hat{\psi}=\log[\hat{p}/(1-\hat{p})]=$

Table 8.1.1 .95 Confidence Intervals for $S(10+0)$ and $S(20+0)$

		$S(10+0)$	$S(20+0)$
Sample 1	1. $\hat{p}$	(.17,.59)	(.0,.22)
	2. $\hat{\psi}$	(.20,.60)	(.024,.31)
	3. LR	(.20,.59)	(.016,.27)
	4. Exact	(.18,.63)	(.012,.30)
Sample 2	1. $\hat{p}$	(.56,.94)	(.40,.85)
	2. $\hat{\psi}$	(.53,.89)	(.39,.82)
	3. LR	(.54,.90)	(.40,.82)

-0.485; also, since $d\psi/dp = p^{-1}(1-p)^{-1}$, we find that $V_{\hat{\psi}}^{1/2} = V_a^{1/2}\hat{p}^{-1}(1-\hat{p})^{-1} = 0.450$. The approximation $\hat{\psi} \sim N(\psi, 0.450)$ gives the .95 confidence interval $(-1.306, 0.396)$ for ψ, which corresponds to the interval (.20,.60) for p. The confidence limits obtained by the likelihood ratio method are found as described earlier.

Except for estimation of $S(20+0)$ in sample 1, there is no substantial difference among the methods. In that case, methods (2) and (3) provide reasonably good approximations to the exact confidence interval, whereas (1) is considerably poorer. The beta transformation (8.1.3) gives answers close to those for the likelihood ratio and logit transformation methods but is slightly more troublesome to use than the logit transformation. Any of the methods should, of course, be used with caution when $S(a)$ is likely to be close to 0 or 1.

8.1.2 Interval Estimates of Distribution Quantiles

Uncensored Data

Distribution-free confidence intervals for the pth quantile t_p of a continuous lifetime distribution can be readily obtained when the data are complete or Type II censored. Two-sided intervals for t_p are, for example, of the form

$$t_{(r)} \leq t_p \leq t_{(s)} \tag{8.1.11}$$

where $1 \leq r < s \leq n$ and $t_{(1)} < \cdots < t_{(n)}$ are the ordered observations in a random sample of size n from the distribution in question.

Let us determine the confidence coefficient for an interval of the form (8.1.11). To do this let X represent the number of observations in a random sample of size n that are $\leq t_p$; X has a binomial distribution with probability function $\binom{n}{x}p^x(1-p)^{n-x}$. The inequality $t_{(r)} \leq t_p \leq t_{(s)}$ is satisfied by a sample if and only if $r \leq X \leq s-1$, and hence

$$\Pr\left[t_{(r)} \leq t_p \leq t_{(s)}\right] = \sum_{x=r}^{s-1} \binom{n}{x} p^x(1-p)^{n-x}. \tag{8.1.12}$$

By utilizing the well-known relationship between the binomial distribution and the incomplete beta function [see (B19)], we can write (8.1.12) in the alternate form

$$\Pr\left[t_{(r)} \leq t_p \leq t_{(s)}\right] = B_p(r, n-r+1) - B_p(s, n-s+1) \tag{8.1.13}$$

where $B_p(a, b)$ is the incomplete beta function (B18). The interval (8.1.11) is thus distribution-free and has confidence coefficient α given by (8.1.12). Confidence coefficients can be calculated directly or obtained from tables of the binomial distribution or incomplete beta function (see Appendix B).

For a given p and n it will be possible to find confidence intervals (8.1.11) only for certain values of α. If a two-sided .90 confidence interval for t_p is wanted, say, then r, s, and n can be selected to make α as close as possible to .90, though it may not be possible to make it exactly .90. Typically, there will be more than one choice of n, r, and s, though other considerations may point to a particular combination. For example, if the data are to be obtained from a life test and it is desirable to terminate the test as soon as possible, one may want to have s as small as possible. For very precise estimation, on the other hand, it is desirable to choose r and s as close together as possible. It should be noted, incidentally, that although the confidence interval procedure described here is distribution-free, its efficiency properties depend on the underlying distribution, as well as on r, n, and s.

One-sided confidence intervals of the form $t_{(r)} \leqslant t_p$ or $t_p \leqslant t_{(s)}$ can also be included under (8.1.11) by defining $t_{(0)}=0$ and $t_{(n+1)}=\infty$. Then the choice $s=n+1$, for example, gives the confidence interval

$$t_{(r)} \leqslant t_p \tag{8.1.14}$$

with confidence coefficient given by (8.1.12) as

$$\sum_{x=r}^{n} \binom{n}{x} p^x (1-p)^{n-x}. \tag{8.1.15}$$

Lower confidence limits for quantiles are often wanted in life testing applications, and typically one is interested in t_p, for small p. In this case r is sometimes chosen to be unity in (8.1.14), giving $[t_{(1)}, \infty)$ as a confidence interval with confidence coefficient

$$\sum_{x=1}^{n} \binom{n}{x} p^x (1-p)^{n-x} = 1-(1-p)^n.$$

Example 8.1.2 Suppose that we want a lower .90 confidence limit on the .10 quantile of a distribution. Let us consider confidence intervals of the form (8.1.14), with different values of r. With $r=1$ the confidence coefficient will be

$$\alpha = 1 - .90^n$$

and it is easily found that $\alpha \doteq .90$ for $n=22$; the exact value of α is actually .902. Similarly, for other values of r (8.1.15) with $p=.10$ can be used to determine the value of n, making α as close as possible to .90. For $r=1,2,3$ the values of n, and the corresponding α, are

r	n	α
1	22	.902
2	37	.896
3	52	.903

Any one of these combinations of r and n gives a lower .90 confidence limit for $t_{.10}$. The values of r and n one selects will depend upon the precision of estimation required and economic factors such as the number of individuals it is feasible to sample, the amount of time available for testing, and so on. The true underlying lifetime distribution affects these considerations, however, so some subjective judgment is necessary in making a choice. Generally speaking, high precision implies larger values of n (and hence of r). On the other hand, economic considerations may suggest small values of r and n. For example, if the underlying distribution were exponential, the expected time to observe $t_{(r)}$ would be

$$E\left(t_{(r)}\right)=\theta\left(\frac{1}{n}+\cdots+\frac{1}{n-r+1}\right)$$

and the three choices $(r,n)=(1,22)$, $(2,37)$, and $(3,52)$ give $E(t_{(r)})=0.045\theta$, 0.055θ, and 0.059θ, respectively.

For large values of n approximations can be employed to calculate the binomial probabilities in (8.1.12). With the usual normal approximation, the confidence coefficient $\alpha=\Pr(t_{(r)}\leqslant t_p\leqslant t_{(s)})$ is approximated by

$$\Pr\left(\frac{r-.5-np}{\left[np(1-p)\right]^{1/2}}\leqslant Z\leqslant\frac{s-.5-np}{\left[np(1-p)\right]^{1/2}}\right)$$

where Z has a standard normal distribution. With $r=3$ and $p=.10$, for example, we have

$$\alpha=\Pr\left(t_{(3)}\leqslant t_{.10}\right)$$

$$\doteq\Pr\left(\frac{2.5-.10n}{(.09n)^{1/2}}\leqslant Z\right).$$

To make $\alpha = .90$ we need

$$\frac{2.5 - .10n}{(.09n)^{1/2}} = -1.281$$

which gives $n = 53$; the value computed from the exact binomial probabilities was 52.

Censored Data

When the data are Type II censored, the methods just described still apply, provided, of course, that the experiment continue until the necessary order statistics $t_{(r)}$ and $t_{(s)}$ in (8.1.12) are observed. With arbitrarily censored data there is no completely satisfactory approach, though by using the relationship between the quantiles and survivor function of a distribution one can obtain rough confidence intervals for quantiles. To do this recall that if t_p is the pth quantile of a distribution with survivor function $S(t)$, then $\Pr(L \leqslant t_p) = \Pr[S(L) \geqslant 1-p]$. It follows that if $p_L(\text{data}; a)$ is a lower α confidence limit for $S(a)$, then a lower α confidence limit for t_p can be obtained by finding a such that $p_L(\text{data}; a) = 1-p$.

Consider, for example, sample 2 in Example 8.1.1; we shall obtain a lower .95 confidence limit for $t_{.50}$, the median of the remission time distribution, basing confidence limits for $S(a)$ on the pivotal (8.1.7) for illustration. In this case the .95 lower confidence limit for $S(a)$ is $\text{LCL} = \hat{S}(a) - 1.645V_a^{1/2}$, where $\hat{S}(a)$ is the product-limit estimate and

$$V_a = \hat{S}(a)^2 \sum_{j:\, t_{(j)} < a} \frac{d_j}{n_j(n_j - d_j)}$$

as in (8.1.6). The value of a that makes LCL equal to .5 can be found by trial and error. Because of the discreteness of $\hat{S}(a)$, it is not in fact possible to satisfy LCL=.5 exactly, but we can determine values giving $\text{LCL} \doteq .5$. To find a it is necessary to consider only the values of observed lifetimes. We find for $a = 16$ that $\hat{S}(16) = .690$, $V_{16}^{1/2} = .107$, and LCL=.514; for $a = 22$ we have $\hat{S}(22) = .627$, $V_{22}^{1/2} = .114$, and LCL=.439. Consequently $a = 16$ can be taken as a conservative lower .95 confidence limit for $t_{.50}$.

This procedure produces the confidence limits based on (8.1.11) and (8.1.12) in the case in which there is no censoring. For arbitrarily censored data the procedure provides reasonable results if the method upon which the confidence limits for $S(a)$ are based is reliable for the sample size and type of censoring in question.

8.2 RANK TESTS FOR COMPARING DISTRIBUTIONS

8.2.1 Linear Rank Tests for the *m*-Sample Problem

Tests for the equality of two or more lifetime distributions are often required. When it is not convenient or appropriate to adopt a parametric family of models within which to carry out tests, distribution-free methods can be used. Some such procedures were discussed in Chapter 7, in connection with the proportional hazards model. This section examines another class of distribution-free tests, based on the linear model (6.1.9) for log lifetimes. Two tests will be discussed in detail. One is the log rank test, or exponential ordered scores test, considered earlier in Section 7.2.2, and the other is a generalization of the Wilcoxon and Kruskal–Wallis tests (Wilcoxon, 1945; Kruskal and Wallis, 1952). These are only two out of a large class of possible tests, but they warrant discussion because of their widespread use in connection with lifetime data. Both tests can be derived from more than one point of view, but we examine them here as linear rank tests. In order to do this we shall review the basic ideas of rank tests and consider their extension to censored data. A thorough discussion of this area is beyond the scope of the book, but the main problems will be outlined.

Several books contain extended treatments of rank tests (e.g., Hajek and Sidak, 1967; Hajek, 1969; Lehmann, 1975), and reference can be made to these for many details. A few results are sketched below for the problem of testing the equality of m distributions. More specifically, we develop tests for comparing distributions that can differ only with respect to location. In this case two distributions, 1 and 2, are assumed to have p.d.f.'s $g(y-\theta)$ and $g(y)$, and they are identical if and only if $\theta=0$. The tests described are thus appropriate for testing whether two or more distributions have the same location parameter, given that they have the same scale and shape. They can be used to test the equality of any group of distributions, but they will be poor at detecting certain types of departures from equality.

Consider m distributions that can differ only with respect to location. Without loss of generality, the p.d.f.'s of the distributions can be assumed to be of the form

$$g_1(y)=g(y-\theta_1),\ldots, g_{m-1}(y)=g(y-\theta_{m-1}), g_m(y)=g(y). \quad (8.2.1)$$

Equality of the distributions is then tested by testing that $\theta_1=\cdots=\theta_{m-1}=0$. As in Chapter 7, results for the m-sample problem can be concisely expressed in terms of dummy regressor variables. Let $\boldsymbol{\theta}=(\theta_1,\ldots,\theta_{m-1})'$ and let $\mathbf{x}=(x_1,\ldots,x_{m-1})$ be a vector of indicator variables, defined so that individuals from distributions $1,\ldots,m-1$ have $\mathbf{x}$ vectors $(1,0,\ldots,0)$,

$(0,1,\ldots,0),\ldots,(0,\ldots,0,1)$, respectively, and individuals from distribution m have $\mathbf{x}=(0,\ldots,0)$. Consider the model in which an individual with regression vector $\mathbf{x}$ has p.d.f.

$$f(y|\mathbf{x})=g(y-\mathbf{x}\boldsymbol{\theta}). \tag{8.2.2}$$

Then individuals with distributions $1,\ldots,m$ have p.d.f.'s as given in (8.2.1).

Tests With Uncensored Data

The construction of rank tests of $\boldsymbol{\theta}=\mathbf{0}$ in (8.2.2) will be described first for the case of uncensored data, with censoring being taken up later. Briefly put, a rank test is one for which the test statistic is a function of the ranks of the observations and not their actual values. Such a test is distribution-free in the sense that significance levels calculated from the distribution of the ranks are valid for arbitrary distributions. The power of a rank test depends on the alternative hypothesis and the underlying distribution, but tests can be selected to have good power against specific types of alternatives. In the present context let $y_1,\ldots,y_n$ be a sample from (8.2.2), selected as a set of random samples from each of distributions $1,\ldots,m$; let N_i be the number of observations from distribution i $(N_1+\cdots+N_m=n)$. (In life distribution work the y_i's will typically be log lifetimes.) Let $\mathbf{r}=[(1),\ldots,(n)]$ denote the rank vector based on the y_i's; that is, (i) is the label of the individual with the ith smallest y value. The ordered observations $y_{(1)}<\cdots<y_{(n)}$ are assumed to be distinct. This entails no loss of generality under a continuous model. Rank tests of $\boldsymbol{\theta}=\mathbf{0}$ can be constructed by considering the scores test based on the distribution of $\mathbf{r}$. This approach is outlined here; a more detailed presentation can be found, for example, in Hajek and Sidak (1967).

The probability function of $\mathbf{r}=[(1),\ldots,(n)]$ is

$$p(\mathbf{r};\boldsymbol{\theta})=\int\underset{A}{\cdots}\int\prod_{j=1}^{n}g\left[y_{(j)}-\mathbf{x}_{(j)}\boldsymbol{\theta}\right]dy_{(1)}\cdots dy_{(n)}$$

where A is the region $\{(y_{(1)},\ldots,y_{(n)}):\ -\infty<y_{(1)}<\cdots<y_{(n)}<\infty\}$ and $\mathbf{x}_{(j)}$ is the regression vector associated with (j). The first derivatives of the log likelihood based on $p(\mathbf{r};\boldsymbol{\theta})$ are thus

$$U_l(\boldsymbol{\theta})=\frac{\partial\log p(\mathbf{r};\boldsymbol{\theta})}{\partial\theta_l}$$

$$=\frac{-1}{p(\mathbf{r};\boldsymbol{\theta})}\sum_{i=1}^{n}\int\underset{A}{\cdots}\int x_{(i)l}\frac{g'(y_{(i)}-\mathbf{x}_{(i)}\boldsymbol{\theta})}{g(y_{(i)}-\mathbf{x}_{(i)}\boldsymbol{\theta})}\prod_{j=1}^{n}g(y_{(j)}-\mathbf{x}_{(j)}\boldsymbol{\theta})\,dy_{(1)}\cdots dy_{(n)}$$

$$l=1,\ldots,m-1. \tag{8.2.3}$$

A test of $\boldsymbol{\theta}=\mathbf{0}$ can be based on the score statistic

$$\mathbf{U}(\mathbf{0})=\left[U_1(\mathbf{0}),\ldots,U_{m-1}(\mathbf{0})\right]'.$$

When $\boldsymbol{\theta}=\mathbf{0}$, all individuals have the same distribution and $p(\mathbf{r};\mathbf{0})=(n!)^{-1}$ for each possible rank vector $\mathbf{r}$. Thus

$$U_l(\mathbf{0})=-n!\sum_{i=1}^{n}\int\cdots\int_A x_{(i)l}\frac{g'(y_{(i)})}{g(y_{(i)})}\prod_{j=1}^{n}g(y_{(j)})\,dy_{(1)}\cdots dy_{(n)}.$$

But $n!\prod_{j=1}^{n}g(y_{(j)})$ is the joint p.d.f. of $y_{(1)},\ldots,y_{(n)}$ when $\boldsymbol{\theta}=\mathbf{0}$ [see (D2)], thus

$$U_l(\mathbf{0})=\sum_{i=1}^{n}x_{(i)l}\alpha_i \qquad l=1,\ldots,m-1 \tag{8.2.4}$$

where

$$\alpha_i=E\left(-\frac{g'(y_{(i)})}{g(y_{(i)})};\boldsymbol{\theta}=\mathbf{0}\right). \tag{8.2.5}$$

The α_i's are called the scores associated with $y_{(1)},\ldots,y_{(n)}$. Note that if one uses the likelihood function based on the actual observations $y_1,\ldots,y_n$, and not just their ranks, the score statistic at $\boldsymbol{\theta}=\mathbf{0}$ has components

$$\sum_{i=1}^{n}x_{il}\frac{-g'(y_i)}{g(y_i)}$$

so that the effect of considering only the ranks is to replace $g'(y_i)/g(y_i)$ with a score that is a function of the rank of y_i.

The mean and variance of $\mathbf{U}(\mathbf{0})$ can be derived by standard permutation theory arguments, since under $H_0:\boldsymbol{\theta}=\mathbf{0}$ all $n!$ possible rank vectors $\mathbf{r}$ are equally likely. The necessary formulas are given by the following well-known result:

Lemma 8.2.1 Let $\mathbf{x}_1,\ldots,\mathbf{x}_n$ be given vectors with $\mathbf{x}_i=(x_{i1},\ldots,x_{ip})$ and let $\alpha_1,\ldots,\alpha_n$ be given constants such that $\Sigma\alpha_i=0$. Let $[(1),\ldots,(n)]$ be a random permutation of $(1,\ldots,n)$ and define

$$U_l=\sum_{i=1}^{n}x_{(i)l}\alpha_i \qquad l=1,\ldots,p.$$

Then $E(U_l)=0$ and

$$E(U_lU_s)=\left(\sum_{i=1}^{n}\alpha_i^2/(n-1)\right)\left(\sum_{i=1}^{n}x_{il}x_{is}-n\bar{x}_l\bar{x}_s\right)\qquad l,s=1,\ldots,p \tag{8.2.6}$$

where

$$\bar{x}_l=\sum_{i=1}^{n}\frac{x_{il}}{n}\qquad l=1,\ldots,p.$$

Proof First,

$$E(U_l)=\sum_{i=1}^{n}\alpha_iE_p(x_{(i)l})$$

where E_p denotes expectation over the set of permutations of $1,2,\ldots,n$. This gives

$$\sum_{i=1}^{n}\alpha_i\left(\sum_{j=1}^{n}x_{jl}\frac{1}{n}\right)=\left(\sum_{i=1}^{n}\alpha_i\right)\bar{x}_l=0.$$

Also,

$$E(U_lU_s)=\sum_{i=1}^{n}\sum_{j=1}^{n}\alpha_i\alpha_jE_p(x_{(i)l}x_{(j)s})$$

$$=\sum_{i=1}^{n}\alpha_i^2\left(\sum_{k=1}^{n}x_{kl}x_{ks}\frac{1}{n}\right)+\sum_{\substack{i=1\\i\neq j}}^{n}\sum_{j=1}^{n}\alpha_i\alpha_j\left(\sum_{\substack{k=1\\k\neq t}}^{n}\sum_{t=1}^{n}x_{kl}x_{ts}\frac{1}{n(n-1)}\right)$$

$$=\left(\sum_{i=1}^{n}\alpha_i^2/n\right)\sum_{k=1}^{n}x_{kl}x_{ks}-\left(\sum_{i=1}^{n}\alpha_i^2/n(n-1)\right)$$

$$\times\left[\left(\sum_{k=1}^{n}x_{kl}\right)\left(\sum_{t=1}^{n}x_{ts}\right)-\sum_{k=1}^{n}x_{kl}x_{ks}\right]$$

$$=\left(\sum_{i=1}^{n}\alpha_i^2/(n-1)\right)\left(\sum_{k=1}^{n}x_{kl}x_{ks}-n\bar{x}_l\bar{x}_s\right).\quad\blacksquare$$

We observe that for the scores (8.2.5) $\Sigma\alpha_i=0$, since

$$\sum \alpha_i = E\left(-\sum \frac{d}{dy}\log g(y_{(i)})\right) = E\left(-\sum \frac{d}{dy}\log g(y_i)\right) = 0.$$

The mean and covariance matrix for the score vector $\mathbf{U(0)}$ are thus given by Lemma 8.2.1 as

$$E[\mathbf{U(0)}]=\mathbf{0} \quad \text{and} \quad \mathbf{V}=E[\mathbf{U(0)U(0)'}]$$

$$=\left(\sum_{i=1}^{n}\alpha_i^2/(n-1)\right)\mathbf{X}_c'\mathbf{X}_c$$

where $\mathbf{X}_c$ is the $n\times p$ matrix of x's centered about their means. For the m-sample problem the score vector $\mathbf{U(0)}$ and covariance matrix $\mathbf{V}$ have components

$$U_l(0)=\sum_{i\in S_l}\alpha_{(i)} \qquad l=1,\ldots,m-1$$

$$V_{ls}=\left(\sum_{i=1}^{n}\alpha_i^2/(n-1)\right)\left(N_l\delta_{ls}-\frac{N_lN_s}{n}\right) \qquad l,s=1,\ldots,m-1$$

(8.2.7)

where $\delta_{ls}=1$ if $l=s$ and 0 if $l\neq s$ and S_l denotes the individuals in the sample who are from distribution l.

It is feasible to calculate significance levels for the test of $\boldsymbol{\theta}=\mathbf{0}$ from the exact permutation distribution of $\mathbf{U(0)}$ only if sample sizes are small. However, under quite general conditions (e.g., Hajek and Sidak; 1967, p. 159) the distribution of $\mathbf{U(0)}$ is asymptotically normal, and the equality of the m distributions can be tested with the statistic

$$X^2=\mathbf{U(0)'V^{-1}U(0)}, \tag{8.2.8}$$

where $\mathbf{U(0)}$ and $\mathbf{V}$ are given by (8.2.7). Under the hypothesis that the distributions are identical, X^2 is distributed approximately as $\chi^2_{(m-1)}$.

In order to obtain a rank test we have to define scores. Usually scores are selected by basing the α_i's in (8.2.5) on some specific p.d.f. $g(y)$. If the data actually arise from a model (8.2.2) with this p.d.f., then the rank test is asymptotically fully efficient relative to the parametric procedure based on the actual observations y_j. In addition, the rank test generally retains substantially higher efficiency than the corresponding parametric test when the model is of the form (8.2.2), but with a different p.d.f. Finally, significance levels calculated from the rank test are valid regardless of the common underlying distribution of the observations. This is, of course, not true for parametric tests based on a specific model.

Before considering the problem of censored data, we consider two examples of rank tests, both of which will be later extended to the censored data case.

Example 8.2.1 [exponential ordered scores (logrank) test] If scores are generated by letting $g(y)$ in (8.2.5) be the extreme value p.d.f $\exp(y-e^y)$, $-\infty<y<\infty$, then $g'(y)/g(y)=1-e^y$ and

$$\alpha_i=E\left(e^{y_{(i)}}-1\right).$$

Since $v=e^y$ has a standard exponential distribution with p.d.f. e^{-v}, $v\geq 0$, $E(\exp y_{(i)})$ is the expected value of the ith order statistic in a random sample of size n from the standard exponential distribution. By (3.4.5), this gives

$$\alpha_i=\sum_{l=1}^{i}\frac{1}{n-l+1}-1$$

$$=e_{i,n}-1.$$

The α_i's are sometimes called exponential ordered scores (Cox, 1964).

In the two-sample test, for example, the rank statistic and its variance are, from (8.2.7),

$$U(0)=\sum_{i\in S_1}e_{(i),n}-N_1$$

$$V=\frac{N_1N_2}{n(n-1)}\sum_{i=1}^{n}\left(e_{i,n}-1\right)^2.$$

The variance can be simplified slightly by using the easily proved relations

$$\sum_{i=1}^{n}e_{i,n}=n \quad\text{and}\quad \sum_{i=1}^{n}e_{i,n}^2=2n-e_{n,n}$$

to give

$$V=\frac{N_1N_2}{n(n-1)}\left(n-e_{n,n}\right).$$

These results were mentioned earlier in Section 7.2.2. If the data arise from two extreme value distributions differing only with respect to location, the asymptotic relative efficiency of this rank test is unity. If the data come from normal distributions differing only with respect to location, the asymptotic relative efficiency turns out to be .82. The efficiency of the test is discussed further in Section 8.2.4.

***Example** 8.2.2* (Wilcoxon test) If scores are generated by taking the logistic p.d.f. $g(y)=e^y/(1+e^y)^2, -\infty<y<\infty$, the so-called Wilcoxon test (Wilcoxon, 1945) is obtained. In this case $g'(y)/g(y)=1-2e^y/(1+e^y)=1-2G(y)$, where $G(y)$ is the distribution function corresponding to $g(y)$. Since $G(y)$ is uniformly distributed on $(0,1)$, we find

$$\begin{aligned}\alpha_i &= E\left[2G(y_{(i)})-1\right]\\ &= \frac{2i}{n+1}-1 \end{aligned} \tag{8.2.9}$$

using the well-known fact that the mean of the ith order statistic from a random sample from $U(0,1)$ is $i/(n+1)$.

For the two-sample test (8.2.7) gives the rank statistic and its variance as

$$U(0)=\frac{2}{n+1}\sum_{i\in S_1}(i)-N_1$$

$$\begin{aligned}V&=\frac{N_1N_2}{n(n-1)}\sum_{i=1}^{n}\left(\frac{2i}{n+1}-1\right)^2\\ &=\frac{N_1N_2}{3(n+1)}.\end{aligned}$$

The two-sample Wilcoxon statistic is often considered in different but equivalent forms. For a discussion of the statistic and its properties in the two-sample case, see, for example, Kendall and Stuart (1967, p. 494 ff.). The Wilcoxon test is asymptotically fully efficient for detecting location shifts when the underlying distributions are logistic. In view of the similarity of the logistic and normal distributions, the Wilcoxon test would be expected to also have high efficiency when the underlying distributions are normal distributions differing only in mean. In fact, the asymptotic relative efficiency of the Wilcoxon test in this case is .95. If the underlying distributions are extreme value distributions, on the other hand, the asymptotic relative efficiency is .75.

Tests With Censored Data

When the data are censored, some modification of the procedures described above is needed. Rank tests with Type II censored data have been discussed by several authors; Johnson (1974) and Mehrotra et al. (1977) contain numerous references. The construction of rank tests with arbitrarily censored data has been examined by Prentice (1978), Peto and Peto (1972), and others. In addition, generalizations of the log rank and Wilcoxon tests to the case of arbitrarily censored data have been available for some time (Gehan, 1965; Breslow, 1970; Mantel, 1966). Only a brief description of the construction of rank tests for censored data will be given here, followed by a more detailed discussion of the generalizations of the log rank and Wilcoxon tests that have proved useful in practice.

Consider the linear regression model (8.2.2) once again, and suppose that from a sample involving n individuals with regression vectors $\mathbf{x}_1,\ldots,\mathbf{x}_n$ there arise k distinct observed log lifetimes $y_{(1)}<\cdots<y_{(k)}$ and $n-k$ censoring times. In addition, suppose that there are m_i log censoring times falling into the interval $[y_{(i)}, y_{(i+1)})$, for $i=0,1,\ldots,k$, where for convenience we define $y_{(0)}=0$ and $y_{(k+1)}=\infty$. Let $\mathbf{x}_{(i)}$ be the regression vector associated with the individual whose y value is $y_{(i)}$, and let $\mathbf{s}_{(i)}$ be the sum of these vectors for the m_i individuals with log censoring times in $[y_{(i)}, y_{(i+1)})$. To construct rank tests of the hypothesis $H_0: \boldsymbol{\theta}=\mathbf{0}$ in this situation Prentice (1978) and most others propose the use of a "score" statistic that has components of the form

$$U_l(\mathbf{0})=\sum_{i=1}^{k}\left(x_{(i)l}\alpha_i+s_{(i)l}a_i\right)\qquad l=1,\ldots,m-1. \tag{8.2.10}$$

That is, individuals whose lifetimes are censored are given scores a_i that are different from the scores of those whose lifetimes are observed. All individuals censored in $[t_{(i)}, t_{(i+1)})$ are given the same score, regardless of their respective censoring times.

Prentice (1978) suggests a general method of obtaining scores α_i and a_i for (8.2.10) and discusses estimation of the covariance matrix of $\mathbf{U}(\mathbf{0})=[U_1(\mathbf{0}),\ldots,U_{m-1}(\mathbf{0})]'$. This will not be discussed here, except to note that scores can be defined so that $E[\mathbf{U}(\mathbf{0})]=\mathbf{0}$ and that a kind of permutation variance for $\mathbf{U}(\mathbf{0})$ can be obtained, with entries

$$\begin{aligned}V_{ls}&=E\left[U_l(\mathbf{0})U_s(\mathbf{0})\right]\\&=\left(\sum_{i=1}^{k}\left(\alpha_i^2+m_ia_i^2\right)/(n-1)\right)\left(N_l\delta_{ls}-\frac{N_lN_s}{n}\right)\\&\qquad\qquad l,s=1,\ldots,m-1.\end{aligned} \tag{8.2.11}$$

This permutation variance is conditional on the particular assignment of scores in the situation on hand and is thus conditioned on $m_0, m_1, \ldots, m_k$. Consequently, formula (8.2.11) should be used only when it can safely be assumed that censoring is independent of $\mathbf{x}$ (e.g., see Gehan, 1965; Mantel, 1967).

The rank test procedures are not strictly distribution-free when there is censoring: the distribution and properties of the test statistics will depend on the censoring and life distributions involved. In this respect, observe that the assignment of scores in (8.2.10) is not, in general, prespecified, but depends on the observed data. It is prespecified, however, for various types of Type II censoring.

We now consider extension of the log rank and Wilcoxon tests to the censored data situation.

8.2.2 The Exponential Ordered Scores (Log Rank) Test With Censored Data

With uncensored data the exponential ordered scores test for the equality of two or more distributions employs the scores α_i given in (8.2.8). To discuss the case of censored data it is convenient to use the notation of earlier chapters: specifically, suppose that n_i is the total number of individuals at risk across all m distributions just prior to $t_{(i)}$, where $t_{(i)} = \exp y_{(i)}$ is the ith observed lifetime $(i=1,\ldots,k)$. Let d_i be the number of deaths at $t_{(i)}$; for now d_i is taken to be unity, since lifetimes are assumed to be distinct, but we shall later allow d_i to be greater than unity to handle ties in the data. Let S_l be the set of individuals from distribution l and define, for $l=1,\ldots,m$ and $i=1,\ldots,k$,

$$d_{li} = \text{Number of deaths at } t_{(i)} \text{ among individuals in } S_l$$

$$n_{li} = \text{Number of individuals from } S_l \text{ at risk just prior to } t_{(i)}\,.$$

Of course,

$$\sum_{l=1}^{m} d_{li} = d_i \quad \text{and} \quad \sum_{l=1}^{m} n_{li} = n_i.$$

Prentice (1978) and Peto and Peto (1972) have suggested the following scores for use with (8.2.10):

$$\left.\begin{aligned} \alpha_i &= \sum_{j=1}^{i} \frac{1}{n_j} - 1 \\ a_i &= \sum_{j=1}^{i} \frac{1}{n_j} \end{aligned}\right\} \qquad i=1,\ldots,k. \tag{8.2.12}$$

It is easily seen that when there is no censoring, the α_i's in (8.2.12) are identical to those given by (8.2.8). To motivate (8.2.12) in the censored case, note that for the extreme value distribution that generates these scores $-g'(y)/g(y)=e^y-1=H(y)-1$, where $H(y)$ is the distribution's cumulative hazard function. In (8.2.12) α_i is seen to be $\tilde{H}(y_{(i)}+0)-1$, where $\tilde{H}(y)$ is the empirical cumulative hazard function (2.3.13). The score a_i is α_i+1; this is motivated by the observation that led to (6.2.13), namely, that $H(y)$ has a standard exponential distribution, suggesting an adjustment of $+1$ to a censored observation.

For the m-sample problem, (8.2.10) in conjunction with the scores (8.2.12) gives the rank statistic

$$U_l(\mathbf{0})=\sum_{i=1}^{k}\left[d_{li}\alpha_i+(n_{li}-d_{li}-n_{l,i+1})a_i\right]$$

$$=\sum_{i=1}^{k}\left[-d_{li}+(n_{li}-n_{l,i+1})\left(\sum_{j=1}^{i}\frac{1}{n_j}\right)\right]$$

$$=-\sum_{i=1}^{k}d_{li}+\sum_{i=1}^{k}\frac{n_{li}}{n_i}.$$

Since $d_i=1$ $(i=1,\ldots,k)$ here, $U_l(\mathbf{0})$ can be rewritten as

$$U_l(\mathbf{0})=-\sum_{i=1}^{k}\left(d_{li}-\frac{n_{li}d_i}{n_i}\right)\qquad l=1,\ldots,m-1. \tag{8.2.13}$$

This is, aside from sign, the statistic (7.2.19) generated in Section 7.2.2 by the partial likelihood arguments for the proportional hazards model. From the expression (7.2.22) obtained in Chapter 7 the estimated covariance matrix for $\mathbf{U}(\mathbf{0})$ can be taken to have entries

$$V_{ls}=\sum_{i=1}^{k}\frac{d_i(n_i-d_i)n_{li}}{n_i(n_i-1)}\left(\delta_{ls}-\frac{n_{si}}{n_i}\right)\qquad l,s=1,\ldots,m-1. \tag{8.2.14}$$

As discussed in Section 7.2.2, (8.2.13) and (8.2.14) can also be used if there is a small number of ties in the data, in which case some of the d_i's will be greater than unity.

The equality of the m distributions is tested with the statistic

$$X^2 = \mathbf{U}(\mathbf{0})'\mathbf{V}^{-1}\mathbf{U}(\mathbf{0}),$$

which is distributed approximately as $\chi^2_{(m-1)}$ under the hypothesis of equality. The variance estimate (8.2.14) is based on the observed information matrix from the partial likelihood of Section 7.3.2; an alternative is the permutation variance given by (8.2.11). This should be used, however, only if the censoring pattern is roughly the same in each of the m samples. When there is no censoring, the permutation variance is exact, as indicated in Example 8.2.1, and should be used. In general, it appears that unless censoring differs somewhat across the m samples, or samples are rather small, the two variance formulas usually give results that are in close agreement. Prentice (1978) briefly discusses asymptotic results for censored rank tests; a number of points in connection with the log rank test have already been made in Section 7.2.3.

The log rank test has been derived from two somewhat different points of view, first as a test based on the proportional hazards model of Chapter 7 and, here, as a linear rank test for location differences. As noted in Example 8.2.1, when there is no censoring, the test is asymptotically fully efficient for detecting location differences under an extreme value model. Crowley and Thomas (1975) show that this result still holds under a random censorship model in which the same censoring distribution applies to each of the m samples but that there is some loss of efficiency when the censoring distributions differ. This is equivalent to stating that the test is asymptotically fully efficient for testing equality of lifetime distributions in a proportional hazards, or Lehmann, family, which is essentially the result that was noted in Section 7.4. Note that if lifetime T in distribution i has survivor function of the form $S_i(t) = S_0(t)^{\delta_i}$, then

$$\log[-\log S_0(T)] + \log \delta_i = \log[-\log S_i(T)].$$

But $\log[-\log S_i(T)]$ has an extreme value distribution, so the model can be written as $Y + \theta_i = U$, where $Y = \log[-\log S_0(T)]$, $\theta_i = \log \delta_i$, and U has a standard extreme value distribution. The problem of testing $\delta_i = 1$ ($i = 1, \ldots, m$) is therefore that of testing $\theta_i = 0$ ($i = 1, \ldots, m$) in the extreme value model. This applies to the case of known $S_0(T)$, of course, which essentially just says that the log rank is asymptotically fully efficient for testing differences among m exponential distributions. However, since a scale factor can be incorporated without loss of generality, the result also applies to the comparison of Weibull distributions with the same shape parameter. More generally, the log rank test is a locally most powerful rank invariant test within the Lehmann family (Peto, 1972b).

8.2.3 The Generalized Wilcoxon Test With Censored Data

The extension of the Wilcoxon and Krushkal–Wallis tests of Example 8.2.2 to the case of censored data has been discussed by several authors. Prentice (1978) suggests the statistic (8.2.10) in conjunction with scores

$$\left.\begin{aligned} \alpha_i &= 1-2\prod_{j=1}^{i}\frac{n_j}{n_j+1} = 1-2F_i \\ a_i &= 1-\prod_{j=1}^{i}\frac{n_j}{n_j+1} = 1-F_i \end{aligned}\right\} \quad i=1,\ldots,k \qquad (8.2.15)$$

where the n_j's are defined as in Section 8.2.2. As motivation for this note that for the logistic distribution that generates the Wilcoxon scores in the uncensored case, $-g'(y)/g(y)=1-2G(y)$, whereas in (8.2.15) F_i is roughly equal to the product-limit estimate $\prod_{j=1}^{i}(n_j-1)/n_j$ at $t_{(i)}+0$. Peto and Peto (1972) and Efron (1967) suggest similar scores, but with F_i replaced by the empirical hazard function estimate

$$F_i' = \exp\left(-\sum_{j=1}^{i} n_j^{-1}\right).$$

The test based on (8.2.15) reduces to the test given in Example 8.2.2 in the case of uncensored data. To see this note that when there is no censoring and no ties, $F_i = n_i/(n+1)$ and $n_i = n-i+1$, so that $\alpha_i = 1-2(n-i+1)/(n+1) = 2i/(n+1)-1$.

An alternate test has been suggested by Gehan (1965) for the two-sample case and by Breslow (1970) for the m-sample case. The test is based on (8.2.10) with scores

$$\left.\begin{aligned} \alpha_i &= \frac{i-n_i}{n+1} \\ a_i &= \frac{i}{n+1} \end{aligned}\right\} \quad i=1,\ldots,k. \qquad (8.2.16)$$

This gives the tests in Example 8.2.2 when there is no censoring but has some undesirable properties, particularly under unequal censoring patterns in different populations (Latta, 1977; Prentice and Marek, 1979). Although the scores in (8.2.15) require slightly more computation than those in (8.2.16), the former are to be preferred for general use.

The components of the score statistic (8.2.10) can be written in a simple form when (8.2.15) are used. We have

$$U_l(\mathbf{0}) = \sum_{i=1}^{k} \left[d_{li}\alpha_i + (n_{li} - d_{li} - n_{l,i+1})a_i\right]$$

$$= -\sum_{i=1}^{k} F_i d_{li} + \sum_{i=1}^{k} (n_{li} - n_{l,i+1})(1 - F_i)$$

$$= -\sum_{i=1}^{k} F_i d_{li} + \sum_{i=1}^{k} (F_{i-1} - F_i) n_{li}.$$

Now $F_{i-1} - F_i = F_i/n_i$, and if $d_i = 1$ (i.e., no ties) we can write

$$U_l(\mathbf{0}) = -\sum_{i=1}^{k} F_i\left(d_{li} - d_i\frac{n_{li}}{n_i}\right) \qquad l = 1,\ldots,m-1. \tag{8.2.17}$$

It is interesting to compare this with the score vector (8.2.12) for the log rank test. They differ only in the weight given to the terms $d_{li} - d_i n_{li}/n_i$: whereas in the log rank test the terms are given equal weight, in (8.2.17) they are weighted according to the estimate F_i of the survivor function at $t_{(i)} + 0$. The Wilcoxon test thus gives relatively more weight to earlier events than later ones, while the log rank test does not.

An estimate of the covariance matrix $\mathbf{V}$ for $\mathbf{U}(\mathbf{0})$ has entries (Tarone and Ware, 1977; Prentice and Marek, 1979; also see Problem 8.4)

$$V_{jl} = \sum_{i=1}^{k} F_i^2 \frac{d_i(n_i - d_i)}{n_i - 1}\frac{n_{ji}}{n_i}\left(\delta_{jl} - \frac{n_{li}}{n_i}\right) \qquad j,l = 1,\ldots,m-1. \tag{8.2.18}$$

Equality of the m distributions can be tested with the statistic $X^2 = \mathbf{U}(\mathbf{0})'\mathbf{V}^{-1}\mathbf{U}(\mathbf{0})$, which, under suitable conditions, is approximately $\chi^2_{(m-1)}$ under the hypothesis of equality. In addition, although the test has been developed on the assumption that there are no ties in the data, (8.2.17) and (8.2.18) can be employed when there is a small number of ties. In this case one should, however, define F_i as

$$\prod_{j=1}^{i} \frac{n_j - d_j + 1}{n_j + 1}.$$

The scores (8.2.16) give statistics similar to (8.2.17) and (8.2.18), the only difference being that F_i is replaced with $n_i/(n+1)$. This test has been discussed by Gehan (1965), Efron (1967), and Mantel (1967) for the two-sample case and Breslow (1970) for the m-sample case. Breslow shows that $\mathbf{U}(\mathbf{0})$ in this case has mean $\mathbf{0}$ and is asymptotically normal under a fixed (Type I) censoring model and under a random independent censoring time model. He also justifies the variance estimate (8.2.18), with F_i replaced by $n_i/(n+1)$. Prentice (1978) gives a brief general discussion of asymptotic properties of linear rank tests with censored data, but, basically, little rigorous work has been done in this area.

As an alternate to (8.2.18), a permutation variance for $\mathbf{U}(\mathbf{0})$ can be used, provided that the censoring pattern is essentially the same in the different populations. According to the general formula (8.2.11), when there are no ties, terms in the permutation covariance matrix are

$$V'_{jl} = \left(\sum_{i=1}^{n} w_i^2/(n-1) \right) \left(N_j \delta_{jl} - \frac{N_j N_l}{n} \right) \qquad j,l=1,\dots,m-1 \quad (8.2.19)$$

where w_i represents the score (α_i or a_i) assigned to the ith individual in the combined sample and N_j is the number of individuals from population j ($j=1,\dots,m$). Breslow (1970) proves in the two-sample case that (8.2.18) and (8.2.19) are asymptotically equivalent, provided that the same random independent censoring mechanism applies in the two populations. In finite samples the two variance estimates do not usually differ much, assuming that censoring is similar in the different populations.

An illustration of the Wilcoxon test is given in the next section.

8.2.4 Discussion

The log rank and generalized Wilcoxon tests are effective ways of assessing the equality of life distributions in many situations. The log rank test is good at detecting departures within a proportional hazards family of models: it is then asymptotically fully efficient when there is no censoring or when there is random but equal censoring in the m samples (Crowley and Thomas, 1975). There is some loss of efficiency for other types of censoring, but this appears to be relatively slight. The generalized Wilcoxon test is less powerful than the log rank test at detecting proportional hazard departures; it has an asymptotic relative efficiency of .75 in this situation when there is no censoring. On the other hand, it appears to be more powerful than the log rank test for detecting many other types of departures from equality.

For example, when there is no censoring and log lifetimes are normally distributed with the same variance but possibly different means, the Wilcoxon test has an asymptotic relative efficiency of .95, whereas the log rank has an asymptotic relative efficiency of .82. Work by Lee et al. (1975) suggests that roughly the same result holds for the two-sample problem with censored data. Lininger et al. (1979) examine power when there is censoring and stratification and obtain similar results.

There are situations in which neither the log rank or Wilcoxon tests are very effective. To see this and to view the two tests another way, note that for both tests the score statistic components for the m-sample problem are of the form

$$U_l = \sum_{i=1}^{k} w_i \left(d_{li} - \frac{n_{li} d_i}{n_i} \right) \qquad l=1,\dots,m-1 \tag{8.2.20}$$

with the same notation as in (8.2.13) and (8.2.17). Here w_i is a weight attached to $d_{li} - n_{li}d_i/n_i$; the log rank and Wilcoxon tests have $w_i = 1$ and $w_i = F_i$ [see (8.2.15)], respectively. This suggests that the log rank test will be relatively more effective at detecting differences in the right tails of the distributions, whereas the Wilcoxon will be more sensitive to earlier differences. Suppose, however, that two distributions differ, but that their hazard functions or survivor functions cross (e.g., see Problem 8.8). In such situations neither the Wilcoxon nor log rank statistic may be very powerful, and it will be sensible to consider other tests of equality of the distributions. For example, Tarone and Ware (1977) and Morton (1978) discuss general statistics of the form (8.2.20), and Fleming and Harrington (1980) and Fleming et al. (1980) present a two-sample test based on the maximum of a Smirnov-type statistic (see Section 9.1.1) designed to measure the distance between estimates of the two distributions. The latter approach is shown to be considerably more effective than the log rank or Wilcoxon tests when two lifetime distributions differ substantially for some range of t values, but not necessarily elsewhere. More work is needed on this problem, however, to produce simple but effective tests.

In summary, the log rank and Wilcoxon tests are effective in a fairly wide range of situations, particularly when the hazard or survivor functions of the different distributions do not cross. When plots of the estimated survivor functions for the distributions suggest that this may not be the case, other tests can be considered.

Before presenting an example of the use of the log rank and Wilcoxon tests, we remark that the methodology presented in Chapter 7 for the proportional hazards model has the nice feature that other regressor variables are easily incorporated into the analysis. The possibility of doing this also exists for the linear rank methods outlined in Section 8.2.1 (e.g., see Prentice, 1978), but the computation associated with these methods is considerably more involved than for the methods of Chapter 7. It might also be remarked that for the extreme value model the approach of Chapter 7 and the linear rank approach that generalizes the results in Section 8.2.1 do not lead to the same methods of testing nonzero $\boldsymbol{\theta}$'s in (8.2.2). In this case the linear rank procedures are in fact somewhat more efficient than those based on the partial likelihood analysis of Chapter 7. For some comments on this, see Prentice (1978, Sec. 4.2).

An illustration of the two-sample log rank and generalized Wilcoxon tests will conclude our discussion.

Example 8.2.3 (Example 7.2.1 revisited) In Example 7.2.1 some data on remission times for leukemia patients were presented. Patients were given one of two treatments and it was desired to test the hypothesis that the remission time distribution was the same for patients given treatment A as for those given treatment B.

The log rank statistics (8.2.13) and (8.2.14) were calculated there, and it was found that $U(0)=3.323$ and $V=8.1962$, giving $U(0)^2/V=1.35$. This yields a significance level of about .24 on $\chi^2_{(1)}$ and does not provide any evidence that the remission time distributions are not equal.

The generalized Wilcoxon statistics (8.2.17) and (8.2.18) are likewise easily calculated and give $U(0)=2.269$ and $V=3.0105$. This gives $U(0)^2/V=1.71$, which yields a significance level of about .19, in rough agreement with the log rank test. Gehan's generalized Wilcoxon statistic based on (8.2.16) uses $n_i/(n+1)$ in place of F_i in (8.2.17) and (8.2.18) and gives $U^2(0)/V=1.67$, in close agreement with the other result.

8.3 PROBLEMS AND SUPPLEMENTS

8.1 Consider the problem in Example 8.1.3 of obtaining a lower .90 confidence limit on the .10 quantile of a lifetime distribution. Suppose that the distribution is actually exponential with mean θ.

a. Compute and compare $E(t_{.10}-t_{(r)})=E(0.105\theta-t_{(r)})$ for the values $r=1,2,3$ discussed in the example. Also compare these with the values of $E(0.105\theta-L)$, where L is the lower .90 confidence limit for $t_{.10}$ obtained by the parametric methods of Section 3.1 (assume that the sample $t_1,\ldots,t_n$ is uncensored).

b. With the distribution-free procedure of Example 8.1.3 it is necessary only to observe the first r lifetimes. Compare $E(0.105\theta-t_{(r)})$ with

$E(0.105\theta - L)$, where L is based only on the r smallest observations from $(t_1, \ldots, t_n)$.

(Section 8.1.2)

8.2 Consider the ball bearing data discussed in Example 5.3.2. Obtain via (8.1.12) a two-sided confidence interval for the .10 quantile of the lifetime distribution, with confidence coefficient as close to .90 as possible. Compare this with the confidence intervals for $t_{.10}$ obtained in Example 5.3 under various generalized gamma models. In view of the uncertainty about the form of the model, what do you feel is a good estimate of $t_{.10}$?

(Section 8.1.2)

8.3 Consider the two-sample problem discussed in Example 1.1.6 and the five-sample problem discussed in Problem 4.11. Carry out log rank and Wilcoxon tests of distributional equality and compare the results of the two tests in each case.

(Section 8.2)

8.4 Consider m-sample tests based on the statistics U_l given by (8.2.20). Show that the covariance matrix for $\mathbf{U} = (U_1, \ldots, U_{m-1})'$ can be estimated by the matrix $\mathbf{V}$ with entries

$$V_{jl} = \sum_{i=1}^{k} w_i^2 \frac{d_i(n_i - d_i)}{n_i - 1} \frac{n_{ji}}{n_i} \left(\delta_{jl} - \frac{n_{li}}{n_i} \right) \qquad j, l = 1, \ldots, m-1.$$

(Hint: develop $\mathbf{U}$ from a partial likelihood along the lines of the argument in Section 1.4d and then consider terms in $\mathbf{U}$ for the ith interval conditional on n_i, the n_{li}'s, and d_i.)

8.5 The following is another way of looking at the Wilcoxon test. Suppose, for simplicity, that there is no censoring and let Y_{1i} $(i = 1, \ldots, n_1)$ and Y_{2i} $(i = 1, \ldots, n_2)$ be independent random samples from two continuous distributions with survivor functions $S_1(y)$ and $S_2(y)$. Define

$$U_{ij} = \begin{cases} 1 & \text{if} \quad Y_{2j} \leqslant Y_{1i} \\ -1 & \text{if} \quad Y_{2j} > Y_{1i} \end{cases} \qquad i = 1, \ldots, n_1 \quad j = 1, \ldots, n_2$$

and let

$$W = \sum_{i=1}^{n_1} \sum_{j=1}^{n_2} U_{ij}.$$

a. Show that $W=(n+1)U(0)$, where $U(0)$ is the Wilcoxon score statistic given in Example 8.2.2.

b. Show under $H_0: S_1(y)=S_2(y)$ that $E(W)=0$ and $\text{Var}(W)=n_1n_2(n_1+n_2+1)/12$.

Gehan (1965), Efron (1967), and Breslow (1970) discuss generalized Wilcoxon tests for censored data from this point of view.

(Section 8.2)

8.6 Confidence intervals for a location difference. The two-sample rank tests of Section 8.2 can be used to get confidence intervals for the difference in location of two distributions with the same shape. Suppose that two distributions have survivor functions $S_1(y)$ and $S_2(y)$, with $S_2(y)=S_1(y-\theta)$. Consider the hypothesis $H_0: \theta=\theta_0$. This can be tested with the rank procedures of Section 8.2 by considering the "samples" $\{y_i: i\in S_1\}$ and $\{y_j-\theta_0: j\in S_2\}$ and applying a rank test of the equality of the distributions from which the two samples come. This test can be carried out for different values θ_0. An α confidence interval for θ consists of all values θ_0 for which the rank test gives a significance level of $1-\alpha$ or greater.

Use this procedure to find an estimate of location difference for the two distributions involved in Example 8.3.2.

(Section 8.2)

8.7 Consider statistics $\mathbf{U}$ of the form (8.2.20) for testing the equality of m distributions. Examine how you could partition $\mathbf{U}$ into components that would assess certain specific hypotheses, such as that $S_1(t)$ and $S_2(t)$ are equal but different from $S_3(t),\ldots,S_m(t)$. Apply your ideas to the data in Problem 7.13.

(Sections 7.2, 7.3, 7.4)

8.8 The data below are survival times for patients with bile duct cancer who took part in a study to determine whether a combination of radiation treatment (R_0R_X) and the drug 5-fluorouracil (5-FU) prolonged survival (Fleming et al., 1980). Survival times, in days, are given for a group of patients given the radiation–drug therapy and for a control group of patients. Asterisks denote censored observations.

$R_0R_X+5-\text{FU}$	30, 67, 79*, 82*, 95, 148, 170, 171, 176, 193, 200, 221, 243, 261, 262, 263, 399, 414, 446, 446*, 464, 777
Control	57, 58, 74, 79, 89, 98, 101, 104, 110, 118, 125, 132, 154, 159, 188, 203, 257, 257, 431, 461, 497, 723, 747, 1313, 2636

Plot the product-limit estimates of the survivor functions for the two populations and note the form of the plot. Use the Wilcoxon and log rank tests to test the hypothesis that the two survival time distributions are equal. Are these likely to be effective tests in this type of situation? Suggest other methods of assessing possible differences in the two distributions.

(Section 8.2.4)

CHAPTER 9

Goodness of Fit Tests

It is important to check the adequacy of models upon which inferences are based. Some ways of doing this were considered in earlier chapters; in particular, plotting procedures discussed in Sections 2.3, 2.4, and 6.2 provide valuable tools for examining a model's suitability. In this chapter we consider some formal goodness of fit tests and tests for discriminating among models.

Consider a random variable X with distribution function $F(x)$. The main problem discussed here is that of testing hypotheses about $F(x)$ of the form

$$H_0: F(x)=F_0(x) \tag{9.0.1}$$

where $F_0(x)$ is a specified family of models. Usually $F_0(x)$ will involve unknown parameters, but occasionally it is fully specified. Tests of H_0 are frequently referred to as goodness of fit tests. It is sometimes useful to distinguish two types of tests: tests that are designed to be effective against wide classes of alternatives to a given $F_0(x)$ are often called "omnibus tests," and tests that are effective at detecting certain specific types of departures from $F_0(x)$, but not others, are referred to as "directional tests." Well-known general goodness of fit test procedures such as the Kolmogorov–Smirnov or Pearson χ^2 often provide reasonably good omnibus tests. On the other hand, it is usually possible to find tests with substantially more power against specific types of departures from a given model.

Section 9.1 presents procedures that are general in the sense that they can be used to test essentially any family of models. Most of these are omnibus tests, though the so-called smooth tests of Section 9.1.2 are more of a compromise between omnibus and directional tests. In Section 9.2 we consider tests of fit for several important lifetime distributions, including the exponential, Weibull, and log-normal distributions; directional as well as omnibus tests are presented. Goodness of fit tests for regression models and tests for discriminating among models are also considered. Section 9.3 concludes the chapter with a few comments on the problem of model specification.

9.1 SOME GENERAL METHODS OF TESTING FIT

We consider here some general methods of testing hypotheses H_0: $F(x)=F_0(x)$, as in (9.0.1). The best known procedures for this are the classical goodness of fit tests based on the empirical distribution function (EDF) for continuous ungrouped data and the Pearson χ^2 or likelihood ratio tests for discrete or grouped data. We shall consider these as well as less well-known smooth goodness of fit procedures. When $F_0(x)$ is completely specified and data are uncensored, the tests are all distribution-free and percentage points for the various test statistics are generally known. This is no longer the case when data are censored or when $F_0(x)$ involves unknown parameters, however. We shall examine tests for use with continuous data first and then consider tests with discrete or grouped data.

9.1.1 Tests Based on the EDF

Let X be a random variable with continuous distribution function (d.f.) $F(x)$ and consider the hypothesis (9.0.1) that $F(x)=F_0(x)$, where $F_0(x)$ is some family of d.f.'s. We first consider the case in which data are uncensored and $F_0(x)$ is completely specified (i.e., does not contain any unknown parameters). Given a random sample $x_1, \ldots, x_n$ from the distribution for X,

$$\tilde{F}_n(x)=\frac{\text{Number of } x_i\text{'s} \leq x}{n}$$

is the empirical d.f. (EDF) for the sample. A great many statistics that have been proposed for testing H_0 are based on the notion of measuring "distance" between $\tilde{F}_n(x)$ and $F_0(x)$ (see, e.g., Kendall and Stuart, 1968, Ch. 30, or Stephens, 1974). Three are discussed here; large values of the statistics are indicative of evidence against the hypothesized model. The statistics are

1. The Kolmogorov–Smirnov statistics:

$$\begin{aligned} D_n^+ &= \sup_x \left[\tilde{F}_n(x)-F_0(x)\right] \\ D_n^- &= \sup_x \left[F_0(x)-\tilde{F}_n(x)\right] \qquad (9.1.1) \\ D_n &= \sup_x \left|\tilde{F}_n(x)-F_0(x)\right| = \max(D_n^+, D_n^-). \end{aligned}$$

2. The Cramer–von Mises statistic:

$$W_n^2 = n\int_{-\infty}^{\infty} \left[\tilde{F}_n(x) - F_0(x)\right]^2 dF_0(x). \tag{9.1.2}$$

3. The Anderson–Darling statistic:

$$A_n^2 = n\int_{-\infty}^{\infty} \frac{\left[\tilde{F}_n(x) - F_0(x)\right]^2}{F_0(x)\left[1 - F_0(x)\right]} dF_0(x). \tag{9.1.3}$$

$\tilde{F}_n(x)$ is a step function with jumps at the order statistics $x_{(1)} < x_{(2)} < \cdots < x_{(n)}$, and for computational purposes the following alternate expressions (see Problem 9.1) are useful:

$$D_n^+ = \max_{1 \leqslant i \leqslant n}\left(\frac{i}{n} - F_0(x_{(i)})\right) \qquad D_n^- = \max_{1 \leqslant i \leqslant n}\left(F_0(x_{(i)}) - \frac{i-1}{n}\right) \tag{9.1.4}$$

$$W_n^2 = \sum_{i=1}^{n}\left(F_0(x_{(i)}) - \frac{i-.5}{n}\right)^2 + \frac{1}{12n} \tag{9.1.5}$$

$$A_n^2 = -\sum_{i=1}^{n} \frac{2i-1}{n}\left\{\log\left[F_0(x_{(i)})\right] + \log\left[1 - F_0(x_{(n+1-i)})\right]\right\} - n. \tag{9.1.6}$$

Tests based on these three statistics are distribution-free in the sense that the distributions of the statistics under H_0 do not depend on $F_0(x)$. This is obvious from (9.1.4), (9.1.5), and (9.1.6), since under H_0 the $F_0(x_{(i)})$'s are the order statistics in a random sample of size n from the uniform distribution on $(0, 1)$. The exact distributions of D_n^+, D_n^-, and D_n under H_0 are known for all n. The distribution theory for W_n^2 and A_n^2 is more difficult (e.g., Durbin, 1973); asymptotic distributions are known, but for finite n only partial analytical results are available, supplemented by information from Monte Carlo studies. Stephens (1974) surveys some results and provides many references.

A simple table given by Stephens (1974), also contained in Pearson and Hartley (1972, see Table 54), allows easy determination of percentage points for D_n^+, D_n^-, W_n^2, and A_n^2 under H_0. The table gives quantiles for these statistics, modified as a function of n, the beauty of this being that one set of quantiles suffices for all n. Some quantiles obtained from the table are naturally off slightly, but the accuracy attained is adequate for virtually all

Table 9.1.1 Quantiles for Functions of D_n^+, D_n, W_n^2, and A_n^2[a]

	Quantile				
Function	.85	.90	.95	.975	.99
$D_n^+(n^{1/2}+.12+.11n^{-1/2})$	0.973	1.073	1.224	1.358	1.518
$D_n(n^{1/2}+.12+.11n^{-1/2})$	1.138	1.224	1.358	1.480	1.628
$(W_n^2-.4n^{-1}+.6n^{-2})(1+n^{-1})$	0.284	0.347	0.461	0.581	0.743
A_n^2	1.610	1.933	2.492	3.070	3.857

[a] Reprinted, with permission, from Stephens (1974).

practical purposes. Table 9.1.1 gives the relevant information for D_n^+, D_n, W_n^2, and A_n^2; quantiles for D_n^- under H_0 are identical to those for D_n^+.

The tests just given are of limited value because of the requirement that $F_0(x)$ be completely specified (but see problem 9.6). The much more important situation is that in which $F_0(x)$ depends on unknown parameters. The tests can be modified by inserting estimates of parameters in $F_0(x)$ in this case, but their distribution theory is then much more complex. The distributions of test statistics also depend on $F_0(x)$, so the tests are no longer distribution free. Methods exist for determining asymptotic distributions of W_n^2 or A_n^2 with estimated parameters (e.g., Durbin 1973, Durbin et al. 1975; Stephens, 1976). There is generally no satisfactory way to get percentage points for small n except by simulation, though a few special results are available (e.g., Durbin, 1975). Percentage points have been obtained by Monte Carlo methods for the cases in which $F_0(x)$ is an exponential, extreme value, or normal distribution. This is discussed in Section 9.2, along with other tests for these models. We now consider modifications of the EDF tests to handle censored data.

EDF Tests with Censored Data

When data are Type II or singly Type I censored, simple modifications can be made to the EDF goodness of fit statistics, and distribution theory becomes only slightly more complicated than in the corresponding uncensored situation. With arbitrary censoring, things are more difficult, so we shall discuss the two situations in turn.

Single Type I or Type II Censoring

First consider the case in which $F_0(x)$ is completely specified. If the data are Type II censored at $x_{(r)}$, the rth smallest observation in a random sample of

n, then D_n, W_n^2, and A_n^2 can be modified as

$$D_{n,r} = \sup_{-\infty < x \leq x_{(r)}} |\tilde{F}_n(x) - F_0(x)|$$

$$W_{n,r}^2 = n\int_{-\infty}^{x_{(r)}} [\tilde{F}_n(x) - F_0(x)]^2 \, dF_0(x) \qquad (9.1.7)$$

$$A_{n,r}^2 = n\int_{-\infty}^{x_{(r)}} \frac{[\tilde{F}_n(x) - F_0(x)]^2}{F_0(x)[1 - F_0(x)]} \, dF_0(x).$$

For single Type I censoring at the point L analogous statistics are defined:

$$D_{n,p} = \sup_{-\infty < x \leq L} |\tilde{F}_n(x) - F_0(x)|$$

$$W_{n,p}^2 = n\int_{-\infty}^{L} [\tilde{F}_n(x) - F_0(x)]^2 \, dF_0(x) \qquad (9.1.8)$$

$$A_{n,p}^2 = n\int_{-\infty}^{L} \frac{[\tilde{F}_n(x) - F_0(x)]^2}{F_0(x)[1 - F_0(x)]} \, dF_0(x)$$

where $p = F_0(L)$.

Alternate forms of (9.1.7) and (9.1.8) are convenient for computation. It can be shown (see problem 9.1) that

$$W_{n,r}^2 = \sum_{i=1}^{r} \left(F_0(x_{(i)}) - \frac{i - .5}{n} \right)^2 + \frac{r}{12n^2} - \frac{n}{3}\left(\frac{r}{n} - F_0(x_{(r)}) \right)^3. \qquad (9.1.9)$$

For $W_{n,p}^2$ let r be the number of observations less than or equal to L. Then

$$W_{n,p}^2 = \sum_{i=1}^{r} \left(F_0(x_{(i)}) - \frac{i - .5}{n} \right)^2 + \frac{r}{12n^2} - \frac{n}{3}\left(\frac{r}{n} - F_0(L) \right)^3. \qquad (9.1.10)$$

For $A_{n,r}^2$ and $A_{n,p}^2$ corresponding expressions are

$$A_{n,r}^2 = -\sum_{i=1}^{r} \left(\frac{2i-1}{n} \log F_0(x_{(i)}) - \frac{2n-2i+1}{n} \log[1 - F_0(x_{(i)})] \right)$$
$$+ \frac{r^2}{n} \log F_0(x_{(r)}) - \frac{(n-r)^2}{n} \log[1 - F_0(x_{(r)})] - nF_0(x_{(r)}) \qquad (9.1.11)$$

$$A_{n,p}^2 = -\sum_{i=1}^{r}\left(\frac{2i-1}{n}\log F_0(x_{(i)}) - \frac{2n-2i+1}{n}\log[1-F_0(x_{(i)})]\right)$$

$$+\frac{r^2}{n}\log F_0(L) - \frac{(n-r)^2}{n}\log[1-F_0(L)] - nF_0(L). \quad (9.1.12)$$

The statistics $D_{n,r}$ and $D_{n,p}$ have been considered by Barr and Davidson (1973), Koziol and Byar (1975), and Dufour and Maag (1978). Dufour and Maag present tables of percentage points for both the Type I and Type II censored cases for samples sizes up to $n=25$. Koziol and Byar determine the common asymptotic distribution of $D_{n,p}$ in the cases of single Type I and Type II censoring where, with Type II censoring, r and n go to infinity, with $r/n=p$ fixed. Excerpts from Koziol and Byar's tables are given in Table 9.1.2, which contains percentage points for the limiting distribution of $\sqrt{n}\,D_{n,p}$. This table can be used with values of n greater than 25 or so if an empirical adjustment due to Dufour and Maag is employed to correct percentage points slightly. The adjustment is as follows: let d_p^α be the αth quantile corresponding to the value p in Table 9.1.2. Then, if $D_{n,p}^\alpha$ is the αth quantile of $D_{n,p}$, take (1) $\sqrt{n}\,D_{n,p}^\alpha = d_p^\alpha - .19n^{-1/2}$ in the case of Type I censoring, and (2) $\sqrt{n}\,D_{n,r}^\alpha = d_p^\alpha - .24n^{-1/2}$ in the case of Type II censoring (where $p=r/n$). This yields satisfactorily accurate percentage points for $D_{n,p}$ and $D_{n,r}$ when $n \geqslant 25$ and $p \geqslant .25$ for (1) and when $n \geqslant 25$ and $p \geqslant .4$ for (2).

The generalizations of W_n^2 and A_n^2 given in (9.1.7) have been discussed by Pettit and Stephens (1976), who determine their asymptotic distributions and give a table of percentage points. Smith and Bain (1976) provide some small-sample percentage points for a variant of $W_{n,r}^2$ that are obtained by Monte Carlo methods. Michael and Schucany (1979) take a different approach with Type II censoring by transforming the observations $x_{(1)} \leqslant \cdots \leqslant x_{(r)}$ to r ordered observations in a complete sample of size r from the uniform distribution on $(0,1)$. This allows one to apply tests for the uncensored case.

The results above are of limited use because $F_0(x)$ is assumed to be known, and we shall not present any percentage points beyond those in Table 9.1.2. In the more important situation where $F_0(x)$ contains unknown parameters, the statistics (9.1.7) can be modified by the insertion of appropriate parameter estimates. General asymptotic distribution theory for $W_{n,p}^2$ and $A_{n,p}^2$ in this case has been considered by Pettit (1976). This and related work on tests for the normal, exponential, and extreme value distributions are discussed in Section 9.2.

Table 9.1.2 Quantiles of the Limiting Distribution of $\sqrt{n}D_{n,p}$[a]

p	.75	.90	.95	.975	.99
.1	0.4714	0.5985	0.6825	0.7589	0.8512
.2	0.6465	0.8155	0.9268	1.0282	1.1505
.3	0.7663	0.9597	1.0868	1.2024	1.3409
.4	0.8544	1.0616	1.1975	1.3209	1.4696
.5	0.9196	1.1334	1.2731	1.3997	1.5520
.6	0.9606	1.1813	1.3211	1.4476	1.5996
.7	0.9976	1.2094	1.3471	1.4717	1.6214
.8	1.0142	1.2216	1.3568	1.4794	1.6272
.9	1.0190	1.2238	1.3581	1.4802	1.6276
1.0	1.0192	1.2238	1.3581	1.4802	1.6277

[a]Adapted, with permission, from Koziol and Byar (1975).

Arbitrary Censoring

There is no satisfactory approach to EDF goodness of fit tests when data are arbitrarily censored. When $F_0(x)$ in (9.0.1) is fully specified, an obvious modification to (9.1.1), (9.1.2) or (9.1.3) is to replace $\tilde{F}_n(x)$ with $1-\hat{S}(x-0)$, where $\hat{S}(x)$ is the product-limit estimate of the survivor function [see (2.3.2)]. Distribution theory is difficult, however, and the fact that distributions of test statistics will depend on the censoring process makes even the tabulation of percentage points by Monte Carlo methods unattractive. Koziol and Green (1976), for example, generalize the Cramer–von Mises statistic (9.1.2) in this way and obtain its asymptotic distribution under the assumption that censoring times are random variables independent of lifetimes, with survivor function of the form $[1-F_0(x)]^{\beta}$. They give asymptotic percentage points for a few values of β. Since both β and $F_0(x)$ are assumed to be known, these tests are not particularly useful. Hall and Wellner (1980), Gregory (1980), and a few others have proposed similar types of tests. A more feasible approach to tests of fit when data are arbitrarily censored might be to use the ideas of smooth tests; this is discussed in Section 9.1.2.

Discussion

The EDF tests that assume $F_0(x)$ to be completely specified are of very limited use. Sometimes these tests are employed when parameters are actually estimated. This is not to be recommended; significance levels are likely to be far off the correct significance levels obtained if one allows for estimation. Modification of the EDF statistics to handle unknown parameters is discussed in Section 9.2 for several important models. In some instances ways can be devised to reduce a problem in which $F_0(x)$ has unknown nuisance parameters to a problem involving a fully known model. For example, tests for the exponential distribution can be obtained in this way (e.g., Cox and Lewis 1966; Problem 9.6). In fact, by a device known as the half-sample method (Durbin, 1973; 1975; Stephens, 1978) one can apply asymptotic theory for EDF statistics when $F_0(x)$ is completely specified to the case in which unknown parameters are present. This is discussed in Section 9.2.4.

Nothing has yet been said about the power of the EDF tests. Broadly speaking, D_n^+, D_n^-, D_n, and W_n^2 of (9.1.1) and (9.1.2) are more effective at detecting departures in the middle of a distribution, whereas A_n^2 of (9.1.3) is relatively more effective at detecting tail departures. It appears that when unknown parameters are present, W_n^2 and especially A_n^2 are often substantially more powerful than D_n. Some power comparisons of these with other tests are mentioned in Section 9.2, where we consider tests of fit for specific distributions. The W_n^2 and A_n^2 tests generally have good power against broad ranges of alternatives.

9.1.2 Smooth Goodness of Fit Tests

Suppose we wish to test the hypothesis that the p.d.f. $f(x)$ of a random variable X has specified form,

$$H_0: f(x)=f_0(x; \boldsymbol{\theta}) \tag{9.1.13}$$

where $\boldsymbol{\theta}=(\theta_1,\dots,\theta_p)$ is an unknown vector of parameters. One way to test H_0 is to embed $f_0(x; \boldsymbol{\theta})$ in a larger parametric family of models and to test H_0 as a parametric hypothesis. Examples of this are given in Section 9.2 in connection with tests for the normal and extreme value distributions.

Neyman (1937) suggested a general procedure for developing tests by embedding $f_0(x; \boldsymbol{\theta})$ in the model with p.d.f.

$$g(x; \boldsymbol{\theta}, \boldsymbol{\beta})=f_0(x; \boldsymbol{\theta})\exp\left(\sum_{j=1}^{k} \beta_j F_0^j(x; \boldsymbol{\theta})-k(\boldsymbol{\beta})\right) \tag{9.1.14}$$

where $F_0(x; \boldsymbol{\theta})$ is the distribution function corresponding to $f_0(x; \boldsymbol{\theta})$ and $\boldsymbol{\beta}=(\beta_1,\dots,\beta_k)$ is a parameter. The function $k(\boldsymbol{\beta})$ is a normalizing constant and does not depend on $\boldsymbol{\theta}$. With the family (9.1.14), one can test H_0 by testing that $\boldsymbol{\beta}=\mathbf{0}$. The hope is that the family of alternatives embodied in (9.1.14) is rich enough to give tests of H_0 with good power in a fairly wide range of situations. It indeed appears that tests of fit obtained in this way are often fairly good, with k typically chosen to be small.

Kopecky and Pierce (1979a) and Thomas and Pierce (1979) review work in this area and propose tests based on score statistics from (9.1.14). These results are potentially quite useful for developing tests of fit in difficult situations and will be sketched briefly, though we do not use them explicitly in the rest of the chapter. The log likelihood from a complete sample $x_1,\ldots,x_n$ from (9.1.14) is

$$\log L(\boldsymbol{\beta},\boldsymbol{\theta})=\sum_{i=1}^{n}\log f_0(x;\boldsymbol{\theta})+\sum_{i=1}^{n}\left(\sum_{j=1}^{k}\beta_j F_0^j(x_i)-k(\boldsymbol{\beta})\right).$$

Denote the score function as $\mathbf{U}=[\mathbf{U}_\beta(\boldsymbol{\beta},\boldsymbol{\theta}),\mathbf{U}_\theta(\boldsymbol{\beta},\boldsymbol{\theta})]'$, where $\mathbf{U}_\beta$ and $\mathbf{U}_\theta$ have entries

$$\frac{\partial\log L}{\partial\beta_l}=\sum_{i=1}^{n}\left[F_0^l(x_i;\boldsymbol{\theta})-(1+l)^{-1}\right]\qquad l=1,\ldots,k$$

and

$$\frac{\partial\log L}{\partial\theta_l}=\sum_{i=1}^{n}\frac{\partial\log f_0(x;\boldsymbol{\theta})}{\partial\theta_l}\qquad l=1,\ldots,p$$

respectively. In addition, let the Fisher information matrix $\mathbf{I}(\boldsymbol{\beta},\boldsymbol{\theta})$ and its inverse be partitioned as

$$\mathbf{I}(\boldsymbol{\beta},\boldsymbol{\theta})=\begin{pmatrix}\mathbf{I}_{\beta\beta} & \mathbf{I}'_{\theta\beta}\\ \mathbf{I}_{\theta\beta} & \mathbf{I}_{\theta\theta}\end{pmatrix}\qquad \mathbf{I}(\boldsymbol{\beta},\boldsymbol{\theta})^{-1}=\begin{pmatrix}\mathbf{I}^{\beta\beta} & \mathbf{I}^{\theta\beta\prime}\\ \mathbf{I}^{\theta\beta} & \mathbf{I}^{\theta\theta}\end{pmatrix}.$$

Thomas and Pierce (1979, Sec. 3) show that $\mathbf{I}(\boldsymbol{\beta},\boldsymbol{\theta})$ has entries

$$E\left(\frac{-\partial^2\log L}{\partial\beta_l\,\partial\beta_m}\right)=\frac{nlm}{(l+1)(m+1)(l+m+1)}\qquad l,m=1,\ldots,k$$

$$E\left(\frac{-\partial^2\log L}{\partial\beta_l\,\partial\theta_m}\right)=-E\left(\sum_{i=1}^{n}\frac{\partial F_0^l(x_i;\boldsymbol{\theta})}{\partial\theta_m}\right)\qquad l=1,\ldots,k\quad m=1,\ldots,p$$

$$(9.1.15)$$

$$E\left(\frac{-\partial^2\log L}{\partial\theta_l\,\partial\theta_m}\right)=-E\left(\sum_{i=1}^{n}\frac{\partial^2\log f_0(x_i;\boldsymbol{\theta})}{\partial\theta_l\,\partial\theta_m}\right)\qquad l,m=1,\ldots,p.$$

Let $\hat{\boldsymbol{\theta}}_0$ be the m.l.e. of $\boldsymbol{\theta}$ under the hypothesis $H_0: \boldsymbol{\beta}=\mathbf{0}$. The partial scores statistic for testing H_0 is then

$$W_k = \mathbf{U}_\beta(\mathbf{0}, \hat{\boldsymbol{\theta}}_0)' \mathbf{I}_0^{\beta\beta} \mathbf{U}_\beta(\mathbf{0}, \hat{\boldsymbol{\theta}}_0) \tag{9.1.16}$$

where $\mathbf{I}_0^{\beta\beta}$ is $\mathbf{I}^{\beta\beta}$ with $\boldsymbol{\beta}=\mathbf{0}$ and $\boldsymbol{\theta}=\hat{\boldsymbol{\theta}}_0$. Under H_0, W_k is asymptotically $\chi^2_{(k)}$, provided that mild regularity conditions are satisfied. Note that (9.1.16) requires only the m.l.e. $\hat{\boldsymbol{\theta}}_0$ of $\boldsymbol{\theta}$, which is obtained from the hypothesized model $f_0(x; \boldsymbol{\theta})$, and that only expectations with respect to $f_0(x; \boldsymbol{\theta})$ are required. In fact, from (9.1.15), the only expectation required here that is not required in the usual large-sample theory for $f_0(x; \boldsymbol{\theta})$ is

$$-E\left(\sum_{i=1}^{n} \frac{\partial F_0^l(x_i; \boldsymbol{\theta})}{\partial \theta_m}\right) \qquad l=1,\dots,k \quad m=1,\dots,p.$$

Thomas and Pierce (1979a) and Kopecky and Pierce (1979a) compute W_k for $k \leqslant 4$ in the cases in which $f_0(x; \boldsymbol{\theta})$ is an exponential, normal, or Weibull p.d.f. They find the power of the resulting tests to be very good in the situations they examine. Note that the tests actually have a quite simple form, since $\partial \log L/\partial \beta_l$ measures the difference between $F_0^l(x_i; \boldsymbol{\theta})$ and its expectation $(1+l)^{-1}$ under H_0. For example, the test statistic W_2 in the case in which $f_0(x; \boldsymbol{\theta})$ is the normal distribution $N(\mu, \sigma^2)$ is

$$W_2 = \frac{-733.40}{n} \sum_{i=1}^{n} \left[\hat{F}_0(x_i) - \tfrac{1}{2}\right]^2 + \frac{749.71}{n} \sum_{i=1}^{n} \left[\hat{F}_0(x_i)^2 - \tfrac{1}{3}\right]^2 \tag{9.1.17}$$

where $\hat{F}_0(x_i) = \Phi[(x_i - \hat{\mu})/\hat{\sigma}]$ and $\Phi(z)$ is the standard normal distribution function.

Smooth tests can, in principle, also handle censored data. However, calculation of the expected values required for (9.1.16) is difficult and needs to be done for each different censoring mechanism. A more attractive procedure is to develop a test statistic like W_k, but based on the observed information matrix. Another point that needs investigation is the adequacy of the χ^2 approximation to the distribution of W_k or similar statistics when samples are not very large.

9.1.3 Tests of Fit Based on Grouped Data

With grouped uncensored data, tests of fit can be based on the multinomial model, the best-known procedures being the classical Pearson (χ^2) test and the likelihood ratio test. We shall briefly review these, with a view to later accommodating censored data.

Consider the usual grouped data setup: observations can fall into $k+1$ classes $I_j=[a_{j-1}, a_j)$, $j=1,\ldots,k+1$, with $a_0=0$, $a_{k+1}=\infty$, and $a_k=T$ as an upper limit of observation. Let d_j represent the number of observations in a random sample of size n that fall into I_j, let p_j be the probability of an observation falling into I_j, and consider the hypothesis

$$H_0: p_j=p_{j0} \qquad j=1,\ldots,k+1 \tag{9.1.18}$$

where the p_{j0}'s are specified but may involve unknown parameters. Let $\tilde{p}_{j0}$ be the m.l.e. of p_j under H_0, or some other asymptotically fully efficient estimator, and let $e_j=n\tilde{p}_{j0}$. The Pearson statistic for testing H_0 is

$$X^2=\sum_{j=1}^{k+1}\frac{(d_j-e_j)^2}{e_j}. \tag{9.1.19}$$

When the p_{j0}'s are known constants, $e_j=np_{j0}$ and the limiting distribution of X^2 is $\chi^2_{(k)}$. When the p_{j0}'s involve s unknown parameters, the limiting distribution is $\chi^2_{(k-s)}$.

The likelihood ratio test is an alternative test of H_0. The likelihood function for $p_1,\ldots,p_k$ is multinomial,

$$L(p_1,\ldots,p_k)\propto\prod_{j=1}^{k+1}p_j^{d_j}$$

where $p_{k+1}=1-p_1-\cdots-p_k$. The likelihood ratio statistic for testing H_0 against the alternative that the p_j's satisfy only $p_j\geqslant 0, \Sigma p_j=1$, is easily seen to be

$$\Lambda=2\sum_{j=1}^{k+1}d_j\log\left(\frac{d_j}{n}\right)-2\sum_{j=1}^{k+1}d_j\log\tilde{p}_{j0}$$

$$=2\sum_{j=1}^{k+1}d_j\log\left(\frac{d_j}{e_j}\right). \tag{9.1.20}$$

The limiting distribution of Λ under H_0 is $\chi^2_{(k-s)}$ when the p_{j0}'s involve s unknown parameters. The likelihood ratio and Pearson tests have been investigated by many authors (see, e.g., Bishop et al. 1975; Larntz, 1978; Chapman, 1976). Essentially, the limiting χ^2 distributions can be used to calculate significance levels for most situations of practical importance.

When testing a hypothesis $H_0: F(x)=F_0(x)$ about a continuous model, the X^2 and likelihood ratio tests have the advantages of easy computation and the ability to accommodate unknown parameters. They are less powerful than tests based on the EDF in some situations, but are not appreciably less so in most situations of practical importance, provided that intervals are chosen appropriately. In general, the X^2 and likelihood ratio statistics provide good omnibus tests of fit. This is illustrated in Example 9.2.1.

Censored Data

When the data are censored, we face the same difficulties in testing (9.1.18) as are encountered in life table estimation (see Section 2.2). Note first that if censoring is single Type I or Type II, there is essentially no problem: if the data are Type I censored at L, we need only take $a_k = T \leq L$, which forces all censoring times into the last interval, within which all observations not in one of $I_1, \ldots, I_k$ must fall anyway. If the data are Type II censored at $x_{(r)}$, one can choose intervals so that $T \leq x_{(r)}$. When intervals are selected by looking at the data in this way, the distributions of X^2 and Λ are altered to some degree, but the χ^2 approximations are usually still suitable.

The main problem is to provide a test of (9.1.18) with arbitrarily censored data. In this case, consider the life table setup of Section 2.2, where the interval $[a_{j-1}, a_j)$ has in it n_j individuals at risk at a_{j-1} and d_j deaths (lifetimes) and w_j withdrawals (censoring times). To modify either the X^2 or likelihood ratio tests appropriately we need specific assumptions about censoring. One approach would be to assume that censoring times are distributed across intervals and to attempt to modify X^2 or Λ in the spirit of the standard life table estimation methods. This is not particularly attractive, and we assume here that censoring occurs only at the ends of intervals $I_1, \ldots, I_k$. We consider only the likelihood ratio statistic, though an analog to X^2 can also be obtained.

Let $S_j = S(a_j)$ be the probability of survival beyond I_j and let $s_j = S_j / S_{j-1}$ be the probability of survival beyond I_j conditional on survival to the start of I_j. Under suitable assumptions about independence of censoring and lifetimes, the likelihood function is given by (1.4.8),

$$L(s_1, \ldots, s_k) = \prod_{j=1}^{k} (S_{j-1} - S_j)^{d_j} S_j^{w_j}.$$

Assumptions about censoring like those preceding (1.4.8) are sufficient to ensure the validity of this likelihood and asymptotic procedures based on it. Noting that $S_j = s_1 s_2 \cdots s_j$ and letting $q_j = 1 - s_j$, we can rewrite L as

$$L(s_1, \ldots, s_k) = \prod_{j=1}^{k} q_j^{d_j} s_j^{n_j - d_j} \tag{9.1.21}$$

when $n_j = n - \sum_{i=1}^{j-1}(d_i + w_i)$ is the number of individuals at risk at a_{j-1}.

Consider the hypothesis

$$H_0: s_j = s_{j0} \qquad j=1,\dots,k \tag{9.1.22}$$

with the alternative hypothesis that the s_j's satisfy only $0 \leqslant s_j \leqslant 1$. The unrestricted m.l.e.'s of the s_j's are easily found, by maximizing (9.1.21), to be

$$\hat{s}_j = \frac{n_j - d_j}{n_j} \qquad j=1,\dots,k.$$

With the m.l.e.'s of the s_j's under H_0 denoted as $\tilde{s}_{j0}$'s, the likelihood ratio statistic for testing H_0 is

$$\Lambda = -2\log\left(\frac{L(\tilde{s}_{10},\dots,\tilde{s}_{k0})}{L(\hat{s}_1,\dots,\hat{s}_k)}\right).$$

This reduces to

$$\Lambda = 2\sum_{j=1}^{k} d_j \log\left(\frac{d_j}{\tilde{d}_{j0}}\right) + 2\sum_{j=1}^{k}(n_j - d_j)\log\left(\frac{n_j - d_j}{n_j - \tilde{d}_{j0}}\right) \tag{9.1.23}$$

where $\tilde{d}_{j0} = n_j(1 - \tilde{s}_{j0})$ can be thought of as the expected number of deaths in I_j under H_0.

Under suitable assumptions concerning independence of censoring and lifetimes, it can be shown that if the s_{j0}'s involve s unknown parameters, the limiting distribution of Λ under H_0 is $\chi^2_{(k-s)}$. Turnbull and Weiss (1978) prove this for the case in which each of the n individuals in the sample has a random censoring time that takes on possible values $a_1, a_2, \dots, a_{k+1}$, independent of the lifetime of the individual.

When the p_j's in (9.1.18) or the s_j's in (9.1.21) are determined from a continuous model with unknown parameters, we are faced with the problem, discussed in Section 5.5, of computing m.l.e.'s from grouped data. If m.l.e.'s based on the ungrouped data are used in the tests, the limiting distributions of X^2 and Λ are no longer $\chi^2_{(k-s)}$ (e.g., Kendall and Stuart 1967, Ch. 30). The limiting distributions are, however, bounded between $\chi^2_{(k)}$ and $\chi^2_{(k-s)}$; this can be used to bound the exact significance level. An example of the use of X^2 and Λ in a test of fit is given in Section 9.2.1 (see Example 9.2.1).

Sometimes it is undesirable to assume that censoring occurs only at interval end points. There has been very little work on tests of fit for other situations, however, partly because it does not often seem reasonable to make the strong assumptions about censoring that are needed to develop tests. One paper that tries to handle censoring is Gail and Ware (1979), though they consider only the case in which $F_0(x)$ is fully known. Often an *ad hoc* adjustment to the likelihood (9.1.21) will provide a sensible way to handle other types of censoring; the limiting χ^2 distribution for Λ of (9.1.23) will not hold then, though it may provide a reasonable approximation. For example, if censoring times are distributed across intervals, a reasonable adjustment is to replace (9.1.21) with

$$L(s_1,\ldots,s_k)=\prod_{j=1}^{k} q_j^{d_j} s_j^{n_j-d_j-w_j/2}$$

where w_j is the number of censoring times in I_j. This is in the same spirit as the likelihood (7.3.20).

9.2 TESTS OF FIT FOR SPECIFIC DISTRIBUTIONS

Tests of fit for several of the most important lifetime distribution models are considered in this section. Graphical methods are invaluable in assessing models; this has been discussed in Chapter 2. Here we consider only formal tests of fit, though, as in Section 9.1.1, some of the tests have obvious relations to plotting procedures. The exponential, extreme value, and normal distributions all involve location and scale parameters; there are numerous tests of fit for them and hence also for the Weibull and log-normal distributions. We now examine tests for these and other models.

9.2.1 Tests of Fit for the Exponential Distribution

In testing goodness of fit one can distinguish between tests designed to detect quite specific types of departures from the hypothesized model and tests designed with only a broad class of alternatives in mind. We shall consider three types of procedures for the exponential distribution. The first involves embedding the exponential distribution in a more general parametric model such as the Weibull or gamma model; such tests are naturally effective at detecting departures from exponentiality within the more general model, but may or may not be effective at detecting other types of departures. We also consider a test of exponentiality that has good power against a broad class of alternatives, specifically those distributions with monotone hazard functions. Finally, we consider tests of fit formulated with no specific alternatives in mind, employing special versions of some of the omnibus tests of Section 9.1.1.

Tests Based on Specific Parametric Models

We consider tests of fit for the one-parameter exponential distribution, in which case the null hypothesis is that the survivor function of T is exponential,

$$H_0: S(t)=e^{-t/\theta} \qquad t>0 \tag{9.2.1}$$

where $\theta>0$ is an unknown parameter. One way to test H_0 is to embed the exponential model in a parametric family with two or more parameters so that H_0 can be tested as a parametric hypothesis. The two most important tests of this sort are that in which the exponential is embedded in a Weibull distribution and that in which the exponential is embedded in a gamma distribution. For example, if we consider the Weibull family with survivor function

$$S(t)=\frac{\beta}{\theta}\left(\frac{t}{\theta}\right)^{\beta-1} e^{-t/\theta} \qquad t>0$$

then testing H_0 is equivalent to testing that $\beta=1$. Tests for the Weibull shape parameter β are discussed in detail in Section 4.1, so there is no need to go into this here.

A similar procedure is to assume a gamma distribution with survivor function

$$S(t)=\frac{1}{\theta\Gamma(k)}\left(\frac{t}{\theta}\right)^{k-1} e^{-t/\theta} \qquad t>0$$

and to test H_0 by testing that $k=1$. Tests with the gamma model are considered in Section 5.1.

Tests of this kind are effective at detecting departures from exponentiality within or “close” to the assumed family of models. These procedures have the advantage of handling arbitrarily censored data with relative ease; see Sections 4.1 and 5.1 for the situation with the Weibull and gamma models. On the other hand, tests that are effective against a broader class of alternatives are often required, since it may be impossible to precisely specify alternatives. We next consider a test that has good power against alternative distributions with monotone hazard functions.

A Test Effective Against Monotone Hazard Function Alternatives

A wide variety of tests of fit have been proposed for the exponential distribution. Many early references are given in Epstein (1960b) and Cox and Lewis (1966, Ch. 6). Wang and Chang (1977), Lee et al. (1980), and references cited therein refer to many additional tests. A number of tests appear to have good power against alternatives that possess either monotone increasing (IFR) or monotone decreasing (DFR) hazard functions, or failure rates. Only one test is given below: this test has good power against the alternatives mentioned, can handle Type II censored data, and requires only a small table of percentage points. The test is also insensitive to round-off and recording errors, to which some other tests are not. It is not claimed that the test is best in any sense, but it possesses desired attributes and is effective in a broad range of situations. Other tests with similar properties can be found (e.g., Lee et al., 1980), but they are not discussed here.

The test, hereafter referred to as the G test, is based on the so-called Gini statistic and is discussed by Gail and Gastwirth (1978) and others. Consider a random sample $t_1,\ldots,t_n$ of size n. The statistic proposed by Gail and Gastwirth for testing (9.2.1) is

$$G_n=\sum_{i=1}^{n}\sum_{j=1}^{n}|t_i-t_j|/2n(n-1)\bar{t}.$$

An alternate expression that is useful for computation and for later generalization to the case of Type II censored sampling employs the scaled spacings

$$W_i=(n-i+1)(t_{(i)}-t_{(i-1)}) \qquad i=1,\ldots,n$$

where $t_{(0)}=0$ and $t_{(1)}\leqslant\cdots\leqslant t_{(n)}$ are the ordered observations. It is easily shown that

$$G_n=\left(\sum_{i=1}^{n-1} iW_{i+1}\right)\Big/(n-1)\sum_{i=1}^{n} W_i. \tag{9.2.2}$$

G_n takes on values between 0 and 1, with values near 0 or 1 providing evidence against exponentiality. Under the null hypothesis (9.2.1) W_i/θ ($i=1,\ldots,n$) are independent and have standard exponential distributions (see Theorem 3.1.). The distribution of G_n under H_0 has been obtained and tabulated for $n=3,\ldots,20$ by Gail and Gastwirth (1978). Some upper quantiles are given in Table 9.2.1; the distribution of G_n is symmetric about .5, so the table also gives the .01, .025, and .05 quantiles. The mean and

Table 9.2.1 Quantiles of the Test Statistic G_n[a]

	Quantile						
n	.950	.975	.990	n	.950	.975	.990
3	.84189	.88818	.92932	12	.64337	.66992	.70020
4	.77686	.82288	.86951	13	.63725	.66275	.69183
5	.73834	.77997	.82501	14	.63185	.65641	.68448
6	.71307	.75079	.79260	15	.62704	.65076	.67792
7	.69439	.72931	.76831	16	.62273	.64567	.67197
8	.67988	.71252	.74921	17	.61882	.64107	.66659
9	.66821	.69896	.73370	18	.61527	.63688	.66168
10	.65855	.68768	.72070	19	.61201	.63304	.65723
11	.65039	.67816	.70972	20	.60902	.62952	.65308

[a]Adapted, with permission, from Gail and Gastwirth (1978).

variance of G_n under H_0 are .5 and $[12(n-1)]^{-1}$, respectively, and for n larger than 20 the approximation

$$\left[12(n-1)\right]^{1/2}(G_n-.5)\sim N(0,1) \tag{9.2.3}$$

is sufficiently accurate for all practical purposes.

The G test is easily modified to handle Type II censored data. If only $t_{(1)}\leqslant\cdots\leqslant t_{(r)}$ are observed in a random sample of n, we define W_i's as before and consider

$$G_{r,n}=\left(\sum_{i=1}^{r-1} iW_{i+1}\right)\Big/(r-1)\sum_{i=1}^{r} W_i. \tag{9.2.4}$$

Values of $G_{r,n}$ close to 0 or 1 provide evidence against exponentiality. In fact, since W_i/θ $(i=1,\ldots,r)$ are independent and have standard exponential distributions, it is obvious that the distribution of $G_{r,n}$ is exactly the same as that of G_r. Consequently, no new tables or results are required to accommodate Type II censored data.

The G test has good power against IFR or DFR alternatives. Values of G_n close to zero are indicative of DFR, and values close to unity indicative of IFR, among the class of distributions with monotone hazard functions. Gail and Gastwirth (1978) examine the power of the test against certain alternative models. They also show that the asymptotic relative efficiency of the test is fairly high in detecting gamma and Weibull departures from exponentiality, the asymptotic efficiencies being .694 for the gamma distribution and .876 for the Weibull distribution.

Omnibus Tests Based on the EDF

With a complete sample $x_1,\dots,x_n$ the Kolmogorov–Smirnov and Cramer–von Mises statistics of Section 9.1.1 can be modified to test (9.2.1) by replacing $F_0(x)$ in (9.1.1) to (9.1.6) with $\hat{F}_0(x)=1-\exp(-x/\hat{\theta})$, where $\hat{\theta}=\Sigma x_i/n$ is the m.l.e. of θ. Exact percentage points for D_n^+ and D_n^- in this case have been found by Durbin (1975) and Margolin and Maurer (1976). Stephens (1974) gives percentage points for each of D_n, W_n^2, and A_n^2, estimated by Monte Carlo methods. Approximate percentage points are given in Table 9.2.2, in the same format as Table 9.1.1. That is, quantiles are given for statistics modified as a function of n in such a way that a single table suffices for all n. The quantiles in the table appear to be quite accurate for n greater than about 20. For small sample sizes a table given by Durbin (1975) can be consulted in the case of D_n.

For Type II or singly Type I censored data, Pettit (1977) has considered tests based on $W_{n,p}^2$ and $A_{n,p}^2$ of (9.1.7), with $F_0(x)$ replaced by $\hat{F}_0(x)=1-e^{-x/\hat{\theta}}$. We shall give results only for $W_{n,p}^2$. From (9.1.9) and (9.1.10), the appropriate test statistics are

$$W_{n,r}^2=\sum_{i=1}^{r}\left(\hat{F}_0(x_{(i)})-\frac{i-.5}{n}\right)^2+\frac{r}{12n^2}-\frac{n}{3}\left(\frac{r}{n}-\hat{F}_0(x_{(r)})\right)^3 \tag{9.2.5}$$

for a Type II censored sample $x_{(1)}\leqslant\cdots\leqslant x_{(r)}$ $(r\leqslant n)$, and

$$W_{n,L}^2=\sum_{i=1}^{r}\left(\hat{F}_0(x_{(i)})-\frac{i-.5}{n}\right)^2+\frac{r}{12n^2}-\frac{n}{3}\left(\frac{r}{n}-\hat{F}_0(L)\right)^3 \tag{9.2.6}$$

for Type I censored data. In (9.2.6) L is the upper limit of observation on X

Table 9.2.2 Quantiles for Functions of D_n, W_n, and A_n^2, With $F_0(x)=1-\exp(-x/\hat{\theta})$[a]

	Quantile				
Function	.85	.90	.95	.975	.99
$(D_n-.2n^{-1})n^{1/2}+.26+.5n^{-1/2}$	0.926	0.990	1.094	1.190	1.308
$W_n^2(1+.16n^{-1})$	0.149	0.177	0.224	0.273	0.337
$A_n^2(1+.6n^{-1})$	0.922	1.078	1.341	1.606	1.957

[a]Adapted, with permission, from Stephens (1974).

(i.e., the censoring time) and r is a random variable equal to the number of observations with values less than or equal to L.

Pettit (1977) determines the asymptotic distributions of (9.2.5) and (9.2.6). If in (9.2.5) we let r and $n \to \infty$, with $r/n \to p$, and in (9.2.6) let $n \to \infty$ with L defined so that $F_0(L)=p$, then the two statistics have a common limiting distribution. Percentage points for this are given in Table 9.2.3; these appear to provide suitable approximations for samples with $n \geqslant 20$ and $r \geqslant .5n$. Smith and Bain (1976) give some small-sample percentage points for a variant of (9.2.5), obtained by Monte Carlo methods. Their paper can be consulted for small sample sizes. No one has obtained analogous exact or asymptotic results for the Kolmogorov–Smirnov statistics under Type II censoring.

Not much is known about the power of these EDF tests. Stephens (1974) suggests that W_n^2 and A_n^2 are usually preferable to D_n, but there is little conclusive evidence on this. The EDF tests should normally be used when no very specific family of alternatives to exponentiality is suggested. If monotone hazard function or even more specific alternatives are considered, the G test or a test based on a parametric model is usually preferable.

Example 9.2.1 Proschan (1963) has given some data on the time, in hours of operation, between successive failures of air conditioning equipment in 13 Boeing 720 aircraft. The data for plane number 3 are as follows:

90, 10, 60, 186, 61, 49, 14, 24, 56, 20, 79, 84, 44, 59, 29, 118, 25, 156, 310, 76, 26, 44, 23, 62, 130, 208, 70, 101, 208.

There are 29 observations in all. An important question is whether the times between successive failures are i.i.d.; this is discussed in Section 10.2.4. We

Table 9.2.3 Quantiles for Asymptotic Distribution of $W_{r,n}^2$ and $W_{r,L}^2$ [a]

	Quantile				
p	.85	.90	.95	.975	.99
.5	.0531	.0635	.0821	.1015	.1279
.6	.0720	.0857	.1103	.1359	.1710
.7	.0921	.1093	.1401	.1721	.2160
.8	.1126	.1333	.1702	.2082	.2613
.9	.1321	.1561	.1986	.2433	.3033
1.0	.1480	.1745	.2216	.2706	.3376

[a] Taken, with permission, from Pettit (1977).

assume here that the times *are* i.i.d. and examine whether the data could conceivably have arisen from an exponential distribution. That is, we test the hypothesis that the times between successive failures are i.i.d. with d.f. of the form $F_0(x)=1-e^{-x/\theta}$, $x>0$.

Figure 9.2.1 shows a probability plot of the points $(t_{(i)}, \alpha_i)$, where α_i is the expected ith standard exponential order statistic given in Section 2.4. Goodness of fit tests supplement the picture. For the EDF tests we use the m.l.e. of θ, which is $\hat{\theta}=\bar{x}=83.517$. The statistics D_{29}^+, D_{29}^-, W_{29}^2, and A_{29}^2 below are calculated from expressions (9.1.4), (9.1.5), and (9.1.6), with $F_0(x)=1-\exp(-x/\hat{\theta})$. Observed values for these statistics and for G_{29} of (9.2.2) are as follows: $G_{29}=.4411$, $D_{29}^+=.064541$, $D_{29}^-=.143991$, $D_{29}=.143991$, $W_{29}^2=.121626$, and $A_{29}^2=.810039$. Significance levels with these statistics can be obtained from 9.2.3 (for G_{29}), and from Table 9.2.2. The significance level from G_{29} is about .27; those from D_{29}, W_{29}^2, and A_{29}^2 are all greater than .15 and appear to be less than .30 or so. (For example, with W_{29}^2 we calculate the modified statistic $W_{29}^2(1+.16/29)=.122$; Table 9.2.2 indicates a probability somewhat over .15 of observing a value at least this large.) There is consequently no evidence against the hypothesis of exponen-

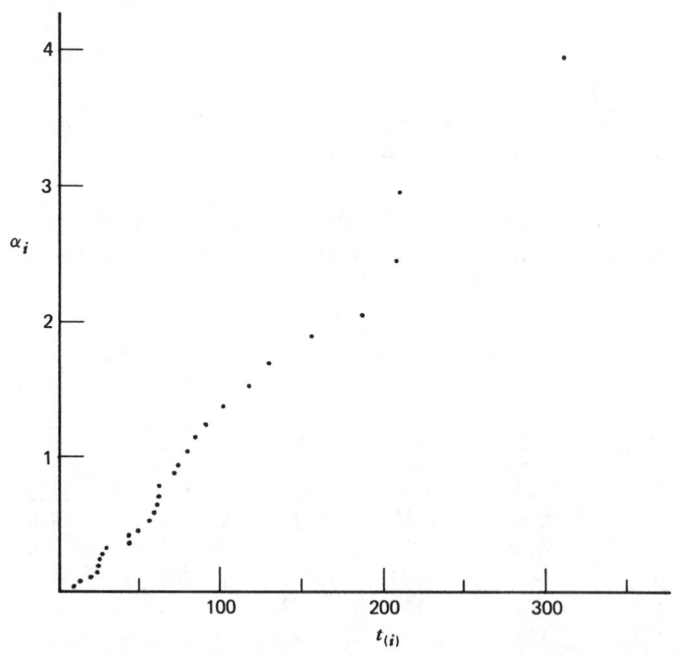

Figure 9.2.1 Probability plot of air conditioning failure data under exponential model (Example 9.2.1)

tiality from any of these tests, and, in addition, the tests give similar significance levels.

A test of fit could also be obtained by grouping the data and using the Pearson X^2 or likelihood ratio statistics of Section 9.1.3. The table below shows observed (d_i) and expected (e_i) frequencies for one particular group of classes:

Number of hours	[0,50)	[50,100)	[100,200)	[200,∞)
d_i	11	10	5	3
e_i	13.57	7.22	5.89	2.32

The expected frequencies are calculated from the m.l.e.'s of the p_j's for the four classes, where $p_j = \Pr$(an observation X is in class j). The multinomial likelihood function for θ is

$$L(\theta) \propto p_1^{11} p_2^{10} p_3^5 p_4^3 = (1-e^{-50/\theta})^{11}(e^{-50/\theta}-e^{-100/\theta})^{10}$$
$$\times (e^{-100/\theta}-e^{-200/\theta})^5 (e^{-200/\theta})^3.$$

If we set $\beta = e^{-50/\theta}$, $L(\theta)$ can be written as

$$L_1(\beta) = \beta^{32}(1-\beta)^{26}(1+\beta)^5.$$

This is easily maximized to give $\hat{\beta}=.532$, and thus $\hat{\theta}=-50/\log\hat{\beta}=79.11$. The expected frequencies in the table now follow; for example, $e_1 = 29(1-e^{-50/79.11}) = 13.57$. The X^2 and Λ statistics given by (9.1.19) and (9.1.20) turn out to equal 1.89 and 1.80. Comparing these with the percentage points of $\chi^2_{(2)}$, one obtains a significance level of over .5 in each case.

Note that the m.l.e. $\hat{\theta}=79.11$ from the multinomial likelihood function is required for this test. If the m.l.e. ($\hat{\theta}=83.52$) from the ungrouped likelihood function is used, close to the same values of X^2 and Λ are obtained, though the limiting distribution of the statistics in this case is not exactly $\chi^2_{(2)}$.

We conclude this section with two additional points. First, tests for one-parameter exponentiality can also be used to test for two-parameter exponentiality. Suppose that the hypothesized model has survivor function $\exp[-(x-\mu)/\theta]$, $x \geqslant \mu$, where μ is a threshold parameter, and that $x_{(1)} \leqslant \cdots \leqslant x_{(r)}$ are the order statistics from a random sample of size n. Then the statistics $x'_{(i)} = x_{(i)} - x_{(1)}$, $i=2,\ldots,n$, can be treated as the order statistics in a random sample of size $n-1$ from the one-parameter exponential distribution and used with the tests for one-parameter exponentiality.

A second point is that there is no completely satisfactory way to test for exponentiality when data are arbitrarily censored. One possible procedure is to group the data and proceed as in Section 9.1.3. Another, very crude, procedure is to adjust censored observations by an amount $\hat{\theta}$, in the spirit of (6.2.11), and to then treat the resulting "sample" as uncensored. If there are more than a few censored observations, one should not, however, rely much on significance levels obtained in this way.

9.2.2 Tests of Fit for the Weibull or Extreme Value Distribution

Although the Weibull model is an important one, relatively little work has been done on goodness of fit tests for it or the equivalent extreme value distribution. We consider here the problem of testing that X has survivor function of extreme value form

$$S(x)=\exp(-e^{(x-\mu)/\sigma}) \qquad -\infty<x<\infty$$

or, equivalently, that the p.d.f. of X is

$$f(x)=\frac{1}{\sigma}\exp\left[\frac{x-\mu}{\sigma}-\exp\left(\frac{x-\mu}{\sigma}\right)\right]. \tag{9.2.7}$$

Four tests are mentioned here: the first is a parametric test of the extreme value model within the three-parameter log-gamma family, the second and third are tests due to Mann et al. (1973) and Tiku (1981), and the fourth is a Cramer–von Mises test. Illustrations of the tests are given at the end of the section.

A Parametric Test Based on the Log-Gamma Model

With log lifetime data and the extreme value distribution, a useful test is obtained by embedding (9.2.7) in the three-parameter log-gamma model of Section 5.3. The p.d.f (5.3.8) of X for the most general model given there is

$$\frac{|\lambda|(\lambda^{-2})^{\lambda^{-2}}}{\sigma\Gamma(\lambda^{-2})}\exp\left(\lambda^{-1}\frac{x-\mu}{\sigma}-\lambda^{-2}e^{\lambda(x-\mu)/\sigma}\right) \qquad -\infty<\lambda<\infty \quad \lambda\neq 0$$

$$\frac{1}{(2\pi)^{1/2}\sigma}\exp\left(\frac{-(x-\mu)^2}{2\sigma^2}\right) \qquad \lambda=0. \tag{9.2.8}$$

A test of the extreme value model within this family is obtained by testing that $\lambda=1$; a likelihood ratio test to do this was described in Section 5.3. An advantage of this test is the ability to handle arbitrarily censored data. A disadvantage is that the test may not be effective for detecting departures from the extreme value model that cannot be approximated by members of (9.2.8).

The Mann–Scheuer–Fertig and Tiku Tests

Mann et al. (1973) present a test of the two-parameter Weibull or extreme value distribution that can accommodate Type II censored data and appears to have reasonably good power against certain types of alternatives. Let $Z_{(i)}$ represent the ith order statistic in a random sample of size n from the standard extreme value distribution. Let $X_{(1)} \leqslant \cdots \leqslant X_{(r)}$ be the r smallest observations in a random sample of size n from the distribution under study, and define normalized spacings

$$l_i = \frac{X_{(i+1)} - X_{(i)}}{E\left(Z_{(i+1)} - Z_{(i)}\right)} \qquad i=1,\ldots,r-1. \tag{9.2.9}$$

The statistic proposed by Mann et al. for testing that the distribution is an extreme value distribution is

$$M = \left[\frac{r}{2}\right] \sum_{i=[r/2]+1}^{r-1} l_i \bigg/ \left[\frac{r-1}{2}\right] \sum_{i=1}^{[r/2]} l_i \tag{9.2.10}$$

where $[m]$ denotes the largest integer less than or equal to m. Large values of M provide evidence against the extreme value model.

It is clear that M is parameter free under the hypothesis H_0, that the underlying distribution is an extreme value distribution. Mann et al. (1973) give percentage points for M found by Monte Carlo methods for samples with n in the range 3 to 25. However, Mann et al. also note that under H_0 the l_i's are very close to being independent standard exponential random variables. Thus the distribution of M under H_0 is closely approximated by a F distribution, as

$$M \sim F_{(2[(r-1)/2],2[r/2])}. \tag{9.2.11}$$

This approximation is quite accurate, especially for $n \geqslant 20$. For small n tables in Mann et al. (1973) or Mann et al. (1974) can be used.

The M test requires the expected value of standard extreme value order statistics or, actually, their successive differences. Mann et al. (1973) tabulate $E(Z_{(i+1)} - Z_{(i)})$ for samples with n from 3 to 25. Alternately, tables of $E(Z_{(i)})$ given by White (1967, 1969) can be consulted. In fact, for most situations an approximation devised by Blom (1958, p. 73 ff.) is sufficiently accurate to obviate the need for tables; this is

$$E(Z_{(i)}) \doteq \log\left[-\log\left(1-\frac{i-.5}{n+.25}\right)\right] \tag{9.2.12}$$

and it is fairly accurate even for n as small as 10. Alternately, approximations given in Appendix D for moments of standardized order statistics can be employed.

Mann et al. (1973) give results of a small power study comparing the M test with the Cramer–von Mises W_n^2 and A_n^2 tests given below when alternatives are normal distributions or log three-parameter Weibull distributions. They find the M test to be somewhat more powerful than the others, though it should be mentioned that the scope of the study was very limited. Littel et al. (1979) come to similar conclusions. Mann and Fertig (1975b) modify the M test by considering

$$M' = \frac{m}{r-1-m} \sum_{i=m+1}^{r-1} l_i \Big/ \sum_{i=1}^{m} l_i$$

where $1 \leqslant m < r-1$. They show that some increase in power against log three-parameter Weibull alternatives is obtained by choosing m roughly equal to $r/3$. Under H_0, M' is approximately $F_{(2(r-1-m),2m)}$. Mann and Fertig also show how to use M' to obtain confidence limits for the threshold parameter in the three-parameter Weibull distribution (see Problem 9.9).

Tiku and Singh (1981) apply the results of Tiku (1981) to devise a test for the extreme value model that is somewhat similar to the Mann–Scheuer–Fertig test. Their test statistic is

$$Z^* = \left(2\sum_{i=1}^{r-2}(r-i-1)l_i\right) \Big/ (r-2)\sum_{i=1}^{r-1} l_i. \tag{9.2.13}$$

Large or small values of Z^* provide evidence against the extreme value model. Let the numerator and denominator of Z^* be D_1 and D_2, and define

$$V = \frac{\operatorname{Var}(D_1)}{E(D_1)^2} + \frac{\operatorname{Var}(D_2)}{E(D_2)^2} - \frac{2\operatorname{Cov}(D_1, D_2)}{E(D_1)E(D_2)}. \tag{9.2.14}$$

Tiku and Singh show that for $n \geqslant 20$ the approximation $Z^* \sim N(1, V)$ provides a very good approximation to the null distribution of Z^*. To calculate V we require means, variances, and covariances of standard extreme value order statistics. For $n \geqslant 25$ the variances and covariances are not readily available; Tiku and Singh suggest using (D10) and (D11) of Appendix D, which are quite accurate for $n \geqslant 20$.

Tiku and Singh discuss results of a small power study that indicates that their test is slightly more powerful than the M test in certain situations. It also performs well in comparison to the EDF tests discussed in this next section.

Tests Based on the EDF

For complete samples $x_1, \ldots, x_n$ from the extreme value distribution Stephens (1977) has used Monte Carlo methods to determine approximate percentage points of the Cramer–von Mises statistics given by (9.1.5) and (9.1.6), with $F_0(x)$ replaced by

$$\hat{F}_0(x) = 1 - \exp(-e^{(x-\hat{\mu})/\hat{\sigma}})$$

where $\hat{\mu}$ and $\hat{\sigma}$ are the m.l.e.'s of μ and σ. Percentage points for W_n^2 and A_n^2 are given in Table 9.2.4, in the same format as Table 9.1.1. The quantiles are for statistics modified by a function of n and they provide good approximations to the quantiles of W_n^2 and A_n^2, even for relatively small values of n.

Littel et al. (1979) have studied the W_n^2 and A_n^2 tests, along with the Kolmogorov–Smirnov test obtained by inserting $\hat{\mu}$ and $\hat{\sigma}$ in (9.1.4). They give percentage points estimated by Monte Carlo methods for sample sizes in the range $n=10$ to $n=40$. For small values of n one can refer to their Table I instead of Table 9.2.4.

The EDF tests have not been extended to handle Type II censored data, though asymptotic distributions of statistics based on (9.1.7), with $\hat{\mu}$ and $\hat{\sigma}$ estimating μ and σ, could be computed from results given in Pettit (1976).

Example 9.2.2 (Example 5.3.1 revisited) The observations below are the ordered log failure times in a sample of size 23, taken from data on the

Table 9.2.4 Quantiles for Functions of W_n^2 and A_n^2[a]

	Quantile				
Function	.75	.90	.95	.975	.99
$(1+.2n^{-1/2})W_n^2$	.073	.102	.124	.146	.175
$(1+.2n^{-1/2})A_n^2$	.474	.637	.757	.877	1.038

[a]Taken, with permission, from Stephens (1977).

endurance of ball bearings:

2.884	3.365	3.497	3.726	3.741	3.820	3.881	
3.948	3.950	3.991	4.017	4.217	4.229	4.229	4.232
4.432	4.534	4.591	4.655	4.662	4.851	4.852	5.156

It has been assumed elsewhere that failure times follow a Weibull distribution; we shall examine this by testing the hypothesis that the log failure times have an extreme value distribution.

To calculate the M statistic (9.2.10), we need the values of $E(Z_{(i+1)}-Z_{(i)})$, $i=1,\ldots,22$, as in (9.2.9). These can be taken from a table in Mann et al. (1973) or approximated by using (9.2.12). Using the exact values in Mann et al., we get

$$M=\sum_{i=12}^{22} l_i \Big/ \sum_{i=1}^{11} l_i = 1.302.$$

The approximation (9.2.12) gives $M=1.323$, which is close to the exact value. The significance level can be obtained from tables in Mann et al.; we find this to be just over .25, indicating that there is no evidence against the hypothesized extreme value family. Alternately, we can use the approximation $M \sim F_{(22,22)}$, which gives the same result.

The EDF tests of fit are also easy to carry out. The m.l.e.'s of the extreme value location and scale parameters were determined in Example 5.3.1 and they are $\hat{\mu}=4.405$ and $\hat{\sigma}=0.476$. Using the Cramer–von Mises statistics (9.1.5) and (9.1.6) with $F_0(x)=1-\exp[-e^{(x-\hat{\mu})/\hat{\sigma}}]$, we get $W_{23}^2=.057931$ and $A_{23}^2=.328844$. From Table 9.2.4, significance levels are seen to be somewhat over .25, in broad agreement with the M test. Finally, consider the parametric test based on the log-gamma model (9.2.8). The likelihood ratio for $\lambda=1$ has been obtained in Example 5.3.1 as $R_{\max}(1)=.485$; this gives the likelihood ratio statistic value $\Lambda=2\log(.485)=1.477$. The significance level obtained by comparing this with $\chi_{(1)}^2$ is about .22.

9.2.3 Tests of Fit for the Normal and Log-Normal Distributions

Many tests of fit have been proposed for the normal distribution. Only a very few are mentioned here, the aim being to provide effective tests of normality without getting into a lengthy discussion of this area. References to numerous tests can be found in Shapiro et al. (1968), Hegazy and Green (1975), Saniga and Miles (1979), and other papers cited.

Parametric Tests of Normality

We consider the hypothesis that a random variable X has a normal distribution with p.d.f. of the form

$$f(x)=\frac{1}{(2\pi)^{1/2}\sigma}\exp\left(\frac{-(x-\mu)^2}{2\sigma^2}\right). \tag{9.2.15}$$

There are numerous families of distributions that include the normal distribution as a special case, and any of these will give a parametric test of normality within that particular family. When X represents log lifetime, asymmetric alternatives to (9.2.15) are often of interest. The log-gamma model (9.2.8) includes (9.2.15) as the special case $\lambda=0$, and for $\lambda\neq0$ it provides asymmetric alternatives. On the other hand, if symmetric long- or short-tailed departures from normality are envisaged, one might consider a family such as the exponential power distributions with p.d.f.

$$f(x;\mu,\sigma,\delta)=\frac{k(\delta)}{\sigma}\exp\left(-\frac{1}{2}\left|\frac{x-\mu}{\sigma}\right|^{\delta}\right) \qquad -\infty<x<\infty$$

where $\delta>0$ is a shape parameter and $k(\delta)$ is a normalizing constant. This includes (9.2.15) as the special case $\delta=2$. Smooth tests of normality can also be obtained from the general approach of Section 9.1.2.

Parametric tests of this kind are easy to obtain and can handle arbitrarily censored data, but they may not be effective at detecting departures from normality that are not closely approximated by a member of the specified family of models. With only vague alternatives in mind, one will usually prefer one of the tests mentioned below.

Some Omnibus Tests of Normality

Many tests of normality have been proposed for general use. Three types will be discussed briefly.

First we consider tests based on the EDF. With complete samples, tests of normality can be obtained from the D_n, W_n^2, and A_n^2 tests represented by (9.1.1) through (9.1.6). We replace $F_0(x)$ in those expressions with

$$\hat{F}_0(x)=\Phi\left(\frac{x-\hat{\mu}}{\hat{\sigma}}\right)$$

where $\Phi(z)$ is the standard normal distribution function and $\hat{\mu}$ and $\hat{\sigma}$ are the m.l.e.'s of μ and σ from a random sample $x_1,\ldots,x_n$. Stephens (1974) gives

Table 9.2.5 Quantiles for Functions of W_n^2 and A_n^2[a]

Function	Quantile .75	.90	.95	.975	.99
$(n^{1/2}-.01+.85n^{-1/2})D_n$	.775	.819	.895	.955	1.035
$(1+.5n^{-1})W_n^2$	.091	.104	.126	.148	0.178
$(1+4n^{-1}-25n^{-2})A_n^2$	.576	.656	.787	.918	1.092

[a]Taken, with permission, from Stephens (1974).

approximate percentage points for D_n, W_n^2, and A_n^2, except that he uses $s=[\Sigma(x_i-\bar{x})^2/(n-1)]^{1/2}$ to estimate σ. These are presented in Table 9.2.5, which is of the same form as Table 9.1.1.: quantiles are given for D_n, W_n^2, and A_n^2, multiplied by a function of n. Percentage points obtained from Table 9.2.5 are sufficiently accurate, for practical purposes, for virtually all n.

Stephens (1974) compares these tests to the Shapiro–Wilk "W" test described next. He finds A_n^2 and W_n^2 to typically be more powerful than D_n, and only slightly less powerful than W, though some other studies (e.g. Lin and Mudholkar, 1980) have found larger differences in power. Pettit (1976) has generalized W_n^2 and A_n^2 to the case of Type II or singly Type I censored data and gives some asymptotic percentage points for the test statistics. Unfortunately, it appears that the asymptotic distributions of the statistics are approached very slowly.

Shapiro and Wilk (1965) proposed a test for normality that has consistently performed well in studies of tests of fit. The test is for complete samples and is based on the statistic

$$W=\left(\sum_{i=1}^{n} a_i x_{(i)}\right)^2 \Big/ \sum_{i=1}^{n} (x_i-\bar{x})^2 \qquad (9.2.16)$$

where $x_1,\ldots,x_n$ is a random sample from the distribution under study. The a_i's are functions of the means, variances, and covariances of the order statistics in a sample of size n from the standard normal distribution. The a_i's are tabulated for $n\leq 50$ by Shapiro and Wilk and are reproduced in Pearson and Hartley (1972, Table 15). *Small* values of W indicate departures from normality. Percentage points of W, obtained by Monte Carlo methods, are presented by Shapiro and Wilk for $n\leq 50$ and reproduced in Table 16 of Pearson and Hartley (1972). These tables are rather extensive, and the reader is referred to one of these references for details about the test and use of the tables. For sample sizes $n>50$, Shapiro and Francia (1972) discuss a modified W statistic.

Shapiro et al. (1968), Pearson et al. (1977), Saniga and Miles (1979), and others have made quite broad power comparisons of tests of normality. The W test is generally found to be very effective and appears to perform better than most other statistics, especially when asymmetric alternatives to normality are considered. Judicious use of other tests, such as those based on sample skewness and kurtosis measures $\sqrt{b_1}=m_3/m_2^{3/2}$ and $b_2=m_4/m_2^2$ (where $m_r=\Sigma(x_i-\bar{x})^r/n$), can also be very effective, however. Lin and Mudholkar (1980) base what appears to be a very good test on a measure of correlation of $\bar{x}$ and s^2. The interested reader is referred to the references for details.

None of the tests mentioned so far handle censored data. Tiku (1981) has given a test that can be used with Type II censored data, based on the same type of statistic used for the test of the extreme value distribution in Section 9.2.2. This is

$$Z^*=\left(2\sum_{i=1}^{r-2}(r-i-1)l_i\right)\Big/(r-2)\sum_{i=1}^{r-1}l_i$$

where

$$l_i=\frac{X_{(i+1)}-X_{(i)}}{E\left(Z_{(i+1)}-Z_{(i)}\right)},$$

$X_{(1)}\leqslant\cdots\leqslant X_{(r)}$ are the first r observations in a random sample of n, and $Z_{(i)}$ is the ith order statistic in a random sample of n from the standard normal distribution. Small or large values of Z^* indicate departures from normality. The normal approximation $Z^*\sim N(1,V)$, with V given by (9.2.14), provides a close approximation to the null distribution of Z^*. For $n\leqslant 20$ exact means, variances, and covariances of the $Z_{(i)}$'s have been tabulated [see, for example Sarhan and Greenberg (1962)]. For $n\geqslant 20$ approximations (D9) through (D11) in Appendix D can be used.

Tiku compares the test based on Z^* with the Shapiro–Wilk and Cramer–von Mises tests of normality and finds that Z^* is generally more powerful against asymmetric alternatives, and less powerful against symmetric alternatives. The statistic Z^* is ineffective against short-tailed symmetric alternatives.

Example 9.2.3 (Example 5.3.1 revisited) Consider once again the ball bearing failure time data discussed in Examples 5.3.1 and 9.2.2, where the data were found to be consonant with a Weibull model. We shall show that the data are also consonant with a log-normal failure time model, that is, a normal model for log failure times.

The Shapiro–Wilk test is easily carried out with the aid of Tables 15 and 16 of Pearson and Hartley (1972). We find, using Table 15, that $(\Sigma a_i x_{(i)})^2 = 6.1544$ and $\Sigma(x_i - \bar{x})^2 = 6.2560$, giving, by (9.2.16), $W = .984$. Remembering that small values of W indicate nonnormality, we see from Table 16 that the significance level is approximately .93. The EDF statistics are also easy to compute: the estimates for μ and σ are $\hat{\mu} = \bar{x} = 4.1504$, $s = 0.5333$, and D_n, W_n^2, and A_n^2 are computed from (9.1.4) to (9.1.6) with $F_0(x) = \Phi[(x - 4.1504)/0.5333]$. We find $D_{23} = .0914$, $W_{23}^2 = .0289$, and $A_{23}^2 = .1867$. From Table 9.2.5 it follows that the significance level is well above .25 in each case. Neither the Shapiro–Wilk nor EDF tests suggest any departure from normality. Finally, we note that the normal model can be tested within the log-gamma family (9.2.8) as $H_0: \lambda = 0$. In Example 5.3.1 it was found that $R_{\max}(\lambda = 0) = .839$, giving $\Lambda = -2 \log R_{\max}(0) = .351$. This corresponds to a significance level on $\chi^2_{(1)}$ of approximately .55.

9.2.4 Other Distributions

Goodness-of-fit tests for families of lifetime distributions other than the exponential, Weibull, and log-normal distributions have not been investigated much. A little work has been done on tests for the gamma (Pettit, 1978; Dahiya and Gurland, 1972) and a few other models (e.g., Smith and Bain, 1976), but for most models no special tests of fit have been developed. The approaches discussed in Sections 9.1.1 and 9.1.2 can be used to get tests in some instances. More generally, if there is enough data, the data can be grouped and the Pearson or likelihood ratio test used.

Another approach to testing the fit of a family of models with unknown parameters is to try to find a device whereby the problem can be treated by using results for the case in which no unknown parameters are present (e.g., Stephens, 1978; Braun, 1980). One such method, often called the half-sample method (Durbin, 1976; Stephens, 1978), is simple to use though it does not give tests with very good power. The basic idea is as follows: consider a random sample of size n to be used to test the hypothesis H_0: $F(x) = F_0(x; \boldsymbol{\theta})$, where $\boldsymbol{\theta}$ is an unknown nuisance parameter. Randomly choose half of the sample and use this to estimate $\boldsymbol{\theta}$ by an efficient procedure such as maximum likelihood. Using the estimate $\tilde{\boldsymbol{\theta}}$ obtained, replace $F_0(x)$ in the usual EDF statistics (9.1.4) to (9.1.6) with $F_0(x; \tilde{\boldsymbol{\theta}})$; the EDF statistic is calculated using the entire sample. It can be shown (Durbin, 1973) that the asymptotic distributions of the EDF test statistics are the same as if $F_0(x)$ were completely known. Thus, at least in large samples, one can obtain percentage points for these statistics from the tabulated asymptotic points available for the $F_0(x)$ known case (see Table 9.1.1). Stephens (1978) checks whether percentage points for small n, given by Table 9.1.1, can be used with the half-sample method. Unfortunately, these appear to be good approximations to the exact points only for certain situations, so at present this method should be reserved for large samples.

Stephens (1978) examines the power of the half-sample method for testing normality or exponentiality. The power is poor compared to that of the corresponding EDF tests that use the m.l.e.'s of parameters estimated from the entire sample (see Sections 9.2.1 and 9.2.3), so one would clearly not use the half-sample tests in these cases. On the other hand, for situations in which no other tests are available, the half-sample method may occasionally be of use.

9.2.5 Tests of Fit With Regression Models

In a regression situation one may want to assess the adequacy of the assumed base, or error, distribution. Consider, for example, linear regression models of the form

$$y_i = \mathbf{x}_i \boldsymbol{\beta} + \sigma e_i \qquad i = 1, \ldots, n \tag{9.2.17}$$

where y_i is the response variable and $\mathbf{x}_i$ is a $1 \times p$ vector of regressor variables. We may want to test, say, that the e_i's are independent random variables with some specified distribution. Probability plots with residuals provide useful assessments of model adequacy (see Section 6.2), but it may be desired to supplement these with formal tests of significance. This is briefly examined below, but only for models of the form (9.2.17) with i.i.d. e_i's.

The adequacy of the error distribution is most easily assessed when there are several observations at each regression vector $\mathbf{x}$. If, for example, there are n_i observations $y_{ij}(j=1,\ldots,n_i)$ from (9.2.17) at $\mathbf{x} = \mathbf{x}_i$, then we essentially examine the fit of the model

$$y_{ij} = \mu_i + \sigma e_{ij} \qquad j = 1, \ldots, n_i \tag{9.2.18}$$

for each value of i. To assess the fit of (9.2.18) one formally requires three or more observations at the value of the i in question. An overall test of the distribution assumed for the e_{ij}'s can be formed by combining statistics from different $\mathbf{x}_i$'s. Some of the goodness of fit statistics in Sections 9.2.1 to 9.2.4 allow easy combination of statistics from independent samples (see

Problem 9.10). In other cases general methods of combining results from independent tests can be used (e.g., Littel and Folks, 1971).

With arbitrary regression data typically not involving repeated observations other methods must be used. Smooth tests obtained by embedding a regression model in a larger model may provide convenient tests. For example, tests of whether the e_i's in (9.2.17) have a standard extreme value distribution or a standard normal distribution can be obtained by embedding these models in the log-gamma regression model given by (6.6.1) and (6.6.3). Parametric tests of this sort are, at least in principle, straightforward.

A second approach to testing distributional assumptions in regression is through test statistics based on suitably defined residuals (e.g., Cox and Snell, 1968; Durbin, 1975). Suppose that the e_i's in (9.2.17) are i.i.d. with distribution function $F(e)$; we also assume that (9.2.17) contains a constant term in $\mathbf{x}_i\boldsymbol{\beta}$. Consider the hypothesis $H_0: F(e)=F_0(e)$, where $F_0(e)$ is a completely specified distribution function. Let $\hat{\boldsymbol{\beta}}$ and $\hat{\sigma}$ be m.l.e.'s of $\boldsymbol{\beta}$ and σ, obtained from the sample $(y_i, \mathbf{x}_i)$, $i=1,\dots,n$, and define residuals

$$\hat{e}_i = \frac{y_i - \mathbf{x}_i\hat{\boldsymbol{\beta}}}{\hat{\sigma}} \qquad i=1,\dots,n. \tag{9.2.19}$$

Tests of H_0 can be based on statistics that are straightforward generalizations of statistics used to test that the model

$$y_i = \mu + \sigma e_i \qquad i=1,\dots,n \tag{9.2.20}$$

has i.i.d. e_i's with distribution function $F_0(e)$. The $\hat{e}_i$'s are not, in general, independent nor identically distributed, and the exact distribution of test statistics will usually be hopelessly complicated. Kopecky and Pierce (1979b) show, however, that statistics that are permutationally invariant functions of the $\hat{e}_i$'s have the same asymptotic distribution under (9.2.17) as under (9.2.20), regardless of the dimension of $\boldsymbol{\beta}$. This means that asymptotic theory concerning tests of fit for location–scale models can be brought to bear on the regression problem (see also Loynes, 1980).

To illustrate these remarks let us consider the problem of testing the hypothesis that the e_i's in (9.2.17) are i.i.d. standard normal random variables. We have

$$H_0: F(e) = \Phi(e) = \int_{-\infty}^{e} \frac{1}{(2\pi)^{1/2}} \exp\left(\frac{-z^2}{2}\right) dz.$$

One of the statistics presented in Section 9.2.3 for testing adequacy of the ordinary normal model (9.2.20), with $e_i \sim N(0,1)$, was the Cramer–von

Mises statistic

$$\hat{W}_n^2 = \sum_{i=1}^{n} \left[\Phi\left(\frac{y_{(i)} - \hat{\mu}}{\hat{\sigma}} \right) - \frac{i - .5}{n} \right]^2 + \frac{1}{12n}.$$

The corresponding statistic in the regression case is

$$\tilde{W}_n^2 = \sum_{i=1}^{n} \left(\Phi(\hat{e}_{(i)}) - \frac{i - .5}{n} \right)^2 + \frac{1}{12n}$$

where $\hat{e}_{(i)}$ is the ith smallest residual from (9.2.19). Kopecky and Pierce's result shows that the limiting distributions of $\tilde{W}_n^2$ and $\hat{W}_n^2$ are the same. Thus for large samples one can use the known limiting distribution of $\hat{W}_n^2$ (see Table 9.2.5) to obtain approximate percentage points for $\tilde{W}_n^2$. One should be cautioned, however, that it is not known how large n should be to make this approximation reasonable.

The $\hat{e}_i$'s are not i.i.d., and it may not always be desirable to use a test statistic that is a symmetric function of them. On the other hand, there is almost nothing else available upon which to base goodness of fit tests. An exception is in the case of the normal distribution, where much work has been done. In particular, to test that the e_i's in (9.2.17) are standard normal, we can first transform the unscaled residuals $\tilde{e}_i = y_i - \mathbf{x}_i\hat{\boldsymbol{\beta}}$ to obtain $n-p$ independent residuals that are identically distributed as $N(0, \sigma^2)$, and then use one of the tests for univariate normality (e.g., Seber 1977, p. 172). This topic is not pursued here.

9.2.6 Model Discrimination and Hypothesis Testing

Sometimes it is desired to establish which one(s) of a small number of families of models are best supported by available data. One way to look at this problem is to assess the fit of the various models and to see if one or more of them are in substantially better agreement with the data than the others. This is useful, though one frequently finds that several models fit the data adequately and that different goodness of fit tests yield different rankings of the models in terms of closeness of fit to the data. Another way that discrimination can be effected is within a family of models that includes the distributions of interest as sub-models. The log-gamma family (9.2.8) and other flexible models (see, e.g., Prentice, 1975) are useful in this respect.

If one's aim is to choose a single model from a set of two or more models, the problem can be treated as a hypothesis testing or discrimination problem. Let us consider the case in which a choice is to be made between two models. To be specific, suppose that two possible models are hypothesized for a random variable X: H_1, that X has p.d.f $f_1(x; \boldsymbol{\theta}_1)$, where $\boldsymbol{\theta}_1 \in \Omega_1$ is an unknown vector of parameters, and H_2, that X has p.d.f. $f_2(x; \boldsymbol{\theta}_2)$, where $\boldsymbol{\theta}_2 \in \Omega_2$ is an unknown vector of parameters. If either f_1 or f_2 is a special case of the other (e.g., f_1 is an exponential p.d.f. and f_2 is a Weibull p.d.f.), hypothesis tests are usually easy to obtain. We shall concentrate here on the case in which H_1 and H_2 represent so-called separate families of hypotheses. This means that an arbitrary member $f_1(x; \boldsymbol{\theta}_1^0)$ in H_1 cannot be arbitrarily closely approximated by a member $f_2(x; \boldsymbol{\theta}_2)$ of H_2.

Likelihood Ratio Procedures

The basic problem is to choose between the following:

$$\begin{aligned} &H_1: X \text{ has p.d.f. } f_1(x; \boldsymbol{\theta}_1) \qquad && \boldsymbol{\theta}_1 \in \Omega_1 \\ &H_2: X \text{ has p.d.f. } f_2(x; \boldsymbol{\theta}_2) \qquad && \boldsymbol{\theta}_2 \in \Omega_2. \end{aligned} \tag{9.2.21}$$

One criterion upon which the choice can be based is the likelihood ratio

$$R_{12} = \frac{\sup\limits_{\boldsymbol{\theta}_1 \in \Omega_1} L_1(\boldsymbol{\theta}_1)}{\sup\limits_{\boldsymbol{\theta}_2 \in \Omega_2} L_2(\boldsymbol{\theta}_2)} \tag{9.2.22}$$

where $L_1(\boldsymbol{\theta}_1)$ and $L_2(\boldsymbol{\theta}_2)$ are the likelihood functions under models H_1 and H_2. If the data are a complete random sample $x_1, \ldots, x_n$, then

$$L_i(\boldsymbol{\theta}_i) = \prod_{j=1}^{n} f_i(x_j; \boldsymbol{\theta}_i) \qquad i = 1, 2.$$

Decision rules will be of the form

$$\begin{aligned} &\text{If } R_{12} \geqslant c \qquad && \text{choose } H_1 \\ &\text{If } R_{12} < c \qquad && \text{choose } H_2\,. \end{aligned} \tag{9.2.23}$$

Two approaches can be mentioned. Sometimes we wish to favor one model by treating it as the null hypothesis and the other as the alternative. Suppose, for example, we test H_1 (null hypothesis) versus H_2 (alternative hypothesis). In this case we usually select c so that the Type I error probability, $\Pr(R_{12} < c;\ H_1 \text{ true})$, is small. The power of the test is then $\Pr(R_{12} < c;\ H_2 \text{ true})$, and this is hopefully large. This approach is appropriate when we wish to retain one of the models (the null hypothesis), unless there is fairly strong evidence against it. On the other hand, it is sometimes desired to merely discriminate between H_1 and H_2, treating the hypotheses symmetrically. This typically involves a larger value of c in (9.2.23) than if one were testing H_1 (null) versus H_2 (alternative hypothesis), the idea being to choose a rule that minimizes some overall risk of making an incorrect decision. Dyer (1973, 1974) examines the hypothesis testing and discrimination approaches for certain families of models.

The main difficulty with the statistic R_{12}, as with other statistics that might be used to choose between H_1 and H_2, is that its distribution under H_1 or H_2 is usually intractable. Even the asymptotic theory is difficult; Cox (1961, 1962) and others have examined these problems, but general theory is not yet at the stage where useful asymptotic approximations to the distribution of R_{12} can be routinely derived. When H_1 and H_2 both represent location–scale parameter families, R_{12} is parameter free (invariant) and in a few situations the distribution of R_{12} has been investigated analytically or by simulation (e.g., Uthoff, 1970; Hogg et al., 1972; Dumonceaux et al., 1973; Bain and Engelhardt, 1980). In an example below results are noted for the case in which H_1 and H_2 represent the families of extreme value and normal distributions. In addition, a few other isolated problems relevant to the analysis of lifetime data have been examined (e.g., Jackson 1968, 1969). Pereira (1977) gives a bibliography of work in this area.

UMPI Procedures

When H_1 and H_2 represent location–scale parameter families, there exists a uniformly most powerful invariant (UMPI) test statistic (Lehmann, 1959). This provides UMPI tests of H_1 (null hypothesis) versus H_2 (alternative hypothesis) or of H_2 versus H_1, and it also has optimal properties in the discrimination context (e.g., Dyer, 1973). Suppose that $f_1(x)$ and $f_2(x)$ are of the forms $\sigma^{-1}g_1[(x-\mu)/\sigma]$ and $\sigma^{-1}g_2[(x-\mu)/\sigma]$, respectively; if the data are a random sample $x_1,\ldots,x_n$, then the UMPI test statistic is (Hajek and Sidak, 1967)

$$M_{12}=\int_0^{\infty}\int_{-\infty}^{\infty}\sigma^{-n}\prod_{j=1}^{n}g_1\left(\frac{x_j-\mu}{\sigma}\right)\frac{d\mu\,d\sigma}{\sigma}\Big/\int_0^{\infty}\int_{-\infty}^{\infty}\sigma^{-n}\prod_{j=1}^{n}g_2\left(\frac{x_j-\mu}{\sigma}\right)\frac{d\mu\,d\sigma}{\sigma}. \tag{9.2.24}$$

This is used in decision rules in the same way R_{12} is used in (9.2.23). Sometimes, but not in general, the statistics M_{12} and R_{12} are equivalent. The exact distribution of M_{12} under H_1 or H_2 is, like that of R_{12}, usually intractable, and asymptotic theory is difficult. UMPI tests can be investigated analytically in only a very few special situations (e.g., see Uthoff, 1970). Like the likelihood ratio test procedures with separate families of hypotheses, use of M_{12} in other situations would depend heavily on the development of asymptotic theory or the determination of percentage points for test statistics via simulation.

The statistic M_{12} has one property that is sometimes useful: if H_1 or H_2 is the correct model, each with probability $\frac{1}{2}$, then the rule "choose $H_1(H_2)$, if $M_{12} \geqslant (<) 1$" minimizes the probability of choosing the incorrect model among the class of all decision rules that are invariant under location–scale transformations (e.g., Dyer, 1973). In a situation where one has no *a priori* reason for preferring H_1 or H_2, this decision rule might be reasonable. Note that this rule is not, in general, the same as the rule "choose H_1 (H_2) if $R_{12} \geqslant (<) 1$."

Tests of separate families of hypotheses are at present of very limited use, but we can consider one problem where these ideas can be employed.

Example 9.2.4 (discrimination between the normal and extreme value distributions) Let us consider the problem of discriminating between the normal and extreme value (or, equivalently, between the log-normal and Weibull) distributions. Consider

$$H_1: X \text{ has p.d.f. } f_1(x) = \frac{1}{\sigma}\exp\left[\frac{x-\mu}{\sigma} - \exp\left(\frac{x-\mu}{\sigma}\right)\right] \qquad -\infty < x < \infty$$

$$H_2: X \text{ has p.d.f. } f_2(x) = \frac{1}{(2\pi)^{1/2}\sigma}\exp\left[-\frac{1}{2}\left(\frac{x-\mu}{\sigma}\right)^2\right] \qquad -\infty < x < \infty$$

where, in both H_1 and H_2, μ $(-\infty < \mu < \infty)$ and σ $(\sigma > 0)$ are parameters.

Suppose we have a complete sample $x_1, \ldots, x_n$. Let $\hat{\mu}_1$ and $\hat{\sigma}_1$ be the m.l.e.'s of μ and σ in $f_1(x)$, obtained as described in Section 4.1.1, and let $\hat{\mu}_2 = \bar{x}$ and $\hat{\sigma}_2 = [\Sigma_{i=1}^n (x_i - \bar{x})^2/n]^{1/2}$ represent the m.l.e.'s of μ and σ in $f_2(x)$. The statistic R_{12} of (9.2.21) in this case reduces to

$$R_{12} = \left(2\pi \sum_{i=1}^n (x_i - \bar{x})^2 \Big/ n\hat{\sigma}_1^2\right)^{n/2} \exp\left(\frac{n}{2} - \sum_{i=1}^n e^{(x_i - \hat{\mu}_1)/\hat{\sigma}_1}\right).$$

The distribution of R_{12} under both H_1 and H_2 has been investigated via simulation by Dumonceaux and Antle (1973). Tables 9.2.6 and 9.2.7 give percentage points as follows:

Table 9.2.6 Quantiles of $R_{12}^{1/n}$ Under Model H_2 (Normal)[a]

	Quantile			
n	.80	.90	.95	.99
20	1.015	1.038	1.082	1.144
30	0.993	1.020	1.044	1.095
40	0.984	1.007	1.028	1.070
50	0.976	0.998	1.014	1.054

[a]Adapted, with permission, from Dumonceaux and Antle (1973).

1. Table 9.2.6 gives percentage points of $R_{12}^{1/n}$ under model H_2. These can be used to test the null hypothesis H_2 versus H_1, with large values of $R_{12}^{1/n}$ providing evidence against H_2.
2. Table 9.2.7 gives percentage points of $R_{21}^{1/n}=R_{12}^{-1/n}$ under model H_1. These can be used to test the null hypothesis H_1 versus H_2, with large values of $R_{21}^{1/n}$ providing evidence against H_1.

The UMPI statistic (9.2.24) is as follows: the numerator is readily found by direct integration to be expressible as

$$\int_0^\infty \frac{\sigma^{-(n-1)}\Gamma(n)\exp\left(\sum_{i=1}^{n} x_i/\sigma\right)}{\left[\sum_{i=1}^{n}\exp(x_i/\sigma)\right]^n}\,\frac{d\sigma}{\sigma}$$

Table 9.2.7 Quantiles of $R_{21}^{1/n}$ Under Model H_1 (Extreme Value)[a]

	Quantile			
n	.80	.90	.95	.99
20	1.008	1.041	1.067	1.120
30	0.991	1.019	1.041	1.088
40	0.980	1.005	1.026	1.063
50	0.974	0.995	1.016	1.045

[a]Adapted, with permission, from Dumonceaux and Antle (1973).

whereas the denominator is (e.g., Uthoff, 1970; Hajek and Sidak, 1967)

$$\frac{\Gamma[(n-1)/2]}{2\sqrt{n}\,\pi^{(n-1)/2}\left[\sum_{i=1}^{n}(x_i-\bar{x})^2\right]^{(n-1)/2}}.$$

The numerator for M_{12} requires numerical integration. Since no one has determined M_{12}'s distribution under either H_1 or H_2, this test is unuseable at present.

Bain (1978, Sec. 4.4) considers another statistic that appears to have essentially the same power as R_{12} or M_{12} in discriminating between H_1 and H_2. This is $S=(\hat{\sigma}_2/\hat{\sigma}_1)^n$, where $\hat{\sigma}_1$ and $\hat{\sigma}_2$ are the m.l.e.'s of σ in $f_1(x)$ and $f_2(x)$. He gives some percentage points for S, obtained by simulation.

Before concluding this section we mention that discrimination between the extreme value and normal distributions can be carried out within the log-gamma model (9.2.8). The situation described above involves testing $H_1:\lambda=1$ versus $H_2:\lambda=0$ within (9.2.8). This problem can be compared with that which tests $H_1:\lambda=1$ versus $H_2:\lambda\neq 1$. The likelihood ratio statistic in this case is

$$R_{\max}(1)=\frac{L_{\max}(1)}{L_{\max}(\hat{\lambda})}$$

and under H_1, $\Lambda=-2\log R_{\max}(1)$ is approximately $\chi^2_{(1)}$. This test is easy to use and under H_1 or H_2 it is asymptotically equivalent to the test based on R_{12}, though for small to moderate n it's power is somewhat less than that of R_{12}.

9.3 MODEL CHOICE AND STATISTICAL METHODS

When inferences are to be based on a statistical model, it is, of course, important to be satisfied as to the appropriateness of the model. As a minimum, the model should be consonant with the data in regard to goodness of fit tests and other assessment procedures. However, we should also recognize the extent to which inferences depend on the assumed model, even if the model fits the data adequately. For example, with small or moderate sample sizes it is often impossible to discriminate between various competing models, and yet inferences may depend rather heavily upon what model is used. Table 9.3.1 shows the power of some size .05 goodness of fit tests for the Weibull or extreme value model against a few log-gamma

Table 9.3.1 Power of Tests of Fit for Extreme Value Model (Size .05)

		Test Statistic		
Alternate Model	n	A_n^2	Z^*	M
Normal	10	.08	.13	.18
	25	.29	.39	.38
	40	.54	.62	.46
Logistic	40	.59	.68	.43
Log-gamma ($k=.5$)	40	.11	.14	.01
Log-gamma ($k=2$)	40	.08	.11	.13

alternatives (Tiku and Singh, 1981). The alternative models are similar to the extreme value model, so fairly large samples are needed in order for the extreme value model to be rejected. For the normal model, which differs the most from the extreme value model, a sample of about 40 is needed for the power to be around .50. On the other hand, inferences about distribution quantiles, for example, can differ markedly under the two models: this has been discussed in some detail with a sample of size 23 in Example 5.3.2.

The main point to be stressed is that it is important to know the effects of departures from assumptions on the inferences one is making. Often data are analyzed under a particular model simply because (1) the model has been used before in similar situations, or (2) it fits the data on hand. This does not imply any absolute validity of the model, and we should ask whether inferences change much if another similarly "plausible" model is used instead. In some situations, such as when quantiles or probabilities in the tail of a distribution are estimated parametrically, results may depend heavily on the model. In others results may be relatively insensitive to changes in the assumed model. In coming to grips with model dependency two approaches are helpful. One is simply the practice of analyzing the data under a number of plausible models, with an examination of the effect of model choice on inferences. This can frequently be done by adopting a "super model" that includes various competing models and allows effects of departures to be examined parametrically. The log-gamma analysis in Example 5.3 is of this type. Other examples of this are given by Dempster (1975), Denby and Mallows (1977), and others. The second approach involves the use of robust procedures, that is, procedures whose validity is maintained under a variety of models. Distribution-free methods like those discussed in Chapters 2, 7, and 8 depend on somewhat weaker model assumptions than many fully parametric procedures, for example, and are usually more robust than the latter.

These comments do not imply that we should not search for good parametric models in various situations. Indeed, this search should go on constantly, for it is only as we build up confidence in the appropriateness of a family of models that we can take full advantage of the increasing precision and power that parametric model-based procedures can give.

9.4 PROBLEMS AND SUPPLEMENTS

9.1 Derive equations (9.1.5) and (9.1.6) by noting that for an uncensored sample $x_1, \ldots, x_n$, the EDF satisfies $\tilde{F}_n(x_{(i)}) = i/n$, $\tilde{F}_n(x_{(i)} - 0) = (i-1)/n$, for $i = 1, \ldots, n$. Generalize to the case of Type II censoring and derive (9.1.9) and (9.1.11).

(Section 9.1.1)

9.2 Consider testing the fit of an exponential distribution with unknown mean θ, using the Kolmogorov–Smirnov statistic D_n [see (9.1.4)]. Suppose we estimate θ by the m.l.e. $\hat{\theta}$ but incorrectly determine the significance level for the test by using Table 9.1.1, which assumes that θ is known, rather than Table 9.2.2. Examine the effect of this for $n = 20, 60, 100$. (Suggestion: determine, for example, the correct .90, .95, and .99 quantiles of D_n from Table 9.2.2 and then see what probabilities these correspond to in Table 9.1.1.)

Carry out a similar examination for W_n^2 and A_n^2.

(Sections 9.1.1, 9.2.1)

9.3 Table 9.4.1 is an adaptation of data given by Gail and Ware (1979) and concerns survival in a cohort of 191 white males who worked in a particular manufacturing plant during the period 1943–1947. The life table shows the number of deaths among individuals in the cohort at different ages; note that there are some withdrawals because of loss to follow-up. The table also shows conditional survival probabilities q_i for each age interval for a comparable cohort from the population at large.

Compare the survival distribution of the cohort of plant workers with that of the overall population cohort. You may want to combine the last two or three age intervals because of the large numbers of withdrawals there. Alternately, you might consider a modification of (9.1.23) to handle withdrawals within intervals. (Gail and Ware present a different test of fit for this situation where the q's are known.)

(Section 9.1.3)

Table 9.4.1 Survival of a Cohort of Manufacturing Plant Workers

Age Interval (Years)	n_i	d_i	w_i	q_i
[40,45)	191	8	0	.02529
[45,50)	183	6	0	.03672
[50,55)	177	11	0	.05709
[55,60)	166	8	0	.08600
[60,62)	158	5	14	.04640
[62,64)	139	8	51	.05470
[64,∞)	80	80	—	1.0

9.4 Jardine (1979) presented the data in Table 9.4.2, concerning the time to failure for 229 sugar centrifuge cloths.

Assess the fit of (1) a two-parameter Weibull distribution and (2) an exponential distribution to these data.

(Section 9.1.3)

9.5 Consider the goodness of fit statistic G_n of Section 9.2.1, written in terms of $W_j=(n-j+1)(t_{(j)}-t_{(j-1)})$, $j=1,\ldots,n$, where $t_{(0)}=0$. Noting that under an exponential distribution the quantities $D_i=W_i/\Sigma_{j=1}^{n}W_j$ $(i=$

Table 9.4.2 Failure Time Frequencies for Centrifuge Cloths

Interval (Weeks)	Number of Failures	Interval (Weeks)	Number of Failures
[0,2)	24	[30,32)	4
[2,4)	36	[32,34)	4
[4,6)	27	[34,36)	5
[6,8)	23	[36,38)	2
[8,10)	15	[38,40)	2
[10,12)	9	[40,42)	2
[12,14)	12	[42,44)	2
[14,16)	11	[44,46)	2
[16,18)	13	[46,50)	0
[18,20)	4	[50,52)	4
[20,22)	12	[52,56)	0
[22,24)	5	[56,58)	1
[24,26)	4	[58,76)	0
[26,28)	4	[76,78)	1
[28,30)	1		

$1,\ldots,n$) have a Dirichlet distribution (see Wilks, 1962, p. 177), express G_n as

$$G_n = \frac{1}{n-1}\sum_{i=1}^{n-1} iD_{i+1}$$

and deduce that $E(G_n) = .5$ and $\text{Var}(G_n) = [12(n-1)]^{-1}$.

(Section 9.2.1; Gail and Gastwirth, 1978)

9.6 Goodness of fit for the exponential distribution. Let $t_{(1)} < \cdots < t_{(r)}$ be the r smallest observations in a random sample of size n that are hypothesized to come from an exponential distribution with unknown mean θ. Let $W_1,\ldots,W_r$ be defined as in Section 9.2 and Problem 9.5, and define

$$\begin{aligned} T_j &= W_1 + \cdots + W_j \\ &= t_{(1)} + \cdots t_{(j-1)} + (n-j+1)t_{(j)} \qquad j=1,\ldots,r. \end{aligned}$$

a. Show that the joint p.d.f. of $T_1,\ldots,T_{r-1}$, given T_r, is

$$\frac{(r-1)!}{T_r^{r-1}} \qquad 0 < T_1 < \cdots < T_{r-1} < T_r.$$

This is the distribution of the order statistics from a random sample of size $r-1$ from the uniform distribution on $(0, T_r)$; equivalently, the quantities $Z_i = T_i/T_r$ $(i=1,\ldots,r-1)$ are distributed like the order statistics in a random sample of size $r-1$ from $U(0,1)$.

b. The result of part (a) removes the nuisance parameter θ and allows one to use the goodness of fit tests of Section 9.1.1, which are for completely specified distributions. Test the hypothesis that the data in Example 9.2.1 come from an exponential model by using each of the EDF statistics D_n, W_n^2, and A_n^2 of Section 9.1.1 (that is, test that the unordered Z_i's are independent $U(0,1)$ random variables). Compare the results of these tests with those carried out in the example.

c. Suppose one plots Z_i against i/r, the expected value of the ith order statistic in a random sample of size $r-1$ from the $U(0,1)$ distribution, for $i=1,\ldots,r-1$. Relate this plot to the tests in part (b). To what types of departures would you expect the tests in part (b) to be sensitive?

(Sections 9.1.1, 9.2.1; Epstein, 1960b)

9.7 Consider the two samples of electrical insulation failure times in Problem 3.11. Test the adequacy of two-parameter exponential models for the distributions from which they come.

(Section 9.2.1)

9.8 Consider the data of Problem 4.9 on failure times for five samples of ball bearing specimens. Test that the data for each of samples I to V is a random sample from a Weibull distribution.

(Section 9.2.2)

9.9 Confidence intervals for a Weibull threshold parameter. Suppose that T has a three-parameter Weibull distribution with threshold parameter λ and p.d.f. as given in Section 4.4.1. Then $X=\log(T-\lambda)$ has an extreme value distribution with p.d.f. (9.2.7). If $T_{(1)}< \cdots <T_{(r)}$ are the first r observations in a random sample of size n, define l_i's and M' as in Section 9.2.2, noting that M' is now a function of λ. One can obtain confidence limits for λ by treating M' as a pivotal quantity.

Consider the data of Example 4.4.1. Treating the two censored observations as though they were uncensored, obtain confidence intervals for λ from M'. Note that this must be done numerically. Compare your results with the assessment of plausible λ values in Example 4.4.1. (Note: the threshold parameter was denoted by μ in the Example 4.4.1). Repeat your calculations, with the two censored observations adjusted upwards to equal the estimated means of lifetimes known to exceed these censoring times.

(Section 9.2.2; Mann and Fertig, 1975b)

9.10 Assessing goodness of fit from several samples. Suppose that it is desired to test that each of m Type II censored samples comes from a Weibull distribution, where the m distributions do not necessarily have the same parameters.

a. Suppose that a statistic Z_j^* defined as in (9.2.13) is obtained from each sample. Tiku and Singh (1981) suggest using the statistic

$$Z^{**}=\sum_{j=1}^{m}(r_j-2)Z_j^* \Big/ \sum_{j=1}^{m}(r_j-2)$$

to test the null hypothesis that each of the samples comes from a Weibull distribution, where the jth sample involves the r_j smallest observations in a random sample of n_j. Determine the mean, variance, and an approximation to the distribution of Z^{**} under the null hypothesis. In what types of situations would you expect a test based on Z^{**} to be effective?

b. Consider the electrical insulating fluid failure time data of Example 4.3.2. Test the adequacy of a Weibull model at each of the voltage levels, using the tests in Section 9.2.2.

c. Assess the overall fit of the Weibull model in part (b) by calculating the significance level α_j corresponding to the test based on Z_j^* at each voltage level $j=1,\ldots,7$. Under the hypothesis that the Weibull model is correct at all voltage levels, the α_j's are independent and uniformly distributed on $(0,1)$ or, equivalently, $w_j=-2\log\alpha_j$ are independent $\chi^2_{(2)}$ random variables. Assess the Weibull fit by plotting the $\alpha_{(j)}$'s against the values $j/8$.

d. Assess the overall Weibull fit by using the statistics (1) Z^{**} of part (a) and (2) $W^*=\Sigma_{i=1}^m w_i$, which is $\chi^2_{(2m)}$ under the hypothesis that each of the m distributions is a Weibull distribution.

(Section 9.2.2)

CHAPTER 10

Multivariate and Stochastic Process Models

There are many other topics that could be profitably examined in connection with lifetime data. Some that come to mind are sequential methods (e.g., Armitage, 1975; Koziol and Petkau, 1978; Jones and Whitehead, 1979), problems in special areas of application such as clinical trials (e.g., Armitage, 1979; Peto et al., 1976, 1977) or reliability (e.g., Mann et al., 1974, Ch. 10; Barlow and Proschan, 1975), Bayesian methods (Mann et al. 1974, Ch. 8; Kalbfleisch and Prentice, 1980, Ch. 8), and robustness. Time and space limitations preclude a discussion of these topics here.

Two additional topics are discussed in this chapter; these concern models that are of fundamental importance in certain problems. Section 10.1 looks at multivariate models, wherein it is desired to examine the joint distribution of two or more variables. The competing risks problem, where individuals are liable to suffer different types of deaths, is examined in conjunction with this. The second topic is a brief discussion of stochastic process models. These arise naturally in many problems involving lifetime data, and in Section 10.2 we survey a few important models and indicate some situations where they might be applied. In both sections the aim is merely to introduce the models and survey basic results; a thorough treatment of statistical methods is not given.

10.1 MULTIVARIATE MODELS

We have till this point in the book mainly considered univariate lifetime data, though covariables are brought into the picture with regression models. In this section we discuss multivariate models for situations where the joint distribution of two or more response variables is of interest. At present our understanding of multivariate lifetime processes is rather limited; further work is needed on the modeling of these processes and on the statistical analysis of multivariate data. Development in this area is hindered to some degree by the fact that multivariate lifetime data are often censored or otherwise structured so as to make examination of certain aspects of the lifetime process impossible. A well-known example of this occurs in the so-called competing risks problem, discussed later.

Before discussing multivariate models, we shall illustrate some of the ways in which multivariate data can arise by giving a few examples.

Example 10.1.1 Let us first consider a simple conceptual model that illustrates several points. Suppose that a system has two components I and II and let T_1 and T_2 be random variables representing time to failure of I and II, respectively, under stated operating conditions. Information about the joint distribution of T_1 and T_2 may come in several forms, depending on the nature of the system. For example, if the system is a series system that fails as soon as either I or II fails, then data from a single system will consist only of $\min(T_1, T_2)$ and which component failed. If the system is a parallel one that fails only when both I and II have failed, however, then both T_1 and T_2 will be observed. Other possibilities also exist: for example, if the system operates only as long as component I operates, then we observe both T_1 and T_2 whenever $T_2 \leqslant T_1$, but only T_1 if $T_2 > T_1$.

Bivariate data may thus be restricted or censored in various ways. This has important consequences, as we see later.

Example 10.1.2 Stone (1978) gives some data from experiments to investigate the failure of epoxy electrical cable insulation specimens under conditions of constant voltage stress. Failures occurred because of a phenomenon known as electrical treeing: in this process there is considered to be an inception, or initiation, stage in which nothing appears to be happening (when viewed under a microscope), but at some point a microscopic defect appears in the material and from then on the defect "grows," eventually causing failure of the insulation. The data consist of the time T_1 to inception of a defect and the subsequent additional elapsed time T_2 to failure. Table 10.1.1 gives results from an experiment in which 20 specimens were subjected to a constant voltage of 55 kilovolts; times are in minutes. For three of the specimens inception had still not occurred when observation ceased. This results in both T_1 and T_2 being censored; T_2 in this case is completely censored.

Table 10.1.1 Cable Insulation Failure Data

Specimen	T_1	T_2	Specimen	T_1	T_2
1	228	30	11	1227	39
2	106	8	12	254	46
3	246	66	13	>2440	—
4	700	72	14	435	85
5	473	25	15	1155	85
6	>1740	—	16	>2600	—
7	155	7	17	195	27
8	414	30	18	117	27
9	1374	90	19	724	21
10	128	4	20	300	96

Example 10.1.3 Faulkner and McHugh (1972) discuss an experiment where mice that have been selectively inbred to yield spontaneous neoplasms are used to test the effect of a treatment hypothesized to provide some resistance against the development of spontaneous carcinomas. For each mouse let T_1 be the age at which a tumor appears in the mouse and let T_2 be the age of the mouse at death. Then T_1 and T_2 are both observed only if a tumor appears in the mouse before death; if a mouse dies without a tumor having appeared, only T_2 is observed.

Example 10.1.4 Hoel (1972) presents mortality data from an experiment in which two groups of mice had received a radiation dose of 300 roentgens at an age of 5 to 6 weeks. The first group of mice lived in a conventional laboratory environment and the second was kept in a germ-free environment. The cause of death for each mouse was determined by autopsy to fall into one of three well-defined categories: thymic lymphoma (C_1), reticulum cell sarcoma (C_2), and all other causes (C_3). The data, which are given in Problem 10.4, consist of a lifetime T and a cause of death (C_1, C_2, or C_3) for each mouse.

Example 10.1.5 Sometimes we wish to consider the joint distribution of lifetimes of two or more "related" individuals. For example, in examining the effects of heredity on lifetime, we may wish to consider the joint distribution of the lifetimes of a father and son in a family. Typical data might then consist of lifetimes T_{1i} and T_{2i} of a father–son pair from each of n unrelated families.

Example 10.1.6 Nelson (1970b) presents data from life tests carried out during the development of a small electrical appliance. Appliances were essentially operated repeatedly until failure, though a few appliances still had not failed after a large number of cycles of use. Failures were classified into several different modes, 18, in fact, in early stages of development. Data consist of a failure time (actually the number of cycles to failure) and a failure mode. A portion of these data has been given in Table 1.1.5 of Example 1.1.9.

These examples show some of the ways in which multivariate lifetime data arise. Two main types of situation are in evidence: (1) those in which there are two or more lifetime variables $T_1,\ldots,T_k$, associated with one or more individuals, and (2) those in which single individuals are susceptible to death from different causes or to deaths of different types, so that associated with each individual is a lifetime T and a cause (or type) of death C. Both situations will be discussed briefly.

10.1.1 Multivariate Lifetime Distributions

Suppose that there are time variables $T_1,\ldots,T_k$, with joint p.d.f. $f(t_1,\ldots,t_k)$ and survivor function

$$S(t_1,\ldots,t_k)=\Pr(T_1\geqslant t_1,\ldots,T_k\geqslant t_k) \tag{10.1.1}$$

for $t_1\geqslant 0,\ldots,t_k\geqslant 0$; when $S(t_1,\ldots,t_k)$ is continuous with respect to $t_1,\ldots,t_k$, $f(t_1,\ldots,t_k)=-\partial^k S(t_1,\ldots,t_k)/\partial t_1\cdots\partial t_k$.

In applications, interest often centers on marginal and conditional distributions associated with (10.1.1) and on the possible independence of various T_i's. A difficulty is that data are often censored in such a way that tests or estimation for certain aspects of the joint distribution of $T_1,\ldots,T_k$ are impossible. A well-known instance of this arises in a common formulation of the competing risks problem, discussed later, where only $\min(T_1,\ldots,T_k)$ is observed for each individual. This makes it impossible to test for independence of $T_1,\ldots,T_k$ without strong assumptions.

Alternate formulations of a multivariate model may be useful in different situations, especially when one is dealing with censoring. We shall consider the model both in terms of the p.d.f. and survivor function and in terms of a hazard function formulation. For convenience we examine only the bivariate case, though the ideas are readily extended to the case where $k>2$. In addition, most of the discussion is general, with only brief references to specific parametric families of distributions.

P.d.f. and Survivor Function Formulations

If data consist of n uncensored pairs of times (t_{1i}, t_{2i}), inference about (10.1.1) poses no problems, either for parametric or nonparametric methods. If T_1 and T_2 have joint p.d.f. $f(t_1, t_2; \boldsymbol{\theta})$ in some parametric family, the likelihood function is

$$L(\boldsymbol{\theta}) = \prod_{i=1}^{n} f(t_{1i}, t_{2i}; \boldsymbol{\theta})$$

and standard parametric methods lead to estimation or tests for $\boldsymbol{\theta}$. Nonparametric methods exist too; for example, the empirical joint survivor function

$$\hat{S}(x, y) = \frac{\text{Number of observations } (t_{1i}, t_{2i}) \text{ with } t_{1i} \geqslant x \text{ and } t_{2i} \geqslant y}{n}$$

provides a nonparametric estimate of $S(x, y)$. Similarly, $\hat{S}_1(x) = $(number of observations with $t_{1i} \geqslant x)/n = \hat{S}(x, 0)$ provides an estimate of the marginal survivor function for T_1, and so on. Nonparametric tests of independence for T_1 and T_2 (e.g., Kendall and Stuart, 1967, Ch. 31; Bhattacharya et al., 1970) are also available. An area where work is needed, however, is in the assessment of the fit of multivariate models.

With censored data matters are more difficult, and many problems need further study. To start, suppose that data are generated in such a way that either t_1 or t_2 can be censored and that censoring is independent of lifetimes, as in Section 1.4. Let us subdivide the n observations into four classes: C_1, both t_{1i} and t_{2i} are observed lifetimes; C_2, t_{1i} is a lifetime and t_{2i} a censoring time (i.e., we know only that $T_{2i} \geqslant t_{2i}$); C_3, t_{1i} is a censoring time and t_{2i} is a lifetime; and C_4, both t_{1i} and t_{2i} are censoring times. The likelihood function for a continuous model is then

$$L = \prod_{i \in C_1} f(t_{1i}, t_{2i}) \prod_{i \in C_2} \frac{-\partial S(t_{1i}, t_{2i})}{\partial t_{1i}} \prod_{i \in C_3} \frac{-\partial S(t_{1i}, t_{2i})}{\partial t_{2i}} \prod_{i \in C_4} S(t_{1i}, t_{2i}). \tag{10.1.2}$$

Estimation of parameters in a parametric family of models is usually straightforward in this situation, but more development of nonparametric methods is needed. For example, Campbell (1979) and others have proposed nonparametric estimates of a bivariate survivor function based on censored data, but further work in this area is needed. Work on tests for independence of T_1 and T_2 is also needed, though Brown et al. (1974) have generalized the rank correlation statistic (Kendall and Stuart, 1967, p. 476) to the case in which just one of T_1 or T_2 is subject to arbitrary right censoring.

Other types of censoring can cause special difficulties, such as when censoring of either T_1 or T_2 is dependent on the other variable. For example, in the competing risks problem observation of an individual continues only until $\min(T_{1i}, T_{2i})$ is observed. The data consist of pairs (δ_i, T_i), where $T_i = \min(T_{1i}, T_{2i})$ and $\delta_i = 1$ if $T_i = T_{1i}$ and 0 if $T_i = T_{2i}$. If we assume that any further censoring involves both T_{1i} and T_{2i} being censored at t_i and that this censoring is independent of (T_{1i}, T_{2i}), the likelihood function is of the form

$$L = \prod_{i \in D} \left(\frac{-\partial S(x, y)}{\partial x} \bigg|_{x=y=t_i} \right)^{\delta_i} \left(\frac{-\partial S(x, y)}{\partial y} \bigg|_{x=y=t_i} \right)^{1-\delta_i} \prod_{i \in \bar{D}} S(t_i, t_i) \tag{10.1.3}$$

where D is the set of individuals for which T_i is observed. As with (10.1.2), inference procedures for parametric models can usually be obtained in standard ways (e.g., Moeschberger, 1974; David and Moeschberger, 1978). However, (10.1.3) is missing one component found in (10.1.2), and it turns out that one cannot test whether T_1 and T_2 are independent nor estimate $S(t_1, t_2)$ nonparametrically, on the basis of (10.1.3). To gain more insight into the difficulties this sort of censoring creates, let us consider a hazard function formulation for bivariate lifetime models.

Hazard Function Formulations

The hazard function concept can be approached in various ways with multivariate distributions. Some authors (e.g., Brindley and Thompson, 1972; Barlow and Proschan, 1975; Johnson and Kotz, 1975) have considered extensions of the monotone hazard function idea to multivariate models (see Problem 1.15). This is useful in connection with system reliability, but it is not of much direct relevance to the development of statistical methods. The hazard function formulation given here is that of Cox (1972a) and it views the bivariate lifetime model as a point process. This is particularly useful when T_1 and T_2 are observed in the same time frame, with censoring taking the form of ceased observation at some point in time.

Define

$$\lambda_i(t) = \lim_{\Delta t \to 0} \frac{\Pr(t \leqslant T_i < t + \Delta t | T_1 \geqslant t, T_2 \geqslant t)}{\Delta t} \qquad i = 1, 2$$

$$\lambda_{12}(t_1|t_2) = \lim_{\Delta t \to 0} \frac{\Pr(t_1 \leqslant T_1 < t_1 + \Delta t | T_1 \geqslant t_1, T_2 = t_2)}{\Delta t} \qquad t_1 > t_2 \qquad (10.1.4)$$

$$\lambda_{21}(t_2|t_1) = \lim_{\Delta t \to 0} \frac{\Pr(t_2 \leqslant T_2 < t_2 + \Delta t | T_1 = t_1, T_2 \geqslant t_2)}{\Delta t} \qquad t_1 < t_2.$$

In terms of the joint survivor function $S(t_1, t_2)$ for T_1 and T_2, it is readily seen that

$$\lambda_1(t) = \left. \frac{-\partial S(t_1, t_2)/\partial t_1}{S(t_1, t_2)} \right|_{t_1 = t_2 = t}$$

$$(10.1.5)$$

$$\lambda_{12}(t_1|t_2) = \frac{-\partial^2 S(t_1, t_2)/\partial t_1 \partial t_2}{\partial S(t_1, t_2)/\partial t_2} \qquad t_1 > t_2$$

with similar expressions for $\lambda_2(t)$ and $\lambda_{21}(t_2|t_1)$.

The functions in (10.1.4) completely specify the joint distribution of T_1 and T_2. The joint p.d.f. for T_1 and T_2 can be shown to be

$$\lambda_2(t_2)\lambda_{12}(t_1|t_2)\exp\left(-\int_0^{t_2}[\lambda_1(u) + \lambda_2(u)]\,du - \int_{t_2}^{t_1}\lambda_{12}(u|t_2)\,du\right) \qquad t_1 \geqslant t_2$$

$$\lambda_1(t_1)\lambda_{21}(t_2|t_1)\exp\left(-\int_0^{t_1}[\lambda_1(u) + \lambda_2(u)]\,du - \int_{t_1}^{t_2}\lambda_{21}(u|t_1)\,du\right) \qquad t_1 \leqslant t_2.$$

$$(10.1.6)$$

This can be seen by viewing the process as a point process: for example, with $t_1 \geqslant t_2$, the probability of having no deaths in $[0, t_2)$ and then the event $T_2 \in [t_2, t_2 + \Delta t_2)$ is

$$\lambda_2(t_2)\Delta t_2 \exp\left(-\int_0^{t_2}[\lambda_1(u) + \lambda_2(u)]\,du\right).$$

Conditional on this, the probability of no further deaths in $[t_2, t_1)$ and the event $T_1 \in [t_1, t_1+\Delta t_1)$ is

$$\lambda_{12}(t_1|t_2)\Delta t_1 \exp\left(-\int_{t_2}^{t_1}\lambda_{12}(u|t_2)\,du\right).$$

Multiplying these probabilities, we get the first line of (10.1.6).

By expressing the likelihood function for a given situation in terms of the hazard functions in (10.1.4), we can determine what can and cannot be estimated. For example, if observation ceases as soon as $T=\min(T_1, T_2)$ is observed, then contributions to the likelihood are of just two types:

$$\begin{aligned} &\lambda_1(t_1)\exp\left(-\int_0^{t_1}[\lambda_1(u)+\lambda_2(u)]\,du\right) && \text{if } T=T_1=t_1 \\ &\lambda_2(t_2)\exp\left(-\int_0^{t_2}[\lambda_1(u)+\lambda_2(u)]\,du\right) && \text{if } T=T_2=t_2. \end{aligned} \tag{10.1.7}$$

The likelihood consequently allows us to estimate only $\lambda_1(t)$, $\lambda_2(t)$, and functions of these, nonparametrically; it is not possible to estimate $f(t_1, t_2)$ or $S(t_1, t_2)$. With a parametric family of distributions, it may be possible to estimate all parameters in the model from data of this type, but we will not be able to assess the validity of the model; this point is examined further later.

Parametric Models

Several parametric families have been suggested as multivariate lifetime distributions. Among these are the multivariate normal (as a model for log-lifetimes) and multivariate generalizations of the exponential and Weibull distributions. These and other models are discussed at length by Johnson and Kotz (1972), Barlow and Proschan (1975), David and Moeschberger (1978), and others. When one of these models is assumed, tests and estimation of parameters is usually straightforward. One might add, however, that at the present time there is not a lot of motivation for such models, either in the form of realistic failure processes that lead to specific models or in the form of bodies of data that certain models can be shown to fit well. We shall not examine any parametric models here (but see Problem 10.1); for further information the interested reader is referred to the references cited.

Independent Lifetimes

When $T_1, \ldots, T_k$ are independent, matters simplify greatly. In the bivariate case T_1 and T_2 are independent if and only if the joint p.d.f. and survivor functions factor: for example, $S(t_1, t_2) = S_1(t_1)S_2(t_2)$, where $S_1(t_1)$ and $S_2(t_2)$ are the marginal survivor functions for T_1 and T_2, respectively. In the hazard function formulation a necessary and sufficient condition for T_1 and T_2 to be independent is that $\lambda_{12}(t_1|t_2) = \lambda_1(t_1)$ and $\lambda_{21}(t_2|t_1) = \lambda_2(t_2)$ (see Problem 10.3).

When T_1 and T_2 are independent, nonparametric estimation of the marginal survivor functions $S_1(t)$ and $S_2(t)$, and hence of $S(t_1, t_2)$, is straightforward in many instances. For example, if censoring is independent of lifetimes and is like that leading to (10.1.2), then (10.1.2) factors into terms for T_1 and T_2 and the sample of (t_{1i}, t_{2i})'s can be treated as two independent censored samples. The situation leading to (10.1.3), where only $T_i = \min(T_{1i}, T_{2i})$ and δ_i ($=1$ if $T_i = T_{1i}$, 0 if $T_i = T_{2i}$) are observed, can be handled in a similar fashion: when T_1 and T_2 are independent, (10.1.3) becomes

$$\prod_{i\in D}[f_1(t_i)S_2(t_i)]^{\delta_i}[f_2(t_i)S_1(t_i)]^{1-\delta_i}\prod_{i\in \bar{D}}S_1(t_i)S_2(t_i)$$

$$= \left(\prod_{i\in D} f_1(t_i)^{\delta_i} S_1(t_i)^{1-\delta_i}\prod_{i\in \bar{D}}S_1(t_i)\right)$$

$$\times\left(\prod_{i\in D} f_2(t_i)^{1-\delta_i} S_2(t_i)^{\delta_i}\prod_{i\in \bar{D}}S_2(t_i)\right). \qquad (10.1.8)$$

The first term involves only $S_1(t)$ and is the likelihood obtained from a censored sample with observed lifetimes at those t_i's for which $\delta_i = 1$. The second term depends only on $S_2(t)$ and is of similar form. Thus, with data of this type, inference about $S_1(t)$ [or $S_2(t)$] is handled as for arbitrarily censored data, with both censoring times and lifetimes for which $\delta_i = 0$ (or $\delta_i = 1$) treated as censoring times.

When only $T_i = \min(T_{1i}, T_{2i})$ and δ_i are observed, it is important to realize that one cannot test the assumption that T_1 and T_2 are independent. This has been pointed out by Cox (1959), Tsiatis (1975), Peterson (1976), and others. The problem is that different bivariate models can give rise to the same hazard functions $\lambda_1(t)$ and $\lambda_2(t)$ of (10.1.4), and hence the same likelihood [see (10.1.7)]. In particular, for any model in which T_1 and T_2 are dependent, there is a model with the same $\lambda_i(t)$'s for which T_1 and T_2 are independent. For example, consider the joint survivor function

$$S(t_1, t_2) = \exp\{1 - \lambda_1 t_1 - \lambda_2 t_2 - \exp[\lambda_{12}(\lambda_1 t_1 + \lambda_2 t_2)]\} \qquad t_1 > 0 \quad t_2 > 0$$

where $\lambda_1 > 0, \lambda_2 > 0$, and $\lambda_{12} > -1$ are parameters. The parameter λ_{12} measures dependence between λ_1 and λ_2, with T_1 and T_2 being independent when $\lambda_{12} = 0$. The $\lambda_i(t)$'s are found from (10.1.5) to be

$$\lambda_i(t) = \lambda_i\left(1 + \lambda_{12} e^{\lambda_{12}(\lambda_1 + \lambda_2)t}\right) \qquad i = 1, 2.$$

It is possible to have a model with these same $\lambda_i(t)$'s in which T_1 and T_2 are independent, however: this is the model for which $\lambda_1(t)$ and $\lambda_2(t)$ are the hazard functions in the marginal distributions of T_1 and T_2. The survivor function for T_i is then

$$S_i(t) = \exp\left(-\int_0^t \lambda_i(u)\, du\right)$$

$$= \exp\left(1 - \lambda_i t - \frac{\lambda_i}{\lambda_1 + \lambda_2} e^{\lambda_{12}(\lambda_1 + \lambda_2)t}\right) \qquad i = 1, 2.$$

Independence of lifetimes is thus a strong assumption that cannot be assessed with certain types of censored data. Note, however, that although it may be unrealistic to assume that T_1 and T_2 are unconditionally independent in a particular situation, it is sometimes possible to condition on regressor variables or to stratify the population under study so that T_1 and T_2 are approximately conditionally independent. This is an important aspect of the treatment of multivariate data.

10.1.2 Models With Competing Causes of Death

A common situation is that in which individuals are liable to die or fail from several different causes (see Examples 10.1.4 and 10.1.6). We shall examine lifetime–cause of death models in some detail. Approaches similar to the one taken here are discussed by Chiang (1968), Altschuler (1970), Desu and Narula (1977), Prentice et al. (1978), Kalbfleisch and Prentice (1980), and others.

We associate with each individual a pair (T, C) where T is lifetime and C the cause of death for that individual; C is assumed to take on values in the set $\{1, 2, \ldots, k\}$. The joint distribution of T and C can be approached through cause-specific hazard, survivor, or probability density functions defined as follows for $j=1,\ldots,k$:

$$h_j(t)=\lim_{\Delta t\to 0}\frac{\Pr(t\leqslant T<t+\Delta t, C=j|T\geqslant t)}{\Delta t} \tag{10.1.9}$$

$$S_j(t)=\Pr(T\geqslant t, C=j) \tag{10.1.10}$$

$$f_j(t)=\frac{-dS_j(t)}{dt}. \tag{10.1.11}$$

Any of these sets of functions specifies the joint distribution of T and C. The marginal distribution of T has hazard and survivor functions and p.d.f.

$$\begin{aligned} h(t)&=\sum_{j=1}^{k} h_j(t) \\ S(t)&=\exp\left(-\int_0^t h(u)\,du\right)=\sum_{j=1}^{k} S_j(t) \\ f(t)&=-S'(t)=\sum_{j=1}^{k} f_j(t). \end{aligned} \tag{10.1.12}$$

The marginal distribution of C has probabilities

$$\pi_j=\Pr(C=j)=S_j(0). \tag{10.1.13}$$

For future reference we note that

$$h_j(t)=\frac{f_j(t)}{S(t)}. \tag{10.1.14}$$

It is also convenient to define pseudocumulative hazard and survivor functions related to the $h_j(t)$'s. Let

$$\begin{aligned} H_j(t)&=\int_0^t h_j(u)\,du \\ G_j(t)&=e^{-H_j(t)}. \end{aligned} \tag{10.1.15}$$

These have the form of cumulative hazard and survivor functions, though they do not represent the distribution of any specific random variable. Note from (10.1.12) and (10.1.15) that

$$S(t)=\prod_{j=1}^{k} G_j(t) \tag{10.1.16}$$

and

$$H(t)=-\log S(t)=\sum_{j=1}^{k} H_j(t)$$

and that the $G_j(t)$'s or $H_j(t)$'s also uniquely specify the distribution of T and C.

Nonparametric Inference Procedures

Suppose that observations are taken on a random sample of n individuals. Assume that censoring is possible and define the usual indicator variable δ_i taking on the value 1 if T_i is observed and 0 if it is censored. The observation on individual i is either of the form $(t_i, C_i, \delta_i=1)$ or $(t_i, \delta_i=0)$, where t_i is a lifetime in the first case and a censoring time in the second. If there is independent censoring of the type discussed in Section 1.4, the likelihood function is

$$L=\prod_{i=1}^{n} f_{C_i}(t_i)^{\delta_i} S(t_i)^{1-\delta_i}. \tag{10.1.17}$$

Under parametric models inference by maximum likelihood methods is straightforward, and we shall discuss only nonparametric estimation procedures. When there is no censoring, estimates of the $S_j(t)$'s are

$$\hat{S}_j(t)=\frac{\text{Number of observations with } T\geqslant t \text{ and } C=j}{n}$$

and $\hat{\pi}_j=\hat{S}_j(0)$. Interval estimates for probabilities can be based on the binomial distribution. If there is censoring, matters are more complicated: we shall develop product-limit estimates of the $G_j(t)$'s of (10.1.15) and then use these to obtain estimates of the $S_j(t)$'s and π_j's. First, (10.1.14) and (10.1.16) allow us to rewrite the likelihood (10.1.17) as

$$\begin{aligned} L&=\prod_{i=1}^{n} h_{C_i}(t_i)^{\delta_i} G_{C_i}(t_i)^{\delta_i} \prod_{l\neq C_i} G_l(t_i) \\ &=\prod_{j=1}^{k}\left(\prod_{i\in D_j} g_j(t_i) \prod_{i\in \overline{D}_j} G_j(t_i)\right) \end{aligned} \tag{10.1.18}$$

where $g_j(t) = -G_j'(t)$ and D_j is the set of individuals observed to die from cause j. The likelihood therefore factors to give a term for each $G_j(t)$, and by proceeding as in the derivation of the product-limit estimate (see Section 2.3), we find the "m.l.e." of $G_j(t)$ to be

$$\hat{G}_j(t) = \prod_{\substack{i: t_i < t \\ C_i = j}} \frac{n_i - d_{ji}}{n_i} \tag{10.1.19}$$

where n_i is the number of individuals at risk just prior to t_i and d_{ji} is the number of individuals dying from cause j at t_i. An alternate estimate of $G_j(t)$ is $\tilde{G}_j(t) = \exp[-\tilde{H}_j(t)]$, where

$$\tilde{H}_j(t) = \sum_{\substack{i: t_i < t \\ C_i = j}} \frac{d_{ji}}{n_i} \tag{10.1.20}$$

is an empirical hazard function.

The marginal survivor function $S(t)$ can be estimated by the product-limit estimate, ignoring cause of death. This is

$$\hat{S}(t) = \prod_{\substack{i: t_i < t \\ \delta_i = 1}} \frac{n_i - d_i}{n_i} \tag{10.1.21}$$

where d_i is the number of deaths at t_i. It is easily seen that $\hat{S}(t) = \hat{G}_1(t) \cdots \hat{G}_k(t)$, provided that there are no tied lifetimes involving different causes of death. Alternately, $H(t) = -\log S(t)$ can be estimated by the empirical hazard function.

Since $S_j(t) = \int_t^\infty h_j(u) S(u)\, du$, a reasonable estimate of $S_j(t)$ is

$$\hat{S}_j(t) = \sum_{\substack{i: t_i \geqslant t \\ C_i = j}} \frac{d_{ji}}{n_i} \hat{S}(t_i). \tag{10.1.22}$$

The π_j's can be estimated as

$$\hat{\pi}_j = \hat{S}_j(0). \tag{10.1.23}$$

When there is no censoring, $\sum_{j=1}^k \hat{\pi}_j = 1$, but in general this is not so. To obtain estimates that add to unity we can replace $\hat{\pi}_j$ with $\hat{\pi}_j' = \hat{\pi}_j / \sum_{i=1}^k \hat{\pi}_i$.

When analyzing cause-specific lifetime data, one often wants to estimate quantities like $\Pr(C = j | T \geqslant t) = S_j(t)/S(t)$. This is readily done with the estimates given above. We may also want to examine the relative behavior of the different causes of death. One way to do this is to compare the cause-specific hazard functions by plotting the $\tilde{H}_j(t)$'s or ($\hat{H}_j(t) = -\log \hat{G}_j(t)$)'s. This shows how the relative risks from the different causes of death vary with time and can also suggest possible parametric models for the $h_j(t)$'s.

Special Case: Proportional Hazards

When the $h_j(t)$'s are proportional, significant simplifications occur. In addition, T and C are independent only in this situation. Suppose that $h_j(t) = w_j h(t)$, where $0 < w_j < 1$ and $\Sigma w_j = 1$. In fact, $w_j = \pi_j$, since

$$\pi_j = S_j(0) = w_j \int_0^\infty h(u) S(u)\, du = w_j.$$

It then follows from (10.1.14) and (10.1.15) that

$$h_j(t) = \pi_j h(t) \qquad S_j(t) = \pi_j S(t) \qquad G_j(t) = S(t)^{\pi_j}. \tag{10.1.24}$$

Nonparametric estimation simplifies in this situation. The likelihood (10.1.17) now becomes

$$L = \prod_{j=1}^{k} \pi_j^{d_{\cdot j}} \prod_{i=1}^{n} f(t_i)^{\delta_i} S(t_i)^{1-\delta_i}$$

where $d_{\cdot j}$ is the total number of deaths due to cause j. Maximization of L gives $\hat{S}(t)$ as the ordinary product-limit estimate (10.1.21), and

$$\hat{\pi}_j = \frac{d_{\cdot j}}{d_{\cdot 1} + \cdots + d_{\cdot k}}. \tag{10.1.25}$$

A check on the possible proportionality of cause-specific hazards can be made by comparing estimates of the $H_j(t)$'s. In particular, plots of $\log \tilde{H}_j(t)$ or $\log \hat{H}_j(t)$ for different j's should be roughly parallel if the $h_j(t)$'s are proportional.

An Example

To illustrate these procedures, let us consider the data given in Example 1.1.9, consisting of failure or censoring times for 36 appliances subjected to an automatic life test. Failures were classified into 18 different modes, though among the 33 observed failures only 7 modes are represented in these data, and only modes 6 and 9 appear more than twice. We shall focus on failure mode 9 by considering just two causes of death: $C=1$ (failure mode 9) and $C=2$ (all other failure modes). The data are given in Table 10.1.2, along with some estimates computed below. Failure times have been ordered; the three censored observations are identifiable by the absence of a cause of death.

The table shows values of $\hat{G}_1(t)$ and $\hat{G}_2(t)$ calculated according to (10.1.19); with the failure times ordered, these are easily calculated recursively. The estimates indicate how the cause-specific hazard functions for the two causes of death vary with time. Quite clearly, the two hazards are not proportional: cause 2 predominates among the early failures and cause 1 predominates among the later ones. Plots of the two $(\log[-\log \hat{G}_i(t)])$'s reflect this.

The $\hat{G}_i(t)$'s facilitate a comparison of the cause-specific hazard functions. In order to estimate survival probabilities of various kinds we need to estimate $S_1(t)$, $S_2(t)$, and $S(t)$. The estimates $\hat{S}(t)$, given by (10.1.21), and $\hat{S}_1(t)$ and $\hat{S}_2(t)$, given by (10.1.22), are shown in Table 10.1.2. From these we can also estimate π_1 and π_2 as $\hat{\pi}_1=\hat{S}_1(0)=.511$ and $\hat{\pi}_2=\hat{S}_2(0)=.451$. These do not sum to unity, and as alternate estimates we might take $\hat{\pi}_1'=.511/(.511+.451)=.531$ and $\hat{\pi}_2'=.469$. We can now estimate conditional probabilities as well. For example, $\Pr(C=1|T\geq 2000)$ is estimated by $\hat{S}_1(2000)/\hat{S}(2000)=.454/.639=.710$. (This agrees with the observation that of the 23 appliances still operating at 2000 cycles, 14 fail from cause 1, 6 from cause 2, and 3 have censored failure times.)

Sometimes it is desired to estimate survival probabilities that would result if one or more causes of death were removed. This is often done by assuming that removal of a cause merely reduces the hazard function for that cause to 0 and leaves the hazards for other causes unchanged. This assumption is, however, often an unrealistic one. In the present situation a check on this is to some extent possible by examining the full data set given by Nelson (1970b), in which new tests are run after certain improvements in the appliance have been made. These show that cause-specific hazard functions are greatly affected by changes made in the appliance during development, and so there is no basis for this assumption here.

Parametric modeling of the $G_i(t)$'s, and hence of the $S_i(t)$'s, is also possible. No particular model is suggested by the background to this problem, though flexible parametric forms such as $G_i(t)=\lambda_i t^{\alpha_i}$ could be considered. There seems to be little point in doing this here, since the situation will most likely change considerably when future modifications to the appliance are made.

Table 10.1.2 Appliance Life Test Data[a]

t_i	C_i	n_i	$\hat{G}_1(t_i+0)$	$\hat{G}_2(t_i+0)$	$\hat{S}(t_i+0)$	$\hat{S}_1(t_i)$	$\hat{S}_2(t_i)$
11	2	36	1.0	.972	.972	.511	.451
35	2	35	1.0	.944	.944	.511	.424
49	2	34	1.0	.917	.917	.511	.396
170	2	33	1.0	.889	.889	.511	.368
329	2	32	1.0	.861	.861	.511	.340
381	2	31	1.0	.833	.833	.511	.312
708	2	30	1.0	.806	.806	.511	.285
958	2	29	1.0	.778	.778	.511	.257
1062	2	28	1.0	.750	.750	.511	.229
1167	1	27	.963	.750	.722	.511	.201
1594	2	26	.963	.721	.694	.482	.201
1925	1	25	.924	.721	.667	.482	.174
1990	1	24	.886	.721	.639	.454	.174
2223	1	23	.847	.721	.611	.426	.174
2327	2	22	.847	.688	.583	.398	.174
2400	1	21	.807	.688	.555	.398	.146
2451	2	20	.807	.654	.528	.371	.146
2471	1	19	.765	.654	.500	.371	.118
2551	1	18	.722	.654	.472	.343	.118
2565	—	17	.722	.654	.472	.315	.118
2568	1	16	.677	.654	.443	.315	.118
2694	1	15	.632	.654	.413	.286	.118
2702	2	14	.632	.607	.384	.256	.118
2761	2	13	.632	.561	.354	.256	.089
2831	2	12	.632	.514	.325	.256	.059
3034	1	11	.574	.514	.295	.256	.030
3059	2	10	.574	.462	.266	.227	.030
3112	1	9	.511	.462	.236	.227	.0
3214	1	8	.447	.462	.207	.197	.0
3478	1	7	.383	.462	.177	.168	.0
3504	1	6	.319	.462	.148	.138	.0
4329	1	5	.255	.462	.118	.108	.0
6367	—	4	.255	.462	.118	.079	.0
6976	1	3	.170	.462	.079	.079	.0
7846	1	2	.085	.462	.039	.039	.0
13403	—	1	.085	.462	Undefined	.0	.0

[a] $C=1$, failure mode 9; $C=2$, all other modes.

Remarks on Competing Risks

The situation in which one observes lifetime and an associated cause of death is often referred to as the "competing risks framework." There is an extensive literature on this; many results and references are given by Chiang (1968), David and Moeschberger (1978), and Kalbfleisch and Prentice (1980), while Seal (1977) presents some historical material. We make two further remarks about this area: the first is a reminder that to estimate hazard or survivor functions that would result if one or more causes of death were removed requires very strong assumptions. A common procedure is to assume that removal of one or more causes of death merely has the effect of nullifying the hazard functions for those causes, while leaving the other hazard functions unchanged (e.g., Chiang, 1968). It is doubtful whether this is very often appropriate, even in special situations such as when T and C are independent. Detailed knowledge of the death process seems necessary if one is to deal realistically with cause removal.

A second remark is that competing risk analysis is often approached by assuming that for an individual there are lifetimes $T_1, \dots, T_k$ associated with the k causes of death. The lifetime of an individual is then $T = \min(T_1, \dots, T_k)$ and the cause of death is the value of i such that $T_i = \min(T_1, \dots, T_k)$. Both parametric and nonparametric methods of analysis can be developed (e.g., Gail, 1975; David and Moeschberger, 1978). A major difficulty with this approach is that, as noted earlier, it is impossible to discriminate among different models giving rise to the same cause-specific hazard functions on the basis of data on T and C alone. It is consequently impossible to assess the adequacy of any such model. By the same token, it is impossible to assess whether $T_1, \dots, T_k$ might be independent. Thus, unless there is physical evidence supporting such a model, one is forced to accept the model on faith.

10.2 SOME STOCHASTIC PROCESS MODELS

The first nine chapters of this book deal mainly with situations in which there are independent univariate observations on individuals, perhaps with covariables also present. Lifetime or time to event data also arise in places where more complicated models, often incorporating some notion of dependence among lifetimes, are necessary. Some examples of this were given in Section 10.1, where multivariate models were discussed. In this section we consider problems where stochastic process models are helpful. Two such instances are (1) when lifetimes arise in a natural order, with the possibility of a time trend of some sort, and (2) when individuals in a population can spend time in different "states," with sojourn times in the states being of interest. As an example of the first type, suppose that "lifetimes" refer to the times between successive failures of a system; each time the system fails, it is repaired and operates until the next failure. Questions arise as to whether times between failures are i.i.d., whether there is correlation between times, whether there is a time trend present, and so on. An example of the second type is a situation in which cancer patients under treatment can be in any of several well-defined states (e.g., in remission, in a progressive disease state, dead). Sojourn times in various states (e.g., time in remission) or times to entry into a state (e.g., time to death) are of obvious interest.

It is beyond the scope of this book to examine in detail the types of stochastic processes that might be useful in connection with lifetime data; this section merely sketches a few important models. More comprehensive discussions can be found in the references cited.

Before presenting a few models, we give some examples of situations where they might be useful.

Example 10.2.1 Proschan (1963) gives the time intervals, in hours of operation, between successive failures of air conditioning equipment in a number of Boeing 720 aircraft; these data have also been discussed by many other authors. The times between failures for the first four planes are given below. Times appear in the order in which they occurred: that is, for plane 1 the first failure was at 194 hours, the second at 209 ($=194+15$) hours, and so on.

Plane 1	194, 15, 41, 29, 33, 181 ($n=6$)
Plane 2	413, 14, 58, 37, 100, 65, 9, 169, 447, 184, 36, 201, 118, 34, 31, 18, 18, 67, 57, 62, 7, 22, 34 ($n=23$)
Plane 3	90, 10, 60, 186, 61, 49, 14, 24, 56, 20, 79, 84, 44, 59, 29, 118, 25, 156, 310, 76, 26, 44, 23, 62, 130, 208, 70, 101, 208 ($n=29$)
Plane 4	74, 57, 48, 29, 502, 12, 70, 21, 29, 386, 59, 27, 153, 26, 326 ($n=15$)

An important question is whether times between failures for a given aircraft can be considered to be i.i.d. or whether time trends of some kind exist. This problem is examined in Section 10.2.4. Another question is whether there are differences among the aircraft.

Example 10.2.2 Berlin et al. (1979), Chiang (1968, 1980), and others describe "illness–death" models in which individuals can suffer from different diseases while alive. If there are k disease classes, an individual is in one of 2^k "alive" states, depending on which combination of diseases he has. A state D, corresponding to death, is also defined. In stochastic process terminology the alive states are transient and D is absorbing, since all individuals are assumed to die eventually.

Berlin et al. (1979) describe animal experiments in which (1) some individuals are observed until natural death, at which time a post mortem shows what alive state they were in just prior to death, and (2) some individuals are sacrificed (killed), again with a post mortem determining their alive state just prior to death. Data consist of sacrifice or natural death times for each individual, along with a list of the diseases present at the time of death. Questions of interest include the possibility of disease dependence, estimation of survival time measured from time of entry to a particular state, and the effects of age upon disease state and survival.

Note that this situation is somewhat similar to the competing risk problem of Section 10.1, where one observes time of death and a "state" at death (namely, the cause of death) for each individual. In the present situation no cause of death is assigned or even considered, however.

Example 10.2.3 Temkin (1978) describes a clinical trial to compare two chemotherapeutic treatments for testicular cancer. Patients were given one of two treatments, A or B. Overall, 48 patients given A had a median survival time of 321 days, and 46 patients given B had a median survival time of 275 days. Other features of the data are also of interest and can be related to a multistate model. We define four states, as follows: (1) an initial state 0, in which each patient is assumed to be at time of entry to the trial, (2) a response state 1, which some patients enter after being given treatment (this state is defined according to whether response to treatment is favorable; in this case "favorable" means a 50% or greater decrease in total tumor area), (3) a progressive disease state 2, into which individuals not responding favorably to treatment can enter (this involves an increase in tumor size and eventual death), and (4) a relapse state 3, in which people who have been in the response state relapse, with an increase in tumor size and eventual death. States 0 and 1 are assumed to be transient, and states 2 and 3 absorbing; transitions are possible only from state 0 to states 1 or 2, and from state 1 to state 3.

A breakdown of data from the trial indicates some interesting features (see Table 10.2.1). For example, the proportion of patients who enter state 1 is higher for treatment B, but patients then spend considerably less time in state 1 than do patients receiving treatment A. Consideration of a multistate model allows a more detailed comparison of the two treatments than a mere examination of survival times.

Table 10.2.1 Comparison of Treatments for Testicular Cancer

	Treatment	
	A (48 patients)	B (46 patients)
Percentage and number entering state 1	19% (9)	67% (31)
Median time to enter state 1	48 days	25 days
Median time spent in state 1	260 days	84 days
Median survival time	321 days	275 days

We shall now sketch a few important models. More details can be found in books on probability and stochastic processes such as Chiang (1968, 1980), Cox and Lewis (1966), Cox and Miller (1965), Feller (1968), Karlin and Taylor (1975), and in other references cited.

10.2.1 The Poisson Process

The Poisson process is treated in detail in most books on stochastic processes (e.g., see Karlin and Taylor, 1975). Consider a point process and let $N(t, t+h)$ denote the number of events occurring in the time interval $(t, t+h]$, where $t \geqslant 0$ and $h > 0$. The process is a Poisson process if

$$\begin{aligned} \Pr[N(t,t+h)=0] &= 1-\lambda h+o(h) \\ \Pr[N(t,t+h)=1] &= \lambda h+o(h) \end{aligned} \tag{10.2.1}$$

as $h \to 0$, where $\lambda > 0$ is a constant, called the intensity of the process, and $o(h)$ denotes a quantity $g(h)$ for which $\lim_{h\to 0} g(h)/h = 0$. The conditions (10.2.1) imply that $\Pr[N(t,t+h) \geqslant 2] = o(h)$ and that the occurrence of events in $(t, t+h]$ is unaffected by what happens prior to time t.

From (10.2.1) it can be shown that for any $s \geqslant 0, t > 0$,

$$\Pr[N(s,s+t)=r] = e^{-\lambda t}\frac{(\lambda t)^r}{r!} \qquad 0,1,2,\ldots \tag{10.2.2}$$

which is a Poisson distribution with mean λt. The process is time homogeneous in the sense that (10.2.2) does not depend on s. The time X between successive events in the Poisson process has an exponential distribution with hazard function λ. To see this note that $\Pr(X \geqslant x) = \Pr[N(t, t+x) = 0] = e^{-\lambda x}$. In addition, times between events are independent, so that a Poisson process can be viewed as a point process in which times between successive events (including time to the first event) are i.i.d. exponential random variables.

Statistical methods developed for the exponential distribution (see Chapter 3) can be used with the Poisson process when data consist of times between events. Cox and Lewis (1966, Ch. 2) discuss methods based on counts of the number of events in fixed time intervals.

10.2.2 Nonhomogeneous (Time-Dependent) Poisson Processes

The assumptions underlying the Poisson process can be modified by allowing the intensity to be time dependent. If λ in (10.2.1) is replaced by $\lambda(t)$, we find that $N(s, s+t)$ now has a Poisson distribution,

$$\Pr[N(s,s+t)=r]=e^{-\mu(s,t)}\frac{\mu(s,t)^r}{r!} \qquad r=0,1,2,\ldots \tag{10.2.3}$$

where

$$\mu(s,t)=\int_s^{s+t}\lambda(u)\,du. \tag{10.2.4}$$

Times between events are neither independent nor identically distributed, and the process is not time homogeneous unless $\lambda(t)=\lambda$. It can be shown, however, that if $t_1<t_2<\cdots$ represent the times at which events occur, then the quantities

$$e_i=\int_{t_{i-1}}^{t_i}\lambda(u)\,du \qquad i=1,2,\ldots \tag{10.2.5}$$

are i.i.d. standard exponential random variables. Here, t_0 is defined for convenience as 0.

Cox and Lewis (1966), Brown (1972), and others discuss statistical procedures for time-dependent Poisson processes. If the process is observed over a fixed time period $[0, t_L)$ and if n events occur at times $0\leqslant t_1<\cdots<t_n\leqslant t_L$, then the likelihood function is

$$\begin{aligned} L&=\left[\prod_{i=1}^{n}\lambda(t_i)\exp\left(-\int_{t_{i-1}}^{t_i}\lambda(u)\,du\right)\right]\exp\left(-\int_{t_n}^{t_L}\lambda(u)\,du\right)\\ &=\left(\prod_{i=1}^{n}\lambda(t_i)\right)\exp\left(-\int_0^{t_L}\lambda(u)\,du\right). \end{aligned} \tag{10.2.6}$$

If the process is observed until the nth event, instead of a fixed time period, the likelihood is given by (10.2.6), with t_L replaced by t_n. These expressions are written down by noting that the data consist of no events until $t_1 - 0$, an event at t_1, no more events until $t_2 - 0$, an event at t_2, and so on.

Two particular time-dependent processes that are frequently used are those with $\lambda(t)=e^{\alpha+\beta t}$ (e.g., Cox and Lewis, 1966) and with $\lambda(t)=\lambda t^{\beta}$ (e.g., Crow, 1974; Lee and Lee, 1978). Time-dependent Poisson processes are sometimes used as models for failures in reparable systems (e.g., Ascher and Feingold, 1978); an application to the failure data of Example 10.2.1 is discussed in Section 10.2.4.

10.2.3 Renewal Processes

A renewal process is a point process in which the times X_i between successive events are i.i.d. Numerous results for these processes are given, for example, by Cox (1962), Cox and Lewis (1966), and Karlin and Taylor (1975). Whether or not a renewal process is appropriate is important, since if times between events are i.i.d., models and methods discussed earlier in this book can often be applied. Cox and Lewis (1966, Ch. 6), Lewis and Robinson (1974), and others discuss tests for a process to be a renewal process. This is not discussed here, but we note that tests of independence of the X_i's are important, as are tests against more specific types of departure from the renewal model. This is addressed briefly in the example of Section 10.2.4.

A potentially useful generalization of the renewal model is the modulated renewal process (Cox, 1972b): at time t let $u(t)$ be the backward recurrence time to the last event and let $z_1(t),\ldots, z_p(t)$ be specified functions. Suppose that given the history of the process up to time t, the probability of an event in $(t, t+h)$ is

$$\{\lambda_0[u(t)]\exp[\beta_1 z_1(t)+ \cdots +\beta_p z_p(t)]\}h+o(h) \qquad (10.2.7)$$

where $\boldsymbol{\beta}=(\beta_1,\ldots, \beta_p)$ is a vector of unknown parameters. The probability of more than one event being in $(t, t+h)$ is $o(h)$. This model is quite flexible and includes the ordinary renewal model as the special case $\boldsymbol{\beta}=\mathbf{0}$. For example, with $p=1$ and $z_1(t)=t$, the intensity at time t is $\lambda_0[u(t)]e^{\beta t}$, which allows us to consider a time trend departure from the renewal model. Statistical methods need to be developed for these models, since not much is currently available. Cox (1972b) gives a partial likelihood procedure for testing that $\boldsymbol{\beta}=\mathbf{0}$.

10.2.4 An Example (Example 10.2.1 Revisited)

Let us examine the data given in Example 10.2.1 on the times between failure (X) of air conditioning equipment in four aircraft. To start, if the

X_i's in each plane can be viewed as i.i.d., the exponential distribution is an acceptable model in each case. For example, the exponential goodness of fit statistic G_n given by (9.2.2) takes on the values .539, .585, .441, and .617 for the data from planes 1 to 4, with associated significance levels .76, .17, .27, and .13, respectively.

There is reason to question the assumption that the X_i's are i.i.d., however; it is plausible that there could be gradual deterioration in the systems over time, with failures becoming more frequent. Or, it may be that nearby observations are correlated as a consequence of the nature of the failures and subsequent repairs. One way to examine the possibility of a time trend is to fit a time-dependent Poisson process model within which the Poisson process (that is, i.i.d. exponential times between failure) can be assessed.

A Time-Dependent Poisson Process

A simple model that allows for a monotone trend in the intensity function is that for which $\lambda(t)=\exp(\alpha+\beta t)$. There are, of course, other models with this property, but this one provides a convenient place to start. From the sentence following (10.2.6) the likelihood function is

$$L(\alpha,\beta)=\exp\left(n\alpha+\beta\sum_{i=1}^{n}t_i-\frac{e^{\alpha}(e^{\beta t_n}-1)}{\beta}\right) \tag{10.2.8}$$

where $t_1<\cdots<t_n$ are the times at which failures occur for a given plane. It follows that Σt_i and t_n are sufficient for α and β; it is easily shown that the m.l.e.'s $\hat{\alpha}$ and $\hat{\beta}$ can be found from the equations

$$\sum_{i=1}^{n}t_i+\frac{n}{\beta}-\frac{nt_n}{1-e^{-\beta t_n}}=0 \tag{10.2.9}$$

$$e^{\alpha}=\frac{n\beta}{e^{\beta t_n}-1}. \tag{10.2.10}$$

Maximum likelihood large-sample theory can be readily developed and is omitted here. One problem of much interest is to test $H:\beta=0$, since under H_0 the process is a Poisson process. This can be done by using a likelihood ratio test or a test based on the large-sample normal approximation to the distribution of $\hat{\beta}$. A simpler procedure (Cox and Lewis, 1966, Sec. 6.3) is to base a scores test on a conditional likelihood for β. To get this we note that the conditional distribution of Σt_i, given t_n, does not depend on α: the p.d.f.

is found to be proportional to

$$L_1(\beta)=\beta^{n-1}\exp\left(\beta\sum_{i=1}^{n-1}t_i\right)\Big/\left(e^{\beta t_n}-1\right)^{n-1}. \qquad (10.2.11)$$

The statistic

$$U=\left.\frac{\partial\log L/\partial\beta}{\left(E(-\partial^2\log L/\partial\beta^2)\right)^{1/2}}\right|_{\beta=0}$$

can be used to test H_0. It can be shown that

$$U=\left(\sum_{i=1}^{n-1}t_i-\tfrac{1}{2}(n-1)t_n\right)\Big/t_n\left(\frac{n-1}{12}\right)^{1/2}. \qquad (10.2.12)$$

Under H_0, U is approximately $N(0,1)$.

If observation of the process continues for a fixed time t_L rather than until the nth failure, the results above need only be modified as follows: (1) in (10.2.8), (10.2.9), and (10.2.10) t_n is replaced by t_L, and (2) in (10.2.11) and (10.2.12) $n-1$ is replaced by n, $\Sigma_{i=1}^{n-1}t_i$ is replaced by $\Sigma_{i=1}^{n}t_i$, and t_n is replaced by t_L. In addition, $\Sigma_{i=1}^{n}t_i$ and n are sufficient for α and β, and the analog to (10.2.11) is obtained by considering the distribution of $\Sigma_{i=1}^{n}t_i$, given n.

Tests of $H_0:\beta=0$ are easily carried out with the data from planes 1 to 4. Note that the data given in Example 10.2.1 are the times X_i between failures and that $t_i=\Sigma_{l=1}^{i}X_l$. For plane 2 we have $n=6$, $\Sigma_{i=1}^{5}t_i=1244$, and $t_6=493$, which by (10.2.12) gives $U=0.04$ and provides no evidence against H_0. The U values for planes 2, 3, and 4 data are 2.24, -1.48, and -0.75, respectively. The largest of these, 2.24, gives a significance level of approximately .025, and plane 2 is the only one for which there is much of a suggestion of trend. On the other hand, 2.24 is not an especially large value for the largest of four independent $|U|$ values under the hypothesis that $\beta=0$ for all planes.

Data from all four planes can be combined to provide an overall test of H_0. Cox and Lewis (1966, Sec. 6.3) show that

$$U=\left(\sum_{l=1}^{k}\sum_{i=1}^{n_l-1}t_{li}-\frac{1}{2}\sum_{l=1}^{k}(n_l-1)t_{ln}\right)\Big/\left(\frac{1}{12}\sum_{l=1}^{k}(n_l-1)t_{ln}^2\right)^{1/2} \qquad (10.2.13)$$

is approximately $N(0,1)$ under the hypothesis that each of the k processes is a Poisson process (possibly with different intensities) when observation in the lth process continues until the n_lth failure. In the case of planes 1 to 4 (10.2.13) gives $U=-0.054$, which provides no evidence against H_0.

There does not appear to be evidence of a time trend, in the form of a time-dependent Poisson intensity, assuming that the time-dependent Poisson model with $\lambda(t)=\exp(\alpha+\beta t)$ is itself suitable. We can assess the model by examining residuals based on the e_i's in (10.2.5),

$$\hat{e}_i=\int_{t_{i-1}}^{t_i} e^{\hat{\alpha}+\hat{\beta}u}\,du$$

$$=\frac{e^{\hat{\alpha}}}{\hat{\beta}}\left(e^{\hat{\beta}t_i}-e^{\hat{\beta}t_{i-1}}\right) \qquad i=1,\ldots,n \ .$$

For example, for plane 2 we find the m.l.e.'s to be $\hat{\beta}=0.000888$ and $\hat{\alpha}=-5.693$, and the 23 $\hat{e}_i$'s are

1.68, 0.07, 0.29, 0.20, 0.56, 0.39, 0.06, 1.14, 3.99, 2.16, 0.47, 2.89, 1.95, 0.60, 0.56, 0.34, 0.34, 1.32, 1.18, 1.36, 0.16, 0.50, 0.80

A plot of the ordered $\hat{e}_{(i)}$'s against the expected standard exponential order statistics

$$\alpha_i=\sum_{l=1}^{i}(n-l+1)^{-1}$$

does not show any unusual features, nor do plots of the $\hat{e}_i$'s against order of appearance. The only point one might remark on is that the large $\hat{e}_i$'s seem to bunch together slightly, as do the original x_i's. The possibility of serial correlation in the X_i's is considered below.

Other Tests

In situations like this it is of interest whether the process can be considered to be a renewal process. The approach taken to this point provides some information on this but supposes the process to be Poisson if it is a renewal process. Other tests for renewal processes are given by Cox and Lewis (1966, Ch. 8), Lewis and Robinson (1974), Ascher and Feingold (1978), and others. We shall briefly examine some of these in the case of the plane 2 data. We have mentioned that large failure times seem to be bunched together slightly, but tests based on estimated serial correlation coefficients do not show any strong evidence of correlation among the X_i's. For example, the estimate of the lag 1 serial correlation coefficient (See Cox and Lewis, 1966,

Sec. 6.4) is $\tilde{\rho}_1 = -.175$; this gives an approximate standard normal variate of $(-.175)\sqrt{22} = -0.82$ for testing that $\rho_1 = 0$. Similarly, the lag 1 rank serial correlation coefficient (Cox and Lewis, 1966, Sec. 6.4) gives an approximate standard normal variate of -1.17 under the hypothesis that $\rho_1 = 0$. Neither test provides evidence against the renewal hypothesis.

Finally, we mention a test of the renewal hypothesis given by Lewis and Robinson (1974). This uses the statistic

$$U_{LR} = \left(\sum_{i=1}^{n-1} t_i - \frac{1}{2}(n-1)t_n \right) \Big/ t_n \left(\frac{n-1}{12} \right)^{1/2} \hat{C}(\mathbf{x}) \qquad (10.2.14)$$

where

$$\hat{C}(\mathbf{x}) = \bar{x} \Big/ \left(\sum_{i=1}^{n} \frac{(x_i - \bar{x})^2}{n-1} \right)^{1/2}$$

is the estimated coefficient of variation for the X_i's and the remaining quantities in (10.2.14) are the same as those in (10.2.12). The statistic (10.2.14) is a generalization of (10.2.12) designed to be effective at detecting departures from the renewal hypothesis when the X_i's are no longer exponential. Under the hypothesis that the X_i's are i.i.d., U is approximately $N(0,1)$. Note that if the X_i's are i.i.d. exponential, then (10.2.14) is asymptotically equivalent to (10.2.12), since the coefficient of variation for the exponential distribution is unity.

For the plane 2 data $\hat{C}(\mathbf{x}) = 0.802$, which gives $U_{LR} = 2.793$. Taken on its own, this provides fairly strong evidence against the renewal hypothesis. On the other hand, we have been led to examine the plane 2 data in part because it has the strongest suggestion of a departure from the Poisson model among the four planes. It is probably fair to conclude that the plane 2 data provide some evidence of a departure from the Poisson (exponential renewal) model, with a trend to more frequent failures over time. However, the evidence is not overwhelming and a longer series of observations would be needed to provide firm conclusions.

10.2.5 Markov Processes

Markov processes are discussed in detail by Feller (1968), Cox and Miller (1965), Karlin and Taylor (1975), and many others. We mention only processes in continuous time with discrete states. Let $X(t)$ represent the state a process is in at time t, where possible states are labeled $0, 1, 2, \ldots,$

and let transition probabilities by represented by

$$P_{ij}(s,s+t)=\Pr[X(s+t)=j|X(s)=i]$$

where $s\geqslant 0$ and $t>0$. When the P_{ij}'s depend only on t, the process is time homogeneous.

The transition probabilities must satisfy the Markov dependence property (e.g., Feller, 1968, Sec. 17.9). The model can be specified in terms of instantaneous transition rate functions defined by the property that as $\delta t\to 0$

$$P_{ij}(t,t+\delta t)=v_{ij}(t)\,\delta t+o(\delta t) \qquad i\neq j.$$

Elapsed time until entry into a state and sojourn times in a state (i.e., elapsed time from entry into a state until departure from the state) are usually of interest in Markov models. Complete data from a situation consists of the sequence of states visited by each individual, along with the sojourn time in each state. Statistical methods for Markov processes are discussed by Bartholomew (1977) for the time-homogeneous case. Berlin et al. (1979) and references therein discuss the general model and apply it to the problem described in Example 10.2.2. A difficulty with non-time-homogeneous models is that it is usually impossible to obtain explicit expressions for transition probabilities and waiting and sojourn time distributions from even fairly simple $v_{ij}(t)$'s.

10.2.6 Semi-Markov Processes

A semi-Markov process is one that passes among states $1,\ldots,N$ according to a Markov chain with transition probability matrix $\mathbf{P}=(p_{ij})$; that is, p_{ij} is the probability of the system entering state j next, having just entered state i. In addition, conditional on the sequence of states entered, sojourn times are independently distributed: given that the system is in state i and then passes to state j, the sojourn time in state i has p.d.f. $f_{ij}(t)$ and survivor function $S_{ij}(t)$. The unconditional distribution for sojourn time in state i has survivor function $S_i(t)=\Sigma_{j\neq i}p_{ij}S_{ij}(t)$.

The semi-Markov process can also be expressed in terms of transition rates or hazard functions. Let us set up the following notation: δ_0 is the initial state of the system, X_0 is the sojourn time in this state, and subsequent states and sojourn times are denoted δ_i and $X_i (i=1,2,\ldots)$. For $n=1,2,\ldots,$ define

$$\lambda_{ij}(t)=\lim_{\delta t\to 0}\frac{\Pr[X_{n-1}\in(t,t+\delta t),\delta_n=j|X_{n-1}\geqslant t,\delta_{n-1}=i]}{\delta t} \qquad i\neq j.$$

This can be rewritten as

$$\lambda_{ij}(t)=$$

$$\lim_{\delta t\to 0}\frac{\Pr[X_{n-1}\in(t,t+\delta t)|\delta_{n-1}=i,\delta_n=j]\Pr(\delta_n=j|\delta_{n-1}=i)\Pr(\delta_{n-1}=i)}{\delta t\Pr(X_{n-1}\geqslant t|\delta_{n-1}=i)\Pr(\delta_{n-1}=i)}$$

$$=\frac{p_{ij}f_{ij}(t)}{S_i(t)}. \qquad (10.2.15)$$

Conversely, if we define

$$\lambda_i(t)=\sum_{j\neq i}\lambda_{ij}(t) \qquad i=1,\dots,N$$

then

$$p_{ij}=\int_0^\infty \lambda_{ij}(t)\exp\left(-\int_0^t\lambda_{ij}(u)\,du\right)dt$$

$$S_i(t)=\exp\left(-\int_0^t\lambda_i(u)\,du\right).$$

Inference for semi-Markov processes is straightforward in some instances; Matthews (1980), Lagakos et al. (1978), and Bartholomew (1977) discuss various problems. Suppose, in particular, that observations on the ith individual in a random sample of m consist of the states visited and corresponding sojourn times, $\delta_0, x_0,\dots,\delta_{n_i-1}, x_{n_i-1}, \delta_{n_i}, x^*_{n_i}$, with the final sojourn time possibly being censored at $x^*_{n_i}$. The contribution to the likelihood from this individual is

$$L_i=\left(\prod_{l=0}^{n_i-1}\lambda_{\delta_l,\delta_{l+1}}(x_l)S_{\delta_l}(x_l)\right)S_{\delta_{n_i}}(x^*_{n_i}). \qquad (10.2.16)$$

This can be seen by noting that the individual survives a time x_l in state δ_l and then makes a transition to state δ_{l+1}.

Inference for parametric models can be based on the likelihood function $L=L_1L_2\cdots L_m$. Nonparametric estimation of transition and survivor functions is also possible (e.g., Matthews, 1980). For example, the product-limit estimate of $S_i(t)$ is

$$\hat{S}_i(t)=\prod_{l:\,t_{il}<t}\frac{n_{il}-d_{il}}{n_{il}}$$

where $t_{i1} < t_{i2} < \cdots < t_{ir_i}$ are the distinct ordered sojourn times in state i of all individuals and n_{il} and d_{il} are the number at risk and the number of sojourn times at t_{il}. When there is no censoring present, p_{ij} can be estimated as the proportion of cases in which individuals who entered state i proceeded next to state j. Matthews (1980) discusses estimation of p_{ij} when censoring is present.

Examples of the application of semi-Markov processes to problems involving disease progression are given by Lagakos et al. (1978) and Temkin (1978).

10.3 PROBLEMS AND SUPPLEMENTS

10.1 Consider the two bivariate lifetime distributions that have survivor functions

$$S(t_1, t_2) = \exp\{1 - \lambda_1 t_1 - \lambda_2 t_2 - \exp[\lambda_{12}(\lambda_1 t_1 + \lambda_2 t_2)]\} \quad (10.3.1)$$

$$S(t_1, t_2) = \exp[-(\lambda_1 t_1 + \lambda_2 t_2 + \lambda_1 \lambda_2 \theta t_1 t_2)] \quad (10.3.2)$$

where in (10.3.1) $\lambda_1 > 0$, $\lambda_2 > 0$, and $\lambda_{12} > -1$, and in (10.3.2) $\lambda_1 > 0$, $\lambda_2 > 0$, and $0 \leqslant \theta \leqslant 1$.

Examine the problem of maximum likelihood estimation for each model if (1) the only possible censoring is that in which it is known only that $T_1 \geqslant t^*$ and $T_2 \geqslant t^*$ and (2) we observe only $T = \min(T_1, T_2)$ and whether $T = T_1$ or $T = T_2$ for each individual.

(Section 10.1.1)

10.2 Examine the electrical insulation failure data of Example 10.1.2. Consider potential models for the distribution of T_1 and for the distribution of T_2 given T_1 from among the models discussed in the first six chapters of this book.

(Section 10.1.1)

10.3 Prove that in the hazard function formulation of a bivariate model represented in (10.1.6) a necessary and sufficient condition for T_1 and T_2 to be independent is that $\lambda_{12}(t_1|t_2) = \lambda_1(t_1)$ and $\lambda_{21}(t_2|t_1) = \lambda_2(t_2)$.

(Section 10.1.1)

10.4 The data in Table 10.3.1 are from Hoel (1972) and give survival times for two groups of mice, all of whom had received a radiation dose of 300

Table 10.3.1 Survival Times and Causes of Death for Laboratory Mice

Control Group	Germ-Free Group
C_1 Deaths	
159, 189, 191, 198, 200, 207, 220, 235, 245, 250, 256, 261, 265, 266, 280, 343, 350, 383, 403, 414, 428, 432	158, 192, 193, 194, 195, 202, 212, 215, 229, 230, 237, 240, 244, 247, 259, 300, 301, 321, 337, 415, 434, 444, 485, 496, 529, 537, 624, 707, 800
C_2 Deaths	
317, 318, 399, 495, 525, 536, 549, 552, 554, 557, 558, 571, 586, 594, 596, 605, 612, 621, 628, 631, 636, 643, 647, 648, 649, 661, 663, 666, 670, 695, 697, 700, 705, 712, 713, 738, 748, 753	430, 590, 606, 638, 655, 679, 691, 693, 696, 747, 752, 760, 778, 821, 986
C_3 Deaths	
40, 42, 51, 62, 163, 179, 206, 222, 228, 252, 259, 282, 324, 333, 341, 366, 385, 407, 420, 431, 441, 461, 462, 482, 517, 517, 524, 564, 567, 586, 619, 620, 621, 622, 647, 651, 686, 761, 763	136, 246, 255, 376, 421, 565, 616, 617, 652, 655, 658, 660, 662, 675, 681, 734, 736, 737, 757, 769, 777, 800, 806, 825, 855, 857, 864, 868, 870, 870, 873, 882, 895, 910, 934, 942, 1015, 1019

roentgens at an age of 5 to 6 weeks. The first group of mice lived in a conventional laboratory environment and the second was kept in a germ-free environment. Cause of death for each mouse was determined by autopsy to be thymic lymphoma (C_1), reticulum cell sarcoma (C_2), or other causes (C_3). The table shows survival times for mice dying from each of the three causes for each group.

Assess these data with the methods of Section 10.1.2; consider the possibility of fitting parametric models for the cause-specific hazards functions. It is of particular interest to compare mortality due to thymic lymphoma in the control and germ-free environments. Do this, paying special attention to whether the relative mortality pattern appears to change with age.

(Section 10.1.2; Hoel, 1972)

10.5 Grouped competing risk data. Suppose that lifetime–cause of death data are grouped in life table form. In particular, let the time axis be partitioned into $m+1$ intervals $[a_{j-1}, a_j)$, where $0=a_0<a_1<\cdots<a_m<a_{m+1}=\infty$. Suppose there is no censoring and that the data consist of the number of individuals who die from each of causes $1,\ldots,k$ in each interval.

a. Determine the likelihood function for the data and show that the m.l.e. of $S_j(a_i)$ is n_{ji}/n, where n_{ji} is the number of deaths due to cause j that occur at time $\geqslant a_i$ and n is the total number of individuals in the sample.

b. Consider the case in which the cause-specific hazard functions are proportional, with $h_j(t)=\pi_j h(t)$ as in (10.1.24). Write down the likelihood in this case and show that the m.l.e.'s of π_j are as given by (10.1.25) and that the m.l.e. of $S(a_i)$ is n_i/n, where n_i is the number of deaths that occur at time $\geqslant a_i$.

c. Give a χ^2 or likelihood ratio test of the hypothesis that the hazards $h_j(t)$ are proportional.

(Section 10.1)

10.6 Mendenhall and Hader (1958) present data on the failure times of radio transmitter receivers. Failures are classified as one of two types, those confirmed on arrival at the maintenance center (Type I), and those unconfirmed (Type II). The data consist of a failure time and type for each receiver, except that when observation ceased after 630 hours, 44 of 369 receivers had still not failed and so have censored failure times. Table 10.3.2 gives a frequency distribution compiled from the original data, and discussed by Cox (1959).

a. Obtain estimates of cause-specific survival probabilities under the general model represented by (10.1.10) and under the proportional hazards model (10.1.24), as described in Problem 10.5. Test the hypothesis that the hazard functions are proportional.

Table 10.3.2 Frequency Distribution of Failure Time and Type for Radio Receivers

Time Interval (Hours)	Type I Failures	Type II Failures	Total Failures
[0,50)	26	15	41
[50,100)	29	15	44
[100,150)	28	22	50
[150,200)	35	13	48
[200,250)	17	11	28
[250,300)	21	8	29
[300,350)	11	7	18
[350,400)	11	5	16
[400,450)	12	3	15
[450,500)	7	4	11
[500,550)	6	1	7
[550,600)	9	2	11
[600,630)	6	1	7
[630,∞)	—	—	44
Total	218	107	309

b. Fit the parametric proportional hazards model in which $h_1(t)=\lambda_1$ and $h_2(t)=\lambda_2$ are exponential hazard functions. Show that in this case the likelihood function for λ_1 and λ_2 is

$$p^{218}(1-p)^{107}(1-\beta)^{318}\beta^{1803.4}(1-\beta^{.6})^7$$

where $p=\lambda_1/(\lambda_1+\lambda_2)$ and $\beta=e^{-50(\lambda_1+\lambda_2)}$. Obtain $\hat{\lambda}_1$ and $\hat{\lambda}_2$ and assess the adequacy of this model.

(Section 10.1.2; Cox, 1959)

10.7 Consider the time-dependent Poisson process with intensity function

$$\lambda(t)=\frac{\beta}{\mu}\left(\frac{t}{\mu}\right)^{\beta-1} \tag{10.3.3}$$

where $\mu>0$ and $\beta>0$.

a.. Suppose that the process is observed up to the nth event and that events occur at times $t_1<t_2<\cdots<t_n$. Obtain the m.l.e.'s of μ and β. Prove that $2n\beta/\hat{\beta}$ has a $\chi^2_{(2n-2)}$ distribution.

b. Suppose that the process is observed to time t and that n events are observed at times $t_1<\cdots<t_n\leqslant t$. Obtain the m.l.e.'s of μ and β in this case. Show that given n, $2n\beta/\hat{\beta}\sim\chi^2_{(2n)}$.

c. Assess the fit of the model (10.3.3) to the four series of failure data discussed in Section 10.2.4.

(Crow, 1974; Lee and Lee, 1978)

10.8 A time-dependent Poisson process that includes both the model with $\lambda(t)=\exp(\alpha+\gamma t)$ and that with $\lambda(t)=\lambda\beta t^{\beta-1}$ is the model with intensity function

$$\lambda(t)=\lambda\beta t^{\beta-1}e^{\gamma t} \tag{10.3.4}$$

where $\lambda>0$, $\beta>0$, and $-\infty<\gamma<\infty$. This also allows nonmonotone intensities. Develop inference procedures for this model, including tests of the hypotheses $\beta=1$ and $\gamma=0$.

(Section 10.2; Lee, 1980)

10.9 The data in Table 10.3.3 show the times of significant maintenance events, in total hours of operation, for a diesel engine on the submarine

U.S.S. Grampus. Asterisks denote times of scheduled engine overhauls. The data are taken from a slightly longer series given by Lee (1980).

Investigate the pattern of maintenance event times. Do things appear to change after engine overhauls?

(Section 10.2)

Table 10.3.3 Maintenance Event Times (in Hours) for a Submarine Engine[a]

860	2,439	4,411	6,137	8,498	10,594*
1,203*	3,197*	4,456	6,221	8,690	11,511
1,258	3,203	4,517	6,311	9,042	11,575
1,317	3,298	4,899	6,613	9,330	12,100
1,442	3,902	4,910	6,975	9,394	12,126
1,897	3,910	5,414*	7,335	9,426	12,368
2,011	4,000	5,676	7,723*	9,872	12,681
2,122	4,247	5,755	8,158	10,191	12,795

[a]Asterisk denotes major engine overhaul.

APPENDIX A

Glossary of Notation, Abbreviations and Other Concepts

NOTATION AND SYMBOLS

$X \sim N(\mu, \sigma^2)$	X has a normal distribution with mean μ and variance σ^2
$\mathbf{X} \sim N_p(\boldsymbol{\mu}, \Sigma)$	$\mathbf{X}$ has a p-dimensional multivariate normal distribution with mean vector $\boldsymbol{\mu}$ and covariance matrix $\boldsymbol{\Sigma}$; the p is often omitted if its value is clear from the context
$T \sim \mathrm{Ga}(k)$	T has a one-parameter gamma distribution (1.3.12)
$\Gamma(x)$, $\psi(x)$, $\psi'(x)$	Gamma, digamma, and trigamma functions (see Appendix B)
$I(r, x)$	Incomplete gamma function [see (B12)]
$B_x(a, b)$	Incomplete beta function [see (B18)]
$\Pr(A)$	Probability of event A
$\|A\|$	Cardinality of the set A
$E[U(X)]$	Expectation of $U(X)$, where X is a random variable
$E_{X\|y}[U(X, Y)]$	Expectation of $U(X, Y)$, given that $Y=y$, where X and Y are random variables
$F(t-0)$	$\lim_{x \to 0^-} F(t+x)$
$F(t+0)$	$\lim_{x \to 0^+} F(t+x)$
$a_n \sim b_n$	$\lim_{n \to \infty} (a_n / b_n) = 1$, where $\{a_n\}$ and $\{b_n\}$ are sequences of real numbers

$X_n \xrightarrow{D} X$	The sequence of random variables $X_n: n=1,2,\ldots,$ converges in distribution to the random variable X (see Rao, 1965, p. 96) as $n \to \infty$
Asvar(X_n)	Variance of the asymptotic (limiting) distribution of X_n
Ascov(X_n, Y_n)	Covariance of the asymptotic (limiting) distribution of (X_n, Y_n)

ABBREVIATIONS

d.f.	Distribution function
EDF	Empirical distribution function (see Section 9.1.1)
ESF	Empirical survivor function [see (2.3.1)]
i.i.d.	Independent and identically distributed
m.l.e.	Maximum likelihood estimate
m.p.l.e.	Maximum partial likelihood estimate
p.d.f.	Probability density function
p.f.	Probability function
PH	Proportional hazards
PL estimate	Product-limit estimate [see (2.3.2)]

MISCELLANEOUS TERMS AND CONCEPTS

The *moment generating function* of a random variable X is defined as $M_X(\theta)=E(e^{\theta X})$, provided that this exists for all θ in some interval $(-h, h)$, where $h>0$.

The *cumulant generating function* of X is $K(\theta)=\log M_X(\theta)$. The mean μ and variance σ^2 of X can be determined as $\mu=K'(0)$, $\sigma^2=K''(0)$.

Pivotal (parameter-free) quantity—A function of random variables and parameters whose distribution does not depend on the values of the parameters.

Ancillary statistic—A statistic whose distribution does not depend on the values of any unknown parameters.

Riemann–Stieltjes integral—We consider only a special form suitable for probability distributions. Let $F(x)$ be a right-continuous distribution function and $g(x)$ a function with at most a finite number of discontinuities in any interval (a, b) and no points of discontinuity in common with $F(x)$. Assume that $F'(x)$ exists and is continuous almost everywhere. The Riemann–Stieltjes integral $\int_a^b g(x)\, dF(x)$ then equals

$$\int_a^b g(x)F'(x)\,dx + \sum_i g(x_i)[F(x_i) - F(x_i - 0)]$$

where the first integral is an ordinary Riemann integral and the sum in the second term is over all points $x_i \in (a, b)$ at which F is discontinuous (i.e., has jumps).

APPENDIX B

The Gamma and Some Related Functions

We summarize some useful results about the gamma function and other functions and probability distributions related to it. Many more details on these topics can be found in the books by Abramowitz and Stegun (1965), Johnson and Kotz (1970), and Mardia and Zemroch (1978). Because these books will be referenced frequently, we shall refer to them as AS, JK, and MZ, respectively.

The gamma function is defined as

$$\Gamma(z)=\int_0^{\infty} u^{z-1}e^{-u}\,du \qquad z>0. \tag{B1}$$

We note the well-known results (see AS, Ch. 6)

$$\Gamma(z+1)=z\Gamma(z) \qquad z>0 \tag{B2}$$

$$\Gamma(\tfrac{1}{2})=\pi^{1/2}=1.77245\ldots \tag{B3}$$

$$\log\Gamma(z)=(z-\tfrac{1}{2})\log z-z+\tfrac{1}{2}\log(2\pi)+\frac{1}{12z}-\frac{1}{360z^3}+\frac{1}{1260z^5}-\cdots \tag{B4}$$

It follows directly from (B1) and (B2) that for z a positive integer, $\Gamma(z+1)=z!$ Extensive tables of $\Gamma(z)$ exist; $\Gamma(z)$ is also easy to compute using (B2) and (B4) together, and most computer installations have on-line subroutines that calculate $\Gamma(z)$.

DIGAMMA AND POLYGAMMA FUNCTIONS

The digamma function is defined as

$$\psi(z)=\frac{d\log\Gamma(z)}{dz}=\frac{\Gamma'(z)}{\Gamma(z)} \qquad z>0. \tag{B5}$$

The polygamma functions are

$$\psi^{(n)}(z)=\frac{d^n\psi(z)}{dz^n} \qquad n=1,2,\dots .$$

Some important results are (AS, Ch. 6)

$$\psi(z+1)=\psi(z)+\frac{1}{z} \qquad z>0 \tag{B6}$$

$$\psi(1)=-\gamma=-0.577215\ldots \tag{B7}$$

(γ is called Euler's constant),

$$\psi(z)=\log z-\frac{1}{2z}-\frac{1}{12z^2}+\frac{1}{120z^4}-\frac{1}{252z^6}+\cdots \tag{B8}$$

$$\psi'(z+1)=\psi'(z)-\frac{1}{z^2} \qquad z>0 \tag{B9}$$

$$\psi'(1)=\frac{\pi^2}{6} \tag{B10}$$

$$\psi'(z)=\frac{1}{z}+\frac{1}{2z^2}+\frac{1}{6z^3}-\frac{1}{30z^5}+\frac{1}{42z^7}+\cdots. \tag{B11}$$

AS contains tables of $\psi(z)$ and $\psi'(z)$, or these can be calculated using (B6), (B8), (B9), and (B11).

THE INCOMPLETE GAMMA FUNCTION AND χ^2 DISTRIBUTION

The incomplete gamma function is defined in this book as

$$I(k,x)=\frac{1}{\Gamma(k)}\int_0^x u^{k-1}e^{-u}\,du \qquad k>0 \quad x>0. \tag{B12}$$

This is the distribution function for the one-parameter gamma distribution (1.3.12), denoted Ga(k), and takes on values between 0 and 1. $I(k, x)$ is related to the distribution function for the χ^2 distribution with ν degrees of freedom (denoted $\chi^2_{(\nu)}$) as follows:

$$F_\nu(x)=\Pr\left(\chi^2_{(\nu)}\leqslant x\right) \qquad x>0$$

$$=\int_0^x \frac{z^{\nu/2-1}e^{-z}}{2^{\nu/2}\Gamma(\nu/2)}\,dz$$

$$=I\left(\frac{\nu}{2},\frac{x}{2}\right). \tag{B13}$$

Extensive tables exist for the incomplete gamma function and the χ^2 distribution function (see, e.g., JK, Ch. 17, or AS, Chs. 6 and 26). MZ (1978, p. 200) give a compact but very useful table of percentage points for $\chi^2_{(\nu)}$. AS (Chs. 6 and 26) give many formulas for computing $I(k, x)$ or $F_\nu(x)$; see also MZ (p. 233). In addition, most computer installations have subroutines that calculate $I(k, x)$ or a related function.

For most situations a computer subroutine or tables like those of MZ are most convenient for getting values of $I(k, x)$ or $F_\nu(x)$, or quantiles of Ga(k) or $\chi^2_{(\nu)}$. For large values of k or ν one of several good approximations can be used. Two simple approximations that are sufficiently accurate for most purposes when $\nu\geqslant 20$ are based on the Wilson–Hilferty transformation (see JK, Ch. 17). Expressed in terms of a $\chi^2_{(\nu)}$ random variable, these are

$$\left(\frac{9\nu}{2}\right)^{1/2}\left[\left(\frac{\chi^2_{(\nu)}}{\nu}\right)^{1/3}+\frac{2}{9\nu}-1\right]\simeq N(0,1) \tag{B14}$$

$$\chi^2_{(\nu),p}\doteq\nu\left[1-\frac{2}{9\nu}+N_p\left(\frac{2}{9\nu}\right)^{1/2}\right]^3 \tag{B15}$$

where $\chi^2_{(\nu),p}$ and N_p are the pth quantiles of the $\chi^2_{(\nu)}$ and standard normal distributions, respectively.

BETA AND INCOMPLETE BETA FUNCTIONS

The beta function is defined as

$$B(a,b)=\int_0^1 t^{a-1}(1-t)^{b-1}\,dt \qquad a>0 \quad b>0. \tag{B16}$$

It is readily shown that

$$B(a,b)=\frac{\Gamma(a)\Gamma(b)}{\Gamma(a+b)}=B(b,a). \tag{B17}$$

The incomplete beta function is

$$B_x(a,b)=\frac{1}{B(a,b)}\int_0^x t^{a-1}(1-t)^{b-1}\,dt \qquad 0\leqslant x\leqslant 1. \tag{B18}$$

$B_x(a,b)$ is the distribution function of the beta distribution with parameters a and b, denoted Be(a,b). AS (Chs. 6 and 26) and JK (Ch. 24) contain many results about $B_x(a,b)$. Its computation is discussed in terms of the F distribution below. A useful special result is that for positive integral n and a $(1\leqslant a\leqslant n)$

$$B_p(a,n-a+1)=\sum_{j=a}^{n}\binom{n}{j}p^j(1-p)^{n-j} \tag{B19}$$

which relates the beta distribution to the binomial distribution.

THE F DISTRIBUTION

The F distribution with (a,b) degrees of freedom, denoted $F_{(a,b)}$, has distribution function

$$\Pr\left(F_{(a,b)}\leqslant x\right)=\int_0^x \frac{(a/b)^{a/2}}{B(a,b)}t^{a/2-1}\left(1+\frac{at}{b}\right)^{-(a+b)/2}dt \qquad x>0. \tag{B20}$$

The F and beta distributions are related by the fact that if $F\sim F_{(a,b)}$, then

$$\frac{aF}{b+aF}\sim \mathrm{Be}\left(\frac{a}{2},\frac{b}{2}\right). \tag{B21}$$

The incomplete beta function can thus be expressed in terms of the F distribution function. AS (Ch. 26) and JK (Ch. 26) discuss the F distribution at length. A very useful set of tables for the F distribution is given by MZ: they present percentage points for a wide range of values of a and b, including noninteger values, and show how to interpolate in the tables for other (a, b) pairs. They also (MZ, p. 227) give algorithms for computing F probabilities.

For a and b each larger than about 15 the Wilson–Hilferty transformation below is suitable for most purposes. If $F \sim F_{(a,b)}$, then

$$\left[F^{1/3}\left(1-\frac{2}{9b}\right)-\left(1-\frac{2}{9a}\right)\right]\left(\frac{2}{9a}+\frac{2}{9b}F^{2/3}\right)^{-1/2} \simeq N(0,1). \quad \text{(B22)}$$

This can be used to approximate (B20). To obtain the pth quantile for $F_{(a,b)}$ we find the value F_p of F that makes the left-hand side of (B22) equal to N_p, the pth quantile of $N(0,1)$. This involves solving the quadratic equation (4.3.5) of Chapter 4.

APPENDIX C

Asymptotic Variance Formulae

The following results are often used in developing large-sample inference procedures. Proofs can be found, for example, in Rao (1965, Ch. 6). Here "$\xrightarrow{D}$" means "converges in distribution to."

THEOREM C1 *Let $T_{1n}, \ldots, T_{kn}$ be statistics such that as $n \to \infty$*

$$\sqrt{n}(T_{1n} - \theta_1, \ldots, T_{kn} - \theta_k) \xrightarrow{D} N(\mathbf{0}, \mathbf{\Sigma})$$

where $\mathbf{\Sigma} = (\sigma_{ij})_{k \times k}$. If $g(x_1, \ldots, x_k)$ is a function whose first derivatives all exist, then as $n \to \infty$

$$\sqrt{n}[g(T_{1n}, \ldots, T_{kn}) - g(\theta_1, \ldots, \theta_k)] \xrightarrow{D} N\left(0, \sum_{i=1}^{k} \sum_{j=1}^{k} \sigma_{ij} \frac{\partial g}{\partial \theta_i} \frac{\partial g}{\partial \theta_j}\right) \quad \text{(C1)}$$

where $\partial g / \partial \theta_i$ means $\partial g(\theta_1, \ldots, \theta_k)/\partial \theta_i$ $(i = 1, \ldots, k)$.

Remarks

1. Often the following terminology is used:

$$\text{Asvar}[g(T_{1n}, \ldots, T_{kn})] = \sum_{i=1}^{k} \sum_{j=1}^{k} \frac{\partial g}{\partial \theta_i} \frac{\partial g}{\partial \theta_j} \text{Ascov}(T_{in}, T_{jn}) \quad \text{(C2)}$$

   where Asvar and Ascov denote variances and covariances in the asymptotic distributions of the indicated variables. (In the statement of Theorem C1, Ascov(T_{in}, T_{jn}) is $n^{-1}\sigma_{ij}$, for example.)

2. The results here are stated for statistics with asymptotic normal distributions. Expressions like (C2) also hold under the weaker conditions

that the variances and covariances of $T_{1n},\ldots,T_{kn}$ are $O(n^{-r})$, where $r>0$.

3. An important special case of Theorem C1 is given by $k=1$: if $\sqrt{n}(T_n-\theta)\xrightarrow{D}N(0,\sigma^2)$ as $n\to\infty$, then if $g(x)$ has first derivative $g'(x)$,

$$\sqrt{n}\left[g(T_n)-g(\theta)\right]\xrightarrow{D}N\left[0,g'(\theta)^2\sigma^2\right]. \tag{C3}$$

This implies that

$$\text{Asvar}\left[g(T_n)\right]=g'(\theta)^2\,\text{Asvar}(T_n). \tag{C4}$$

4. The results above can be proved with what is sometimes referred to as the δ method, based on Taylor series expansions. For example, the function $g(T_{1n},\ldots,T_{kn})$ has expansion

$$g(T_{1n},\ldots,T_{kn})=g(\theta_1,\ldots,\theta_k)+\sum_{i=1}^{k}\delta T_{in}\frac{\partial g}{\partial\theta_i}+\text{higher order terms}$$

where $\delta T_{in}=T_{in}-\theta_i$. The results follow from this and simple convergence results for random variables.

Theorem C1 can be generalized to the case in which there are several functions of $T_{1n},\ldots,T_{kn}$, as follows.

THEOREM C2 *Let $(T_{1n},\ldots,T_{kn})$ be statistics defined as in Theorem C1 and let $g_i(x_1,\ldots,x_k)$, $i=1,\ldots,q$, be functions, all of whose first derivatives exist. Then the joint distribution of $\sqrt{n}[g_i(T_{1n},\ldots,T_{kn})-g_i(\theta_1,\ldots,\theta_k)]$, $i=1,\ldots,q$, is asymptotically q-variate normal with mean $\mathbf{0}$ and covariance matrix $\mathbf{G}\boldsymbol{\Sigma}\mathbf{G}'$, where $\mathbf{G}$ has (i,j) entry $G_{ij}=\partial g_i/\partial\theta_j$.*

Remark When there are two functions $g_1(x_1,\ldots,x_k)$ and $g_2(x_1,\ldots,x_k)$, the theorem gives

$$\text{Ascov}\left[g_1(T_{1n},\ldots,T_{kn}),g_2(T_{1n},\ldots,T_{kn})\right]=\sum_{i=1}^{k}\sum_{j=1}^{k}\frac{\partial g_1}{\partial\theta_i}\frac{\partial g_2}{\partial\theta_j}\text{Ascov}\left(T_{in},T_{jn}\right). \tag{C5}$$

APPENDIX D

Order Statistics

A few results about order statistics are given here. Extended treatments can be found in the books by David (1970) and Sarhan and Greenberg (1962).

Suppose that X has continuous p.d.f. $f(x)$ and distribution function $F(x)$ and that $X_1, \ldots, X_n$ is a random sample from this distribution. The X_i's, rearranged in order of magnitude, and denoted

$$X_{(1)} \leqslant X_{(2)} \leqslant \cdots \leqslant X_{(n)}$$

are called the order statistics of the sample. The joint p.d.f. of $X_{(l_1)}, \ldots, X_{(l_k)}$, where $1 \leqslant l_1 < l_2 < \cdots < l_k \leqslant n$ and $1 \leqslant k \leqslant n$, can be shown to be

$$\left(n! \Big/ \prod_{i=1}^{k+1} (l_i - l_{i-1} - 1)! \right) \prod_{i=1}^{k+1} \left[F(x_{(l_i)}) - F(x_{(l_{i-1})}) \right]^{l_i - l_{i-1} - 1} \prod_{i=1}^{k} f(x_{(l_i)}) \tag{D1}$$

where $x_{(l_1)} \leqslant x_{(l_2)} \leqslant \cdots \leqslant x_{(l_k)}$ and where, for convenience, we define $l_0 = 0$, $l_{k+1} = n+1$, $x_{(l_0)} = -\infty$, and $x_{(l_{k+1})} = +\infty$.

Important special cases of (D1) are the following.

1. The joint p.d.f. of $X_{(1)}, \ldots, X_{(r)}$ $(r \leqslant n)$ is

$$\frac{n!}{(n-r)!} \left(\prod_{i=1}^{r} f(x_{(i)}) \right) \left[1 - F(x_{(r)}) \right]^{n-r}. \tag{D2}$$

2. The p.d.f. of $X_{(i)}$ $(1 \leqslant i \leqslant n)$ is

$$\frac{n!}{(i-1)!(n-i)!} f(x_{(i)}) F(x_{(i)})^{i-1} \left[1 - F(x_{(i)}) \right]^{n-i}. \tag{D3}$$

3. The joint p.d.f. of $X_{(i)}$ and $X_{(j)}$, for $i<j$, is

$$\frac{n!}{(i-1)!(j-i-1)!(n-j)!} f(x_{(i)}) f(x_{(j)}) F(x_{(i)})^{i-1}$$

$$\times\left[F(x_{(j)})-F(x_{(i)})\right]^{j-i-1}\left[1-F(x_{(j)})\right]^{n-j}. \qquad \text{(D4)}$$

These expressions are straightforward to obtain. For example, (D4) can be found from the probability that of the n values $X_1, \ldots, X_n$, $i-1$ are less than $x_{(i)}$, one is in $(x_{(i)}, x_{(i)}+\Delta)$, $j-i-1$ are between $x_{(i)}+\Delta$ and $x_{(j)}$, one is in $(x_{(j)}, x_{(j)}+\Delta)$, and $n-j$ are above $x_{(j)}+\Delta$.

MOMENTS

Moments of order statistics, especially means, variances, and covariances, are important in many applications. Means, variances, and covariances can be obtained from (D3) and (D4) for any given distribution, though it is usually not possible to get simple analytical expressions for them. Two exceptions are for the uniform and exponential distributions, for which the following results are easily established.

1. For the uniform distribution on (0, 1), with p.d.f. $f(x)=1$ $(0\leqslant x\leqslant 1)$,

$$E(X_{(i)})=\frac{i}{n+1}$$
$$\text{Var}(X_{(i)})=\frac{i^2}{(n+1)^2(n+2)}. \qquad \text{(D5)}$$

2. For the standard exponential distribution, with p.d.f. $f(x)=e^{-x}$ $(x\geqslant 0)$,

$$E(X_{(i)})=\sum_{l=1}^{i}(n-l+1)^{-1}$$
$$\text{Var}(X_{(i)})=\sum_{l=1}^{i}(n-l+1)^{-2}. \qquad \text{(D6)}$$

Results for the uniform and exponential distributions are important, since other problems can often be transformed to problems for these distributions. For example, if X has distribution function $F(x)$, then the random variables $F(X_{(i)})$, $i=1,\ldots,n$, are order statistics in a sample of n from the uniform distribution on $(0,1)$, since $F(X)$ is uniformly distributed on $(0,1)$.

For location–scale parameter distributions we need only consider moments of order statistics for the standard distribution. In particular, if X has p.d.f. of the form $b^{-1}g[(x-u)/b]$, $-\infty<x<\infty$, where $-\infty<u<\infty$ and $b>0$, then $Z=(X-u)/b$ has the standard form of the distribution, with p.d.f. $g(z)$, $-\infty<z<\infty$, and results for Z's order statistics yield results for X's order statistics. For the standard normal distribution Sarhan and Greenberg (1962) have tabulated means, variances, and covariances of order statistics for samples of size up to $n=20$. Tietjen et al. (1977) have extended this to $n=50$. White (1967, 1969) gives means and variances of order statistics for the standard extreme value distribution for sample sizes up to $n=100$; Mann (1967b) discusses computation of covariances. Some moments have been tabulated for other distributions as well (e.g., see Harter, 1969), but these are of less direct relevance to the topics in this book.

Moments of order statistics can also be approximated by asymptotic expansions. These are often suitable, provided that n is moderately large.

ASYMPTOTIC RESULTS

Several types of asymptotic results can be established for order statistics. First, we consider $X_{(i)}$, where $i=np$ and $0<p<1$ as $n\to\infty$. The pth quantile of the distribution of X is

$$\xi_p=F^{-1}(p).$$

It can then be shown (e.g., Cramer, 1946) that if $f(x)=F'(x)$ is continuous at ξ_p and $f(\xi_p)\neq 0$, then $X_{(i)}$ is asymptotically normal with mean ξ_p and variance

$$\frac{p(1-p)}{n[f(\xi_p)]^2}. \tag{D7}$$

{More precisely, $[nf(\xi_p)^2/p(1-p)]^{1/2}(X_{(i)}-\xi_p)\xrightarrow{D}N(0,1)$.} There is also a multivariate generalization of this (Mosteller, 1946) that states that for $0<p_1<\cdots<p_k<1$ and $i_j=np_j$ ($j=1,\ldots,k$), as $n\to\infty$ the limiting distribution of $X_{(i_1)},\ldots,X_{(i_k)}$ is multivariate normal with mean $(\xi_{1p},\ldots,\xi_{kp})$ and covariance matrix given by

$$\text{Ascov}\left[X_{(i_j)}, X_{(i_l)}\right] = \frac{p_j(1-p_l)}{nf(\xi_j)f(\xi_l)} \qquad j \leqslant l \tag{D8}$$

where $\xi_j = \xi_{p_j}$ $(j=1,\ldots,k)$.

The results above apply when $0<p_1<\cdots<p_k<1$. There are also many asymptotic results about extreme order statistics such as $X_{(1)}$ or $X_{(n)}$ (e.g., Gumbel, 1958; Galambos, 1978), adjacent order statistics (e.g., Pyke, 1965), and linear combinations of order statistics (e.g., Chernoff et al., 1967; Stigler, 1974).

Asymptotic approximations to the moments of order statistics can be given. In connection with a random sample of size n, let i and j be such that $1 \leqslant i < j \leqslant n$ and define $p_i = i/(n+1)$, $q_i = 1-p_i$, and $\xi_i = \xi_{p_i} = F^{-1}(p_i)$. Then (see David, 1970, p. 65, or Johnson and Kotz, 1970, Ch. 12)

$$E(X_{(i)}) \sim \xi_i + \frac{p_i q_i \xi_i''}{2(n+2)} + \frac{p_i q_i}{(n+2)^2}\left(\frac{(q_i-p_i)\xi_i'''}{3} + \tfrac{1}{8}p_i q_i \xi_i''''\right) \tag{D9}$$

$$\text{Var}(X_{(i)}) \sim \frac{p_i q_i}{n+2}(\xi_i')^2 + \frac{p_i q_i}{(n+2)^2}\left\{2(q_i-p_i)\xi_i'\xi_i'' + p_i q_i\left[\xi_i'\xi_i''' + \tfrac{1}{2}(\xi_i'')^2\right]\right\} \tag{D10}$$

$$\text{Cov}(X_{(i)}, X_{(j)}) \sim \frac{p_i q_j}{n+2}\xi_i'\xi_j' + \frac{p_i q_i}{n+2}\Big[(q_i-p_i)\xi_i''\xi_j' + (q_j-p_j)\xi_i'\xi_j'' + \tfrac{1}{2}p_i q_i \xi_i'''\xi_j' + \tfrac{1}{2}q_j p_j \xi_i'\xi_j''' + \tfrac{1}{2}p_i q_j \xi_i''\xi_j''\Big]. \tag{D11}$$

Here $\xi_i' = d\xi_i/dp_i$, $\xi_i'' = d^2\xi_i/dp_i^2$, and so on.

Other simple approximations are often useful. For example, Blom (1958) examines approximations of the form

$$E(X_{(i)}) \doteq F^{-1}\left(\frac{i+a}{n+2b+1}\right)$$

where a and b are suitably chosen constants. For the standard normal distribution $a=b=-.375$ gives a good approximation, and for the standard extreme value distribution $a=-.5$ and $b=-.375$ is good. An application of this is given in Section 9.2.2.

APPENDIX E

Maximum Likelihood Large-Sample Theory

We give here only a brief survey of important results. Books, for example, Cramer (1946, Ch. 7) and Cox and Hinkley (1974, Ch. 9) contain more details and references.

Consider first the case in which observations $x_1,\dots,x_n$ are a random sample from a distribution with p.d.f. $f(x;\boldsymbol{\theta})$, where $\boldsymbol{\theta}=(\theta_1,\dots,\theta_k)'$ is a vector of unknown parameters taking on values in a set Ω. The x_i's can be vectors, but for simplicity we shall write them as scalars. The likelihood function for $\boldsymbol{\theta}$ is defined as

$$L(\boldsymbol{\theta})=\prod_{i=1}^{n} f(x_i;\boldsymbol{\theta}). \tag{E1}$$

Actually, $L(\boldsymbol{\theta})$ depends on $x_1,\dots,x_n$, but it is customary to suppress this in the notation. Let $\hat{\boldsymbol{\theta}}$ be a point in Ω at which $L(\boldsymbol{\theta})$ is maximized; $\hat{\boldsymbol{\theta}}$ is called a maximum likelihood estimate (m.l.e.) of $\boldsymbol{\theta}$. In most simple models $\hat{\boldsymbol{\theta}}$ exists and is unique. Methods of obtaining $\hat{\boldsymbol{\theta}}$ are discussed in Appendix F. It is often convenient to work with $\log L(\boldsymbol{\theta})$, which is also maximized at $\hat{\boldsymbol{\theta}}$, and in many cases $\hat{\boldsymbol{\theta}}$ can be readily found by solving the so-called maximum likelihood equations $U_i(\boldsymbol{\theta})=0$ $(i=1,\dots,k)$, where

$$U_i(\boldsymbol{\theta})=\frac{\partial \log L(\boldsymbol{\theta})}{\partial \theta_i} \qquad i=1,\dots,k. \tag{E2}$$

The $U_i(\boldsymbol{\theta})$'s are called scores, and the $k\times 1$ vector $\mathbf{U}(\boldsymbol{\theta})=[U_1(\boldsymbol{\theta}),\dots,U_k(\boldsymbol{\theta})]'$ is called the score vector.

The score vector is a sum of i.i.d. random variables, since $\log L(\boldsymbol{\theta})=\Sigma \log f(x_i;\boldsymbol{\theta})$, and under mild conditions (e.g., see Cox and Hinkley, 1974, Sec. 9.2) it is asymptotically normally distributed. In addition, $\mathbf{U}(\boldsymbol{\theta})$ has mean $\mathbf{0}$ and covariance matrix $\mathbf{I}(\boldsymbol{\theta})$, with entries

$$I_{ij}(\theta)=E\left(\frac{-\partial^2 \log L(\boldsymbol{\theta})}{\partial\theta_i\,\partial\theta_j}\right) \qquad i,j=1,\ldots,k. \tag{E3}$$

The matrix $\mathbf{I}(\boldsymbol{\theta})$ is called the Fisher (or expected) information matrix.

Another important quantity is the $k\times k$ observed information matrix $\mathbf{I}_0$. This has (i,j) entry

$$I_{0,ij}=\left.\frac{-\partial^2 \log L(\boldsymbol{\theta})}{\partial\theta_i\,\partial\theta_j}\right|_{\boldsymbol{\theta}=\hat{\boldsymbol{\theta}}}. \tag{E4}$$

Under mild conditions $\mathbf{I}_0$ is a consistent estimator of $\mathbf{I}(\boldsymbol{\theta})$.

ASYMPTOTIC RESULTS AND LARGE-SAMPLE METHODS

Under mild regularity conditions $\hat{\boldsymbol{\theta}}$ is a consistent estimator of $\boldsymbol{\theta}$. In addition, several other asymptotic results hold that lead to useful inference procedures.

First, with a somewhat casual wording, $\mathbf{U}(\boldsymbol{\theta})$ is asymptotically $N_k[\mathbf{0},\mathbf{I}(\boldsymbol{\theta})]$. [The strictly correct statement is that $\mathbf{BU}(\boldsymbol{\theta})$ is asymptotically standard k-variate normal, where $\mathbf{B}$ is a matrix such that $\mathbf{BB}'=\mathbf{I}(\boldsymbol{\theta})$, but for convenience we shall use the casual wording.] This means that under the hypothesis $H_0: \boldsymbol{\theta}=\boldsymbol{\theta}_0$

$$\mathbf{U}'(\boldsymbol{\theta}_0)\mathbf{I}(\boldsymbol{\theta}_0)^{-1}\mathbf{U}(\boldsymbol{\theta}_0) \tag{E5}$$

is asymptotically $\chi^2_{(k)}$. This can be used to test H_0 and to obtain confidence regions for $\boldsymbol{\theta}$ consisting of those $\boldsymbol{\theta}_0$ that give a significance level greater than a specified value. Tests and estimates for a subset of the θ_i's can also be obtained: suppose that $\boldsymbol{\theta}$ is partitioned as $\boldsymbol{\theta}=(\boldsymbol{\theta}_1,\boldsymbol{\theta}_2)'$, where $\boldsymbol{\theta}_1$ is $p\times 1$ and $\boldsymbol{\theta}_2$ is $(k-p)\times 1$. Partition $\mathbf{U}(\boldsymbol{\theta})$, $\mathbf{I}(\boldsymbol{\theta})$, and, $\mathbf{I}^{-1}(\boldsymbol{\theta})$ in a corresponding way:

$$\mathbf{U}(\boldsymbol{\theta})=\begin{pmatrix}\mathbf{U}_1(\boldsymbol{\theta})\\ \mathbf{U}_2(\boldsymbol{\theta})\end{pmatrix} \qquad I(\boldsymbol{\theta})=\begin{pmatrix}\mathbf{I}_{11}(\boldsymbol{\theta}) & \mathbf{I}_{12}(\boldsymbol{\theta})\\ \mathbf{I}_{21}(\boldsymbol{\theta}) & \mathbf{I}_{22}(\boldsymbol{\theta})\end{pmatrix} \tag{E6}$$

$$\mathbf{I}^{-1}(\boldsymbol{\theta})=\begin{pmatrix}\mathbf{I}^{11}(\boldsymbol{\theta}) & \mathbf{I}^{12}(\boldsymbol{\theta})\\ \mathbf{I}^{21}(\boldsymbol{\theta}) & \mathbf{I}^{22}(\boldsymbol{\theta})\end{pmatrix}. \tag{E7}$$

It is convenient to note that $\mathbf{I}^{11}(\boldsymbol{\theta})$ can be shown to equal $\mathbf{I}_{11}(\boldsymbol{\theta})-\mathbf{I}_{12}(\boldsymbol{\theta})\mathbf{I}_{22}^{-1}(\boldsymbol{\theta})\mathbf{I}_{21}(\boldsymbol{\theta})$. For a given $\boldsymbol{\theta}_1=\boldsymbol{\theta}_{10}$, let $\tilde{\boldsymbol{\theta}}_2(\boldsymbol{\theta}_{10})$ be the m.l.e. of $\boldsymbol{\theta}_2$, obtained by maximizing $L(\boldsymbol{\theta}_{10},\boldsymbol{\theta}_2)$. Denote $[\boldsymbol{\theta}_{10},\tilde{\boldsymbol{\theta}}_2(\boldsymbol{\theta}_{10})]'$ as $\tilde{\boldsymbol{\theta}}$. Then, under $H_0:\boldsymbol{\theta}_1=\boldsymbol{\theta}_{10}$,

$$\mathbf{U}_1(\tilde{\boldsymbol{\theta}})'\left[\mathbf{I}^{11}(\tilde{\boldsymbol{\theta}})\right]^{-1}\mathbf{U}_1(\tilde{\boldsymbol{\theta}}) \tag{E8}$$

is asymptotically $\chi^2_{(p)}$. Tests based on (E5) are often referred to as scores tests, and those based on (E8) as partial scores tests. A test based on (E5) has the convenient property of not requiring calculation of $\hat{\boldsymbol{\theta}}$.

Tests and estimates can also be based on $\hat{\boldsymbol{\theta}}$. Under mild regularity conditions $\hat{\boldsymbol{\theta}}$ is asymptotically $N_k[\boldsymbol{\theta},\mathbf{I}^{-1}(\boldsymbol{\theta})]$. Thus under $H_0:\boldsymbol{\theta}=\boldsymbol{\theta}_0$

$$(\hat{\boldsymbol{\theta}}-\boldsymbol{\theta}_0)'\mathbf{I}(\boldsymbol{\theta}_0)(\hat{\boldsymbol{\theta}}-\boldsymbol{\theta}_0) \tag{E9}$$

is asymptotically $\chi^2_{(k)}$. This can be used to test H_0 and to get confidence regions for $\boldsymbol{\theta}$. Since $\hat{\boldsymbol{\theta}}$ is a consistent estimator of $\boldsymbol{\theta}$ and $\mathbf{I}_0$ is a consistent estimator of $\mathbf{I}(\boldsymbol{\theta})$, two other test statistics that are asymptotically equivalent to (E9) are

$$(\hat{\boldsymbol{\theta}}-\boldsymbol{\theta}_0)'\mathbf{I}(\hat{\boldsymbol{\theta}})(\hat{\boldsymbol{\theta}}-\boldsymbol{\theta}_0) \tag{E10}$$

and

$$(\hat{\boldsymbol{\theta}}-\boldsymbol{\theta}_0)'\mathbf{I}_0(\hat{\boldsymbol{\theta}}-\boldsymbol{\theta}_0). \tag{E11}$$

Tests and estimates about a portion of $\boldsymbol{\theta}$ are also possible. Partition $\boldsymbol{\theta}=(\boldsymbol{\theta}_1,\boldsymbol{\theta}_2)'$, $\mathbf{U}(\boldsymbol{\theta})$, $\mathbf{I}(\boldsymbol{\theta})$, and $\mathbf{I}^{-1}(\boldsymbol{\theta})$ as in (E6) and (E7). Then under $H_0:\boldsymbol{\theta}_1=\boldsymbol{\theta}_{10}$

$$(\hat{\boldsymbol{\theta}}_1-\boldsymbol{\theta}_{10})'\mathbf{I}^{11}(\boldsymbol{\theta})^{-1}(\hat{\boldsymbol{\theta}}_1-\boldsymbol{\theta}_{10}) \tag{E12}$$

is asymptotically $\chi^2_{(p)}$. This can be used to test H_0 or to obtain confidence regions for $\boldsymbol{\theta}_1$, but only if $\boldsymbol{\theta}_2$ is known. Otherwise, we can use an asymptotically equivalent statistic to (E12) obtained by replacing $\boldsymbol{\theta}$ in $\mathbf{I}^{11}(\boldsymbol{\theta})^{-1}$ with $\tilde{\boldsymbol{\theta}}$ or $\hat{\boldsymbol{\theta}}$ or by replacing $\mathbf{I}^{11}(\boldsymbol{\theta})^{-1}$ with the corresponding observed information submatrix.

A third procedure for obtaining tests and estimates is the likelihood ratio method. It can be shown that under $H_0:\boldsymbol{\theta}=\boldsymbol{\theta}_0$

$$\Lambda=-2\log\left(\frac{L(\boldsymbol{\theta}_0)}{L(\hat{\boldsymbol{\theta}})}\right) \tag{E13}$$

is asymptotically $\chi^2_{(k)}$. Similarly, if $\boldsymbol{\theta}$ is partitioned as $\boldsymbol{\theta}=(\boldsymbol{\theta}_1, \boldsymbol{\theta}_2)'$ and we consider $H_0: \boldsymbol{\theta}_1 = \boldsymbol{\theta}_{10}$, then

$$\Lambda = -2\log\left(\frac{L[\boldsymbol{\theta}_{10}, \tilde{\boldsymbol{\theta}}_2(\boldsymbol{\theta}_{10})]}{L(\hat{\boldsymbol{\theta}}_1, \hat{\boldsymbol{\theta}}_2)}\right) \tag{E14}$$

is asymptotically distributed as $\chi^2_{(p)}$, where, as above, $\boldsymbol{\theta}_1$ is $p \times 1$ and $\tilde{\boldsymbol{\theta}}_2(\boldsymbol{\theta}_{10})$ is the m.l.e. of $\boldsymbol{\theta}_2$ under H_0.

Several large-sample inference procedures have been mentioned. By parameterizing a model in various ways it is possible to obtain tests and estimates for more or less arbitrary functions of parameters. However, the procedures rely on asymptotic approximations to distributions of statistics, and the adequacy of these approximations in finite samples must be considered. The situation changes from model to model and it is difficult to make general statements, but the distributions of the likelihood ratio statistics (E13) and (E14) often appear to approach their limiting distributions considerably more rapidly than the distribution of $\hat{\boldsymbol{\theta}}$. It is also important that whereas the likelihood ratio and score statistics are not affected by reparameterization of the model, statistics based on the limiting normality of $\hat{\boldsymbol{\theta}}$ [see (E9), (E10), and (E11)] are. If the latter approach is used, one should attempt to find a parameterization that is convenient and for which the distribution of $\hat{\boldsymbol{\theta}}$ approaches normality as rapidly as possible. Choice of parameter transformations in this context has been discussed by Anscombe (1961, 1964), Sprott (1973), and others. There are no strong reasons for favoring either (E9), (E10), or (E11) over the others, though there is evidence that in some situations use of the observed information matrix and (E10) is to be preferred (e.g., Efron and Hinkley, 1978; Hinkley 1978).

Procedures based on the asymptotic normality of $\hat{\boldsymbol{\theta}}$ are often used because they are the simplest computationally. Likelihood ratio and scores procedures for finding confidence intervals usually require more computation, for example. However, they often produce intervals with closer to the nominal coverage probabilities because of the superior approximations that are provided by their asymptotic distributions. Sections 3.2 and 4.1 provide comparisons of the different methods in two specific situations.

LIKELIHOOD METHODS IN OTHER SAMPLING SITUATIONS

The summary of likelihood results given so far is for when observations are i.i.d. Much of the discussion can be extended to other sampling situations. We make only a few brief comments regarding places in the book where this is necessary. Further research is needed to clarify the picture of asymptotic properties in some of the areas mentioned.

First, the results given apply without modification when observations are independent but not identically distributed, though slight changes in the regularity conditions underlying the methods are involved. This covers two important problems examined in this book, namely, regression situations and situations where independent observations are subject to independent random or fixed censoring mechanisms. A second place where the results apply is to the case of Type II censored sampling (Halperin, 1952). Another, more difficult problem concerns asymptotic properties of likelihood methods when data are arbitrarily censored (see Section 1.4). Further work is needed in this area.

Two other places where special treatment is necessary have to do with stochastic process models and nonparametric methods. An important situation where likelihood methods can be applied in basically the usual way is for sampling from Markov processes (see Section 10.2, Berlin et al., 1979, and Cox and Hinkley, 1974, p. 141, for examples). With nonparametric problems special difficulties are created by the presence of infinite-dimensional parameter spaces. An example of this is given in Section 2.3, where the product-limit estimate is viewed as an m.l.e.

MARGINAL, CONDITIONAL, AND PARTIAL LIKELIHOOD

Sometimes, especially when there are nuisance parameters present, it is appropriate or convenient to use only a portion of the likelihood function. This leads to the concepts of marginal, conditional, and partial likelihoods.

Suppose that $\boldsymbol{\theta}=(\boldsymbol{\theta}_1,\boldsymbol{\theta}_2)$ and that the data, denoted simply as $\mathbf{x}$, are transformed into (S,T), where one or both of S and T may be vectors. Suppose that the distribution of S depends on $\boldsymbol{\theta}_1$, but not $\boldsymbol{\theta}_2$; then the likelihood $L_S(\boldsymbol{\theta}_1)$ obtained from the marginal distribution of S is termed a "marginal likelihood." Similarly, suppose that the conditional distribution of T, given S, depends on $\boldsymbol{\theta}_1$, but not $\boldsymbol{\theta}_2$. In this case the likelihood $L_{T|S}(\boldsymbol{\theta}_1)$ obtained from this conditional distribution is referred to as a conditional likelihood. Marginal and conditional likelihoods have been discussed by several authors (e.g., Fraser, 1968; Kalbfleisch and Sprott, 1970, 1974). A main problem is in assessing whether there is a substantial loss of information entailed in using a marginal or conditional likelihood in a given situation. If there is not, these provide convenient methods of making inferences about $\boldsymbol{\theta}_1$ in the absence of knowledge of $\boldsymbol{\theta}_2$. Asymptotic properties have not been thoroughly examined for marginal and conditional likelihood, though in many situations it is clear that the usual asymptotic results apply. In the places in this book where they are used it can be shown that large-sample theory is valid for them.

Marginal and conditional likelihood are special cases of the more general concept of partial likelihood (Cox, 1975). The basic idea behind this is as follows: suppose that the data $\mathbf{x}$ have p.d.f. $f(\mathbf{x}; \boldsymbol{\theta}_1, \boldsymbol{\theta}_2)$, where $\boldsymbol{\theta}_1$ is of interest and $\boldsymbol{\theta}_2$ is a nuisance parameter. Suppose that $\mathbf{x}$ can be transformed into $S_1, T_1, S_2, T_2, \ldots, S_m, T_m$, where some or all of the S_i's and T_i's might be vectors. The joint p.d.f. of $S_1, T_1, \ldots, S_m, T_m$ can be written as

$$\prod_{i=1}^{m} f_{s_i|s^{(i-1)}, t^{(i-1)}}\left(s_i|s^{(i-1)}, t^{(i-1)}; \boldsymbol{\theta}\right) \prod_{i=1}^{m} f_{t_i|s^{(i)}, t^{(i-1)}}\left(t_i|s^{(i)}, t^{(i-1)}; \boldsymbol{\theta}\right)$$

where $s^{(i)}=(s_1, \ldots, s_i)$ and $t^{(i)}=(t_1, \ldots, t_i)$.

If the second term depends just on $\boldsymbol{\theta}_1$, this is termed a partial likelihood for $\boldsymbol{\theta}_1$. There will typically be some loss of information involved in using a partial likelihood, though this may be difficult to assess. In addition, to use partial likelihoods we need to know that asymptotic results of the kind described for ordinary likelihoods hold. Cox (1975) outlines conditions under which this would be so, but it is often difficult to check these in individual problems. Further research is needed to delineate the properties of partial likelihood.

APPENDIX F

Optimization Methods for Maximum Likelihood

Many statistical problems require the determination of maximum likelihood estimates (m.l.e.'s), that is, points $\boldsymbol{\theta}=(\theta_1,\ldots,\theta_k)'$ at which a (log) likelihood function is maximized. Many optimization methods are available, and we shall merely review several of the most useful ones. In a few situations m.l.e.'s can be found analytically; we consider problems where this is not possible. The books by Chambers (1977, Ch. 6) and Walsh (1975) can be consulted for further details and references on optimization.

NEWTON–RAPHSON ITERATION

Denote the likelihood function as $L(\theta_1,\ldots,\theta_k)=L(\boldsymbol{\theta})$, defined over the parameter space Ω. We shall consider situations in which the point $\hat{\boldsymbol{\theta}}$ at which $L(\boldsymbol{\theta})$, and also $\log L(\boldsymbol{\theta})$, is maximized is a local maximum and satisfies the likelihood equations

$$U_i(\boldsymbol{\theta})=\frac{\partial \log L}{\partial \theta_i}=0 \qquad i=1,\ldots,k. \tag{F1}$$

The $k\times 1$ vector $\mathbf{U}(\boldsymbol{\theta})=[U_1(\boldsymbol{\theta}),\ldots,U_k(\boldsymbol{\theta})]'$ is called the score vector at $\boldsymbol{\theta}$. Suppose that $\boldsymbol{\theta}_0$ is a first guess at $\hat{\boldsymbol{\theta}}$ and expand each of the functions $U_i(\boldsymbol{\theta})$ in a Taylor series about $\boldsymbol{\theta}_0$. To first order this gives

$$\mathbf{U}(\boldsymbol{\theta})\doteq\mathbf{U}(\boldsymbol{\theta}_0)+\mathbf{G}(\boldsymbol{\theta}_0)(\boldsymbol{\theta}-\boldsymbol{\theta}_0) \tag{F2}$$

where $\mathbf{G}(\boldsymbol{\theta})$ is the $k\times k$ matrix with entries

$$G_{ij}(\boldsymbol{\theta})=\frac{\partial^2 \log L}{\partial \theta_i \partial \theta_j}.$$

Since $\hat{\boldsymbol{\theta}}$ satisfies $\mathbf{U}(\hat{\boldsymbol{\theta}})=\mathbf{0}$, (F2) yields the approximation

$$\hat{\boldsymbol{\theta}} \doteq \boldsymbol{\theta}_0 - G(\boldsymbol{\theta}_0)^{-1}\mathbf{U}(\boldsymbol{\theta}_0). \tag{F3}$$

In practice, (F3) is used to define an iteration scheme for obtaining $\hat{\boldsymbol{\theta}}$, with the right-hand side of (F3) producing a second approximation $\boldsymbol{\theta}_1$ to $\hat{\boldsymbol{\theta}}$; this is in turn inserted in the right-hand side of (F3) to produce a third approximation, and so on. This is usually called the Newton–Raphson iteration procedure. Provided that $\log L(\boldsymbol{\theta})$ is well behaved at $\hat{\boldsymbol{\theta}}$ and that the first guess $\boldsymbol{\theta}_0$ is not too far from $\hat{\boldsymbol{\theta}}$, the sequence of approximations generated will converge to $\hat{\boldsymbol{\theta}}$. "Well behaved" here means essentially that $\log L(\boldsymbol{\theta})$ can be approximated by a quadratic in the neighborhood of $\hat{\boldsymbol{\theta}}$. If the likelihood satisfies the regularity conditions for standard asymptotic normal theory (see Appendix E), then this is usually so, at least in large samples. Newton–Raphson iteration generally performs well with such likelihoods. Sometimes problems are encountered with the iteration procedure, but these can be remedied by a simple modification of (F3); Chambers (1977, Ch. 6) discusses this. Common difficulties are that if $\boldsymbol{\theta}_0$ is not sufficiently close to $\hat{\boldsymbol{\theta}}$, the next point produced may lie outside of Ω or have a smaller value of $\log L(\boldsymbol{\theta})$ than $\boldsymbol{\theta}_0$.

A minor adjustment to the Newton–Raphson procedure that is sometimes used in statistical problems is known as the "scoring method." This consists of replacing $\mathbf{G}(\boldsymbol{\theta})$ in (F3) with $-\mathbf{I}(\boldsymbol{\theta})$, the negative of the expected information matrix (E3). The scoring method has convergence properties similar to Newton–Raphson iteration, and on occasions gives a simpler looking algorithm than does the Newton–Raphson procedure.

Both the Newton–Raphson and scoring procedures conveniently yield an estimate of the information matrix, and hence the asymptotic covariance matrix for $\hat{\boldsymbol{\theta}}$. In the former case $-\mathbf{G}(\hat{\boldsymbol{\theta}})$ is the observed information matrix, and in the latter case $\mathbf{I}(\hat{\boldsymbol{\theta}})$ is the estimated expected information matrix.

EXAMPLE: EXPONENTIAL REGRESSION MODEL

The exponential regression model (6.3.1) gives the log likelihood (6.3.5) with censored data,

$$\log L(\boldsymbol{\beta}) = \sum_{i \in D} (y_i - \mathbf{x}_i\boldsymbol{\beta}) - \sum_{i=1}^{n} \exp(y_i - \mathbf{x}_i\boldsymbol{\beta}) \tag{F4}$$

where $\mathbf{x}_i = (x_{i1}, \ldots, x_{ip})$ is a vector of regressor variables and $\boldsymbol{\beta} = (\beta_1, \ldots, \beta_p)'$

is a vector of parameters. The likelihood equations and matrix $\mathbf{G}(\boldsymbol{\beta})$ have components

$$U_r(\boldsymbol{\beta}) = \frac{\partial \log L}{\partial \beta_r} = -\sum_{i \in D} x_{ir} + \sum_{i=1}^{n} x_{ir} \exp(y_i - \mathbf{x}_i \boldsymbol{\beta})$$

$$= 0 \qquad r = 1, \ldots, p$$

$$G_{rs}(\boldsymbol{\beta}) = \frac{\partial^2 \log L}{\partial \beta_r \, \partial \beta_s} = -\sum_{i=1}^{n} x_{ir} x_{is} \exp(y_i - \mathbf{x}_i \boldsymbol{\beta}) \qquad r, s = 1, \ldots, p.$$

The Newton–Raphson procedure involves the following steps:

1. Obtain an initial estimate $\boldsymbol{\beta}_0$ of $\hat{\boldsymbol{\beta}}$.
2. Calculate $\mathbf{U}(\boldsymbol{\beta}_0)$ and $\mathbf{G}(\boldsymbol{\beta}_0)$.
3. Calculate the next approximation $\boldsymbol{\beta}_1$ to $\hat{\boldsymbol{\beta}}$ using (F3), as

$$\boldsymbol{\beta}_1 = \boldsymbol{\beta}_0 - \mathbf{G}(\boldsymbol{\beta}_0)^{-1} \mathbf{U}(\boldsymbol{\beta}_0).$$

   This involves inverting $\mathbf{G}(\boldsymbol{\beta}_0)$ or, at least, solving the system of linear equations $\mathbf{G}(\boldsymbol{\beta}_0)(\boldsymbol{\beta}_1 - \boldsymbol{\beta}_0) = -\mathbf{U}(\boldsymbol{\beta}_0)$ for $\boldsymbol{\beta}_1$.
4. Repeat steps (2) and (3), replacing $\boldsymbol{\beta}_0$ with $\boldsymbol{\beta}_1$. Continue repeating this until convergence is (hopefully) achieved. One stops when $\boldsymbol{\beta}_0$ and $\boldsymbol{\beta}_1$ are close together and $\mathbf{U}(\boldsymbol{\beta}_1)$ is close to $\mathbf{0}$.

When the log likelihood (F4) in this problem is based on an uncensored sample, the scoring method produces a slightly simpler procedure than the Newton–Raphson algorithm. With no censoring we have (see Section 6.3.1)

$$I_{rs}(\boldsymbol{\beta}) = E\left(\frac{-\partial^2 \log L}{\partial \beta_r \, \partial \beta_s} \right)$$

$$= \sum_{i=1}^{n} x_{ir} x_{is}.$$

Since $\mathbf{I}$ does not depend on $\boldsymbol{\beta}$, if $-\mathbf{I}$ replaces $\mathbf{G}(\boldsymbol{\beta}_0)$ in steps (2) and (3), it is necessary to compute and invert only one matrix over all, rather than one per iteration.

QUASI-NEWTON METHODS

The Newton–Raphson algorithm is a good procedure in many maximum likelihood problems. However, it requires second derivatives of $\log L$, and sometimes these may be inconvenient or expensive to complete. Quasi-Newton methods (see Chambers, 1977, Sec. 6c) have properties that are somewhat similar to those of the Newton–Raphson procedure but require only first derivatives of $\log L$. Many procedures of this kind have been proposed in the numerical analysis literature, and several good ones have been implemented in widely available computer programs. Chambers (1977) can be consulted for details.

SEARCH PROCEDURES

When the likelihood function is not well behaved or when even first derivatives of $\log L$ are inconvenient to compute, a direct search procedure for finding the maximum of $L(\boldsymbol{\theta})$ may be the best approach. The simplex search procedure of Nelder and Mead (1965) has proved successful in many problems, particularly when there are not too many parameters present. It has been implemented in several computer programs that are widely available; Chambers (1977, Sec. 6d) and Olsson and Nelson (1975) provide references.

Other search procedures such as those of Powell (1964) and Fletcher and Reeves (1964) are also widely available in computer programs; Chambers (1977) gives references. Some of these techniques perform well when there are many parameters, in which case the Nelder–Mead procedure can be unsuitably slow.

OTHER CONSIDERATIONS

Several general optimization methods have been mentioned. We supplement the discussion with some general remarks about maximizing likelihood functions.

1. Sometimes, for example, in certain distributions with only two or three parameters, it is possible to partially solve the maximization problem analytically and to then use numerical methods on the remainder. In obtaining m.l.e.'s $\hat{b}$ and $\hat{u}$ in the two-parameter extreme value distribution (see Section 4.1.1), for example, one can solve a single equation numerically for $\hat{b}$ and then determine $\hat{u}$ from this.

2. One optimization method may work well with respect to part of a problem, whereas another may be preferable on other parts. For example, in the three-parameter generalized gamma model (5.3.6) any differentiation of $\log L$ with respect to k is inconvenient and to be avoided. A good method for obtaining the m.l.e. $(\hat{\mu}, \hat{\sigma}, \hat{k})$ is to use Newton–Raphson iteration to get m.l.e.'s $\hat{\mu}(k)$ and $\hat{\sigma}(k)$ for specified k values, combined with a search procedure to obtain $\hat{k}$, the value of k that maximizes $\log L_{max}(k) = \log L[\hat{\mu}(k), \hat{\sigma}(k), k]$.
3. Good initial estimates are very important in most procedures, and due consideration should be given to methods for obtaining them. In problems with a small number of parameters use of an interactive computing facility often facilitates initial exploration of a likelihood function and the selection of initial estimates.
4. Alternate parameterizations in a problem can improve convergence properties of a method and remove restrictions on parameters. Parameter transformations that improve the asymptotic normal approximations for m.l.e.'s will often help convergence of algorithms as well, since they make the log likelihood more nearly quadratic near its maximum.
5. For many problems special iteration procedures can be devised, with both speed of convergence and simplicity in mind. An example is the Sampford–Taylor algorithm based on (5.2.11) of Section 5.2, which finds the m.l.e.'s in the normal model from a censored sample.
6. Problems where m.l.e.'s are on the boundary of Ω and problems in which $\log L(\boldsymbol{\theta})$ is unbounded can occur. These require special care and are often difficult to handle. Problems of this kind arise in connection with maximum likelihood estimation of the three-parameter Weibull or log-normal distribution (see Sections 4.4 and 5.2, respectively), for example.

APPENDIX G

Inference in Location–Scale Parameter Models

A univariate location–scale parameter distribution is one with p.d.f. of the form

$$f(y; u, b) = \frac{1}{b} g\left(\frac{y-u}{b}\right) \qquad -\infty < y < \infty \qquad \text{(G1)}$$

where u $(-\infty < u < \infty)$ is a location parameter, b $(b > 0)$ is a scale parameter, and $g(\ )$ is a fully specified p.d.f. defined on $(-\infty, \infty)$. The survivor function corresponding to (G1) is $G[(y-u)/b]$, where

$$G(x) = \int_x^\infty g(z)\, dz.$$

Tests and interval estimation for u and b, and for parameters in related regression models, are outlined here. The ideas involved actually apply to a wider class of models (e.g., Hora and Buehler, 1966; Fraser, 1968), but this is not needed for this book.

EQUIVARIANT STATISTICS

We allow Type II censoring and suppose that $y_1 \leqslant \cdots \leqslant y_r$ are the r smallest observations in a random sample of size n from (G1). The results here also apply when there is progressive Type II censored sampling, but for simplicity we shall not examine this explicitly. Suppose that $\tilde{u} = \tilde{u}(y_1, \ldots, y_r)$ and $\tilde{b} = \tilde{b}(y_1, \ldots, y_r)$ are statistics with the following properties:

$$\tilde{u}(dy_1 + c, \ldots, dy_r + c) = d\tilde{u}(y_1, \ldots, y_r) + c \qquad \text{(G2)}$$

$$\tilde{b}(dy_1 + c, \ldots, dy_r + c) = d\tilde{b}(y_1, \ldots, y_r) \qquad \text{(G3)}$$

for any real constants c $(-\infty<c<\infty)$ and d $(d>0)$. Then $\tilde{u}$ and $\tilde{b}$ are termed equivariant statistics; they are also commonly referred to as equivariant estimators of u and b (e.g., Zacks, 1971, Ch. 7). The requirements (G2) and (G3) are natural ones for estimators of location and scale parameters and most, if not all, of the common types of estimators satisfy them. For example, we have

THEOREM G1 *Let $\hat{u}$ and $\hat{b}$ be maximum likelihood estimators of u and b in (G1), based on a Type II censored sample. Then $\hat{u}$ and $\hat{b}$ are equivariant.*

Proof The likelihood function based on $y_1 \leqslant \cdots \leqslant y_r$ is [see (1.4.1)]

$$L_{\mathbf{y}}(u,b)=\frac{1}{b^r}\left[\prod_{i=1}^{r} g\left(\frac{y_i-u}{b}\right)\right]\left[G\left(\frac{y_r-u}{b}\right)\right]^{n-r} \tag{G4}$$

where we write $\mathbf{y}=(y_1,\ldots,y_r)$. It is easily seen that

$$L_{\mathbf{y}}(u,b)=d^r L_{\mathbf{y}'}(u',b')$$

where $y_i'=dy_i+c$, $\mathbf{y}'=(y_1',\ldots,y_r')$, $u'=du+c$, and $b'=db$. If $L_{\mathbf{y}}(u,b)$ is maximized for u and b at $\hat{u}(\mathbf{y})$ and $\hat{b}(\mathbf{y})$, then $L_{\mathbf{y}'}(u,b)$ is maximized at $\hat{u}(\mathbf{y}')=d\hat{u}(\mathbf{y})+c$ and $\hat{b}(\mathbf{y}')=d\hat{b}(\mathbf{y})$. This proves the result. ■

PIVOTALS AND ANCILLARIES

The next theorem follows easily from the definition of equivariant estimators.

THEOREM G2 *Let $\tilde{u}$ and $\tilde{b}$ be equivariant estimators, based on a Type II censored sample from (G1). Then*

(i) $Z_1=(\tilde{u}-u)/\tilde{b}$, $Z_2=\tilde{b}/b$, *and* $Z_3=(\tilde{u}-u)/b$ *are pivotal quantities.*

(ii) *The quantities $a_i=(x_i-\tilde{u})/\tilde{b}$, $i=1,\ldots,r$, form a set of ancillary statistics, only $r-2$ of which are functionally independent.*

Proof From (G1) the random variable $W=(Y-u)/b$ has p.d.f. $g(w)$, $-\infty<w<\infty$, not depending on u or b. Thus the joint p.d.f. of $w_1=(y_1-u)/b,\ldots,w_r=(y_r-u)/b$ does not depend on u or b, and consequently neither does the distribution of $\tilde{u}(w_1,\ldots,w_r)$ or $\tilde{b}(w_1,\ldots,w_r)$. But since $\tilde{u}$ and

$\tilde{b}$ are equivariant, it follows from (G2) and (G3) that

$$\tilde{u}(w_1,\ldots,w_r)=\frac{\tilde{u}(y_1,\ldots,y_r)-u}{b}=Z_3$$

$$\tilde{b}(w_1,\ldots,w_r)=\frac{\tilde{b}(y_1,\ldots,y_r)}{b}=Z_2\,.$$

This proves that Z_2 and Z_3 are pivotal; Z_1 is thus also pivotal, since $Z_1=Z_3/Z_2$.

Regarding (ii), the a_i's are clearly ancillary, since $a_i=(y_i-\tilde{u})/\tilde{b}=(w_i-Z_3)/Z_2$, and hence their distribution does not depend on u or b. Finally, the a_i's satisfy two restrictions, since

$$\tilde{u}(a_1,\ldots,a_r)=[\tilde{u}(y_1,\ldots,y_r)-\tilde{u}]/\tilde{b}=0$$

and

$$\tilde{b}(a_1,\ldots,a_r)=\tilde{b}(y_1,\ldots,y_r)/\tilde{b}=1.$$

Thus there are just $r-2$ functionally independent a_i's. ■

The following theorem concerning the distribution of the pivotals and ancillaries also provides an alternate proof of Theorem G2.

THEOREM G3 *Let $\tilde{u}$ and $\tilde{b}$ be any equivariant estimators of u and b under the conditions of Theorem G2. Then the joint p.d.f. of $Z_1, Z_2, a_1,\ldots,a_{r-2}$ is of the form*

$$k(\mathbf{a},r,n)z_2^{r-1}\left(\prod_{i=1}^{r}g(a_iz_2+z_1z_2)\right)\left[G(a_rz_2+z_1z_2)\right]^{n-r} \tag{G5}$$

where $k(\mathbf{a},r,n)$ is a function of $a_1,\ldots,a_{r-2},r$, and n only. The conditional p.d.f. of Z_1 and Z_2, given $\mathbf{a}=(a_1,\ldots,a_r)$, is also of the form (G5).

Proof The joint p.d.f. of $y_1,\ldots,y_r$ is

$$\frac{n!}{(n-r)!}b^{-r}\left[\prod_{i=1}^{r}g\left(\frac{y_i-u}{b}\right)\right]\left[G\left(\frac{y_r-u}{b}\right)\right]^{n-r}.$$

Make the change of variables from $(y_1,\ldots,y_r)$ to $(\tilde{u},\tilde{b},a_1,\ldots,a_{r-2})$; this transformation can be written as

$$y_i = \tilde{b}a_i + \tilde{u} \qquad i=1,\ldots,r$$

where we note that a_r and a_{r-1} can be expressed in terms of $a_1,\ldots,a_{r-2}$. The Jacobian $\partial(y_1,\ldots,y_r)/\partial(\tilde{u},\tilde{b},a_1,\ldots,a_{r-2})$ is of the form $\tilde{b}^{r-2}k'(\mathbf{a},r,n)$, where k' is a rather complicated function. The joint p.d.f. of $\tilde{u},\tilde{b},a_1,\ldots,a_{r-2}$ is therefore

$$\frac{n!}{(n-r)!}k'(\mathbf{a},r,n)\frac{\tilde{b}^{r-2}}{b^r}\left[\prod_{i=1}^{r} g\left(a_i\frac{\tilde{b}}{b}+\frac{\tilde{u}-u}{b}\right)\right]\left[G\left(a_r\frac{\tilde{b}}{b}+\frac{\tilde{u}-u}{b}\right)\right]^{n-r}.$$

Making the further change of variables from $(\tilde{u},\tilde{b},a_1,\ldots,a_{r-2})$ to $(z_1,z_2,a_1,\ldots,a_{r-2})$, we get (G5). Finally, the conditional p.d.f. of Z_1 and Z_2, given $\mathbf{a}$, is of the same form [though the function $k(\mathbf{a},r,n)$ is different], since dividing (G5) by the joint p.d.f. of $a_1,\ldots,a_{r-2}$ gives a new function of the same form. ■

CONFIDENCE INTERVALS

The pivotals Z_1 and Z_2 can be used to obtain confidence intervals for u and b, respectively. The pth quantile of (G1) is $y_p = u + w_p b$, where w_p satisfies $G(w_p) = 1-p$. Confidence intervals for y_p can be based on the pivotal $Z_p = [w_p b - (\tilde{u}-u)]/\tilde{b} = w_p Z_2^{-1} - Z_1$. This can also be used to get confidence intervals for the survivor function $G[(y-u)/b]$.

There are two main approaches to the construction of confidence intervals. One is to base intervals on the unconditional distributions of pivotals. For example, the probability statement $\Pr(l_1 \leqslant Z_1 \leqslant l_2) = \gamma$ gives $(\tilde{u}-l_2\tilde{b},\ \tilde{u}-l_1\tilde{b})$ as a γ confidence interval for u. Any equivariant estimators $\tilde{u}$ and $\tilde{b}$ can be used, but the properties of the intervals will depend heavily on the estimators. A second approach, first suggested by Fisher (1934), is to base confidence intervals on the conditional distributions of Z_1, Z_2, and Z_p, given the observed value of the ancillary statistic $\mathbf{a}$. For example, the probability statement $\Pr(l_1 \leqslant Z_1 \leqslant l_2 | \mathbf{a}) = \gamma$ gives the confidence interval $(\tilde{u}-l_2\tilde{b}, \tilde{u}-l_1\tilde{b})$ for u. Note that l_1 and l_2 are functions of $\mathbf{a}$ in this case.

A great deal has been written about the two approaches (e.g., see Lawless, 1978, for some comments and references). It seems agreed upon by most that from a purely logical point of view the conditional approach is more appropriate. However, there are sometimes advantages to the unconditional approach. We note a few points of practical importance. First, for all distributions other than the normal the unconditional distributions of pivotals are impossible to obtain analytically; note that to do this involves integrating $a_1, \ldots, a_{r-2}$ out of (G5). The conditional distributions are computationally tractable, however, since $k(\mathbf{a}, r, n)$ in (G5) is a normalizing constant that can be obtained by two-dimensional integration (numerical, if necessary) from the fact that for any $\mathbf{a}$

$$\int_0^\infty \int_{-\infty}^\infty h(z_1, z_2 | \mathbf{a})\, dz_1\, dz_2 = 1$$

where $h(z_1, z_2|\mathbf{a})$ is (G5), the joint p.d.f. of Z_1 and Z_2, given $\mathbf{a}$. Fraser (1979) and Lawless (1978) discuss the computation of probabilities for (G5) in specific situations; see also Section 4.1.1.

Note, however, that if pivotals with good properties are available, and if percentage points for their unconditional distributions can be found (this is often possible by Monte Carlo methods), then these can be tabulated once and for all to provide simple, convenient inference procedures. The conditional approach, on the other hand, requires fresh computation for each new sample.

Three other properties of the conditional approach should be noted.

Property 1 Confidence intervals obtained by the conditional approach are unconditional confidence intervals in the usual sense. We will demonstrate this property for confidence intervals for b. Suppose $[d_1(\mathbf{x}), d_2(\mathbf{x})]$ is a confidence interval for b, obtained from the conditional probability statement $\Pr[l_1 \leqslant \tilde{b}/b \leqslant l_2 | \mathbf{a}] = \gamma$. That is, $d_1(\mathbf{x}) = \tilde{b}/l_2$ and $d_2(\mathbf{x}) = \tilde{b}/l_1$; note that l_1 and l_2 are functions of $\mathbf{a}$. Then

$$\begin{aligned} \Pr[d_1(\mathbf{x}) \leqslant b \leqslant d_2(\mathbf{x})] &= E_{\mathbf{a}}\{\Pr[d_1(\mathbf{x}) \leqslant b \leqslant d_2(\mathbf{x}) | \mathbf{a}]\} \\ &= E_{\mathbf{a}}(\gamma) \\ &= \gamma. \end{aligned}$$

Property 2 Level γ confidence intervals for u, b, or x_p constructed by the conditional method are numerically equivalent to level γ Bayes posterior probability intervals, obtained with the improper prior distribution $b^{-1} du\, db$ for u and b. This is readily established; see, for example, Hora and Buehler (1966) and Lawless (1973).

Property 3 Different equivariant estimators lead to the same confidence intervals, with the conditional approach. This follows from Property 2, since any conditional confidence limit produced is numerically equivalent to a unique Bayes posterior probability limit. This property does not hold for the unconditional method, where properties of confidence intervals depend heavily on the estimators used to form the pivotals.

REGRESSION MODELS

These results generalize in a straightforward way for linear regression models (e.g., Verhagen, 1961; Fraser, 1979). Suppose that Y_i is a response variable, $\mathbf{x}_i=(x_{i1},\dots,x_{ik})$ is a regressor variable, and the p.d.f. of Y_i, given $\mathbf{x}_i$, is of the form

$$f(y_i|\mathbf{x}_i;\boldsymbol{\beta},b)=\frac{1}{b}g\left(\frac{y_i-\mathbf{x}_i\boldsymbol{\beta}}{b}\right) \qquad -\infty<y_i<\infty \tag{G6}$$

where $\boldsymbol{\beta}=(\beta_1,\dots,\beta_k)'$ is a vector of unknown regression coefficients. Suppose that a random sample $y_1,\dots,y_n$ is taken from (G6), corresponding to fixed regressor variable vectors $\mathbf{x}_1,\dots,\mathbf{x}_n$. Let $\mathbf{X}$ be the $n\times k$ matrix with rows $\mathbf{x}_1,\dots,\mathbf{x}_n$ and let $\mathbf{y}=(y_1,\dots,y_n)'$; estimators $\tilde{\boldsymbol{\beta}}=\tilde{\boldsymbol{\beta}}(\mathbf{y})$ and $\tilde{b}=\tilde{b}(\mathbf{y})$ of $\boldsymbol{\beta}$ and b are now called equivariant if for any real vector $\mathbf{c}=(c_1,\dots,c_k)'$ and scalar d ($d>0$)

$$\tilde{\boldsymbol{\beta}}(d\mathbf{y}+\mathbf{X}\mathbf{c})=d\tilde{\boldsymbol{\beta}}(\mathbf{y})+\mathbf{c} \tag{G7}$$

$$\tilde{b}(d\mathbf{y}+\mathbf{X}\mathbf{c})=d\tilde{b}(\mathbf{y}). \tag{G8}$$

The least squares and maximum likelihood estimators are readily shown to be equivariant (e.g., Verhagen, 1961).

Theorem G2 generalizes for uncensored samples under the regression model (G6) to give

THEOREM G4 *Let $\tilde{\boldsymbol{\beta}}$ and $\tilde{b}$ be any equivariant estimators of $\boldsymbol{\beta}$ and b, based on a random sample as defined above. Then* (1) $\mathbf{Z}_1=(\tilde{\boldsymbol{\beta}}-\boldsymbol{\beta})/\tilde{b}$ *and* $Z_2=\tilde{b}/b$ *are pivotals, and* (2) *the quantities* $a_i=(y_i-\mathbf{x}_i\tilde{\boldsymbol{\beta}})/\tilde{b}$, $i=1,\dots,n$ *are ancillaries, only* $n-k$ *of which are functionally independent.*

In addition, the final part of Theorem G3 generalizes to

THEOREM G5 *Under the same conditions as in Theorem G4, the joint p.d.f. of* $\mathbf{Z}_1$, Z_2, *and* $\mathbf{a}=(a_1,\dots,a_n)$ *is of the form*

$$k(\mathbf{a},\mathbf{X},n)z_2^{n-1}\prod_{i=1}^{n}g(a_iz_2+\mathbf{x}_i\mathbf{z}z_2). \tag{G9}$$

The conditional p.d.f. of $\mathbf{Z}_1$ *and* Z_2*, given* $\mathbf{a}$*, is of the same form.*

As in the ordinary location–scale case, inferences about $\boldsymbol{\beta}$ and b can be based on the unconditional distribution of $\mathbf{Z}_1$ and Z_2 or on the conditional distribution of $\mathbf{Z}_1$ and Z_2, given $\mathbf{a}$. Except when (G6) is a normal model, the conditional approach is usually not feasible, however, because of the necessity of performing two- or higher-dimensional numerical integration.

Bibliography

Aalen, O. (1976). Nonparametric inference in connection with multiple decrement models. *Scand. J. Stat.*, **3**, 15–27.

Aalen, O. (1978). Nonparametric inference for a family of counting processes. *Ann. Stat.*, **6**, 701–726.

Abramowitz, M., and I. A. Stegun, Eds. (1965). *Handbook of Mathematical Functions*. New York: Dover.

Aitkin, M., and D. G. Clayton (1980). The fitting of exponential, Weibull and extreme value distributions to complex censored survival data using GLIM. *Appl. Stat.*, **29**, 156–163.

Aitkin, M., and G. T. Wilson (1980). Mixture models, outliers and the EM algorithm. *Technometrics*, **22**, 325–331.

Altschuler, B. (1970). Theory for the measurement of competing risks in animal experiments. *Math. Biosci.*, **6**, 1–11.

Anscombe, F. J. (1961). Estimating a mixed-exponential response law. *J. Am. Stat. Assoc.*, **56**, 493–502.

Anscombe, F. J. (1964). Normal likelihood functions. *Ann. Inst. Stat. Math (Tokyo)*, **16**, 1–19.

Aranda-Ordaz, F. J. (1980). Extensions of the proportional hazards model. Paper presented at the 1980 American Statistical Association Meetings, Houston, Texas, August 11–14, 1980.

Armitage, P. (1975). *Sequential Medical Trials*, 2nd ed. New York: Wiley.

Armitage, P. (1979). The design of clinical trials. *Aust. J. Statist.*, **21**, 266–281.

Armitage, P., and E. A. Gehan (1974). Statistical methods for the identification and use of prognostic factors. *Int. J. Cancer*, **13**, 16–36.

Aroian, L. A. (1976). Applications of the direct method in sequential analysis. *Technometrics*, **18**, 301–306.

Ascher, H., and H. Feingold (1978). Is there repair after failure? In *Proceedings of the 1978 Annual Reliability and Maintainability Symposium*, 190–197. New York: Institute of Electrical and Electronics Engineers.

Ascher, H., and H. Feingold (1979). The air conditioner data revisited. In *Proceedings of the 1979 Annual Reliability and Maintainability Symposium (IEEE)*, 153–159.

Atiqullah, M. (1962). The estimation of residual variance in quadratically balanced least squares problems and the robustness of the F-test. *Biometrika*, **49**, 83–91.

Bailey, K. R. (1979). The general maximum likelihood approach to the Cox regression model. Unpublished Ph.D. Thesis. University of Chicago, Illinois.

Bain, L. J. (1972). Inferences based on censored sampling from the Weibull or extreme value distribution. *Technometrics*, **14**, 693–702.

Bain, L. J. (1974). Analysis for the linear failure rate distribution. *Technometrics*, **16**, 551–559.

Bain, L. J. (1978). *Statistical Analysis of Reliability and Life-Testing Models*. New York: Dekker.

Bain, L. J., and M. Engelhardt (1975). A two-moment chi-square approximation for the statistic $\log(\bar{x}/\tilde{x})$. *J. Am. Stat. Assoc.*, **70**, 948–950.

Bain, L. J., and M. Engelhardt (1980). Probability of correct selection of Weibull versus gamma based on likelihood ratio. *Commun. Statist.*, **A9**, 375–381.

Barlow, R. E., and F. Proschan (1965). *Mathematical Theory of Reliability*. New York: Wiley.

Barlow, R. E., and F. Proschan (1975). *Statistical Theory of Reliability and Life Testing*. New York: Holt, Rinehart & Winston.

Barlow, R. E., A. W. Marshall, and F. Proschan (1963). Properties of probability distributions with monotone failure rate. *Ann. Math. Stat.*, **34**, 375–389.

Barlow, R. E., E. Proschan, A. Madansky, and E. M. Scheuer (1968). Estimation procedures for the "burn-in process." *Technometrics*, **10**, 51–62.

Barnett, V. (1975). Probability plotting methods and order statistics. *Appl. Stat.*, **24**, 95–108.

Barr, D. R., and T. Davidson (1973). A Kolmogorov–Smirnov test for censored samples. *Technometrics*, **15**, 739–757.

Bartholomew, D. J. (1957). A problem in life testing. *J. Am. Stat. Assoc.*, **52**, 350–355.

Bartholomew, D. J. (1963). The sampling distribution of an estimate arising in life testing. *Technometrics*, **5**, 361–374.

Bartholomew, D. J. (1977). The Analysis of Data Arising From Stochastic Processes. In *The Analysis of Survey Data*, Vol. 2, C. A. O'Muircheartaigh and C. Payne, Eds. New York: Wiley.

Bartlett, M. S. (1937). Properties of sufficiency and statistical tests. *Proc. R. Soc. London A*, **160**, 268–282.

Bartlett, M. S., and D. G. Kendall (1946). The statistical analysis of variance-heterogeneity and the logarithmic transformation. *J.R. Stat. Soc. Suppl.*, **8**, 128–138.

Bartolucci, A., and J. M. Dickey (1977). Comparative Bayesian and traditional inference for gamma-modelled survival data. *Biometrics*, **33**, 343–354.

Basu, A. P., and J. K. Ghosh (1980). Asymptotic properties of a solution to the likelihood equation with life-testing applications. *J. Am. Stat. Assoc.*, **75**, 410–414.

Batchelor, J. R., and M. Hackett (1970). HLA matching in treatment of burned patients with skin allografts. *Lancet*, **2**, 581–583.

Berkson, J., and R. P. Gage (1950). Calculation of survival rates for cancer. *Proc. Staff Meet. Mayo Clin.*, **25**, 270–286.

Berlin, B., J. Brodsky, and P. Clifford (1979). Testing disease dependence in survival experiments with serial sacrifice. *J. Am. Stat. Assoc.*, **74**, 5–14.

Berretoni, J. N. (1964). Practical applications of the Weibull distribution. *Ind. Qual. Control*, **21**, 71–79.

Bhattacharya, G. K., R. A. Johnson, and H. R. Neave (1970). Percentage points of some nonparametric tests for independence and empirical power comparisons. *J. Am. Stat. Assoc.*, **65**, 976–983.

Billmann, B., C. Antle, and L. J. Bain (1972). Statistical inferences from censored Weibull samples. *Technometrics*, **14**, 831–840.

Bishop, Y. M. M., S. E. Fienberg, and P. W. Holland (1975). *Discrete Multivariate Analysis*. Cambridge, Massachusetts: Massachusetts Institute of Technology Press.

Blom, G. (1958). *Statistical Estimates and Transformed Beta-Variables*. New York: Wiley.

Boag, J. W. (1949). Maximum likelihood estimates of the proportion of patients cured by cancer therapy. *J.R. Stat. Soc. B*, **11**, 15–53.

Bogdanoff, D., and D. A. Pierce (1973). Bayes-fiducial inference for the Weibull distribution. *J. Am. Stat. Assoc.*, **68**, 659–664.

Braun, H. (1980). A simple method for testing goodness of fit in the presence of nuisance parameters. *J.R. Stat. Soc. B.*, **42**, 53–63.

Breslow, N. E. (1970). A generalized Kruskal–Wallis test for comparing k samples subject to unequal patterns of censorship. *Biometrika*, **57**, 579–594.

Breslow, N. E. (1974). Covariance analysis of censored survival data. *Biometrics*, **30**, 89–99.

Breslow, N. E. (1975). Analysis of survival data under the proportional hazards model. *Int. Stat. Rev.*, **43**, 45–58.

Breslow, N. E., and J. Crowley (1974). A large sample study of the life table and product limit estimates under random censorship. *Ann. Stat.*, **2**, 437–453.

Breslow, N., and C. Haug (1972). Sequential comparison of exponential survival curves. *J. Am. Stat. Assoc.*, **67**, 691–697.

Brindley, E. C., and W. A. Thompson (1972). Dependence and aging aspects of multivariate survival. *J. Am. Stat. Assoc.*, **67**, 822–830.

Brown, B. W., M. Hollander, and R. M. Korwan (1974). Nonparametric tests of independence for censored data with applications to heart transplant studies. In *Reliability and Biometry* (F. Proschan and R. J. Serfling, Eds.). Philadelphia: SIAM.

Brown, M. (1972). Statistical analysis of non-homogeneous Poisson processes. In *Stochastic Point Processes*, P. A. W. Lewis, Ed. New York: Wiley.

Bryant, C. M., and J. Schmee (1979). Confidence limits of MTBF for sequential test plans of MIL-STD 781. *Technometrics*, **21**, 33–42.

Bryson, M. C., and M. M. Siddiqui (1969). Some criteria for aging. *J. Am. Stat. Assoc.*, **64**, 1472–1483.

Buckland, W. R. (1964). *Statistical Assessment of the Life Characteristics*. London: Griffin.

Buckley, T., and I. James (1979). Linear regression with censored data. *Biometrika*, **66**, 429–436.

Campbell, G. (1979). Nonparametric estimation with randomly censored data. Unpublished manuscript.

Canfield, R. V., and L. E. Borgman (1975). Some distributions of time to failure for reliability applications. *Technometrics*, **17**, 263–268.

Carbone, P. O., L. E. Kellerhouse, and E. A. Gehan (1967). Plasmacytic myeloma: a study of the relationship of survival to various clinical manifestations and anomolous protein type in 112 patients. *Am. J. Med.*, **42**, 937–948.

Chambers, J. M. (1977). *Computational Methods for Data Analysis*. New York: Wiley.

Chao, M., and R. E. Glaser (1978). The exact distribution of Bartlett's test statistic for homogeneity of variances with unequal sample sizes. *J. Am. Stat. Assoc.*, **73**, 422–426.

Chapman, J. A. (1976). A comparison of the X^2, $-2\log r$, and multinomial probability criteria for significance tests when expected frequencies are small. *J. Am. Stat. Assoc.*, **71**, 854–863.

Chernoff, J., J. Gastwirth, and M. Johns (1967). Asymptotic distribution of linear combinations of functions of order statistics with applications to estimation. *Ann. Math. Stat.*, **38**, 52–72.

Chhikara, R. S., and J. L. Folks (1977). The inverse Gaussian distribution as a lifetime model. *Technometrics*, **19**, 461–468.

Chiang, C. L. (1960a). A stochastic study of the life table and its applications: I. Probability distributions of the biometric functions. *Biometrics*, **16**, 618–635.

Chiang, C. L. (1960b). A stochastic study of the life table and its applications: II. Sample variance of the observed expectation of life and other biometric functions. *Hum. Biol.*, **32**, 221–238.

Chiang, C. L. (1968). *Introduction to Stochastic Processes in Biostatistics*. New York: Wiley.

Chiang, C. L. (1980). *An Introduction to Stochastic Processes and Their Applications*. New York: Krieger.

Cohen, A. C. (1961). Tables for maximum likelihood estimates: singly truncated and singly censored samples. *Technometrics*, **3**, 535–541.

Cohen, A. C. (1965). Maximum likelihood estimation in the Weibull distribution based on complete and censored samples. *Technometrics*, **7**, 579–588.

Cohen, A. C. (1975). Multicensored sampling in the three-parameter Weibull distribution. *Technometrics*, **17**, 347–351.

Cohen, A. C., and B. J. Whitten (1980). Estimation in the three-parameter lognormal distribution. *J. Am. Stat. Assoc.*, **75**, 399–404.

Cox, D. R. (1953). Some simple approximate tests for Poisson variates. *Biometrika*, **40**, 354–360.

Cox, D. R. (1959). The analysis of exponentially distributed lifetimes with two types of failure. *J.R. Stat. Soc. B*, **21**, 411–421.

Cox, D. R. (1961). Tests of separate families of hypotheses. *Proc. Fourth Berkeley Symp.*, **1**, 105–123.

Cox, D. R. (1962). Further results on tests of separate families of hypotheses. *J.R. Stat. Soc. B*, **24**, 406–423.

Cox, D. R. (1964). Some applications of exponential ordered scores. *J.R. Stat. Soc. B*, **26**, 103–110.

Cox, D. R. (1970). *The Analysis of Binary Data*. London: Methuen.

Cox, D. R. (1972a). Regression models and life tables (with discussion). *J. R. Stat. Soc. B*, **34**, 187–202.

Cox, D. R. (1972b). The statistical analysis of dependencies in point processes. In *Stochastic Point Processes*, P. A. W. Lewis, Ed. New York: Wiley.

Cox, D. R. (1975). Partial likelihood. *Biometrika*, **62**, 269–276.

Cox, D. R. (1978). Some remarks on the role in statistics of graphical methods. *Appl. Stat.*, **27**, 4–9.

Cox, D. R., and D. V. Hinkley (1968). A note on the efficiency of least squares estimates. *J.R. Stat. Soc. B*, **30**, 284–289.

Cox, D. R., and D. V. Hinkley (1974). *Theoretical Statistics*. London: Chapman & Hall.

Cox, D. R., and P. A. W. Lewis (1966). *The Statistical Analysis of Series of Events*. London: Methuen.

Cox, D. R., and H. D. Miller (1965). *The Theory of Stochastic Processes*. London: Methuen.

Cox, D. R., and E. J. Snell (1968). A general definition of residuals. *J.R. Stat. Soc. B*, **30**, 248–275.

Cramer, H. (1946). *Mathematical Methods of Statistics*. Princeton, New Jersey: Princeton University Press.

Crow, L. H. (1974). Reliability analysis for complex, repairable systems. In *Reliability and Biometry*, F. Proschan and R. J. Serfling, Eds. Philadelphia, Pennsylvania: SIAM.

Crowley, J. (1970). A comparison of several life table methods. Unpublished M. S. Thesis. University of Washington.

Crowley, J. (1974). A note on some recent likelihoods leading to the log rank test. *Biometrika*, **61**, 533–538.

Crowley, J., and N. E. Breslow (1975). Remarks on the conservatism of $\Sigma(0-E)^2/E$ in survival data. *Biometrics*, **31**, 957–961.

Crowley, J., and M. Hu (1977). Covariance analysis of heart transplant data. *J. Am. Stat. Assoc.*, **72**, 27–36.

Crowley, J., and D. R. Thomas (1975). Large sample theory for the log rank test. University of Wisconsin, Department of Statistics, Technical Report No. 415.

D'Agostino, R. B. (1971). Linear estimation of the Weibull parameters. *Technometrics*, **13**, 171–182.

Dahiya, R. C., and J. Gurland (1972). Goodness of fit tests for the gamma and exponential distributions. *Technometrics*, **14**, 791–801.

David, H. A. (1970). *Order Statistics*. New York: Wiley.

David, H. A., and M. L. Moeschberger (1978). *Theory of Competing Risks*. London: Griffin.

Davis, D. J. (1952). An analysis of some failure data. *J. Am. Stat. Assoc.*, **47**, 113–150.

Davis, H. T., and M. L. Feldstein (1979). The generalized Pareto law as a model for progressively censored survival data. *Biometrika*, **66**, 299–306.

Dempster, A. P. (1975). A subjectivist looks at robustness. *Bull. Int. Stat. Inst.*, **46**, 349–374.

Dempster, A. P., N. M. Laird, and D. B. Rubin (1977). Maximum likelihood from incomplete data via the EM algorithm (with discussion). *J. R. Stat. Soc. B*, **39**, 1–38.

Denby, L., and C. L. Mallows (1977). Two diagnostic displays for robust regression analysis. *Technometrics*, **19**, 1–13.

Desu, M. M., and S. C. Narula (1977). Reliability estimation under competing causes of failure. In *The Theory and Applications of Reliability*, Vol. 2, C. P. Tsokos and I. N. Shimi, Eds. New York: Academic.

Dickinson, J. P. (1974). On the resolution of a mixture of observations from two gamma distributions by the method of maximum likelihood. *Metrika*, **21**, 133–141.

Dixon, W. J. (1960). Simplified estimation from censored normal samples. *Ann. Math. Stat.*, **31**, 385–391.

Draper, N. R., and H. Smith, Jr. (1966). *Applied Regression Analysis*. New York: Wiley.

Drolette, M. E. (1975). The effect of incomplete followup. *Biometrics*, **31**, 135–144.

Dufour, R., and U. Maag (1978). Distribution results for modified Kolmogorov–Smirnov statistics for truncated or censored samples. *Technometrics*, **20**, 29–32.

Dumonceaux, R., and C. E. Antle (1973). Discrimination between the log-normal and the Weibull distributions. *Technometrics*, **15**, 923–26.

Dumonceaux, R., C. E. Antle, and G. Haas (1973). Likelihood ratio test for discrimination between two models with unknown location and scale parameters. *Technometrics*, **15**, 19–27.

Dunsmore, I. R. (1978). Some approximations for tolerance factors for the two-parameter exponential distribution. *Technometrics*, **20**, 317–318.

Durbin, J. (1973). *Distribution Theory for Tests Based on the Sample Distribution Function*. Philadelphia, Pennsylvania: SIAM.

Durbin, J. (1975). Kolmogorov–Smirnov tests when parameters are estimated with applications to tests of exponentiality and tests on spacings. *Biometrika*, **62**, 5–22.

Durbin, J. (1976). Kolmogorov–Smirnov tests when parameters are estimated. In *Empirical Distributions and Processes*, P. Ganssler and P. Revesz, Eds. Lecture Notes in Mathematics No. 566. Berlin: Springer-Verlag.

Durbin, J., M. Knott, and C. C. Taylor (1975). Components of Cramer–von Mises statistics II. *J.R. Stat. Soc. B*, **37**, 216–237.

Dyer, A. R. (1973). Discrimination procedures for separate families of hypotheses. *J. Am. Stat. Assoc.*, **68**, 970–974.

Dyer, A. R. (1974). Hypothesis testing procedures for separate families of hypotheses. *J. Am. Stat. Assoc.*, **69**, 140–145.

Dyer, D. D., and J. P. Keating (1980). On the determination of critical values for Bartlett's test. *J. Am. Stat. Assoc.*, **75**, 313–319.

Efron, B. (1967). The two sample problem with censored data. *Proc. Fifth Berkeley Symp.*, **4**, 831–853.

Efron, B. (1977). The efficiency of Cox's likelihood function for censored data. *J. Am. Stat. Assoc.*, **72**, 555–565.

Efron, B., and D. V. Hinkley (1978). Assessing the accuracy of the maximum likelihood estimator: Observed versus expected Fisher information (with discussion). *Biometrika*, **65**, 457–488.

Elandt-Johnson, R. C. (1977). Various estimators of conditional probabilities of death in followup studies: summary of results. *J. Chronic Dis.*, **30**, 246–256.

Elveback, L. (1958). Estimation of survivorship in chronic disease: The "actuarial" method. *J. Am. Stat. Assoc.*, **53**, 420–440.

Engelhardt, M. (1975). On simple estimation of the parameters of the Weibull or extreme value distribution. *Technometrics*, **17**, 369–374.

Engelhardt, M., and L. J. Bain (1973). Some complete and censored sampling results for the Weibull or extreme value distribution. *Technometrics*, **15**, 541–549.

Engelhardt, M., and L. J. Bain (1974). Some results on point estimation for the two-parameter Weibull or extreme value distribution. *Technometrics*, **16**, 49–56.

Engelhardt, M., and L. J. Bain (1975). Tests of two-parameter exponentiality against three-parameter Weibull alternatives. *Technometrics*, **17**, 353–356.

Engelhardt, M., and L. J. Bain (1977a). Simplified statistical procedures for the Weibull or extreme value distribution. *Technometrics*, **19**, 323–331.

Engelhardt, M., and L. J. Bain (1977b). Uniformly most powerful unbiased tests on the scale parameter of a gamma distribution with a nuisance shape parameter. *Technometrics*, **19**, 77–81.

Engelhardt, M., and L. J. Bain (1978a). Construction of optimal unbiased inference procedures for the parameters of the gamma distribution. *Technometrics*, **20**, 485–489.

Engelhardt, M., and L. J. Bain (1978b). Tolerance limits and confidence limits on reliability for the two-parameter exponential distribution. *Technometrics*, **20**, 37–39.

Engelhardt, M., and L. J. Bain (1979). Prediction limits and two-sample problems with complete or censored Weibull data. *Technometrics*, **21**, 233–237.

Epstein, B. (1954). Truncated life tests in the exponential case. *Ann. Math. Stat.*, **25**, 555–564.

Epstein, B. (1958). The exponential distribution and its role in life-testing. *Ind. Qual. Control*, **15**, 2–7.

Epstein, B. (1960a). Statistical life test acceptance procedures. *Technometrics*, **2**, 435–446.

Epstein, B. (1960b). Testing for the validity of the assumption that the underlying distribution of life is exponential. *Technometrics*, **2**, 83–101, 167–183.

Epstein, B., and M. Sobel (1953). Life testing. *J. Am. Stat. Assoc.*, **48**, 486–502.

Epstein, B., and M. Sobel (1954). Some theorems relevant to life testing from an exponential distribution. *Ann. Math. Stat.* **25**, 373–381.

Epstein, B., and M. Sobel (1955). Sequential life tests in the exponential case. *Ann. Math. Stat.*, **26**, 82–93.

Epstein, B., and C. K. Tsao (1953). Some tests based on ordered observations from two exponential populations. *Ann. Math. Stat.*, **24**, 456–466.

Falls, L. W. (1970). Estimation of parameters in compound Weibull distributions. *Technometrics*, **12**, 399–407.

Farewell, V. T., and R. L. Prentice (1977). A study of distributional shape in life testing. *Technometrics*, **19**, 69–75.

Farewell, V. T., and R. L. Prentice (1980). The approximation of partial likelihood with emphasis on case-control studies. *Biometrika*, **67**, 273–278.

Faulkner, J. E., and R. S. McHugh (1972). Bias in observable cancer age and life-time of mice subject to spontaneous mammary carcinomas. *Biometrics*, **28**, 489–498.

Feigl, P., and M. Zelen (1965). Estimation of exponential survival probabilities with concomitant information. *Biometrics*, **21**, 826–838.

Feinleib, M. (1960). A method of analyzing lognormally distributed survival data with incomplete follow-up. *J. Am. Stat. Assoc.*, **55**, 534–545.

Feller, W. (1968). *An Introduction to Probability Theory and Its Applications*, Vol. 1, 3rd ed. New York: Wiley.

Fertig, K. W., and N. R. Mann (1980). Life test sampling plans for two-parameter Weibull populations. *Technometrics*, **22**, 165–177.

Fisher, R. A. (1934). Two new properties of mathematical likelihood. *Proc. R. Soc. A*, **144**, 285–307.

Fleming, T. R., and D. P. Harrington (1980). A class of hypothesis tests for one and two sample censored survival data. School of Engineering and Applied Science, University of Virginia. DAMACS Technical Report No. 80-9. To appear in *Commun. Statist. A*, 1981.

Fleming, T. R., J. R. O'Fallon, P. C. O'Brien, and D. P. Harrington (1980). Modified Kolmogorov–Smirnov test procedures with application to arbitrarily right censored data. *Biometrics*, **36**, 607–626. Also Technical Report Series, Section of Medical Research Statistics, Mayo Clinic.

Fletcher, R., and C. M. Reeves (1964). Function minimization by conjugate gradients. *Comput. J.*, **7**, 149–154.

Folkes, E. B. (1979). Some methods for studying the mixture of two normal (log-normal) distributions. *J. Am. Stat. Assoc.*, **74**, 561–575.

Fraser, D. A. S. (1968). *The Structure of Inference*. New York: Wiley.

Fraser, D. A. S. (1979). *Inference and Linear Models*. New York: McGraw-Hill.

Freireich, E. O. et al. (1963). The effect of 6-mercaptopurine on the duration of steroid induced remission in acute leukemia. *Blood*, **21**, 699–716.

Fryer, J. G., and D. Holt (1970). On the robustness of the standard estimates of the exponential mean to contamination. *Biometrika*, **57**, 641–648.

Fryer, J. G., and D. Holt (1976). On the robustness of the power function of the one-sample test for the negative exponential distribution. *Commun. Statist.*, **A5**, 723–734.

Furth, J., A. C. Upton, and A. W. Kimball (1959). Late pathologic effects of atomic detonation and their pathogenesis. *Radiat. Res. Suppl.*, **1**, 243–264.

Gail, M. H. (1975). A review and critique of some models used in competing risk analysis. *Biometrics*, **31**, 209–222.

Gail, M. H., and J. L. Gastwirth (1978). A scale-free goodness of fit test for the exponential distribution based on the Gini statistic. *J.R. Stat. Soc. B*, **40**, 350–357.

Gail, M. H., and J. Ware (1979). Comparing observed life table data with a known survival curve in the presence of random censorship. *Biometrics*, **35**, 385–391.

Galambos, J. (1978). *The Asymptotic Theory of Extreme Order Statistics*. New York: Wiley.

Gehan, E. A. (1965). A generalized Wilcoxon test for comparing arbitrarily singly-censored samples. *Biometrika*, **52**, 203–223.

Gehan, E. A. (1969). Estimating survival functions from the life table. *J. Chronic Dis.*, **21**, 629–644.

Gehan, E. A., and M. M. Siddiqui (1973). Simple regression methods for survival time studies. *J. Am. Stat. Assoc.*, **68**, 848–856.

Gehan, E. A., and D. G. Thomas (1969). The performance of some two-sample tests in small samples with and without censoring. *Biometrika*, **56**, 127–132.

Glaser, R. E. (1976a). The ratio of the geometric mean to the arithmetic mean for a random sample from a gamma distribution. *J. Am. Stat. Assoc.*, **71**, 480–487.

Glaser, R. E. (1976b). Exact critical values for Bartlett's test for homogeneity of variances. *J. Am. Stat. Assoc.*, **71**, 488–490.

Glaser, R. E. (1980). Bathtub and related failure rate characterizations. *J. Am. Stat. Assoc.*, **75**, 667–672.

Glasser, M. (1965). Regression analysis with dependent variable censored. *Biometrics*, **21**, 300–307.

Glasser, M. (1967). Exponential survival with covariance. *J. Am. Stat. Assoc.*, **62**, 561–568.

Goldthwaite, L. (1961). Failure rate study for the lognormal lifetime model. *Proc. Seventh Nat. Symp. Reliab. Qual. Control*, 208–213.

Govindarajulu, Z. (1964). A supplement to Mendenhall's bibliography on life testing and related topics. *J. Am. Stat. Assoc.*, **59**, 1231–1291.

Greenberg, R., S. Bayard, and D. Byar (1974). Selecting concomitant variables using a likelihood ratio step-down procedure and a method of testing goodness of fit of an exponential survival model. *Biometrics*, **30**, 601–608.

Greenwood, M. (1926). The natural duration of cancer. *Reports of Public Health and Medical Subjects*, Vol. 33. London: Her Majesty's Stationery Office.

Gregory, G. C. (1980). Goodness-of-fit with randomly censored data: tests based on the conditional distribution. Unpublished manuscript.

Griffiths, D. A. (1980). Interval estimation for the three-parametric lognormal distribution via the likelihood function. *Appl. Stat.*, **29**, 58–68.

Gross, A. J., and V. A. Clark (1975). *Survival Distributions: Reliability Applications in the Biomedical Sciences*. New York: Wiley.

Grubbs, F. E. (1971). Fiducial bounds on reliability for the two-parameter negative exponential distribution. *Technometrics*, **13**, 873–876.

Guenther, W. C., S. A. Patil, and V. R. R. Uppuluri (1976). One-sided β-content tolerance factors for the two-parameter exponential distribution. *Technometrics*, **18**, 333–340.

Gumbel, E. J. (1958). *Statistics of Extremes*. New York: Columbia University Press.

Gupta, S. S. (1960). Order statistics from the gamma distribution. *Technometrics*, **2**, 243–262.

Gupta, S. S., and P. A. Groll (1961). Gamma distribution in acceptance sampling based on life tests. *J. Am. Stat. Assoc.*, **56**, 942–970.

Hager, H. W., and L. J. Bain (1970). Inferential procedures for the generalized gamma distribution. *J. Am. Stat. Assoc.*, **65**, 1601–1609.

Hager, H. W., L. J. Bain, and C. E. Antle (1971). Reliability estimation for the generalized gamma distribution and robustness of the Weibull model. *Technometrics*, **13**, 547–558.

Hajek, J. (1969). *A Course in Nonparametric Statistics*. San Francisco: Holden-Day.

Hajek, J., and Z. Sidak (1967). *Theory of Rank Tests*. New York: Academic.

Hall, W. J., and J. A. Wellner (1980). Confidence bands for a survival curve from censored data. *Biometrika*, **67**, 133–143.

Halperin, M. (1952). Maximum likelihood estimation in truncated samples. *Ann. Math. Stat.*, **23**, 226–238.

Harter, H. L. (1967). Maximum-likelihood estimation of the parameters of a four-parameter generalized gamma population for complete and censored samples. *Technometrics*, **9**, 159–165.

Harter, H. L. (1969). *Order Statistics and Their Use in Testing and Estimation*. Washington, D.C.: U.S. Government Printing Office.

Harter, H. L., and A. H. Moore (1965). Maximum likelihood estimation of the parameters of gamma and Weibull populations from complete and from censored samples. *Technometrics*, **7**, 639–643.

Harter, H. L., and A. H. Moore (1967). Asymptotic variances and covariances of maximum likelihood estimators, from censored samples, of the parameters of Weibull and gamma distributions. *Ann. Math. Stat.*, **38**, 557–570.

Harter, H. L., and A. H. Moore (1968). Maximum likelihood estimation, from doubly censored samples, of the parameters of the first asymptotic distribution of extreme values. *J. Am. Stat. Assoc.*, **63**, 889–901.

Harter, H. L., and A. H. Moore (1976). An evaluation of exponential and Weibull test plans. *IEEE Trans. Reliab.*, **R25**, 100–104.

Hegazy, Y. A. S., and J. R. Green (1975). Some new goodness of fit tests using order statistics. *Appl. Stat.*, **24**, 299–308.

Hensler, G. L., K. G. Mehrotra, and J. E. Michalek (1977). A note on some exact efficiency calculations for survival distributions. *Biometrika*, **64**, 635–637.

Hill, B. M. (1963a). The three-parameter lognormal distribution and Bayesian analysis of a point-source epidemic. *J. Am. Stat. Assoc.*, **58**, 72–84.

Hill, B. M. (1963b). Information for estimating the proportions in mixtures of exponential and normal distributions. *J. Am. Stat. Assoc.*, **58**, 918–932.

Hinkley, D. V. (1978). Likelihood inference about location and scale parameters. *Biometrika*, **65**, 253–262.

Hjorth, J. (1980). A reliability distribution with increasing, decreasing, constant, and bathtub-shaped failure rates. *Technometrics*, **22**, 99–108.

Hoel, D. G. (1972). A representation of mortality by competing risks. *Biometrics*, **28**, 475–488.

Hogg, R. V. (1956). On the distribution of the likelihood ratio. *Ann. Math. Stat.*, **27**, 529–532.

Hogg, R. V., V. A. Uthoff, R. H. Randles, and A. S. Davenport (1972). On the selection of the underlying distribution and adaptive estimation, *J. Am. Stat. Assoc.*, **67**, 597–600.

Holford, T. R. (1976). Life tables with concomitant information. *Biometrics*, **32**, 587–597.

Holt, J. D. (1978). Competing risk analysis with special reference to matched pair experiments. *Biometrika*, **65**, 159–166.

Holt, J. D., and R. L. Prentice (1974). Survival analysis in twin studies and matched pair experiments. *Biometrika*, **61**, 17–30.

Hora, R. B., and R. J. Buehler (1966). Fiducial theory and invariant estimation. *Ann. Math. Stat.*, **37**, 643–656.

Irwin, J. O. (1942). The distribution of the logarithm of survival times when the true law is exponential. *J. Hyg.*, **42**, 328–333.

Jackson, O. A. Y. (1968). Some results on tests of separate families of hypotheses. *Biometrika*, **55**, 355–363.

Jackson, O. A. Y. (1969). Fitting a gamma or log-normal distribution to fibre-diameter measurements on wool tops. *Appl. Stat.* **18**, 70–75.

Jardine, A. K. S. (1979). Solving industrial replacement problems. *Proceedings of the 1979 Annual Reliability and Maintainability Symposium*, 136–141. New York: Institute of Electrical and Electronics Engineers.

Johansen, S. (1978). The product limit estimate as a maximum likelihood estimate. *Scand. J. Stat.*, **5**, 195–199.

Johnson, N. L. (1977). Approximate relationships among some estimators of mortality probabilities. *Biometrics*, **33**, 542–545.

Johnson, N. L., and S. Kotz (1969). *Discrete Distributions*. Boston, Massachusetts: Houghton Mifflin.

Johnson, N. L., and S. Kotz (1970). *Continuous Univariate Distributions*, Vols. 1 and 2. Boston, Massachusetts: Houghton Mifflin.

Johnson, N. L., and S. Kotz (1972). *Distributions in Statistics: Continuous Multivariate Distributions*. New York: Wiley.

Johnson, N. L., and S. Kotz (1975). A vector multivariate hazard rate. *J. Multivar. Anal.*, **5**, 53–66.

Johnson, R. A. (1974). Some optimality results for one and two sample procedures based on the smallest *r* order statistics. In *Reliability and Biometry*, F. Proschan and R. J. Serfling, Eds. Philadelphia, Pennsylvania: SIAM.

Jones, D. R., and J. Whitehead (1979). Sequential forms of the log rank and modified Wilcoxon tests for censored data. *Biometrika*, **66**, 105–113.

Kalbfleisch, J. D. (1974). Some efficiency calculations for survival distributions. *Biometrika*, **61**, 31–38.

Kalbfleisch, J. D., and R. J. MacKay (1978). Censoring and the immutable likelihood. University of Waterloo, Department of Statistics, Technical Report 78-09.

Kalbfleisch, J. D., and R. J. MacKay (1980). Time-dependent covariates. Unpublished manuscript.

Kalbfleisch, J. D., and A. McIntosh (1977). Efficiency in survival distributions with time-dependent covariables. *Biometrika*, **64**, 47–50.

Kalbfleisch, J. D., and R. L. Prentice (1973). Marginal likelihoods based on Cox's regression and life model. *Biometrika*, **60**, 267–279.

Kalbfleisch, J. D., and R. L. Prentice (1980). *The Statistical Analysis of Failure Time Data*. New York: Wiley.

Kalbfleisch, J. D., and D. A. Sprott (1970). Application of likelihood methods to models involving a large number of parameters (with discussion). *J.R. Stat. Soc. B*, **32**, 175–208.

Kalbfleisch, J. D., and D. A. Sprott (1974). Marginal and conditional likelihoods. *Sankhya A*, **35**, 311–328.

Kao, J. H. K. (1959). A graphical estimation of mixed Weibull parameters in life testing of electron tubes. *Technometrics*, **1**, 389–407.

Kao, P., E. P. C. Kao, and J. M. Mogg (1979). A simple procedure for computing performance characteristics of truncated sequential tests with exponential lifetimes. *Technometrics*, **21**, 229–232.

Kaplan, E. L., and P. Meier (1958). Nonparametric estimation from incomplete observations. *J. Am. Stat. Assoc.*, **53**, 457–481.

Karlin, S., and H. M. Taylor (1975). *A First Course in Stochastic Processes*. New York: Academic.

Kay, R. (1977). Proportional hazard regression models and the analysis of censored survival data. *Appl. Stat.*, **26**, 227–237.

Kay, R. (1979). Some further asymptotic efficiency calculations for survival data regression models. *Biometrika*, **66**, 91–96.

Kempthorne, O. (1952). *Design and Analysis of Experiments*. New York: Wiley.

Kempthorne, O., and J. L. Folks (1971). *Probability, Statistics and Data Analysis*. Ames, Iowa: Iowa State University Press.

Kendall, M. G., and A. Stuart (1967). *The Advanced Theory of Statistics*, Vol. 2, 2nd ed. London: Griffin.

Kendall, M. G., and A. Stuart (1968). *The Advanced Theory of Statistics*, Vol. 3, 2nd ed. London: Griffin.

Kendall, M. G., and A. Stuart (1969). *The Advanced Theory of Statistics*, Vol. 1, 3rd ed. London: Griffin.

Kimball, A. W. (1960). Estimation of mortality intensities in animal experiments. *Biometrics*, **16**, 505–521.

Klimko, L., C. Antle, A. Rademaker, and H. Rockette (1975). Upper bounds for the power of invariant tests for the exponential distribution with Weibull alternatives. *Technometrics*, **17**, 357–360.

Kodlin, D. (1967). A new response time distribution. *Biometrics*, **23**, 227–239.

Kopecky, K. J., and D. A. Pierce (1979a). Efficiency of smooth goodness of fit tests. *J. Am. Stat. Assoc.*, **74**, 393–397.

Kopecky, K. J., and D. A. Pierce (1979b). Testing goodness of fit for the distribution of errors in regression models. *Biometrika*, **66**, 1–5.

Koziol, J. A., and D. P. Byar (1975). Percentage points of the asymptotic distributions of one and two sample $K-S$ statistics for truncated or censored data. *Technometrics*, **17**, 507–510.

Koziol, J. A., and S. B. Green (1976). A Cramer–von Mises statistic for randomly censored data. *Biometrika*, **63**, 465–474.

Koziol, J. A., and A. J. Petkau (1978). Sequential testing of the equality of two survival distributions using the modified Savage statistic. *Biometrika*, **65**, 615–623.

Krall, J., V. Uthoff, and J. Harley (1975). A step-up procedure for selecting variables associated with survival. *Biometrics*, **31**, 49–57.

Kramer, H. C., and M. Paik (1979). A central t approximation to the noncentral t distribution. *Technometrics*, **21**, 357–360.

Krane, S. A. (1963). Analysis of survival data by regression techniques. *Technometrics*, **5**, 161–174.

Kruskal, W. H., and W. A. Wallis (1952). Use of ranks in one-criterion analysis of variance. *J. Am. Stat. Assoc.*, **47**, 583–621.

Kumar, S., and H. I. Patel (1971). A test for the comparison of two exponential distributions. *Technometrics*, **13**, 183–189.

Kuzma, J. W. (1967). A comparison of two life table methods. *Biometrics*, **23**, 51–64.

Lagakos, S. W. (1979). General right censoring and its impact on the analysis of survival data. *Biometrics*, **35**, 139–156.

Lagakos, S. W., C. J. Sommer, and M. Zelen (1978). Semi-Markov models for partially censored data. *Biometrika*, **65**, 311–317.

Larntz, K. (1978). Small-sample comparisons of exact levels for chi-squared goodness of fit statistics. *J. Am. Stat. Assoc.*, **73**, 253–263.

Latta, R. B. (1977). Generalized Wilcoxon statistics for the two-sample problem with censored data. *Biometrika*, **64**, 633–635.

Lawless, J. F. (1971). A prediction problem concerning samples from the exponential distribution, with application in life testing. *Technometrics*, **4**, 725–730.

Lawless, J. F. (1972). Confidence interval estimation for the parameters of the Weibull distribution. *Utilitas Math.*, **2**, 71–87.

Lawless, J. F. (1973). On the estimation of safe life when the underlying life distribution is Weibull. *Technometrics*, **15**, 857–865.

Lawless, J. F. (1975). Construction of tolerance bounds for the extreme value and Weibull distributions. *Technometrics*, **17**, 255–261.

Lawless, J. F. (1976). Confidence interval estimation in the inverse power law model. *Appl. Stat.*, **25**, 128–138.

Lawless, J. F. (1978). Confidence interval estimation for the Weibull and extreme value distributions. *Technometrics*, **20**, 355–364.

Lawless, J. F. (1979). Statistical inference for Type I censored data from the exponential and Weibull distributions. Unpublished manuscript.

Lawless, J. F. (1980). Inference in the generalized gamma and log gamma distributions. *Technometrics*, **22**, 409–419.

Lawless, J. F., and N. R. Mann (1976). Tests for homogeneity for extreme value scale parameters. *Commun. Stat.*, **A5**, 389–405.

Lawless, J. F., and K. Singhal (1978). Efficient screening of nonnormal regression models. *Biometrics*, **34**, 318–327.

Lawless, J. F., and K. Singhal (1980). Analysis of data from life-test experiments under an exponential model. *Nav. Res. Log. Q.*, **27**, 323–334.

Lee, L. (1980). Testing adequacy of the Weibull and log linear rate models for a Poisson process. *Technometrics*, **22**, 195–200.

Lee, L, and S. K. Lee (1978). Some results on inference for the Weibull process. *Technometrics*, **20**, 41–45.

Lee, E. T., M. M. Desu, and E. A. Gehan (1975). A Monte Carlo study of the power of some two-sample tests. *Biometrika*, **62**, 425–432.

Lee, S. C. S., C. Locke, and J. D. Spurrier (1980). On a class of tests of exponentiality. *Technometrics*, **22**, 547–554.

Lehmann, E. L. (1959). *Testing Statistical Hypotheses*. New York: Wiley.

Lehmann, E. L. (1975). *Nonparametrics: Statistical Methods Based on Ranks*. San Francisco, California: Holden-Day.

Lemon. G. (1975). Maximum likelihood estimation for the three-parameter Weibull distribution, based on censored samples. *Technometrics*, **17**, 247–254.

Lewis, P. A. W., and D. Robinson (1974). Testing for a monotone trend in a modulated renewal analysis. In *Reliability and Biometry*, F. Proschan and R. J. Serfling, Eds. Philadelphia, Pennsylvania: SIAM.

Lieblein, J., and M. Zelen (1956). Statistical investigation of the fatigue life of deep groove ball bearings. *J. Res. Nat. Bur. Stand.*, **57**, 273–316.

Lin, C., and G. S. Mudholkar (1980). A simple test for normality against asymmetric alternatives. *Biometrika*, **67**, 455–461.

Ling, R. F. (1978). A study of the accuracy of some approximations for t, chi^2 and F tail probabilities. *J. Am. Stat. Assoc.*, **73**, 274–283.

Linhart, H. (1965). Approximate confidence limits for the coefficient of variation of gamma distributions. *Biometrics*, **21**, 733–738.

Lininger, L., M. Gail, S. Green, and D. Byar (1979). Comparison of four tests for equality of survival curves in the presence of stratification and censoring. *Biometrika*, **66**, 419–428.

Littell, A. S. (1952). Estimation of the T-year survival rate from followup studies over a limited period of time. *Hum. Biol.* **24**, 87–116.

Littel, R. C., and J. L. Folks (1971). Asymptotic optimality of Fisher's method of combining independent tests. *J. Am. Stat. Assoc.*, **66**, 802–806.

Littel, R. C., J. T. McClave, and W. W. Offen (1979). Goodness of fit tests for two-parameter Weibull distribution. *Commun. Stat.*, **B8**, 257–269.

Liu, P. Y., and J. Crowley (1978). Large sample properties of the maximum partial likelihood estimate based on Cox's regression model for censored survival data. University of Wisconsin, Department of Biostatistics, Wisconsin Clinical Career Center, Technical Report No. 1.

Lloyd, E. H. (1952). Least-squares estimation of location and scale parameters using order statistics. *Biometrika*, **39**, 88–95.

Locks, M. O., M. S. Alexander, and B. J. Byars (1963). New Tables of the Noncentral t-Distribution. Wright-Patterson Air Force Base, Ohio, ARL Technical Report ARL63-19.

Louis, T. A. (1977). Sequential allocation in clinical trials comparing two exponential curves. *Biometrics*, **33**, 627–634.

Loynes, R. M. (1980). The empirical distribution function of residuals from generalized regression. *Ann. Stat.*, **8**, 285–298.

MacKay, R. J. (1979). A comparison of some experimental designs for the one-parameter exponential distribution. Unpublished manuscript.

Mann, N. R. (1967a). Tables for obtaining the best linear invariant estimates of parameters of the Weibull distribution. *Technometrics*, **9**, 629–645.

Mann, N. R. (1967b). Results on location and scale parameter estimation with application to the extreme value distribution. Wright-Patterson Air Force Base, Ohio, ARL Technical Report ARL 67-0023.

Mann, N. R. (1968a). Point and interval estimation procedures for the two-parameter Weibull and extreme value distributions. *Technometrics*, **10**, 231–256.

Mann, N. R. (1968b). Results on statistical estimation and hypothesis testing with application to the Weibull and extreme value distributions. Wright-Patterson Air Force Base, Ohio, ARL Technical Report ARL 68-0068.

Mann, N. R. (1969). Optimum estimators for linear functions of location and scale parameters. *Ann. Math. Stat.*, **40**, 2149–55.

Mann, N. R. (1972). Design of over-stress life-test experiments when failure times have the two-parameter Weibull distribution. *Technometrics*, **14**, 437–451.

Mann, N. R. (1977). An *F* approximation for two-parameter Weibull and log-normal tolerance bounds based on possibly censored data. *Nav. Res. Log. Q.*, **24**, 187–196.

Mann, N. R. (1978). Calculation of small-sample Weibull tolerance bounds for accelerated testing. *Commun. Stat.*, **A7**, 97–112.

Mann, N. R., and K. W. Fertig (1973). Tables for obtaining confidence bounds and tolerance bounds based on best linear invariant estimates of parameters of the extreme value distribution. *Technometrics*, **15**, 87–101.

Mann, N. R., and K. W. Fertig (1975a). Simplified efficient point and interval estimators for Weibull parameters. *Technometrics*, **17**, 361–368.

Mann, N. R., and K. W. Fertig (1975b). A goodness-of-fit test for the two parameter vs. three parameter Weibull; confidence bounds for the threshold. *Technometrics*, **17**, 237–245.

Mann, N. R., and K. W. Fertig (1977). Efficient unbiased quantile estimators for moderate-size complete samples from extreme value and Weibull distributions: confidence bounds and tolerance and prediction intervals. *Technometrics*, **19**, 87–93.

Mann, N. R., and K. W. Fertig (1980). An accurate approximation to the sampling distribution of the studentized extreme value statistic. *Technometrics*, **22**, 83–98.

Mann, N. R., K. W. Fertig, and E. M. Scheuer (1971). Confidence and tolerance bounds and a new goodness of fit test for the two-parameter Weibull or extreme value distribution with tables for censored samples of size 3(1)25. Wright-Patterson Air force Base, Ohio, ARL Technical Report ARL 71-0077.

Mann, N. R., E. M. Scheuer, and K. W. Fertig (1973). A new goodness of fit test for the two-parameter Weibull or extreme value distribution. *Commun. Stat.*, **2**, 383–400.

Mann, N. R., R. E. Schafer, and N. D. Singpurwalla (1974). *Methods for Statistical Analysis of Reliability and Lifetime Data*. New York: Wiley.

Mantel, N. (1963). Chi-square tests with one degree of freedom: extensions of the Mantel–Haenszel procedure. *J. Am. Stat. Assoc.*, **58**, 690–700.

Mantel, N. (1966). Evaluation of survival data and two new rank order statistics arising in its consideration. *Cancer Chemother. Rep.*, **50**, 163–170.

Mantel, N. (1967). Ranking procedures for arbitrarily restricted observations. *Biometrics*, **23**, 65–78.

Mantel, N., and W. Haenszel (1959). Statistical aspects of the analysis of data from restrospective studies of disease. *J. Nat. Cancer Inst.*, **22**, 719–748.

Mantel, N., and M. Myers (1971). Problems of convergence of maximum likelihood iterative procedures in multiparameter situations. *J. Am. Stat. Assoc.*, **66**, 484–491.

Mardia, K. V., and P. J. Zemroch (1978). *Tables of the F- and Related Distributions with Algorithms*. New York: Academic.

Margolin, B. H., and W. Maurer (1976). Tests of the Kolmogorov–Smirnov type for exponential data with unknown scale and related problems. *Biometrika*, **63**, 149–160.

Matthews, D. E. (1980). Some observations on semi-Markov models for partially censored data. Unpublished manuscript.

McCool, J. I. (1970). Inferences on Weibull percentiles and shape parameter for maximum likelihood estimates. *IEEE Trans. Reliab.*, **R19**, 2–9.

McCool, J. I. (1974). Inferential techniques for Weibull populations. Wright-Patterson Air Force Base, Ohio, ARL Technical Report ARL-74-0180.

McCool, J. I. (1975a). Multiple comparisons for Weibull parameters. *IEEE Trans. Reliab.*, **R24**, 186–192.

McCool, J. I. (1975b). Inferential techniques for Weibull populations II. Wright-Patterson Air Force Base, Ohio. ARL Technical Report ARL-75-0233.

McCool, J. I. (1979). Analysis of single classification experiments based on censored samples from the two-parameter Weibull distribution. *J. Stat. Plan. Infer.*, **3**, 39–68.

McCool, J. I. (1980). Confidence limits for Weibull regression with censored data. *IEEE Trans. Reliab.*, **R29**, 145–150.

Meeker, W. Q., and W. Nelson (1975). Estimation of Weibull percentiles from censored data. *IEEE Trans. Reliab.*, **R25**, 20–24.

Meeker, W. Q., and W. Nelson (1977). Weibull variances and confidence limits by maximum likelihood for singly censored data. *Technometrics*, **19**, 473–476.

Mehrotra, K. G., R. A. Johnson, and G. K. Bhattacharya (1977). Locally most powerful rank tests for multiple-censored data. *Commun. Stat.*, **A6**, 459–469.

Meier, P. (1975). Estimation of a distribution function from incomplete observations. In *Perspectives in Probability and Statistics*, J. Gani, Ed. Sheffield, England.: Applied Probability Trust.

Mendenhall, W. (1958). A bibliography on life testing and related topics. *Biometrika*, **45**, 521–543.

Mendenhall, W., and R. J. Hader (1958). Estimation of parameters of mixed exponential distributed failure times from censored life test data. *Biometrika*, **45**, 504–520.

Michael, J. R., and W. R. Schucany (1979). A new approach to testing goodness of fit with censored samples. *Technometrics*, **21**, 435–442.

Miller, R. G. (1976). Least squares regression with censored data. *Biometrika*, **63**, 449–464.

MIL-STD 781C (1977). Military standard reliability qualification on production acceptance tests: exponential distribution. Washington, D.C.: U.S. Government Printing Office.

Moeschberger, M. L. (1974). Life tests under competing causes of failure. *Technometrics*, **16**, 39–47.

Morton, R. (1978). Regression analysis of life tables and related nonparametric tests. *Biometrika*, **65**, 329–333.

Mosteller, F. C. (1946). On some useful "inefficient" statistics. *Ann. Math. Stat.*, **17**, 377–408.

Murthy, V. K., G. Swartz, and K. Yuen (1973). Realistic models for mortality rates and their estimation, I and II. University of California at Los Angeles, Department of Biomathematics, Technical Report.

Nelder, J. A., and R. Mead (1965). A simplex method for function minimization. *Comput. J.*, **7**, 380–313.

Nelson, W. B. (1970a). Statistical methods for accelerated lifetest data—the inverse power law model. General Electric Co. Technical Report 71-C-011. Schenectady, New York.

Nelson, W. B. (1970b). Hazard plotting methods for analysis of life data with different failure modes. *J. Qual. Technol.*, **2**, 126–149.

Nelson, W. B. (1972a). Graphical analysis of accelerated life test data with the inverse power law model. *IEEE Trans. Reliab.*, **R21**, 2–11.

Nelson, W. B. (1972b). Theory and applications of hazard plotting for censored failure data. *Technometrics*, **14**, 945–965.

Nelson, W. B., and G. J. Hahn (1972). Linear estimation of a regression relationship from censored data. Part I—Simple methods and their applications. *Technometrics*, **14**, 247–269.

Nelson, W. B., and G. J. Hahn (1973). Linear estimation of a regression relationship from censored data. Part II—Best linear unbiased estimation and theory. *Technometrics*, **15**, 133–150.

Nelson, W. B., and J. Schmee (1979). Inference for (log) normal life distributions from small singly censored samples and blue's. *Technometrics*, **21**, 43–54.

Neyman, J. (1937). 'Smooth' tests for goodness of fit. *Skand. Aktuar.*, **20**, 149–199.

Oakes, D. (1977). The asymptotic information in censored survival data. *Biometrika*, **64**, 441–448.

Olsson, D. M., and L. S. Nelson (1975). The Nelder–Mead simplex procedure for function minimization. *Technometrics*, **17**, 45–51.

Owen, D. B. (1962). *Handbook of Statistical Tables*. Reading, Massachusetts: Addison-Wesley.

Owen, D. B. (1963). *Factors for One-Sided Tolerance Limits and for Variables Sampling Plans*. Sandia Corporation Monograph SCR-607. Albuquerque, New Mexico.

Owen, D. B. (1968). A survey of properties and applications of the noncentral t-distribution. *Technometrics*, **10**, 445–478.

Owen, D. B., and T. A. Hua (1977). Tables of confidence limits on the tail area of the normal distribution. *Commun. Statist.* **B6**, 285–311.

Parr, V. B., and J. T. Webster (1965). A method for discriminating between failure density functions used in reliability predictions. *Technometrics*, **7**, 1–10.

Peace, K., and R. Flora (1978). Size and power assessments of tests of hypotheses on survival parameters. *J. Am. Stat. Assoc.*, **73**, 129–132.

Pearson, E. S., and H. O. Hartley (1966). *Biometrika Tables for Statisticians*, Vol. 1, 3rd ed. Cambridge, England: Cambridge University Press.

Pearson, E. S., and H. O. Hartley (1972). *Biometrika Tables for Statisticians*, Vol. 2. Cambridge, England: Cambridge University Press.

Pearson, E. S., R. B. D'Agostino, and K. O. Bowman (1977). Tests for departure from normality: comparison of powers. *Biometrika*, **64**, 231–246.

Pereira, B. (1977). Discriminating among separate models: A bibliography. *Int. Stat. Rev.*, **45**, 163–172.

Persson, T., and H. Rootzen (1977). Simple and highly efficient estimators for a Type I censored normal sample. *Biometrika*, **64**, 123–128.

Peterson, A. V. (1976). Bounds for a joint distribution function with fixed subdistribution functions: application to competing risks. *Proc. Nat. Acad. Sci. U.S.A.*, **73**, 11–13.

Peterson, A. V. (1977). Expressing the Kaplan–Meier estimator as a function of empirical subsurvival functions. *J. Am. Stat. Assoc.*, **72**, 854–858.

Peto, R. (1972a). Discussion of paper by D. R. Cox. *J. R. Stat. Soc. B*, **34**, 205–207.

Peto, R. (1972b). Rank tests of maximal power against Lehmann-type alternatives. *Biometrika*, **59**, 472–475.

Peto, R., and P. Lee (1973). Weibull distributions for continuous carcinogensis experiments. *Biometrics*, **29**, 457–470.

Peto, R., and J. Peto (1972). Asymptotically efficient rank invariant procedures (with discussion). *J. R. Stat. Soc. A*, **135**, 185–206.

Peto, R., and M. C. Pike (1973). Conservatism of the approximation $(0-E)^2/E$ in the log rank test for survival data or tumor incidence data. *Biometrics*, **29**, 579–584.

Peto, R., P. N. Lee, and W. S. Paige (1972). Statistical analysis of the bioassay of continuous carcinogens. *Br. J. Cancer*, **26**, 258–261.

Peto, R., M. C. Pike, P. Armitage, N. E. Breslow, D. R. Cox, S. V. Howard, N. Mantel, K. McPherson, J. Peto, and P. G. Smith (1976). Design and analysis of randomized clinical trials requiring prolonged observation of each patient. Part I: Introduction and design. *Br. J. Cancer*, **34**, 585–612.

Peto, R., M. C. Pike, P. Armitage, N. E. Breslow, D. R. Cox, S. V. Howard, N. Mantel, K. McPherson, J. Peto, and P. G. Smith (1977). Design and analysis of randomized clinical trials requiring prolonged observation of each patient. Part II: Analysis and examples. *Br. J. Cancer*, **35**, 1–39.

Pettit, A. N. (1976). Cramer–von Mises statistics for testing normality with censored samples. *Biometrika*, **63**, 475–481.

Pettit, A. N. (1977). Tests for the exponential distribution with censored data using Cramer–von Mises statistics. *Biometrika*, 629–632.

Pettit, A. N. (1978). Generalized Cramer–von Mises statistics for the gamma distribution. *Biometrika*, **65**, 232–235.

Pettit, A. N., and M. A. Stephens (1976). Modified Cramer–von Mises statistics for censored data. *Biometrika*, **63**, 291–298.

Pierce, D. A. (1973). Fiducial, frequency and Bayesian inference on reliability for the two-parameter negative exponential distribution. *Technometrics*, **15**, 249–253.

Pierce, D. A., W. H. Stewart, and K. J. Kopecky (1979). Distribution-free analysis of grouped survival data. *Biometrics*, **35**, 785–793.

Pike, M. C. (1966). A method of analysis of a certain class of experiments in carcinogenesis. *Biometrics*, **22**, 142–161.

Powell, M. J. D. (1964). An efficient method for finding the minimum of a function of several variables without calculating derivatives. *Comput. J.*, **1**, 155–162.

Prentice, R. L. (1973). Exponential survival with censoring and explanatory variables. *Biometrika*, **60**, 279–288.

Prentice, R. L. (1974). A log gamma model and its maximum likelihood estimation. *Biometrika*, **61**, 539–544.

Prentice, R. L. (1975). Discrimination among some parametric models. *Biometrika*, **62**, 607–614.

Prentice, R. L. (1978). Linear rank tests with right-censored data. *Biometrika*, **65**, 167–179.

Prentice, R. L., and L. A. Gloeckler (1978). Regression analysis of grouped survival data with application to breast cancer data. *Biometrics*, **34**, 57–67.

Prentice, R. L., and J. D. Kalbfleisch (1979). Hazard rate models with covariates. *Biometrics*, **35**, 25–39.

Prentice, R. L., and P. Marek (1979). A qualitative discrepancy between censored data rank tests. *Biometrics*, **35**, 861–886.

Prentice, R. L., and R. Shillington (1975). Regression analysis of Weibull data and the analysis of clinical trials. *Utilitas Math.*, **8**, 257–276.

Prentice, R. L., J. D. Kalbfleisch, A. V. Peterson, N. Flournoy, V. T. Farewell, and N. E. Breslow (1978). The analysis of failure times in the presence of competing risks. *Biometrics*, **34**, 541–554.

Proschan, F. (1963). Theoretical explanation of observed decreasing failure rate. *Technometrics*, **5**, 375–383.

Pyke, R. (1965). Spacings (with discussion). *J. R. Stat. Soc. B*, **27**, 395–449.

Quandt, R. E., and J. B. Ramsey (1978). Estimating mixtures of normal distributions and switching regressions (with discussion). *J. Am. Stat. Assoc.*, **73**, 730–752.

Rao, C. R. (1965). *Linear Statistical Inference and Its Applications*. New York: Wiley.

Regal, R. (1980). The F test with time-censored data. *Biometrika*, **67**, 479–481.

Resnikoff, G. J., and G. J. Lieberman (1957). *Tables of the Non-Central t-Distribution*. Stanford, California: Stanford University Press.

Rockette, H., C. Antle, and L. Klimko (1974). Maximum likelihood estimation with the Weibull model. *J. Am. Stat. Assoc.*, **69**, 246–249.

Sacher, G. A. (1956). On the statistical nature of mortality, with special reference to chronic radiation mortality. *Radiation*, **67**, 250–257.

Sampford, M. R., and J. Taylor (1959). Censored observations in randomized block experiments. *J. R. Stat. Soc. B.*, **21**, 214–237.

Saniga, E. M., and J. A. Miles (1979). Power of some standard goodness of fit tests of normality against asymmetric stable alternatives. *J. Am. Stat. Assoc.*, **74**, 861–865.

Sarhan, A. E., and B. G. Greenberg (1962). *Contributions to Order Statistics*. New York: Wiley.

Savage, I. R. (1956). Contributions to the theory of rank order statistics—the two sample case. *Ann. Math. Stat.*, **27**, 590–615.

Schafer, R. E., and T. S. Sheffield (1976). On procedures for comparing two Weibull populations. *Technometrics*, **18**, 231–235.

Schmee, J., and W. Nelson (1976). Confidence intervals for parameters of (log) normal life distributions from small singly censored samples by maximum likelihood. General Electric Co. R. & D. TIS Report 76CRD218. Schenectady, New York.

Schmee, J., and W. Nelson (1977). Estimates and approximate confidence limits for (log) normal life distributions from singly censored samples by maximum likelihood. General Electric Co. R. & D. TIS Report 76CRD250. Schenectady, New York.

Schoenfeld, D. (1980). Chi-squared goodness-of-fit tests for the proportional hazards regression model. *Biometrika*, **67**, 145–153.

Seal, H. L. (1977). Studies in the history of probability and statistics XXXV. Multiple decrements or competing risks. *Biometrika*, **64**, 429–439.

Seber, G. A. F. (1977). *Linear Regression Analysis*. New York: Wiley.

Shaked, M. (1977). Statistical inference for a class of life distributions. *Commun. Stat.*, **A6**, 1323–1339.

Shapiro, S. S., and R. S. Francia (1972). An approximate analysis of variance test for normality. *J. Am. Stat. Assoc.*, **67**, 215–216.

Shapiro, S. S., and M. B. Wilk (1965). An analysis of variance test for normality (complete samples). *Biometrika*, **52**, 591–611.

Shapiro, S. S., M. B. Wilk, and J. Chen (1968). A comparative study of various tests of normality. *J. Am. Stat. Assoc.*, **63**, 1343–1372.

Shooman, M. L. (1968). *Probabilistic Reliability: An Engineering Approach*. New York: McGraw-Hill.

Singhal, K. (1978). Topics in exponential regression models. Unpublished Ph.D. Thesis. University of Waterloo, Waterloo, Ontario, Canada.

Smith, R. M. (1977). Some results on interval estimation for the two-parameter Weibull or extreme value distribution. *Commun. Stat.*, **A6**, 1311–1322.

Smith, R. M., and L. J. Bain (1975). An exponential power life-testing distribution. *Commun. Stat.*, **A4**, 469–481.

Smith, R. M., and L. J. Bain (1976). Correlation type goodness of fit tests with censored sampling. *Commun. Stat.*, **A5**, 119–132.

Sprott, D. A. (1973). Normal likelihoods and relation to a large sample theory of estimation. *Biometrika*, **60**, 457–465.

Stacy, E. W. (1962). A generalization of the gamma distribution. *Ann. Math. Stat.*, **33**, 1187–1192.

Stacy, E. W., and G. A. Mihram (1965). Parameter estimation for a generalized gamma distribution. *Technometrics*, **7**, 349–358.

Stephens, M. A. (1974). EDF statistics for goodness of fit and some comparisons. *J. Am. Stat. Assoc.*, **69**, 730–737.

Stephens, M. A. (1976). Asymptotic results for goodness of fit statistics with unknown parameters. *Ann. Stat.*, **4**, 357–369.

Stephens, M. A. (1977). Goodness of fit for the extreme value distribution. *Biometrika*, **64**, 583–588.

Stephens, M. A. (1978). On the half-sample method for goodness of fit. *J. R. Stat. Soc. B*, **40**, 64–70.

Stigler, S. M. (1974). Linear functions of order statistics with smooth weight functions. *Ann. Stat.*, **2**, 676–693.

Stone, G. C. (1978). Statistical analysis of accelerated aging tests on solid electrical insulation. Unpublished M.A.Sc. Thesis, University of Waterloo, Waterloo, Ontario, Canada.

Stone, G. C., and J. F. Lawless (1979). The application of Weibull statistics to insulation aging tests. *IEEE Trans. Electr. Insul.*, **EI14**, 233–239.

Sukhatme, P. V. (1937). Tests of significance for samples of the χ^2 population with two degrees of freedom. *Ann. Eugen.*, **8**, 52–56.

Tallis, G. M., and R. Light (1968). The use of fractional moments for estimating the parameters of a mixed exponential distribution. *Technometrics*, **10**, 161–175.

Tarone, R. E. (1975). Tests for trend in life table analysis. *Biometrika*, **62**, 679–682.

Tarone, R. E., and J. Ware (1977). On distribution-free tests for equality of survival distributions. *Biometrika*, **64**, 156–160.

Taylor, J. (1973). The analysis of designed experiments with censored observations. *Biometrics*, **29**, 35–43.

Temkin, N. R. (1978). An analysis for transient states with application to tumor shrinkage. *Biometrics*, **34**, 571–580.

Thoman, D. R., and L. J. Bain (1969). Two-sample tests in the Weibull distribution. *Technometrics*, **11**, 805–816.

Thoman, D. R., L. J. Bain, and C. E. Antle (1969). Inferences on the parameters of the Weibull distribution. *Technometrics*, **11**, 445–460.

Thoman, D. R., L. J. Bain, and C. E. Antle (1970). Reliability and tolerance limits in the Weibull distribution. *Technometrics*, **12**, 363–371.

Thomas, D. R., and G. L. Grunkemeier (1975). Confidence interval estimation of survival probabilities for censored data. *J. Am. Stat. Assoc.*, **70**, 865–871.

Thomas, D. R., and D. A. Pierce (1979). Neyman's smooth goodness of fit tests when the hypothesis is composite. *J. Am. Stat. Assoc.*, **74**, 441–445.

Thomas, D. R., and W. M. Wilson (1972). Linear order statistic estimation for the two-parameter Weibull and extreme value distributions from Type II progressively censored samples. *Technometrics*, **14**, 679–691.

Thompson, W. A., Jr. (1977). On the treatment of grouped observations in life studies. *Biometrics*, **33**, 463–470.

Tietjen, G. L., R. Kahaner, and R. J. Beckman (1977). In *Selected Tables in Mathematical Statistics*, Vol. 5, D. B. Owen and R. E. Odeh, Eds. Providence, Rhode Island: Institute of Mathematical Statistics.

Tiku, M. L. (1967). Estimating the mean and standard deviation from a censored normal sample. *Biometrika*, **54**, 155–165.

Tiku, M. L. (1970). Monte Carlo study of some simple estimators in censored normal samples. *Biometrika*, **57**, 207–210.

Tiku, M. L. (1981). Goodness of fit statistics based on the spacings of complete or censored samples. To appear in *Aust. J. Stat.*

Tiku, M. L., and M. Singh (1981). Testing the two parameter Weibull distribution. To appear in *Commun. Stat. A*.

Tsiatis, A. A. (1975). A nonidentifiability aspect of the problem of competing risks. *Proc. Nat. Acad. Sci. U.S.A.*, **72**, 20–22.

Tsiatis, A. A. (1978a). A large sample study of the estimate for the integrated hazard function in Cox's regression model for survival data. University of Wisconsin, Department of Statistics, Technical Report No. 526.

Tsiatis, A. A. (1978b). A heuristic estimate of the asymptotic variance of the survival probability in Cox's regression model. University of Wisconsin, Department of Statistics, Technical Report No. 524.

Turnbull, B. W., and L. Weiss (1978). A likelihood ratio statistic for testing goodness of fit with randomly censored data. *Biometrics*, **34**, 367–375.

Uthoff, V. A. (1970). An optimum test property of two well-known statistics. *J. Am. Stat. Assoc.*, **65**, 1597–1600.

Verhagen, A. M. W. (1961). The estimation of regression and error-scale parameters when the joint distribution of the errors is of any continuous form and known apart from a scale parameter. *Biometrika*, **48**, 125–132.

Walsh, G. R. (1975). *Methods of Optimization*. New York: Wiley.

Wang, Y. H., and S. A. Chang (1977). A new approach to the nonparametric tests of exponential distribution with unknown parameters. In *The Theory and Applications of Reliability*, Vol. 1, C. P. Tsokos and I. N. Shimi, Eds. New York: Academic.

Ware, J. H., and D. P. Byar (1979). Methods for the analysis of censored survival data. In *Perspectives in Biometrics*, Vol. 2, R. M. Elashoff, Ed. New York: Academic Press.

Watson, G. S., and W. R. Wells (1961). On the possibility of improving the mean useful life of items by eliminating those with short lines. *Technometrics*, **3**, 281–298.

Weibull, W. (1951). A statistical distribution function of wide applicability. *J. Appl. Mech.*, **18**, 293–297.

Wetherill, G. B. (1977). *Sampling Inspection and Quality Control*. London: Chapman and Hall.

White, J. S. (1967). The moments of log-Weibull order statistics. Research Publication GMR-717, Research Laboratories, General Motors Corporation, Warren, Michigan.

White, J. S. (1969). The moments of log-Weibull order statistics. *Technometrics*, **11**, 373–386.

Whittemore, A., and B. Altschuler (1976). Lung cancer incidence in cigarette smokers: further analysis of Doll and Hill's data for British physicians. *Biometrics*, **32**, 805–816.

Wilcoxon, F. (1945). Individual comparisons by ranking methods. *Biometrics*, **1**, 80–83.

Wilk, M. B., R. Gnanadesikan, and M. J. Huyett (1962a). Probability plots for the gamma distribution. *Technometrics*, **4**, 1–20.

Wilk, M. B., R. Gnanadesikan, and M. J. Huyett (1962b). Estimation of parameters of the gamma distribution using order statistics. *Biometrika*, **49**, 525–545.

Wilks, S. S. (1962). *Mathematical Statistics*. New York: Wiley.

Williams, J. S. (1978). Efficient analysis of Weibull survival data from experiments on heterogeneous patient populations. *Biometrics*, **34**, 209–222.

Wolynetz, M. S. (1974). Analysis of Type I censored normally distributed data. Unpublished Ph.D. Thesis, University of Waterloo, Waterloo, Ontario, Canada.

Wolynetz, M. S. (1979). Statistical algorithms AS138 and AS139. *Appl. Stat.*, **28**, 185–206.

Woodall, R. C., and B. M. Kurkjian (1962). Exact operating characteristics for truncated sequential life tests in the exponential case. *Ann. Math. Stat.*, **33**, 1403–1412.

Wright, F. T., M. Engelhardt, and L. J. Bain (1978). Inferences for the two-parameter exponential distribution under Type I censored sampling. *J. Am. Stat. Assoc.*, **73**, 650–655.

Wyckoff, J., and M. Engelhardt (1979). Inferential procedures on the shape parameter of a gamma distribution from censored data. *J. Am. Stat. Assoc.* **74**, 866–871.

Zacks, S. (1971). *The Theory of Statistical Inference*. New York: Wiley.

Zelen, M. (1959). Factorial experiments in life testing. *Technometrics*, **1**, 269–288.

Zelen, M. (1960). Analysis of two-factor classifications with respect to life tests. In *Contributions to Probability and Statistics*, I. Olkin, Ed. Stanford, California: Stanford University Press.

Zelen, M., and M. Dannemiller (1961). The robustness of life testing procedures derived from the exponential distribution. *Technometrics*, **3**, 29–49.

Zippin, C., and P. Armitage (1966). Use of concomitant variables and incomplete survival information in the estimation of an exponential survival parameter. *Biometrics*, **22**, 665–672.

Author Index

Subject Index